Modern Approaches in Solid Earth Sciences

Volume 15

More information about this series at http://www.springer.com/series/7377

Harald G. Dill

The Hagendorf-Pleystein Province: the Center of Pegmatites in an Ensialic Orogen

Springer

Harald G. Dill
Gottfried-Wilhelm-Leibniz University
Hannover, Germany

Responsible Series Editor: F. Pirajno

ISSN 1876-1682 ISSN 1876-1690 (electronic)
Modern Approaches in Solid Earth Sciences
ISBN 978-3-319-18805-8 ISBN 978-3-319-18806-5 (eBook)
DOI 10.1007/978-3-319-18806-5

Library of Congress Control Number: 2015950906

Springer Cham Heidelberg New York Dordrecht London

Printed on acid-free paper

Springer International Publishing AG Switzerland is part of Springer Science+Business Media (www.springer.com)

Dedicated to Berthold Weber (1959–2013)

Berthold was a focused man and he was more than a friend, who you cannot substitute for. As a software engineer by profession he tried to become a leading figure in his field of electronic printing, a frontrunner in his business. But never had he allowed his business to become the dominant place and absorb all his time. His personal and time management was great. He still had time enough for his fishpond, well kept and being located near Pleystein our "scientific field camp". I could not imagine to see him without a judo mat close by, either as a sportsman fighting himself or conveying his experience and knowledge to younger athletes as a trainer. In both ways, Berthold was superb and gained the highest honors. I got in contact with him,

sharing the same passion for minerals and the NE Bavarian region. The Oberpfalz was his home and my adopted country. There, he was born, raised, he was drafted to the armed forces and he began collecting minerals and practicing his sports. It was not up to him travelling across the world to pack his shelves and showcases with minerals from distant countries, he was focused on the Oberpfalz, the heartland for enthusiasts in rocks and minerals (German: Steinpfalz = Stony Palatinate).

To cooperate with him meant to have a friend at your side you can have full confidence in. Even if his desk was cramped with work and he was almost drowned with the daily routine, he did not postpone my questions. I did not receive an answer from him, he provided me with a solution. His motto was: "Carpe diem" (seize the day). He adhered to this motto even during the last months of his life when we all knew that he is going to lose this fight, even if he never gave up. But once we knew he has to leave the team forever, a few days before the last "Glückauf" at the hospital he sent me the photographs of what he called the best of Hagendorf, a series of micro mounts, accompanied by several cartoons illustrating the ideal morphology of these minerals. But he was not given time to send me the scale bar for each photograph. So I left his legacy as it stands. Those images (if some readers may find some are not of the best quality, it was my fault by making a selection too much biased by science with too little sense of mine for beauty) will keep his memory and work, alive, and bear witness that this work could not be done to completeness.

Glückauf forever!
Harald

Preface

Pegmatitic rocks are very coarse-grained rocks, generally of granitic composition; they contain as major constituents the three rock-forming minerals feldspar, quartz and mica, which in places develop mega-crystals or show up in a graphic intergrowth. On the opposite end of the grain-size scale, their little sisters are placed, finer-grained than most granites with which they share the mineralogical composition these white rocks are called aplites. Apart from the size of their crystals, be aplitic or pegmatitic, it is the varied spectrum of rare elements, e.g., Nb, Ta, Be or Li and the plethora of extraordinary minerals resultant from these elements, that draw the attention of mineralogists and mineral collectors, alike, to these felsic rocks. It has to be noted that the lion share of raw material extracted from pegmatites is feldspar, quartz and mica, and only a tiny fraction of pegmatites contains rare elements at a level so as to render mining of these rare-element pegmatites feasible. The traditional mining of pegmatitic and aplitic rocks in the Bohemian Massif, which is shared by Germany, the Czech Republic, Poland and Austria, has been focused on these industrial minerals, mainly for ceramic purposes.

There will be hardly a crustal section to match with this part of Central Europe under consideration with regard to the number, the various types and chemical and mineralogical variability of pegmatitic and aplitic rocks. In light of more than 1000 years of mining and an intensive study of the deeper parts of the basement during the recent past, using not only geological, mineralogical and geochemical methods but also all the principal techniques of deep geophysics, this region enables us to look at the origin of pegmatites in their natural habit from whatever angle you would like to do.

Hagendorf has been mined for feldspar and the deposit is not yet exhausted as far as the high-purity quartz at the core of the pegmatite is concerned. In comparison with other pegmatite deposits in the area, however, the Hagendorf-Pleystein Pegmatite Province (HPPP) is not only the largest concentration of feldspar and quartz of its kind in Central Europe but also a mineralogical treasure box with more than 250 minerals, some of which were described for the first time from this locality. It almost goes without saying that such a huge amount of minerals attracts first and foremost mineralogists who have been doing a lot of fine work, particularly at

Hagendorf-South which is second to none in the HPPP. Other geoscientific disciplines lived a bit in the shadow of mineralogy which was governed by the late Professor Strunz. Not far away from his place of birth, at Weiden he had Hagendorf-South always in his focus and as an outsider you might have come to the conclusion, that each and everything has been told about this prominent pegmatite. Professor Strunz was a generous and open-minded person, who handed over his books to young recruits like me and still left some open questions to be answered by later generations.

In the course of a renewed investigation that went way beyond mineralogy, four new occurrences (Trutzhofmühle Aploid, Pleystein New Aplite, Miesbrunn Pegmatite Aplites Swarm West and East, Reinhardsrieth Aplite) have been found and several abandoned mines and outcrops were re-sampled together with the "nigrine" placer deposits. "Nigrine", not a mineral accepted by IMA, is an intergrowth of ilmenite and rutile rife with mineral inclusions indicative of a pegmatitic source area. Taking into account these mineralogical features, these dull black heavy mineral aggregates can successfully be used as proximity indicators for pegmatitic rocks. It is a case in point where mineralogy, sedimentology and applied geomorphology worked hand in hand well together in practice. In the majority of cases geological and mineralogical results published or unpublished were combined here with an extensive examination of open-file reports and geophysical studies forming a broad and solid basis to shed some light on the economic geology of pegmatite deposits.

The current book project has been written using the term economic geology still in its traditional meaning to find new mineral deposits and enhance the exploitation of existing ones. That is why, "pegmatites and economic geology" is not used to fill an appendix at the end of the book but is placed ahead of all the other sections, even if much of it has today a historical touch. The perspectives to find new profitable mineral deposits in the region are not very promising and mining in Germany is obviously a profession of the past, but learning from nature in this part of the world may help people elsewhere in the world to be successful, or at least raise awareness of these extraordinary and still enigmatic rocks called pegmatites.

The geological evolution of the Central European Variscides is described from the geodynamic and metallogenetic point of view with the pegmatites embedded into this regional economic geology or in other words forming the centerpiece of economic geology, while often they are trailing behind VMS-, porphyry copper or epithermal mineral deposits. Finding pegmatites may open up new avenues and possibilities also for the exploration of non-pegmatitic mineral deposits, and vice versa; metallic deposits of a certain ore type can provide a clue where exploration for pegmatites is done with a chance of success. The HPPP is viewed as if looking through a magnifying glass from the small-scale overview of the economic geology of Central Europe, closely related to the various geodynamic zones, to the individual outcrops of the HPPP, and investigated during underground mapping or by means of drill core examination.

A classification scheme addressing different structural types of metapegmatites/ metaaplites, pegmatoids/aploids, pegmatites/aplites, granite pegmatites and pseudopegmatites is put forward. It is branch of the "Chessboard classification scheme of mineral deposits" and, similarly, based upon those geological and mineralogical features that experienced geologists who work nose on rock can determine in the field or in their study with the routine techniques at hand. It is a classification scheme which is open for amendments and designed for the practical use, following the tripartite subdivision in (economic) geology: describe, interpret and recommend.

The host environment of pegmatitic rocks, the ensialic orogen is the most favorable crustal section to bring about pegmatites of calc-alkaline affiliation, attractive to mineralogists and become an operational target for mining geologists. In an idealistic transect through the crust this orogen takes the central position and, given the Variscides as reference type, it is called Variscan-type in this discussion of pegmatites. On one side the Variscan type passes into the Alpine-type, where pegmatites get reworked, while on the opposite side the transect ends up in the Rift-Type where pegmatites of alkaline affiliation form.

Pegmatitic rocks in Central Europe whether they are of metamorphic or magmatic origin are not only part of a geodynamic environment but also constitute an integral part of the chronological evolution of the Variscides from the early to the late Paleozoic. Radiometric data are a key element to the understanding of pegmatites. The geochemical atlas of the F.R. of Germany, albeit not covering all rare elements of relevance for the formation of pegmatites, is supplemented by local chemical surveys so as to get a full-blown picture of the distribution pattern of rare elements on a regional scale along the western edge of the Bohemian Massif. While these chemical surveys provide information on the surface geology, seismic, gravimetric, geoelectric and magnetic surveys penetrate deeper into the crystalline basement and thereby help to reduce the field of speculation on the existence of ultra deep structures underneath the pegmatites and granites or even help to shape the various pegmatitic and aplitic bodies.

A great deal of the book is devoted to the mineralogy, considering all mineral groups present in the HPPP. In the run-up to this book, all techniques from the scanning electron microscope to the stereomicroscope were applied during mineralogical investigations, but whenever it deemed necessary, the colorful hand specimens and micro mounts were given preference over back-scattered electron images or photographs of polished sections. The latter had their merits by providing background information to establish a sequence of mineralization for the HPPP and it is prevalently their role as a marker for the physical-chemical conditions that counts when comes to the process-oriented investigation.

Large-scale mapping involving underground work on structural geology and lithology leads to a fine-tuning of the "minero-stratigraphy" and provides a clue to the pathways that were opened up for the melt and solutions and singles out the structures most favorable in terms of accommodation space for the stock-like and tabular pegmatites.

Broadly speaking, the pegmatite, aplites and granites are brothers and sisters. In the Variscides, very soon they stroke out on their own, developing their individual characteristics in terms of composition and structure so that the various lithological processes leading to these felsic intrusive rocks can more easily be studied, each in its own compartment, than in complex pegmatites (pseudopegmatites) which are obviously the result of multi-stage emplacement and alteration processes with only the most recent of its kind preserved to the present. The notoriously raised question which granite spawned this pegmatite is a question each reader will be able to answer by himself after passing through the book.

Hannover, Germany Harald G. Dill

Contents

Chapter 1
Pegmatitic Rocks and Economic Geology

Abstract Three minerals contribute to the built-up of Earth's crust by almost ¾ and predominate in the group of industrial minerals. Feldspar is the top scorer with 56 vol.%; quartz is second most in abundance and accounts for 12 vol.% while phyllosilicates of the mica group are fifth among the top 10 with 4 vol.%. They are the major rock-forming constituents of the granite suite and the pegmatitic and aplitic rocks. Pegmatitic rocks similar in the gross mineralogical composition to granites but conspicuously different in size and quality of their minerals can deliver much better to the requirements of consumers from the industry using feldspar, quartz and mica as a raw material. In addition, these rocks may contain specific elements present in quantity so that they attain even ore grade as it is the case with Nb, Ta, Li, Mo, Sn, W, Be, Sc, Cs, Rb, REE, Y and U or they are targeted upon for their gemstones such as B, Be and Li. Although operated mainly for their quartz and feldspar contents, some of these rocks in Central Europe are also strongly enriched in rare elements such as Li, Nb, Ta, Be, REE and U and therefore have not only attracted the attention of mining engineers interested in industrial minerals but also encouraged mineralogists to investigate the wealth of minerals, with much success. The Hagendorf-Pleystein mining district consists of several pegmatites and aplites, including the largest pegmatite stock in Central Europe at Hagendorf-South, with more than 200 minerals. In the "Chessboard classification scheme of mineral deposits" (Dill, Earth Sci Rev 100:1–420, 2010) the outstanding economic importance of pegmatitic and aplitic rocks, particularly for future technologies is on a world-wide scale in context with other mineral deposits hosted by magmatic and metamorphic rocks. Several attempts have been made to squeeze pegmatites into classification schemes, yet with different success. A classification scheme should cater for the extractive and genetic part of economic geology alike. A new classification scheme, which is based on the **C**hemical composition, the **M**ineral assemblage and the **S**tructural geology of pegmatitic rocks and for which the acronym CMS has been coined is put forward as "CMS classification scheme of pegmatitic and aplitic rocks" (Dill, Ore Geol Rev 69, 2015). It can be used in a short and long version and has two major columns the ore body and ore composition, it can be correlated with the "Chessboard classification scheme of mineral deposits", and as such open for amendments and applicable also in a digital way. The extractive geology of various types of pegmatitic and rocks is depicted from the past until today.

H.G. Dill, *The Hagendorf-Pleystein Province: the Center of Pegmatites in an Ensialic Orogen*, Modern Approaches in Solid Earth Sciences 15,
DOI 10.1007/978-3-319-18806-5_1

1.1 Mining Is Life – From Past to Modern Exploitation of Raw Materials

To subsist on hunting and gathering, our ancestors roamed through the African savannah approximately 200,000 ago. It took some time to change their behavior. Archaeological records from Swaziland revealed mining to be among the first professions exercised by humans during prehistoric times, 43,000 years ago, besides hunting and gathering (Swaziland Natural Trust Commission 2007). Although not being the cradle of extractive geology, people in Central Europe have contributed very much to the enhancement of modern mining techniques, as they were sinking new shafts, drifting galleries or constructing new dewatering systems for underground mines, particularly in the Harz Mountains, Germany (Förderverein Rammelsberg Bergbaumuseum 2011/2012). The renown base metal-iron sulfide-barite deposit Rammelsberg near Goslar, a sedimentary-hosted massive sulfide deposit, was opened in 968 AD and only shut down in 1988, looking back on more than 1000 years of mining and processing ore, in the course of which its miners wrote mining history and encouraged economic geologists to use the site's name as a reference-type for syndiagenetic to syngenetic sulfide ores in fine-grained siliciclastics in marine sediments of rift basins (Dill et al. 2008a, b, c). During the initial period of mining Pb, Cu and Ag, the metals extracted from the mine helped the people of Goslar to accumulate wealth and contributed much to the prosperity of the town.

Not only in this famous mining site in Central Europe (and elsewhere in the world too) ore mining was the reason for people to settle in a region but likewise the focal point of geoscientists to investigate these mineral deposits and create new models on how they have formed, as shown in the classical textbook by Schneiderhöhn (1962). Non-metallic deposits or industrial minerals were all too often eclipsed by vein-type Pb-Zn, Sn-W and U deposits in the literature and did not draw the attention on themselves to the extent which they deserved in the files of raw material supply. This is especially true for quartz and feldspar which used to be exploited for the domestic market only and often were cast aside by research workers from universities on account of the rather monotonous mineral assemblage of their deposits. Today non-metallic commodities and construction raw materials play a vital role, for the people living in Germany with more than 100,000 white- and blue-collar employees directly or indirectly related to its exploitation, processing and manufacturing of the final products, a situation no very much different from many other countries in the western world (Dill and Röhling 2007; Dill and Lohmann 2011).

Three minerals contribute to the built-up of Earth's crust by almost ¾ and predominate in the group of industrial minerals. Feldspar is the top scorer with 56 vol.%, quartz is second most in abundance and accounts for 12 vol.% while phyllosilicates of the mica group are fifth among the top 10 with 4 vol.%. They are the major rock-forming constituents of the granite suite which is won as a decoration and building stone and used, e.g., for gladding or manufacturing tiles but also as crushed rocks to supply the construction industry with aggregates of different grain size (Prentice 1990; Lorenz 1991, 1995). If these commodities are ranked by cumulative production values, these construction raw materials occupy the top places.

Feldspar, quartz and mica are found in a more or less homogeneous distribution, spanning a rather narrow grain size range in granitic raw materials largely exposed in the crystalline basement on a worldwide basis. It is the distance between the place of quarrying and the place of final use which plays a crucial part when the salesman calculates the price for the consumer. Granitic aggregates are as typical as sand for construction purposes of the so-called high-place-value commodity group. If the three minerals have to meet higher quality requirements such as feldspar for ceramic purposes, quartz shipped to the electronic industry or muscovite demanded on account of its chemically inertness, elastic and insulating properties, granites can no longer comply with these diverse and high-level requirements.

Pegmatitic rocks similar in the gross mineralogical composition to granites but conspicuously different in size and quality of their minerals can deliver much better to the most standards set by the consumer from the industry and by doing so shift into the high-unit value commodity segment, the most well-known representative of which is diamond. At this point it becomes evident that classification schemes separating non-metallic and metallic deposits from each other are often difficult to handle in practice as the boundary between typical ceramic pegmatites and rare metal pegmatites are not as sharp as may be suggested by the large Hagendorf-South pegmatite. The economic term high-unit value commodity becomes more relevant all the more so as minor elements improve quality (Li ceramic-electronic grade). Some of these specific elements may increase in quantity so that they attain even ore grade as it is the case with Nb, Ta, Li, Mo, Sn, W, Be, Sc, Cs, Rb, REE, Y and U or they are targeted upon for their gemstones such as Be and Li (Černý 1989; Linnen and Cuney 2005; Selway et al. 2006; Dill 2010; Borisova et al. 2012). Lithium is a good example in the field of pegmatites to demonstrate what a porous barrier the boundary between non-metallic/industrial minerals and metallic commodities is. It covers such a wide range of semi-finished and final products from ceramics, through batteries, solders, lubricants to pharmaceuticals and propellants and also can attain jeweler's quality as kunzite or hiddenite, the precious modifications of spodumene.

The Central European Variscides form the geological basis of what is called today politically Belgium, Germany, Luxembourg, France, The Netherlands, Poland, the Czech Republic and Austria. Together with their southern extension represented by some isolated crystalline massifs within the Alpine-Carpathian Mountain Belt, extending from Switzerland, through Austria into Slovakia, this crustal section is host to a great variety of pegmatitic rocks from the early through the late Paleozoic, each of which nicely separated from each other by their geodynamic setting and age of formation. Although operated mainly for their quartz and feldspar contents, some are also strongly enriched in rare elements such as Li, Nb, Ta, Be, REE and U and therefore have not only attracted the attention of mining engineers interested in industrial minerals but have also encouraged mineralogists to investigate the wealth of minerals, with much success. In the center of that pegmatite-prone area the HPPP is located in the NE Bavarian Basement (Oberpfalz/SE Germany) at western edge of the Bohemian Massif. This mining district consists of several pegmatites and aplites, including the largest pegmatite stock in Central Europe at Hagendorf-South. More than 200 minerals are found only in this rare-element pegmatite stock (Figs. 1.1a and 1.1b).

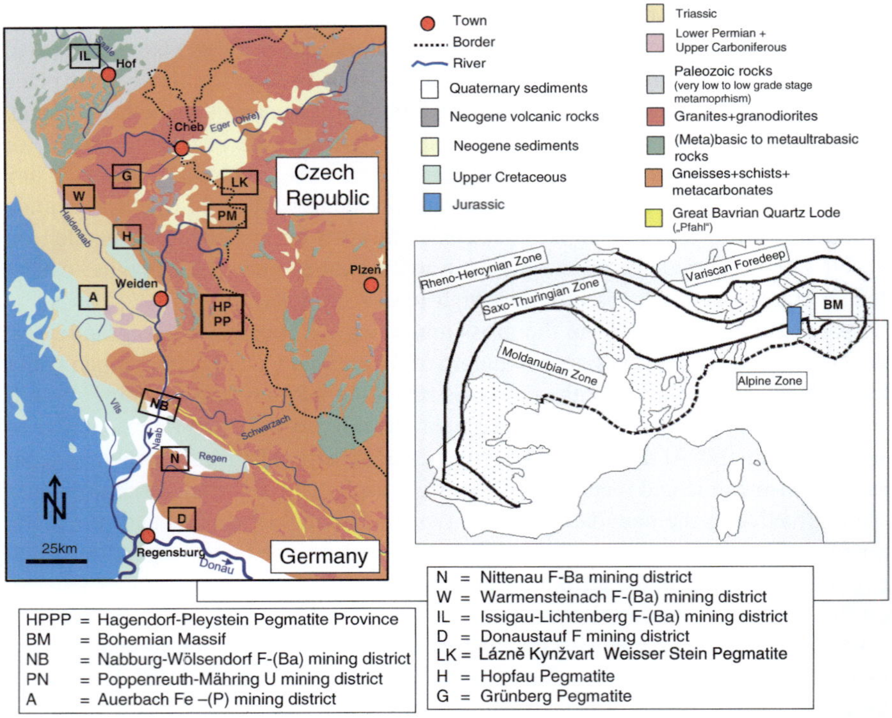

Fig. 1.1a Topographic position and geological setting of the Hagendorf-Pleystein Pegmatite Province (HPPP) at the western edge of the Bohemian Massif (BM) in Central Europe. The position of this large-scale geological map along the western edge of the Bohemian Massif covering the boundary between the Saxo-Thuringian and Moldanubian Zones is given in the index map that illustrates the various geodynamic realms of the European Variscides from Portugal to Poland. In addition to the Hagendorf-Pleystein Pegmatite Province (HPPP) further mineralized sites referred to in the text are shown in the geological map

1.2 Pegmatites and Classification Schemes

1.2.1 The Pegmatites in the "Chessboard Classification Scheme of Mineral Deposits"

In the following section a brief overview of pegmatitic rocks *sensu lato* will be given. Some examples underscore their outstanding economic importance particularly for future technologies which are in need of elements often designated with different names such as "critical elements", "strategic elements" or "electronic elements", the latter term is certainly the most precise of its kind as to application and final use. In view of the current statistics available on critical raw materials, it is predominantly the concentration of Nb, Ta, Li, REE, W and Be in pegmatitic rocks that can alleviate the current pressure on the supply of these elements which is not

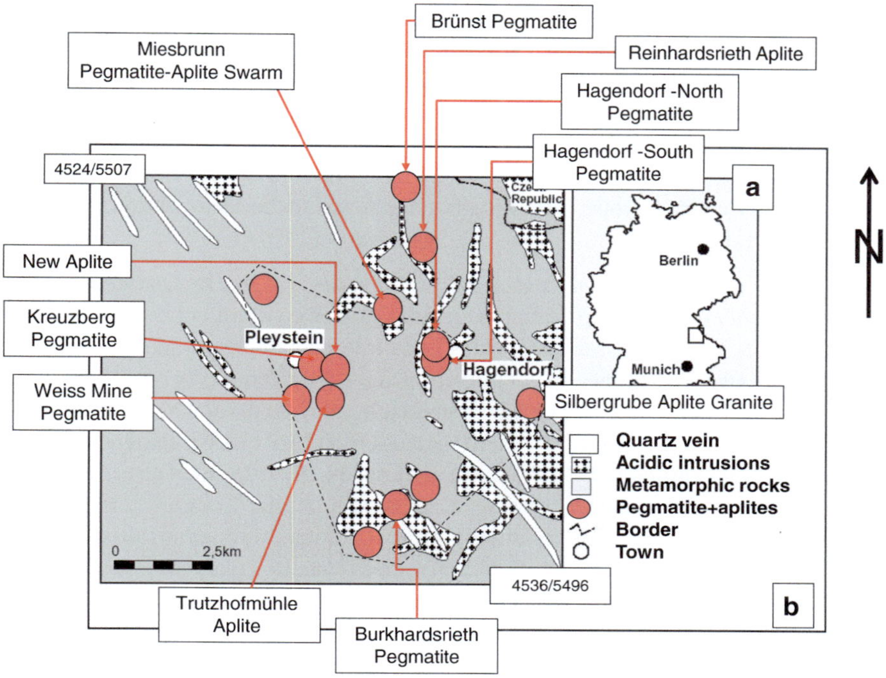

Fig. 1.1b The pegmatitic and aplitic rocks of the HPPP referred to in this book and their geological setting

in any case based on geogenic factors, in other words a shortage of deposits but too often manmade. Too often industrial minerals like fluorite and graphite, also ranked as critical raw materials, are found in a position to be sidelined during reviews of pegmatites as potential sources of rare elements and critical minerals. The term critical, is in my opinion, mainly used from the consumer's point of view to keep the wheels in the western world turning. What about artisanal mining or small-scale operations targeted upon colored gemstones in pegmatites? The number of pits and shafts exceeds by some orders of magnitude those of Be or Li in pegmatites operated by opencast or ramp mining. For the small-scale miners colored gemstones are critical for the daily life and the only source of income.

The "Chessboard classification scheme of mineral deposits" elaborated by Dill (2010) is used here as a red thread through the economic geology of pegmatites and consequently its coding comes into effect during classification.

Sn-W-Ta-Nb Pegmatites (12a D) The Manono-Kitolo deposits is the most well-known representative of a Sn-W-Nb-Ta-Be-Li province along the eastern boundary of the DR Congo with Burundi, Rwanda and Uganda. Further pegmatites of this type are known from Rondonia, Brazil, Pitkäranta, Russia, and Cornwall, Great Britain Metalliferous pegmatites to be coded 12a D carry arsenopyrite, loellingite, wolframite with minor cassiterite associated with tourmaline, zinnwaldite, apatite, chlorite, stilbite, stokesite, topaz and triplite (Fe phosphate) and thoreaulite. They are closely related to the so-called stockscheider, a name coined by miners from the Erzgebirge, Germany, and applied to the apical parts of the tin granites there, displaying a conspicuous pegmatitic texture. There is another type of alteration zone rich in silica, topaz (pyknite) and zinnwaldite, called greisen, with which pegmatitically textured host rocks are closely related in time and space. Referring to our study area in Central Europe, this type is well exposed in the Saxo-Thuringian Zone (see Sect. 2.1.4.4).

Sn-Ta-Nb-Sc in Calc-alkaline and Alkaline Pegmatites (13b D) One can already deduce the chemical kinship of this pegmatite-hosted element concentration from the coding. It is a pegmatite where the last three elements prevail over tin. The zoned petalite-subtype Tanco pegmatite at Bernic Lake, Canada, is one of the main producers of Ta in the world. The Tanco pegmatite may be quoted as an example of an almost unrivalled study of pegmatite zonation and its marker minerals (Černý 1972; Černý and Ferguson 1972; Černý and Simpson 1977, 1978). The ore minerals of Ta are commonly manganotantalite, manganocolumbite, wodginite, and microlite; Ta-rich cassiterite is also commonly present. Tantalum mineralization tends to occur in albitic aplite, mica-rich (lepidolite, cleavelandite (platy albite), ± lepidolite), and spodumene/petalite pegmatite zones. The internal zonation of granitic pegmatites is still debated very controversially (Jahns and Burnham 1969; London 2008). The classical model on zonation claims pegmatites owe their distinctive textural and zonal characteristics to the buoyant separation of aqueous vapor from silicate melt, giving rise to a K-rich pegmatitic upper portion and Na-rich aplitic lower zone of individual pegmatites. More recent theories indicate that pegmatite-forming melts cool quickly, or in any case, more quickly than crystallization can keep pace with. A significant contribution to the cooling behavior of pegmatites can also be made by the numerous mineral ages collected in and around the pegmatites along the western edge of the Bohemian Massif (Sect. 3.1). Often these pegmatites show a complex mineral zonation with only a peculiar zone strongly mineralized with those elements giving their name to the type of deposits. This is the case with the Be-Li-Ta-Nb-REE-U pegmatites in the Alto Ligonha province, Mozambique whose deposits may be subdivided into (1) Na-Li pegmatite (beryl, tantalite, microlite, tourmaline lepidolite, topaz), (2) K pegmatite (beryl, feldspar, mica, COLTAN, REE, monazite, U minerals), (3) amazonite pegmatite with amazonite and tourmaline. A peculiarity is the presence of scandium in this type of pegmatite, with minerals such as cascandite and scandiobabingtonite to be found in the Heftetjern pegmatite in Norway (Raade et al. 2002). One of the few mines operated only for scandium is located within the Befanamo pegmatite, Madagascar, where 40 kg of thortveitite were extracted. The pegmatite belongs to the monazite-thortveitite subtype of the NYF pegmatite clan *sensu* Pezzotta (2001). Pretulite and kolbeckite occur in quartz veins and in the alteration zone of pegmatites (Bernhard et al. 1998; Dill et al. 2006a, b, c). The Trutzhofmühle aplite represented the first discovery of a complex Sc mineralization in the HPPP. It has to anticipate in this paragraph, that these complex pegmatites, whose zonation can hardly be approximated with any of the nice cartoons often found in one or the other textbook on mineral deposits, are true pegmatites *sensu stricto*. They are rather called pseudopegmatites, strikingly different from those dealt with in the HPPP as a consequence of multiphase reworking during the Alpine-Type orogenies (Sects. 2.2 and 2.3). There primary structure and element content in the original pegmatite has been wiped out or at least been obliterated so that the hardly qualify as reference types. What has been preserved in a metamorphosed porphyry-like deposit in the Scandinavian Precambrian terrains resembling features as being compared with the prototypes along the "Ring of Fire?"

Beryl-Emerald-Euclase-Hambergite-Bearing Granite Pegmatites (14a D) Beryllium is an element typical of pegmatites, also in the area under consideration in Germany (section 4.2.3). In this type of host rocks it is not the element beryllium that counts as a source of the metal but as may be concluded form the subheading, the quality or aesthetic value of its minerals that attain gemological quality in places. Gem-quality emeralds fill veins in the immediate surroundings of the pegmatites and granites and replacement of primary or pegmatitic beryl may end up in the formation of secondary beryllium minerals. At high temperature (~550 °C), gem-quality aquamarine precipitates in vugs surrounded by alteration haloes of albite, muscovite, cassiterite and fluorite (Markl and Schumacher 1997). Černý (1991) used beryl as the marker mineral to distinguish the so-called beryl- or LCT-type pegmatites from NYT or REE pegmatites. Aquamarine mineralization in the Mzimba district, Malawi, is obviously unrelated to granites and supposed to have been derived from metamorphic devolatilization and partial melting of basement rocks. These pegmatites are transitional between granitic metapegmatites and pegmatitic mobilizates and corroborate what has been stated at the end of the previous paragraph (Dill 2007). As we see later beryl is one of the few minerals crossing the boundary between the different types of pegmatites and aplites *sensu lato*. Grew (1988) investigated such Be-bearing pegmatites in granulite-facies rocks and published P-T conditions at the time of pegmatite emplacement at 800–900 °C and 7–8 kbar. Beryl in pegmatites decomposes in the course of hydration to a wide range of Be mineral such as euclase and hambergite (O'Donoghue 2006). Cesium-bearing Be oxide was found within the alkaline late stage parts of the LCT granite pegmatites at Antandrokomby and Antsongombato, Madagascar. Beryl deposits are bound to highly-fractionated granites (leucocratic muscovite-bearing quartz-rich granites) enriched in W, Sn, Mo and U. Fluorine used to transport beryllium in the fluid or vapor and shows up as (1) fluorite in quartz veins, (2) fluorine-enriched granites and greisens and (3) topaz mineralization.

Chrysoberyl in Rare Element Pegmatites 14c E Chrysoberyl forms together with beryl in some complex rare-element pegmatites (Walton 2004).

Beryl-bearing Pegmatites 14d D This kind of beryl enrichment is a bit ambiguous and bridge the gap between metamorphic rocks/migmatites and magmatic rocks; they evolved along cleavage planes of crystalline schists in several provinces in Argentina. Compared with the afore-mentioned pegmatites from Africa and North

America, these pegmatites in the Andes are poor in rare-element minerals and beryl, although very widespread among the rock-forming minerals only attains industrial quality. In that context, they reveal similarities to the pegmatitic mobilizates of the Münchberg Gneiss Complex (Sect. 1.3.2.2). Maybe there is reciprocal relationship, encouraging research workers in Argentina to revisit and re-evaluate their hypothesis in the light of the data gathered in the Münchberg Gneiss Complex which is peculiar body emplaced along prominent suture zone (Sects 1.3.2.2, 2.1.4.2, and 6.3)

Chrysoberyl in Pegmatitic Mobilizates in the Contact Zone to Metaultrabasic Rocks (14b A) and Metapelites (14b J) These pegmatites are hard to be linked up with granites in the metamorphic terrains (Walton 2004). It is a metasomatic contact zone between Be-enriched pegmatites and silica-poor wall rocks that lead to the finest gemstones, including chrysoberyl may be associated with emerald and phenakite. Existing hypothesis are focused on a desilification of a pegmatite which intruded into SiO_2-poor country rocks and assimilation of Al_2O_3-rich country rocks. Neither hypothesis can fully account for the mechanism of this Be enrichment.

Li-Cs-Rb Pegmatites (15a D) Pegmatites are important sources of Li and Cs (Bessemer City, USA, Greenbushes, Australia, and Bikita, Zimbabwe). Normally these pegmatitic deposits are exploited for Nb and Ta and the light metals Li and Cs won as byproducts (Bernic Lake, Canada Manono-Kitotolo, DR Congo). Spodumene, Li mica and pollucite are extracted from these deposits to produce Cs and Li concentrates. The Li-Cs-(Rb) pegmatite, Bikita, Zimbabwe, is besides Bernic Lake (Lac-du-Bonnet), Canada, the only site where another alkaline element rubidium was found enriched to such a high level so as to make its recovery from the ore feasible. The Hagendorf-South pegmatite province has also been investigated for Li and Li ore was extracted mainly from triphylite as a byproduct of the running feldspar exploitation. This chain of Variscan Li-bearing pegmatites may be extended through the Czech Republic into Poland with lepidolite pegmatites near Rožná and Dobrá Voda (Novák and Černý 2001). Hiddenite and amblygonite are exploited at Governador Valadares from a swarm of unzoned spodumene-rich K feldspar pegmatites with montebrasite and cookeite (Romeiro and Pedrosa-Soares 2005). Spodumene-rich pegmatites within amphibolites and mica schists in Carinthia, Austria, are the southernmost Variscan pegmatites within the Alpine Mountain range (Göd 1978).

REE-U-Nb Pegmatites (24a D) These pegmatite deposits are in places, transitional into intragranitic deposits with Mo-W-U-Be and known from the Tien Shan Mts. of Kyrgyzstan, where a REE deposits occur in the Aktiuz area, in Achaean gneisses and amphibolites intruded by subalkaline leucocratic granites. Niobium, fluorine and yttrium act as marker elements and enable us to attribute this sort of pegmatites to the so-called NYF (Nb-Y-F) pegmatite clan, a technical term used here only to hook it up with the traditional classification schemes. Fluid separation took place during an initial phase with concentration of F. Subsequently, Nb-, Ta-, Be- and Zr granite-related mineralization was emplaced. The latest phase has seen convectively circulating fluids depleting the lower parts in REE and causing redeposition of them at a shallower level. Columbite is host of Nb and the REE association consists of bastnaesite, parisite, synchysite and REE-bearing thorite. In this case, the granite is not the "Father of all things" resulting at the very end in a distinct pegmatite but is obviously the other way round.

U Pegmatites (25b D) Uraniferous pegmatites were discovered among others at Bancroft, Canada. A great deal of pegmatites and aplites contain U in form of uraninite or uraniferous Ta-Nb oxides but mostly at a subeconomic level so that only very few of them may be ranked among the uranium deposits.

Tourmaline- Danburite- Dumortierite- Kornerupine Pegmatites (30a D) The assemblage of boron minerals used as common denominator reveals that these pegmatites have been drawn attention to for the aesthetic value of their B minerals and are not exploited as a source of boron. The best gem-quality tourmaline mines of this kind are located in Brazil, in the State of Minas Gerais, Espirito Santo, Bahia and Paraiba (Delaney 1992; Adusumilli et al. 1994). Tourmalines are exploited from pegmatites that are intrusive into schistose or granitoid rocks. Other sites of high-quality tourmaline are known from, e.g., Nepal (Hyakule, Phakuwa), Russia (Mursinska Mts. in the Ural), Kenya (Voi-Taveta) and Zambia (Chipata). An extraordinary tourmaline called Paraiba Tourmaline was discovered near Salgadinho, Brazil, in pegmatites which were intruded into Proterozoic muscovite-quartzites. At Quintos Mine-Pernambuco, Brazil, this Cu-bearing elbaite is exploited. Crystallization of slender tourmaline starts off in the siliceous core forming some kind of a band around it and stop in a zone made up of clay minerals, apatite, morganite, spessartite and uranium minerals. Danburite and dumortierite in gem quality are observed in granites and pegmatites. Jeremejevite is a rare constituent of the late stage hydrothermal alteration of pegmatites.

F-(Sc) Pegmatites (32a D) The pegmatite at Crystal Mountain in Montana, USA is one of the few fluorite-bearing pegmatites. In addition to fluorite, this pegmatite also contains Be accommodated in the lattice of euclase and Sc in thortveitite.

Li Phosphate Pegmatite (38a D) The wording already aimed at the final use of these pegmatites. It may to some extent used as source of Li and, in places, for the aesthetic value of the Li-Fe-Mn-Mg phosphate minerals provided they attain of gem-quality, what for the majority of cases may only be applied to the lazulite-scorzalite s.s.s.

Si-, Quartz- and Semi-precious Gemstone Pegmatites (40d D) In zoned pegmatites, a quartz core is common and druses in granite pegmatites, which are partially filled with quartz modifications such as smoky quartz, rock crystal or amethyst, may be operated for ceramic purposes together with feldspar and mica or selectively handpicked for showcase silica minerals Quartz concentrations in simple or zoned granitic pegmatites are numerous worldwide, but only a few are mined as deposits for high-purity quartz as at Drag, Norway, and Spruce Pine, USA. Quartz of pegmatites may gain a high to ultra-high purity in the course of its splitting apart from feldspar and mica. But nature often did not fulfill the strict requirements laid down by the consumers on high-purity quartz (<100 ppm non-Si impurities) and ultrahigh-purity quartz (<10 ppm non-Si impurities). In addition to the above pegmatites mined for industrial quartz in Brazil, India, Australia, China, Norway, Russia and the USA, there are numerous sites where pegmatitic quartz is mined for its aesthetic value. The pegmatites at Vorondolo, Madagascar, are named in place of many other sites worldwide, for their excellent quality of rose quartz which shows asterism due to some dumortierite included in the lattice (Pezzotta 2001). The Kreuzberg rose quartz pegmatite in the study area belongs to this type of pegmatite.

Feldspar-Quartz Pegmatites and Aplites (41d D) The major source of feldspar seldom lies in the granitic bodies but in the pegmatites or aplites some of which are closely related with those zoned pegmatite also worked for silica deposits. Numerous granite pegmatites worked for feldspar have been listed by Harben and Kužvart (1996). It may be a bit pre-emptive in this place, but keeping to the rules of economic geology it has to be stated explicitly, that the numerous pegmatitic rocks mentioned from the Bohemian Massif have been right on target for their feldspar and not for the wealth of minerals. But even a battle-hardened business man can hardly take away oneself from the charm of the colorful and well-shaped minerals scattered across many of these feldspar pegmatites there are two main contenders, sodium feldspar or albite-enriched plagioclase and potash feldspar, mainly present as microcline. The granular feldspar is used for glass making, while the milled products is shipped to the ceramic industry. K feldspar is welcomed for the high-quality glass laboratories and households, TV appliances and high-voltage insulators. In contrast, Na-enriched plagioclase can be used for the inferior glass-qualities, cosmetics and in the production of fiberglass insulations. In addition to these branches were a selection of feldspar is crucial prior to the glassmaking or producing the ceramic body, there is a wide range of manufactures who base their products upon feldspar such as those producing frits, ceramics, paints, fillers, plastics, abrasives, to mention only a few of them.

Amazonite-Moonstone-Sunstone Pegmatites (41e D-41f D) Some feldspar varieties also gained recognition in the field of gemology. Although not grouped among the top tree (diamond-precious corundum-emerald), amazonite, moonstone and sunstone find their lovers and thereby gave local craftsmen and traders an opportunity as breadwinners. Amazonite develops in pegmatites emplaced in a wide range of environments. In the pegmatites related to the Pb–Zn mineralization of the ore deposit at Broken Hill, N.S.W., Australia, and amazonite has originated by the same metamorphic processes which produced the surrounding gneisses and which were also responsible for the metamorphism of the Broken Hill ore deposit (Čech et al. 1971). In the Luc Yen region, Vietnam, the amazonite pertains to a group of minerals typical of the rare alkali metal pegmatite type (Tuyet et al. 2006). At Anjahamiary, Madagascar, in a complexly zoned pegmatite deposit giant amazonite crystals associated with lepidolite and pink elbaite were discovered in the core of the pegmatite as further north in the Embu District and in the Sultan Hamud area, Kenya (DuBois and Walsh 1970; Pezzotta 2001). The lapidary term moonstone describes an optical effect that is observed in K feldspar varieties such as adularia and peristerites of the plagioclase s.s.s. The application of the term sunstone was restricted to colorless oligoclase with oriented inclusions of hematite and goethite but now it has been extended also to labradorite and all aventurescent plagioclase.

Feldspar-Quartz Metapegmatites (41h D) These rocks are widespread in part of the study area in Germany (Teuscher and Weinelt 1972). Differentiation of felsic constituents during the pegmatitic state might have reached an economic level but suffered, locally, a blow during subsequent metamorphism and deformation, when the differentiation was undone by the metamorphic processes superimposed on the pegmatites and lead to a comminution of the previous large mineable crystals of feldspar and quartz produced during the initial pegmatitic stage. They will be dealt with in Sects. 1.3.2.3 and 6.2.

Feldspar-Quartz Pegmatoids (41g D) In contrast to metapegmatites, dealt with in the succeeding paragraph, these feldspar-quartz rocks did not suffer from an intensive metamorphism that might have had a detrimental effect on the quality and quantity of the industrial minerals in the pegmatoid. This type of pegmatite *sensu lato* is not only of widespread occurrence in the study area but found worldwide and often exploited for the domestic market. They will be dealt with in Sects. 1.3.2.2 and 6.3.

K Feldspar-Bearing Marble-Pegmatoid Series (41g K) Near Itrongay, Madagascar, K feldspar of gem-quality is exploited from calcite marble that also contains considerable amounts of diopside and phlogopite. In general, marble is not a common host of K-feldspar deposits but in this particular case pegmatites ramified in these metacarbonates and were responsible for this extraordinary gemstone deposit (Pollner et al. 1997). The coloring of these golden transparent gemstones is caused by a significant amount of Fe^{3+} substituting for Al^{3+}.

Scapolite Pegmatites (42a D) In Tajikistan, pegmatites with gem-quality scapolite occur in the Kukurt antiform in the Rangkul (Kukurtskoe) district in the East Pamir Mountains (Kievlenko 2003). The Kukurt anticline is cored by Precambrian quartzite, marble, granite-gneiss and migmatite. The gem-bearing pegmatites are related to leucocratic biotite granite and two-mica granite of the Mesozoic-Cenozoic Shatput Complex.

Nephelinized Gneisses and Pegmatites (42c E) The close intertonguing of nephelinized gneisses and pegmatites lead to corundum accumulation, e.g., at Thambani, Malawi, Kishangarh, Rajasthan, India, the Wolfe nepheline belt, Ontario, Canada, and the Amazonian gneiss belt, Brazil. The true nature of the processes responsible for this concentration of Al oxide is not yet fully known.

Zeolite Pegmatites (43a DE) Tschernich (1992) gave a detailed account on zeolitization within the endo- and the exocontact of pegmatites. Xenoliths falling into the molten magma may react to form zeolites. The ijolitic pegmatites of La Madera, Argentine, formed dykes running through volcanic olivine melanephelinite of Late Cretaceous age (Galliski et al. 2004). The ijolitic pegmatites were formed by H2O-undersaturated, P_2O_5–, CO_2– and incompatible-element-bearing melts derived by fractional crystallization of a parent olivine melanephelinite. Pegmatites have brought about rather uncommon zeolites together with semi-precious gemstones such as at Ambatovita, Madagascar. Chiavennite a rare Be-bearing zeolite was reported to be associated with amazonite, spodumene, Cs-bearing beryl and pezzottaite, Cs-rich muscovite-lepidolite (Laurs et al. 2003; Warin and Jacques 2003). Syenite pegmatites of the Larvik pluton, Norway, contain apart from many other minerals zeolites such as natrolite (Petersen 1978).

Fe-Mn Garnet Pegmatites (47b D) The Laghman pegmatites, a series of Be-Li-pegmatites in Nuristan, Afghanistan, may serve as an example where Mn-enriched garnet from pegmatites attains gem-quality. The host pegmatite contains Li-tourmalines, pink and blue beryl, spodumene (including kunzite) and spessartite (Bariand and Poullen 1978). Spessartite-enriched garnet is concentrated in Na-Li pegmatites, in marginal, aplitic and quartz-muscovite zones, albitized zones and also in the quartz cores of some pegmatites (Kievlenko 2003). The author also reports gem garnets to have been recovered from pegmatite deposits from Brazil, Madagascar, Myanmar and Sri Lanka. Manganiferous garnet, although not attaining gem quality, are also very widespread in the HPPP.

Corundum Pegmatites (50b E) Corundum and felsic rocks are not a common association in nature, as aluminum and silica tends to go together and develop silicates, the most primitive ones are andalusite, sillimanite and kyanite. Therefore geological settings where both elements go together but keep separated from each other are rare and fulfill all requirements for a gemstone deposit, provided there are chromophores at hand to enhance the aesthetic value of the aluminum oxide. The desilification of the environment normally is expressed by the presence of syenites instead of granites. Gem-quality sapphires in Canada are recorded from syenite pegmatites of the Haliburton-Bancroft alkaline complex. The pendant to the Haliburton-Bancroft alkaline complex is the Ilmeny Gory in the Chelyabinsk Oblast, Urals Region, Russia (Popov and Popova 2006).

Corundum Pegmatites (Plumasite and Marundite) (49c CD-50a CD) Corundum is a constituent of diorite-plumasite pegmatites and found together with margarite in marundites. Occurrences of sub-precious corundum of these types were exploited in Canada, Italy; Malawi and the Pamir Mts. Range, but only the Soutpansberg industrial-grade corundum concentration in South Africa is mined (Rossovskiy and Konovalenko 1977). As silica-bearing pegmatites intrude rocks undersaturated with silica, the silica is extracted or exchanged from the pegmatite and reacts with the undersaturated host rocks producing new minerals that contain silica. Pegmatites may have been desilicified by their fluids interacting with silica-undersaturated country rocks e.g. ultramafic country rocks. At Mahenge, Tanzania, rubies are found in pegmatites bearing green tourmaline as well as in marbles (Hauzenberger et al. 2005). In plumasite and marundite, sapphires, rubies, showcase- quality or industrial grade corundum may come into existence. At Barauta, Zimbabwe, in Kashmir, India and in Moneragala and Okkampiitya, Sri Lanka, sapphire is encountered within pegmatites (Hughes 1990). Some Al silicate-bearing pegmatites in the study area may come close to this type of rock.

Mica Pegmatite (59b D) Muscovite occurs in booklets and large plates in many pegmatites and also forms the fourth group of pegmatites besides LCT zoned pegmatite (Li-Cs-Ta), NYF pegmatites (Nb-Y-F) and feldspar pegmatites (Černý and Ercit 2005). Often mica cannot be won economically as a stand-alone commodity and is only being sold as a by-product of quartz, feldspar. Muscovite books were found predominantly in the border zone close to the country rocks.

Graphite Pegmatite (52a D) Graphite occurs in alkaline pegmatites at Hackman Valley, Mt. Yukspor and Chibina Massif, Russia. It is associated with minerals including aegirite, apatite, albite, nepheline and natrolite (Jaszczak et al. 2007). Graphite-bearing pegmatitic dikes with abundant CO_2-rich inclusions occur side-by-side with wollastonite-bearing calcsilicates and gneiss–charnockite horizons in the supracrustal terrain of the Kerala Khondalite Belt. This close association attests to the transfer of carbonic fluids through magmatic conduits (Satish-Kumar and Santosh, 1998).

Kaolinite Group Minerals in Pegmatitic Rocks 55a CD Kaolinite-group (kandite-group) minerals have three polymorphs kaolinite (triclinic), dickite (monoclinic) and nacrite (monoclinic). Kaolinite is often associated with halloysite which it is going to replace as temperature of formation increases in high sulfidation-type deposits. There is a continuum from metalliferous high-sulfidation-type deposits into non-metallic deposit with the kaolinite-alunite deposits barren as to gold and copper at the opposite end of the mineralizing sequence. Kaolinite may form on top of felsic intrusive rocks acting as the major rock-forming phyllosilicate in relic kaolin or being part of the hypogene alteration which granites as well as pegmatites suffer from during their later stage of evolution. This type is described in detail in Sect. 5.1.2, and exemplified by the Pleystein Trend in the HPPP.

A detailed monographic study on the economic geology of pegmatites and aplites worldwide is given in Dill (2015).

1.2.2 The Pegmatites and Their Classification Schemes in the Scientific Discourse

Several attempts have been made to squeeze pegmatites into classification schemes, yet with different success as shown by the list of classification schemes and their producers. Niggli (1920) was one of the first who addressed this issue. To come to the point, his rather descriptive pioneer work was well ahead of some classification scheme introduced subsequently, because of his omission of genetic connotations. Fersman (1930) represents the state-of-the-art during the 1920s and 1930s in terms the temperature of formation of pegmatites. Landes (1933) again favored the style elaborated by Niggli (1920). Bjørlykke (1937) used index-minerals for his classification. Ginsburg and Rodionov (1960) again played the "genetic ball" and elaborated a classification based on the depth zonation of pegmatites.

One of the most comprehensive studies on pegmatites was published by Schneiderhöhn (1961) who not only provided an overview of the pegmatites known across the world at that time but also dealt with the general issues while crossing the boundary also to types of pegmatites different from what we call today granitic pegmatites ("pseudopegmatites"). People interested in pegmatites mainly in their natural habitat will have difficulties to find a paper even today being a match for this two-volume book, written in German. Unlike this book by Schneiderhöhn (1961), the book by Ginsburg et al. (1979), locally, is cited in the Anglo-Saxon literature, despite being written in Russian. It was not until 1991, that Černý published a classification scheme of granitic pegmatites consisting of four classes which subsequently was added up with some amendments in another publication by Černý and Ercit in 2005. Four lines, five in the amended classification scheme and five columns make up the spreadsheet of classification. Along the y-axis you may find a textural term "miarolitic", indicative of cavities in granites, a chemical term "rare-element", covering the entire range from REE to P, which is among the most common elements in the Earth crust, a phyllosilicate "muscovite", which is almost present in each pegmatite *sensu lato* and last but not least "abyssal", an adjective normally used to describe the depth of the sea and the level of intrusions in the Earth crust. It was certainly the merit of these authors to deal with this very complex category of rocks where almost each outcrop delivers a type of its own. For the user in the field it is hard to recognize a red line between the various terms put forward for the principal classification of pegmatites. The family clan can be subdivided into LCT (= Li-Cs-Ta) and NYT (=Nb-Y-Ta), but its subclasses can only be applied to two of the three classes. Only in the revised version multiple applications of those terms have been allowed for classification. The third column labeled "environment" enters already the field of interpretation as to pressure and temperature but cannot cover the full physical regime. The last three columns address the position of the pegmatite relative to the granite and its shape. The "rare element class" has been split up into such a huge number of subtypes denominated with some marker minerals that it can no longer be considered as close to reality. Ecrit (2005) made some fine-tuning of this classification scheme but did not really help clarify the interrelationship. Apart from the incoherence of the terms used for classification there are another set of problems that make its application difficult. It is first and foremost the focus on granites as the parental material from which the pegmatites are said to originate and from which the ensuing pegmatites acquired their trace elements as a result of fractionation. This may be true for some cases where the granite and pegmatites intertonguing with other or the felsic igneous rocks hosting vugs and druses rife with pegmatitically-textured minerals. One of the most recent papers making us aware of questioning a great deal of this approach was published by Tkachev (2011).

Another classification scheme elaborated by Pezzotta (2001) set aside genetic allusions. Wise (1999) uses chemical parameters such as peralkaline, metaaluminous and peraluminous as basic criteria to subdivide the pegmatites, yet further subdivision is drowned in a plethora of minerals which can hardly be used in the field for any classification. The classification put forward by Zagorsky et al. (1999)

is a relapse to the first mentioned scheme of classification combining descriptive terms from geochemistry and mineralogy with physical terms referring to three different levels of pressure, high, moderate and low.

All of these pegmatite classification schemes have an odd structure and too often try and mix up description of features from different camps with genetic ideas. By all accounts these classification schemes render difficult any application during practical work, and the genetic value I avoid to assess. Many are trained to work "nose-on- rock" following a strict "chain of command" in geosciences: (1) description (2) interpretation (3) recommendation (for the applied geosciences). This sounds like commonplace or transporting coal to Newcastle, but having a closer look at the current "paper world" it has to be repeated and we are obliged to rethink our way of handling these issues in accordance with nature. Otherwise geology will become the main adversary to all our models created in the laboratory and the office, while sitting on the swivel chair in front of the screen and creating a pseudo-world of geology. It is one more reason to show the outcrop, the rock and the mineral – see sections 4 and 5. Cross plots and diagrams which are necessary are given in the basic studies, quoted in each place.

1.2.3 The CMS Classification Scheme of Pegmatitic and Aplitic Rocks

A classification scheme should cater for the extractive and genetic part of economic geology alike and ought to be applicable first and foremost in the field, due to the fact that pegmatites like other geological bodies need accommodation space to get emplaced, and are related in time and space to crystalline country rocks in a tripartite fashion. The criteria for classification must be descriptive and measurable in the field as far as the three-dimensional representation of pegmatites is concerned and the composition needs to be determined by visual inspection in the field for first-hand information. It goes without saying that the second issue referring to the composition of the ore body has to be backed and supplemented by mineralogical and chemical methods in the laboratory so as to achieve some fine-tuning of the class of pegmatites.

All classification schemes of pegmatites put forward so far are focused on the pegmatite itself, with obviously no structural or field geologists or even economic geologists consulted. Another classification scheme of pegmatitic and aplitic rocks has been designed here, stimulated by the ideas of the "Chessboard classification scheme of mineral deposits" with the same basic principle of the concept, simple in the design, using terms well entrenched in geosciences, open for amendments and directed towards improving the exploration of new and exploitation of preexisting deposits as well as genetic interpretation.

The classical definition of pegmatites given in Bates and Jackson (1987) can still today be agreed with in general, removing, however, some phrases that in my opinion overestimate the role of granites as the father of all things or parents of pegmatites what certainly cannot be claimed for all felsic rocks of this texture and mineralogical composition-see overview of pegmatites *sensu lato* (Sect. 1.2.1).

Today we evaluate the role of granites, especially in the field of economic geology a bit more realistic and see a lot of other sources of elements besides granites. The same may also be applied to the family relationship between granites and pegmatites both of which are rather brethren than part of a parent-child relationship. In geoscientific terms, they are two sides of the same coin, a different physical-chemical response to heat and/or pressure events in the crystalline basement at a certain level of structural deformation. The pressure obviously has less impact on the development of pegmatites than temperature but rock mechanics and tectonics do have. All geoscientific investigations in the study area along the western edge of the Bohemian Massif, dealing with the temporal and spatial relationship between pegmatites s.l. and granitic rocks corroborate this metaphorical language used above The studies attest to different ways of how pegmatites formed and consider the derivation of pegmatites (coarser-grained than granites s.s.) and aplites (finer-grained than granites s.s.) from granites (in rare places also granodiorites, diorites and gabbros) only as one among many ways. This multiple-choice approach is not really a surprise, taking into account that melting in the crust is anything but a byproduct of metamorphism. This calls for a fundamental re-assessment and with the aforementioned constraints in mind, I put forward a classifications scheme that is descriptive in essence and open for interpretation and recommendation/application (Table. 1.1). Like other rocks, pegmatitic and aplitic rocks are defined by the body (1st and 2nd order terms) and the composition (3rd and 4th order terms).

1st Order Term of Classification – Type of Aplitic and Pegmatitic Rock All aplitic and pegmatitic rocks occur in crystalline host rock lithologies, either metamorphic in origin or of magmatic-intrusive derivation. Against this background, factual data have to be collected in the field and the timing of the emplacement of these felsic rocks relative to the surrounding country rocks has to be established at the contact between both rock types. The analysis of the structural geology and metamorphic processes conduces to the type of aplitic and pegmatitic rock in the classification scheme; a wealth of information how to come to grips with that kinematic issue was published in the books of Davis and Reynolds (1996), Van der Pluijm and Marshak (2004) and Fossen (2010). For those who want it small, neat and handy, paperback versions for the fieldwork is recommended (Fry 1991; Lisle et al. 2011). Based upon descriptive mineralogical and structural features, metapegmatites/metaaplites which evolved prekinematic/premetamorphous can be distinguished from pegmatoids/aploids which are synkinematic/synmetamorphous and from pegmatites/aplites which formed in the aftermaths of these regional- and dynamometamorphic processes or after the emplacement and lithification of magmatic rocks (Table 1.1). The term pseudopegmatite was introduced for the metamorphic realm to avoid classification of pegmatite-like mobilizates and jumping into genetic conclusion which afterwards prove to be premature. This is especially true in the central European area under study when pegmatites re-appear south of the border of

Table 1.1 Classification scheme of pegmatitic and aplitic rocks for applied and genetic economic geology

ORE BODY						
Host rock lithology	Metamorphic rocks		Metamorphic rocks	Metamorphic and magmatic rocks	Magmatic rocks	Remarks
1 st order term **Type of pegmatitic/ aplitic rock**	**Pseudopegmatite/ pseudoaplite**	**Metapegmatite/ metaaplite**	**Pegmatoid/ aploid**	**Pegmatite/ aplite**	**Plutonic pegmatite/ aplite**	**Mandatory**
	Aplitic: grain size $\leq$ host rock and homogeneous Pegmatitic : grain size >> host rock and heterogeneous				Aplitic: grain size < host rock Pegmatitic : grain size >> host rock host rock and heterogeneous	
Specific type of host rock	*Gneiss, amphibolite , eclogite* *e.g., pegmatoid- (cordierite-sillimanite-gneiss)*				*Granite, syenite, granodiorite*	*Optional*
Determination	Mapping in the field the ore-host rock relation and measuring the grain size by visual examination					
2 th order term **Shape and structure**	**Tabular, schlieren, stock-like, pockets, vein-type, pipes, chimneys, floors** e.g. Sc-Nb aplite tabular, quartz-albite pegmatoid tabular				**Miarolitic, pod-like, pockets, vein-type, schlieren** e.g. topaz granite pegmatite miarolitic	**Mandatory**
Internal structure	Unzoned - rimmed-complex/ ungraded-complex/ graded (e.g. UST)					
Size (thickness)	*cm-sized, dm-sized , meter-sized*					*Optional*
Determination	Mapping in the field the shape by visual examination and measuring the morphological increments and size with a yardstick					
3 th order term **Chemical qualifier** **Rare metal pegmatites sensu lato**	**Sn, W, Ta, Nb, Sc, Be, Li, Cs, Rb, REE, Y, U, Th, B, F, P, Zr** **e.g. Nb-Li pegmatite, (Sc-U)-Nb-P aploid**					**Mandatory** Can be linked to the "Chessboard classification scheme of mineral deposits", using alpha-numerical codes
Determination	By visual inspection of rock-forming minerals, including hand lens for accessory minerals (put in brackets) Rock-forming and accessory minerals $\Rightarrow$ chemical symbols (e.g. beryl, euclase$\Rightarrow$ Be, Li mica$\Rightarrow$ Li, allanite $\Rightarrow$ REE, if necessary LREE)					
4 th order term **Mineralogical qualifier** **Industrial mineral**	**quartz, feldspar, foid, garnet, zeolite, mica, corundum, graphite** **e.g. (andalusite)-quartz-feldspar metapegmatite, graphite-feldspar-quartz pegmatite**					**Mandatory** Can be linked to the "Chessboard classification scheme of mineral
pegmatites						deposits", using alpha-numerical codes
Specific type of minerals for gemstone-bearing pegmatites, fine-tuning and genetic interpretation	*e.g. Alpegmatite (ruby), F-Sn-W granite pegmatite (topaz>fluorite), Li-Nb-P pegmatite (triphylite), Li-Nb-P pegmatite (amblygonite)*					*Optional - Composite of level 3 and 4*
Determination	See 3 nd order term for determination an					
ORE COMPOSITION						

the Variscides within the Alpine mountain belt. Terms like "remobilized pegmatite" or "reworked pegmatite" seem to be reasonable, yet they would violate the scheme of classification to be descriptive and field-oriented *ab initio*. To clarify and elucidate this classification problem, two images from the Alpine mountain range are on display, one showing scheelite and amphibole from the western mining camp of the Felberthal tungsten deposit, Austria, the other image displays spodumene intergrown with amphibole from the Koralpe lithium deposit (Figs. 1.2a, 1.2b, and 1.2c) (Göd 1989). The pseudopegmatites from the Austrian Li deposits have much in common as to the outward appearance with the Greenbushes deposit, Australia (Kippenberger et al. 1988; Partington et al. 1995; Dill 2010,). Both pegmatitic rocks are intercalated among metabasic igneous and metasedimentary rocks while there was no granite nearby as the pseudopegmatites were emplaced. This is especially striking in Greenbushes where the emplacement of the granites predates the formation of the pegmatite by about 90 m.y.

Lens-shaped intercalations of green amphibole in an matrix of coarse-grained scheelite from the Felbertal W deposit (western ore shoot) in the Austrian Alps closely resembles that of the amphibolite-hosted pseudopegmatite from the Koralpe Li deposit, Austria (Fig. 1.2d). This W mineralization in the Alps may be correlated with the late Variscan W mineralization from the Central European Variscides and the scheelite mineralization is consequently a metallogenetic pendant to the re-activated Li pegmatites in the Bohemian Massif, now being present at pseudopegmatites.

Fig. 1.2a Overview of the lithium pegmatites of the Koralpe Li deposit (Courtesy of R. Göd 1989). Adit exposing a lit-par-lit structure of aplitic and pegmatitic layers (*white*) and amphibolite (*dark*)

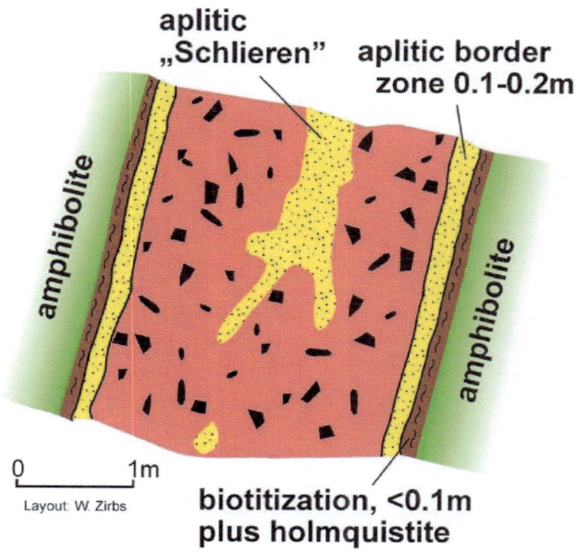

Fig. 1.2b Schematic cross section though the amphibolite-hosted lithium-bearing pseudopegmatite of the Koralpe deposit (Courtesy of R. Göd 1989)

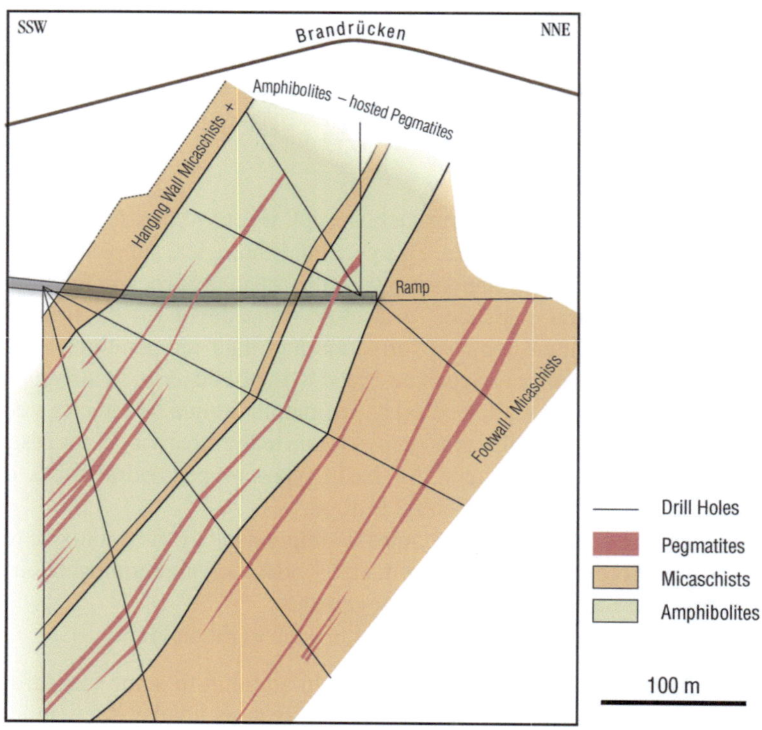

Fig. 1.2c Pseudopegmatites and pegmatites at the Brandrücken in the lithium deposit Koralpe, Austria (Redrawn from R. Göd 1989)

Fig. 1.2d Lens-shaped intercalations of green amphibole in a matrix of coarse-grained scheelite from the Felbertal W deposit (western ore shoot). The ore texture closely resembles that of the amphibolite-hosted pseudopegmatite from the Koralpe Li deposit. The W mineralization in the Alps may be at least in parts correlated with the late Variscan W mineralization known from the Central European Variscides and the scheelite mineralization considered as a re-activated pendant to the Li pseudopegmatites among the granite -related mineralizations (See also schematic cross section of Fig. 2.2a)

Depending upon the type of intrusive host rock we may classify the pegmatites as granite or granodioritic pegmatite. This peculiar lithological supplement can only be applied if the paragenetic relationship with the felsic host rock is for sure based on structural and textural evidence. To disclose more details of the origin of the pegmatitic rocks as to their depth of formation and their temperature regime, the type of metamorphic country rock can be added to the basic term, as an option, e.g., metaaplite (cordierite-sillimanite gneiss) or pegmatoid (amphibole gneiss – eclogite amphibolite). There are no limitations set by system to the creativity of the researcher handling this classification system. All basic types of pegmatitic rocks can be encountered in this study area in Central Europe.

The distinction of aplites and pegmatites is simply done by the grain-size relation between host and ore body: Aplitic: grain size ≤ host rock and homogeneous, pegmatitic : grain size >> host rock and heterogeneous.

The determination of the various features is done in the field.

2nd Order Term of Classification – Shape In addition to some standard terms that illustrate the shape of the pegmatitic rocks, such as tabular, schlieren, stock-like, or vein-type, the applicant can define further qualifiers. Some however, will be restricted to the granite pegmatites, such as pod-like, vein-type, schlieren or miarolitic, the latter adjective has also been used by Černý (1991) and by Černý and Ercit (2005). In their classification scheme the above authors used this term to address the pegmatites at the shallowest level of intrusion. From tabular to miarolitic pegmatites all sorts may be mapped in the study area.

Some may criticize this classification scheme becomes to industry-minded when geometric terms are added to a classification of pegmatites. This only half the truth. Engineers will be fond of finding some general terms as to the shape of pegmatitic rocks, because it may ease trenching, drilling and drifting and last but least help prepare a (pre)feasibility study of a pegmatite project.

The other half of the truth is that even geoscientists focusing on the genetic part of pegmatites may benefit from structural and geometrical terms. For those who can read the book of structural geology in the field the shape can tell them, e.g., where the material came from and what was the pathway, whether the material has totally come out of the surrounding rocks and they might get some ideas about the timing of the various pegmatites in a certain pegmatite province (Table 1.1). This is especially true when mineralized shear zones and host anticlines crop out and are accessible to mapping in the field. The pegmatitic rocks can be assigned a certain style and phase of deformation so that the various pegmatite bodies are no longer erratic rocks but architectural elements of a dynamo-metamorphic plan. Consequently these felsic intrusive rocks are amenable for an incorporation into an exploration plan for pegmatitic deposits (Sect. 5.1.4). Browsing the pertinent literature this part seems to be least well stocked one and a lot of study needs to be done to provide the information necessary for the pegmatite classification.

The internal structure of pegmatites is rather varied and it is difficult to draw a line between what is a "must" for the classification of these rocks and what is only to be disclosed to the specialist (Cameron et al. 1949). To make the classification scheme comprehensible also for geoscientists outside the field of pegmatology I have decided to do not overload this level with too many terms and facts only known to an insider. Some of the terms are self-explanatory such as unzoned and rimmed. Complex/ungraded means a sequence of zones evolves either in a tabular or miarolitic body, showing neither a trend in the size variation nor in the composition. By contrast complex/graded pegmatites may show conspicuous trends and textures which reveal a cross evolution bottom-up or may be comb-like crystal layers as they were described by Shannon et al. (1982) as unidirectional solidification texture (UST). Terms like wall zone, border zone, intermediate zone etc. sometimes used in publications may be helpful but they are not very elucidating to the morphology and structure of the ore body. Therefore preference was given to these few technical terms mentioned above.

The terms in use for the description of the "Ore Body" were selected so as to be applicable also for geophysical surveys during exploration for pegmatites – see Figs. 3.9 and 3.10.

3rd Order Term of Classification – Chemical Qualifier The 3rd order term starts the description of the second part called the "Ore Composition" (Table 1.1). While many publications lack any structural details so as to describe the "Ore Body", a wealth of mineralogical data on pegmatites exist and enable us to conduct a classification of many well-studied but no longer accessible pegmatites.

Feldspar, quartz and mica form the main constituents of granitic rocks as well as of pegmatitic and aplitic rocks. Thus these ubiquitous rocks need not be explicitly named in a classification scheme dealing with granitic and pegmatitic rocks focused for its rare metal contents. It is the group of minor constituents next in abundance to the afore-mentioned rock-forming silicates and the commodities, which the pegmatite is operated for, that play a decisive role in a more detailed classification scheme. All elements concentrated during the emplacement of these pegmatitic rocks such as Sn, W, Ta, Nb, Sc, Be, Li, Cs, Rb, REE, U, (Th), B, F, and P can be used to specify the pegmatitic rocks. Five to six qualifier may be sufficient in practice to render the classification scheme manageable, even though there is no real limitation to the number of qualifiers. Sometimes even minor elements warrant mentioning in the classification as chemical qualifiers to predict, e.g., element scavenging by certain minerals, as it might be the case with tourmaline (B) and columbite (Nb) in pegmatites which are crucial as to the Fe-Mn ratio in columbite evolving in the aftermaths of schorl or dravite. The elements are arranged in the order of decreasing abundance with the most widespread marker-element next to the 1st order term, e.g., Sn-Sc-Nb pegmatite that is a pegmatite most strongly enriched in Nb while Sc prevails over Sn. Chemical qualifiers are placed in front of the type of pegmatite. Pegmatite-forming minerals such as spodumene or beryl, irrespective of their economic or subeconomic grades are listed by the chemical symbol Li and Be, respectively. If beryl is a minor constituent and only present on a microscopic scale the chemical symbol is put in curved brackets.

4th Order Term of Classification – Mineralogical Qualifier There may be two possibilities that need to be mentioned in this classification scheme. In addition to the typomorphic elements of the afore-mentioned group (3rd order term of classification), there may be minerals in the pegmatitic rocks others than feldspar, quartz and mica, containing major elements such as Al, Fe, Mg, and Mn accommodated in the lattice of andalusite, almandine-enriched garnet, cordierite or spessartite. Irrespective of the fact whether they are exploited for a special purpose or not, they need to be addressed for genetic reasons (Table 1.2). Similar to the chemical qualifier which are decisive for the classification of pegmatites, mineralogical qualifiers are also placed in front of the type of pegmatite, with minor constituents put in curved brackets, as exemplified: (andalusite)-quartz-feldspar metapegmatite or graphite- feldspar-quartz pegmatite. Under certain physical-chemical conditions it might happen that no rare elements giving rise either to cassiterite or beryl and minerals others than these rare-element minerals show up in the pegmatitic rocks as diagnostic minerals. In this case, you can skip the step referring to the chemical qualifiers and directly move to the next step, where the mineralogical qualifier plays the major part in naming the rock, e.g., andalusite pegmatoid. Quartz, feldspar, foid, garnet, zeolite, mica, corundum, and graphite are the common mineralogical qualifier, leading immediately to the rock name, such as feldspar pegmatoid, quartz pegmatite or graphite pegmatite. For reasons of final use, it may sometimes be advisable

Table 1.2 A NW-SE transect through the Central European Variscides and the Alpine Mountain range with focusing on the pegmatitic rocks

Geodynamic unit	Pegmatitic/aplitic deposits	Mineral association	Host rocks	Chemical/ mineralogical qualifier	Type	Geodynamic setting
Mid-German Crystalline Rise (Ref.: Sect. 2.1.3)	Dahlem's Buckel (Aftholder), Glattbach, Grauenstein, Huthberg near Aschaffenburg, Goldbach, Dürrmorsbach	Li-bearing biotite, tourmaline, garnet (spessartite), magnetite, sillimanite, muscovite, beryl, apatite, gadolinite, orthite	Gneiss	B-(Li)	Pegmatoid	Suture zone extending across present-day continents and resulting from the late Variscan closure of the Rheic Ocean between Gondwana and Laurussia with remnants of an arc-related plutonism
				P		
				REE-U		
				Be		
Saxothuringian Zone- Münchberg Gneiss Complex (Ref.: Sect. 2.1.4.2	Autengrün, Bärenbühl, Bärlas, Böseneck, Bucheck, Falls, Friedmannsdorf, Gefrees, Grosslossnitz, Hampelhof, Kesselberg, Lübnitz, Marktschorgast, Mechlenreuth, Metzlendorf, Münchberg, Oberpferd, Schlegel	Albite-oligoclase, muscovite, quartz, kyanite, zoisite	Eclogite, eclogite amphibolite, gneiss, metabasite, metabasite-gneiss	Feldspar-quartz	Pegmatoid (metapegmatite)	Allochtonous metamorphic complex, nappe
	Schwarzenbach near Münchberg, Seulbitz, Stammbach, Streitau, Tennersreuth, Weissdorf, Weissenstein Mt., Witzleshofen, Wulmersreuth, Wundenbach, Zettlitz					

(continued)

Table 1.2 (continued)

Geodynamic unit	Pegmatitic/aplitic deposits	Mineral association	Host rocks	Chemical/ mineralogical qualifier	Type	Geodynamic setting
Saxothuringian Zone- Fichtelgebirge- Erzgebirge Anticline (Ref.: Sect. 2.1.4.3)	Karches aplite granite, Waldstein aplitgranite, Holenbrunn, Stemmas, Epprechtstein, Kleiner Kornberg, Gregnitzgrund near Nagel, Grünberg, Rudolfstein, Fuchsbau, Selb	herderite, tourmaline, fluorite, topaz, apatite, cassiterite, zinnwaldite, lepitholite, arsenopyrite, wolframite, scheelite, euclase, beryl, bertrantite, bavenite, phenakite, hematite, albite, smoky quartz, muscovite, andalusite, anatase, arsenopyrite, autunite, meta-autunite, torbnerite, metatorbernite, amblygonite garnet (spessartite), andalusite	granite, calcsilicate, marble, mica schist, phyllite	B	pegmatite granite-pegmatite (miarolitic) pegmatite-aplite pegmatoid	Subfluence zone
				Be		-continent-continent collision and thickening of the crust
				Be-B		
				Be-B-F-Li-Sn		
				Be-F-B-Li-U		-attenuation of the thickened crust
	Wintersberg-Katharinenberg near Wunsiedel, Schönlindt Zeche (mine), Röslau Schönlindt Castlehill			Sn-F-P-As		-exhumation
	Dillenberg (Tillenberg) near Waldsassen					
Moldanubian Zone- Contact zone between Zone of Erbendorf Vohenstrauß and Moldanubian s.str. (Ref.: Sect. 2.1.5) NE Oberpfälzer Wald/Zone of Tirschenreuth-Mähring	Hopfau near Erbendorf, Leichau near Stein, Schirnbrunn, Sägmühle near Tirschenreuth, Marchaney, Ahornberg, Lauterbach, Ellenfeld, Dippersreuth, Poppenreuth, Großkonreuth, Grün SE of Tirschenreuth, Schwarzenbach SE Tirschenreuth, Honnersreuth Mine near Liebenstein, Pilmersreuth, Beidl, Plößberg, Püchersreuth "Auf der Wacht"	apatite, laueite, rockbridgeite, leucophosphite, schoonerite, mitridatite, beraunite, cacoxenite, stewartite, strengite, eosphorite, pharmacosiderite, strunzite, pseudolaueite, triphylite, triplite, phosphosiderite, wavellite, heterosite, vivianite, lazulite, metavivianite, santabarbaraite, reddingite, topaz, arsenopyrite, scorodite, tourmaline, beryl, columbite, zircon, muscovite, durmortierite, schorl, garnet, staurolite, tourmaline, U mica, native Bi	granite, gneiss-granite, mica schist, metabasite	B	pegmatite-aplite pegmatite	Strong diaphtoresis and shearing in the contact zone between the Saxothuringian and Moldanubian zones *sensu lato*.
				B-P-Be-Nb-As-Zr		
				B-Be-Nb		
				Be-B-Nb		
				Be		
				Be-Nb		
				P-F-As		
				P-B		

Moldanubian Zone-Zone of Erbendorf Vohenstrauß (Ref.: Sect. 2.1.5) NW Oberpfälzer Wald	Wildenreuth, Püllersreuth (Grube Maier), Püllersreuth II (Trial mining), Wilma Mine near Wendersreuth, Gertrude Mine (+2 satellite mines) near Klobenreuth, Lenkermühle, Menzlhof (Grube Gertrud), Menzelhof (Gube Klara), Obersdorf (Grube Mandt), Zeßmannrieth, Ach Mine Ödenthal, Irchenrieth, Blockhütte, Muglhof	beryl, columbite, zircon, uraninite, sphalerite, tourmaline, siderite, marcasite, pyrrhotite, pyrite, rockbridgeite, cacoxenite, vivianite, spessartite, zeolite, titanite, clinozoisite, diopside, prehnite, wurtzite, garnet, autunite, molybdenite	mica schist, biotite gneiss, biotite-silimanite-kyanite gneiss, amphibolite	Be-Nb	meta-pegmatite	Suboceanic mantle, ocean ridge to within- plate-magmatism with subsequent rifting in an allochthonous position
				P	pegmatoid	
				feldspar-quartz	pegmatite	
Moldanubian Zone- (Ref.: Sect. 2.1.5) Central Oberpfälzer Wald-Hagendorf-Pleystein-Pegmatite Province	Brünst Mine, Reinhardsrieth, Peugenhammer, Pleystein Kreuzberg, Pleystein "New Aplite", Pleystein Trutzhofmühle Aplite, Pleystein Rehbühl, Hagendorf-North, Hagendorf-South, Silbergrube Aplite near Waidhaus, Burkhardsrieth, Miesbrunn West and East Pegmatite-Aplite Swarm, Kreuth, Eulenberg, Schwand, Kalvarienberg, Steinach near Vohenstrauss, Zessmannsrieth, Walthurn, Weissenstein, Kühbühl, Galgenberg, Kreen, Hasenbühl (all of the near Pleystein), Gröbenstedt, Lohma-Premhof, Isgier, Dreihöf, Fuchsbach- Zengerhöf, Pingermühle, Kupferbühl, Vogelherd, Nassgalle, Neuenhammer, Schafbruck, Zottbachtal, Georgenberg, Neukirchen	Acanthite, alanite, albite, alluaudite, anatas, apatite-(MnF), arrojadite-(BaFe), arrojadite-(KFe), arrojadite-(Na), arrojadite-(NaFe), arrojadite-(SrFe), arsenopyrite, autunite, barbosalite, baryte, bassetite, benyacarite, beraunite, bermanite, berthierine, bertrandite, beryl, beudandite, biotite, bismite, bismuth, bismuthinite, brazilianite, brochantite, cacoxenite, calcioferrite, calcite, carlhintzeite, cassiterite, cerite, chalcanthite, chalcocite, chalcophanite, chalcopyrite, chalcosiderite, chamosite, cheralite, chernikovite, childrenite, chlorite group, chrysocolla, churchite-(Y), coffinite, columbite-(Fe), ("niobite"), connellite, copper, covellite, crandallite, cryptomelane, cubanite, cuprite, cuprobismutite, cyrilovite, devilline, diadochite, dickinsonite-(NaMn)	biotite-sillimanite gneiss, calcsilicate, granite	B B-P-REE P-B P-B-Nb-Zr P-Nb P-Nb-Li P-Nb P-Nb-Sc P-Li-Nb P-Nb-Li-Zn-Be	pegmatite aplite	High grade metamorphic rocks in an autochtonous position with a protolith mainly of Proterozoic age. At the margin overtrusted onto adjacent geodynamic units (Moravo-Silesicum) and penetrated by multiple intrusions

(continued)

Table 1.2 (continued)

Geodynamic unit	Pegmatitic/aplitic deposits	Mineral association	Host rocks	Chemical/ mineralogical qualifier	Type	Geodynamic setting
		dickinsonite-(KMn), digenite, djurleite, dufrénite, earlshannonite, emplectite, eosphorite, fairfieldite, ferriallanite, ferrisicklerite, ferristrunzite, ferrolaueite			pegmatoid	The Hagendorf-Pleystein Pegmatite Province is the root zone for the nappe complexes thrusted onto the north-western geodynamic realms

		ferrostrunzite, fluellite, fluoroapatite, fluorite, frondelite, gahnite, galena, garnet, goethite, gordonite, goyazite, graftonite, graphite, greenockite, gypsum, hagendorfite, hematite, hemimorphite, heterosite, hopeite, hureaulite, ilmenite, jahnsite-(CaMnFe), jarosite, jungite, kaolinite, kastningite, keckite, kingsmountite, kidwellite, kolbeckite, kryzhanovskite, landesite, langite, laueite, lazulite, lehnerite, lermontovite, lepidocrocite, leucophosphite, libethenite, 'limonite', lipscombite, ludlamite, mackinawite, magnetite, malachite, mangangordonite, mantienneite, marcasite, matildite, mesolite, matulaite, messelite, metaautunite, metavivianite, metaswitzerite, metatorbernite, meurigite-K, microcline, mitridatite, molybdenite, monazite, morinite, muscovite, natrodufrenite, nontronite, nordgauite, orthoclase, oxiberaunite, pachnolite, parahopeite, parascholzite, paravauxite, pavonite, perloffite, petitjeanite, petscheckite, pharmakosiderite, phosphoferrite, phosphophyllite, phosphoscorodite, phosphosiderite, posnjakite, pseudolaueite, pseudomalachite, psilomelane, purpurite, pyrite, pyrolusite, pyrosmalite, pyrrhotite, quartz, reddingite, rhodochrosite, riomarinaite, rittmannite, robertsite, rockbridgeite				

(continued)

Table 1.2 (continued)

Geodynamic unit	Pegmatitic/aplitic deposits	Mineral association	Host rocks	Chemical/ mineralogical qualifier	Type	Geodynamic setting
		rosasite, rutile, romanechite, rozenite, santabarbaraite, samarskite-(Y) var. Uransamarskite, sarcopside, scholzite, schoonerite ("oxi-schoonerit"), scorodite, scorzalite, siderite, sillimanite, silver, sphalerite, stannite, sternbergite, stewartite, stibnite, stilpnomelane, strengite, stromeyerite, strunzite, sulphur, switzerite, tainiolite, tavorite, titanite, torbernite, tourmaline, triphylite, triplite, triploidite, turquoise, uraninite, uranocircite, uranophane, uranosphaerite uranpyrochlore, vandendriesscheite, variscite, vivianite, vochtenite, vyacheslavite (?), 'wad', wagnerite, wavellite, waylandite, whiteite-(CaMnMn), whitmoreite, wilhelmvierlingite, wittichenite, wolfeite, xanthoxenite, xenotime-(Y), zairite, zircon, zwieselite				
Moldanubian Zone- (Ref.: Sect. 2.1.5) Southern Oberpfälzer Wald-Vorderer Bayerischer Wald	Weiding near Schwarzenfeld, Unterauerbach, Kemnath –Fuhrn, Döfering, Waldmünchen, Rötz, Lengau – Herzogau, Katzberg, Roßbach, Erlbach, Schwarzenweiher/Fußenberg	triplite, andalusite, tourmaline, cordierite, apatite, fluorite, chalcopyrite, galena, torbernite	gneiss, granite, quartz lode, metabasite, granite-mylonite, diorite	P	pegmatite	See above
				B	aplite	

Moldanubian Zone- (Ref.: Sect. 2.1.5) Southern Bayerischer and Böhmer Wald	Schwarzeck near Lam, Hörlberg, Schmelz, Blötz, Schneiderberg, Arnbruck, Frath Mine, Drachselrieth "In der Frathau", Bärenloch Arber Mts., Hühnerkobel, Zwiesel Kahlenberg, Zwiesel –Birkhöhe, Taferlhöhe near Oberfrauenau, Hirschbach – Frauenau, Reitenberg-Kaitersberg, Schöllnach, Wimhof	corundum, andalusite, apatite, tourmaline, topaz, tourmaline, garnet (spessartite), triplite, uraninite, vesuvianite, dumortierite, cordierite, kyanite, samarskite, autunite, zwieselite, quartz, albite-oligoclase, muscovite, biotite, autunite, meta-autunite, torbernite, pyrite, limonite, pyrolusite, uranocircite, löllingite, columbite-(Fe), beryl, bertrandite, triplite, jahnsite, scorzalite, arrojadit, keckite, sicklerite, kidwellite, childrenite, triphylite, diadochite, eosphorite, montgomeryite, mitridatite, kingsmountite, rockbridgeite, alluaudite, anapaite, cacoxene, strunzite, ludlamite, laueite, beraunite, leukophosphite, stewartite, strengite, manganosegeleritecassiterite, scorodite, arsenopyrite, tantalite, columbite, uranophane, uranophane-beta, sillimanite, triploidite, lazulite, scorzalite, vivianite, fluorite, pyrochlore, allanite, bazzite, bavenite, helvine, bertrandite, bityite, triphylite, monazite, arrojadite, hureaulite, zeolite, anatase, titanite, ilmenorutile, titanite, magnetite	marble, calcsilicate, mica schist, granite, cordierite-sillimanite gneiss, granite-gneiss-mylonite, granodiorite	As	pegmatite	See above
				B-F-P	aplite	
				B-P-U	pegmatite skarn	
				B-P-Nb-REE		
				B-P		
				B		
				B-P –As		
				B-Nb		
				B-U		
				P-Nb-Be-U -F		
				P-Nb-B		
				P-B-Be		
				P-Be-REE-U-B-F-Li-Sc		
				P-Be-B		

(continued)

Table 1.2 (continued)

Geodynamic unit	Pegmatitic/aplitic deposits	Mineral association	Host rocks	Chemical/ mineralogical qualifier	Type	Geodynamic setting
Variscan Massifs in the Austrian Alpine Mountain Range (Ref.: Sect. 2.2.3) Kärnten-Steiermark	Spittal an der Drau, granite pegmatite near Villach, Erling near Spittal, Wildbachgraben, Koralpe-Weinebene deposits, St. Radegund	allanite-(Ce), andalusite, apatite-(F), arsenopyrite, axinite, autunite, bavenite, bertrandite, beryl, beryl (aquamarine), cassiterite, chessboard albite, columbite, chrysoberyl, fairfieldite, ferroniobite, ferrosicklerite, fluorite, garnet, greifensteinite, heterosite, holmquisite, hydroxylherderite, ilmenorutile (niobian rutile), lithiophilite, kyanite, ludlamite, meta-autunite, meta-torbernite, microcline, monazite, montebrasite, muscovite, phenakite, purpurite, pyrochlore, roscherite, scheelite, schorl, spodumene, staurolite, tapiolite, tourmaline, triphylite, uralolite, uraninite, uranophane, vivianite, weinebeneite	garnet-bearing mica schist, mica schist, quartzites, granite, amphibolite	B-Be-P-Nb-U- Be-B-P-F-As Be-Li-B Li Li-Be-Nb- Sn-U Li-Be-P-REE-U As-B-Nb	granitic pegmatite, meta-pegmatite, pegmatoid, pseudo-pegmatite	Variscan lithologies incorporated into the Alpine orogen and reactivated during the Alpine orogeny

The various sites are arranged as a function of the geodynamic evolution of both orogens. For references see the various sections in the text

The **Geodynamic unit** is given by the name used by geoscientists in Central Europe. In rounded bracket the section in this book is given for further information. The **Pegmatitic/aplitic deposits** were mined at various sites in the past and only a few are still under operation. The **Mineral association** is a compilation of minerals found in the pertinent mining district or pegmatite province. Feldspar and quartz are only mentioned for the so-called barren, pegmatites and aplites devoid of rare metals. The **Host rocks** refer to those rocks giving host to the pegmatitic and aplitic rocks. The **Chemical/mineralogical qualifier** and the **Type** are used according to the classification scheme proposed in Table 1.1. The **Geodynamic setting** summarizes the results of the geodynamic interpretations done by various researchers referred to in the text (Sects. 2.1 and 2.2)

Note for reasons of comparison the mineralogical and chemical qualifier are listed in the opposite direction as used to be done in the CMS scheme. The prevailing element is placed on the left-hand side of the column to provide full particulars as to the composition of feldspar, e.g., albite pegmatoid, rose quartz pegmatite, scapolite-sapphire pegmatoid. It is not only the mining engineer or gemologists who may reap the benefit of this more detailed mineral description but it enables geoscientists also to fine-tune the physical-chemical regime of pegmatitization. The classification scheme also allows for a combined use of the terms from level 3 and level 4 in Table 1.1. In case of lithium pegmatites, the element may be accommodated in the lattice of spodumene or phyllosilicates such as zinnwaldite or lepidolite, and thus it is advisable to use one of the Li minerals as qualifier, e.g., Li-Nb-P pegmatite (amblygonite) versus Li-Nb-P pegmatite (triphylite). In this case amblygonite and triphylite do not act as qualifiers but as specification for the Li and P components. As such they were placed behind type of pegmatite. This mineralogical setting is common at the passage from the Moldanubian into the Saxo-Thuringian Zone. A term like that is more in harmony with nature and tells you much more than the word LCT granite pegmatite, when cesium cannot be pinpointed mineralogically, the columbite s.s.s. is enriched in Nb rather than Ta and no granite is close by.

Ab initio, a pegmatite is defined by its chemical/mineralogical composition and by its three-dimensional shape. This is valid irrespective of the depth of emplacement, the temporal relation between formation and deformation. As this classification scheme is based on the **C**hemical composition, the **M**ineral assemblage and the **S**tructural geology of pegmatitic rocks the acronym CMS has been coined and hence it called the "**CMS classification scheme of pegmatitic and aplitic rocks**". Whether you see the pegmatitic rocks as a source of raw materials or as an objective for genetic study, this descriptive classification scheme is a sound basis for both camps of economic geology to live with. The third strong point of the **CMS classification**, lies in its use as a key element or legend for mapping projects. A similar approach has already been taken and tested for the metallogenetic map 1:2,500,000 available on CD (Dill et al. 2008a, b).

1.3 Pegmatites and Economy

1.3.1 Pegmatites in NE Bavaria a Source of all Kinds of Everything

It has already emphasized in the previous chapter by the variable associations of elements and minerals in the CMS classification that minerals from all three commodity groups ore minerals, industrial minerals, and gemstones can be exploited for a profit from the various type of pegmatites. In the study area in Central Europe, almost all mining operations on pegmatites were/are being concentrated on industrial minerals and only in one case the light metal lithium was won as a byproduct for a short period of time at Hagendorf-South. Therefore at the beginning of this discussion on use and final products of pegmatite-related mineral raw materials, the industrial minerals have to be placed much emphasis with feldspar ranking first. It has to be noted, that much of what is going to be dealt with in this section, the reader can also find in the "Chessboard classification scheme of mineral deposits" where a direct link to the US Geological Survey Database at the end of each chapter enables the user to update his/her information on an individual commodity on an annual basis.

Feldspar The feldspar-group minerals albite to albite-oligoclase and K feldspar is not only the silicate groups that is second to none in the built-up of the Earth crust but also dominating the list of applications with glassmaking and manufacturing fine and coarse ceramic goods by as much as 80–90 %. Second in the row, are classical products such as paint fillers, abrasive fillers, welding rods and cosmetics. It is a point of particular note, that not only the run-off mine product is what counts on this market of siliceous raw material but also the processing of the feldspar. While the mining is still done in a rather conventional way with little change in the basic tools throughout the last decades, processing of the siliceous raw material has to keep pace with the very latest developments in order to provide material of a precise particle size and highest purity. Ground feldspar is delivered to manufacturers of ceramic goods and granular raw material to those making glass. Feldspar improves the quality of glass by increasing its resistance to devitrification, making it more durable and amenable to forming. It has also an impact on the workability and thermal expansion. Natural feldspar from pegmatitic rocks are in fierce competition with recycled glass from households and the industry. Processing this raw material will become more and more important relative to the exploitation of the natural raw material.

In the ceramic industry the feldspar's position is rather unrivaled because of the more complex manufacturing process, which includes the production of the ceramic body, ceramic glazes and frits, a mixture of fused ceramic raw materials. Alkaline feldspar which is one component besides quartz and kaolinite in the production of chinaware has positive as well as negative impacts on the ceramic body being fired at temperatures of as much as 1,200 °C. It lowers the firing temperature in the kiln but on the other hand decreases the refractoriness. Higher feldspar content goes along with lower absorption of water that reaches its minimum high-strength ceramics.

The finest grain sizes of feldspar used as a filler in paints where this silicate improves the durability and resistivity. One of the main operation targeted upon albite pegmatoids in the study area to make the perfect cleavage of feldspar available as an abrasive utilized in households and the industry.

The lion share of feldspar is used as filler in plastics, paint, sealants, adhesives, abrasives, flame damper in the match industry and especially for ceramics (porcelain) and glass but a small fraction is also used as gemological products, mainly by local lapidarists. The main producers of feldspar are Italy, Turkey, China, Japan, Thailand and the USA.

Quartz In the Earth's crust, quartz is second in abundance behind feldspar and so does it in pegmatites when it comes to the application of industrial minerals. It has become clear that impurities much more than in feldspar should be removed from quartz that is used for the production of semiconductor chips and special glass devices. The common pegmatite quartz is exploited for a wide range of applications in the glass industry, for ceramics, abrasives, fillers and special chemical compounds such as Si carbides. High-purity or even ultra-high-purity quartz find application in the chip industry, for the manufacture of solar cells, and special ceramics. In Table 4.4, in Sect. 4.2 the trace element contents of some quartz pegmatites and quartz dikes straddling the Czech-German border are listed. They were listed in this section because the trace elements can also shed some light on the physical-chemical conditions under which the host pegmatites formed. Apart from the Hagendorf-North and Hagendorf-South pegmatite, there is only the Kreuzberg pegmatite in the town of Pleystein that fulfills the requirements of a high-purity quartz deposit (Fig. 1.3). The latter, however, will never come into the reaches of exploitation since its quartz reef has to be called an "aesthetic deposit" rather than a quartz deposit in its own rights exploited by the tourists only. The reserves of this deposits are inexhaustible.

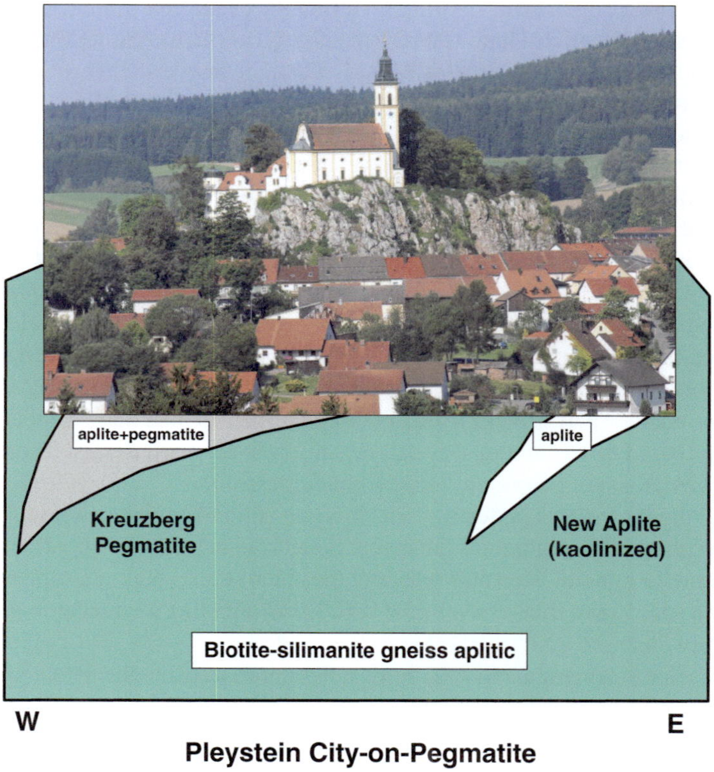

Fig. 1.3 The quartz reef of the Kreuzberg stands out from the city center of Pleystein ("City on Pegmatite"). The top of the pinnacle is topped with the Salesian monastery and its church. The depth penetration of the Kreuzberg Pegmatite and the New Aplite has schematically been illustrated underneath the photograph by a sickle-shaped wedge. The viewing direction is facing northward (See also Fig. 5.3a for more details on the lithological setting)

Mica It is the white mica muscovite and to a lesser extent the Fe-enriched dark biotite, which are mined from pegmatites for a wide range of products, mainly for their excellent electrical properties, chemical inertness, plasticity and low thermal expansion. Thin sheets of white mica are transparent, a reason why this sheet silicate was used as window in the fireplace of ovens as "Glass from Moscow". Today this mineral is no longer in use for that purpose, that accounts for its name, muscovite, but mainly used as filler, coating and lubricant. Cosmetics and insect repellents also make use of muscovite as a basic ingredient. Mica is mainly extracted from pegmatites as a by-product in the course of mining other commodities.

Lithium A wide range of elements such as Sn, W, Ta, Nb, Sc, Be, Li, Cs, Rb, REE, Y, U, Th, and F can be concentrated in pegmatites and reach economic grade rendering their exploitation feasible (Sect. 1.2.1). In NE-Bavaria, only lithium attained economic importance. While in most pegmatites spodumene, petalite and Li-bearing micas such as lepidolite and zinnwaldite are the number one Li ore minerals in hard rocks, at Hagendorf-South triphylite, a Li-bearing phosphate [$LiFePO_4$] was mined and shipped to Langelsheim (Northern Germany) where the lithium ore was processed at Hans-Heinrich-Hütte (Hütte = metallurgical plant), the first metallurgical plant for Li salts established in 1922/1923. There is an ever-increasing demand in lithium driven by the so-called "Green Technologies" where solid-state batteries play a center part in the automobile industry. Even though large salars in Argentina, Bolivia and Chile offer vast brine sources easy to access, the pegmatites did not lose their significance as a source of lithium. The smaller size and, in places, costly underground mining operations are fully compensated for by Li concentrations some orders of magnitude higher than in the sedimentary deposits (low-grade-large- tonnage deposits). The energy issue is so high up on the agenda in politics and plays such a prominent role in the public awareness that lithium electronic applications overshadow everything else in the ceramic and pharmaceutical industry also based upon lithium. In this field of ceramic application spodumene and petalite are the frontrunners. Apart from its use as grease, lubricant, solder and synthetic rubber, new drugs based on Li carbonate and its organic salts are often the last resort to mitigate the symptoms of people suffering from depression.

For all those elements, industrial minerals and colored gemstones, widely known to be concentrated in pegmatites the reader is referred to the "Chessboard classification scheme of mineral deposits" where at the end of each chapter a supplementary section on supply and final use is found with a link directing the reader to the USGS commodity database (Dill 2010).

Africa, South America and Asia have much more to offer for gemologists than the old world. Nevertheless, even the western edge of the Bohemian Massif cannot completely sidelined in this case as it has also a small share in this gemological application. Topaz, euclase and phenakite have been spotted in cavities of granite pegmatites of the Fichtelgebirge Mountains in the Saxo-Thuringian Zone. None attained more than showcase quality and the attention of mineral collectors.

1.3.2 *Extractive Geology Pegmatite Deposits at the Western Edge of the Bohemian Massif – A Historical Perspective*

1.3.2.1 The Extractive Geology of Pegmatites

Hagendorf South

Exploration and Mining It was no really a surprise that mining engineers very late became aware of the quality and size of the feldspar-quartz deposits of the HPPP which is located about 200 km NNE of München (Munich) and 100 km N of Regensburg, close to the Czech-German border in a highly reliefed terrain of the German low mountain range densely forested mainly with softwood trees (Fig. 1.1b). The HPPP in the Oberpfalz-SE/Germany, ranks among the largest concentrations of pegmatitic and aplitic rocks in Europe. The largest pegmatite of this mining district, named Hagendorf-South, totals 4.4 million tons of pegmatitic ore (Forster et al. 1967). When the first rare minerals, mainly phosphates were discovered, geologists and mineralogist from academia as well as rock hounds paid more and more attention to these pegmatites and aplites and after the first papers appeared on these pegmatites, sparked a run for new minerals in the HPPP which is still going on (Scholz 1925; Forster 1965; Forster and Kummer 1974; Strunz 1961; Strunz et al. 1975; Uebel 1975; Mücke 1987, 1988, 2000; Mücke et al. 1990).

Hagendorf-South is the largest pegmatite body in this pegmatite province, attracting geoscientists and those interested in the regional history, alike. Its exploration and exploitation provides an impressive insight into what has happened during mining raw materials, in a region mainly cultivated by farmers and forest workers and far off the large mining districts, such as the Coal Measures in the Ruhr District or the deep salt mines in northwestern and central Germany. In both mining areas, high steel head frames of the collieries still today remind us of the mining activities in the past and huge mining dumps of waste salt and useless excavation material will be an everlasting hint even if the salt domes are mined out and the sites have been converted into a derelict mining area (Fig. 1.4a and 1.4b).

The history of mining of the Hagendorf-South pegmatite was summarized by Keck (1990). Hagendorf-South is also known among locals in the Northern Oberpfalz as "Wildenauer Mine". Taking the name of the first owner to denominate the newly sunken shaft or mine has become common practice over the centuries in Germany. So did the Pleystein resident Michael Wildenauer who started mining Hagendorf-South in 1900. Yet he was not the one who discovered the deposit which happened to be found in 1894 when the country road from the hamlet Spielhof to Hagendorf was going to be constructed.

It was not until 1964 that the Amberger Kaolinwerke took over the mining area and began mining and exploration in the immediate vicinity on a larger scale. Prior to that stage of operation, the ownership at Hagendorf –South changed several times. Periods of operation were followed by periods when the mine was temporally abandoned (1919 Schmidt, Retsch & Co, Wunsiedel). During the Second World

Fig. 1.4a Immediately after mining ceased in 1982, the open pit has not yet been flooded and still allows a look at the underground works with the mined out chambers and pillars stabilizing the roof. The mining edifice with the head frame of the inclined shaft still visible on the western edge of the open pit (Courtesy of G. Rank)

Fig. 1.4b Oblique aerial view of the flooded open pit with the overburden and dumps in the background at Hagendorf-South already covered with a slash green vegetation of hardwood trees and shrubs. The mining area has be reclaimed and reused by a metal-working company while the mining relics have been converted into a nature-sanctuary. Sampling is no longer an easy task

War, unlike many other mining districts in the region, such as the Nabburg-Wölsendorf fluorite mining district, Hagendorf-South was idle until 1945. Feldspar and quartz were not considered strategic minerals and of utmost importance for the German armament industry.

After reopening the mine in 1945, the Bayerische Feldspatwerke Döllner & Wildenauer continued mining operations until 1964, when the deposit was sold to Amberger Kaolinwerke at Hirschau. Through 1986, Hagendorf-South did not only saw an intensification of the mining operation at the "Cornelia-Schacht" (=shaft) but also an extensive ground follow-up geophysical exploration mainly radiometry surveys, deep drilling and trenching in the immediate vicinity of the mining site. Cheaper raw materials and rising labor cost forced the company to close down the mining operations in 1986 and after having been cleared by the regional mining authority, nature conservationists took over the mining site, including the vegetated waste dumps nearby so that sampling is almost impossible now. The mining site, formerly occupied by the head frame and the processing plant has been reclaimed and it forms now part of the premise of a metal-working plant (Fig. 1.4b). The open cut is fenced all around and declared a prohibited zone, exclusive to wild animals with no entrance to mineral collectors anymore.

It did not take very long for the entrepreneurs to realize that a felsic intrusive body like Hagendorf-South could not be operated for feldspar and quartz on an economic basis by opencast mining only. Consequently, the miners began sinking a shaft as early as 1900 which reached the 34 m level in 1911. Gradually they went deeper and deeper with haulage levels at 17, 34, 45, 60, 67, 76 and 93 m below ground. When they started mining the central quartz core another level undercut the 93-m level at a depth of 115 m, while pump sump was located at the final depth of 135 m below ground – for a cross section see Sect. 5.1.2.

The various sublevels listed in the previous paragraph were worked for feldspar using the room- and-pillar mining technique (Fig. 1.4c). For roof support and for the suspension to prevent from collapse the mined-out parts of the pegmatite were backfilled with discard brought into the mine from outside. There did not exist any incline up to the surface to allow for LHD (load-haul and dump) operations. To increase the recovery of feldspar, pillars left in place by the ancient miners for security reasons formed part of the won minerals and were step by step mined out in the most recent mining period from 1965 to 1983. The fluctuation in ownership and a shortage of money caused some deviations from the well-known and adopted mining techniques and locally conduced to a rather liberal design of the underground operations which by and large did neither add very much to the good of the mining safety nor did it increase the output of minerals. During the initial stages of mining between haulage level 54 and 93 the influx of water was measured to be as much as 50 m^3/h, it gradually increased to 70 m^3/h and eventually raised to as much as 120 m^3/h when the miners targeted upon the quartz core from the 115-m-level. When the mine drainage station was faced with these problems at the brink from the feldspar to the quartz mining operational period, the closure of the mine was looming. The inclined shaft "Cornelia" was an overall improvement of the mining activities and gave a temporary relief to the workforce and enhanced the working

Fig. 1.4c Viewing the pillars and the chambers in the feldspar zone at the 87 m-level of Hagendorf-South Mine. For scale see the overhead- loader at the foot of a pillar and the flood light near the pillar on the left-hand side (Courtesy of E. Keck 1990)

conditions of the blue collar staff as well as the output in ceramic raw materials. The mining operations suffered a setback when mining operations focusing on the quartz at the deepest levels of the mine had to be grinded to halt for economic reasons. The end of the mining story at Hagendorf-South was not written or dictated simply by the exhaustion of the deposit, there is still a huge mass of the high-quality quartz down in the mine, but by economic problems whose essence to talk about would go far beyond such a monographic study of a mining district. Miners, especially those outside the large Ruhr District in the western part of Germany with its collieries, have never been in the limelight and rarely attracted the attention of the media or the politics as elsewhere in the world, where miners sometimes have taken to streets and contribute to oust the government. The mining site is located near the Czech-German border, which was still the nicknamed the "Iron Curtain" during that time, far off the city of München, where the local government resides and far more distant from the then provisional capital of the Federal Republic of Germany in Bonn. This deposit which once formed part of the mining activities which brought Germany back from the knees after the Second World War disappeared quietly from the mining stage similar to its neighbor Nabburg-Wölsendorf in the Oberpfalz, which made history among the word class fluorite deposits (Dill and Weber 2011).

Quantity and Quality of Run-of Mine Raw Material The feldspar has been a staple for almost the entire period of mining at Hagendorf-South. The quantity of quartz sold during entire lifetime of the mine was less than 20,000 t and the amount of lithium produced from triphylite, averaging 8.6 wt% Li_2O was approximately

1500 t. Reserve calculations made by H. Schmid yielded 1.6 mio to of feldspar, 1.4 mio of quartz and 2200 t of accessory minerals such as the Li phosphates mentioned above. Professor A. Forster, who revisited these reserve calculations came to a different result with 1.8 mio tons of feldspar and 2.7 mio tons of quartz. He listed the phosphate minerals as a separate entity making up approximately 1800 t together with 6,000 t of accessory minerals, among others columbite-(Fe). Until 1964 the output of feldspar has totaled 500,000 t. Between 1964 and 1984 276,849 t of powdered and grained feldspar has been sold. Loss on mining and processing stood at 32 %.

Processing was done by hand, picking the 20- to 40-mm and 40- to 180-mm sized fragments of feldspar from the conveyor belt (Fig. 1.4d). On increasing demands of the consumers for high-quality feldspar the Amberger Kaolinwerke were forced to implement sorting of the minerals in the magnetic field at 23,000 G which was the strongest magnetic field ever achieved at that time in mineral processing.

The company offered three different classes of feldspar named "Spezial", "H" and "I" which are arranged in decreasing order of purity grade:

"Spezial":	K feldspar 69 %, Na feldspar 25 % Quartz 6 %	Fe content 0.05 %
"H":	K feldspar 68 %, Na feldspar 24 % Quartz 8 %	Fe content 0.10 %
"I":	K feldspar 67 %, Na feldspar 23 % Quartz 10 %	Fe content 0.18 %

Fig. 1.4d Female workers sorting the high-quality feldspar by hand and picking the trash on the conveyor belt at Hagendorf-South (Courtesy of E. Keck 1990)

The chemical composition of feldspar from Hagendorf-South is listed and compared with feldspar form other pegmatites and aplites along the western edge of the Bohemian Massif in Table 1.3.

70 % of the total production went into the domestic market, the remaining part to European countries. The lion share of 40 % was used for bodies and glazes of sanitary ceramic final products, 30 % for high-quality tableware and 13 % for frits, ready-made glazes and enamel. 9 % were used for electro technical and technical porcelain, 1 % for abrasives (abrasive wheels) and the remainder was used for a wide range of unspecified ceramic final products.

The high purity of the quartz qualified this industrial mineral for the production of silica and ferro-silica (Table 4.4,)

Hagendorf -North

The Cornelia Mine operating the large Hagendorf-South pegmatite was not the only mining activity centered on feldspar in this region. As early as 1860, at the western edge of Hagendorf village a much smaller feldspar-quartz pegmatite was exploited by opencast mining operations down to a depth of approximately 10 m (Müllbauer 1925) (Fig. 1.1b) The mine was named after its owner as Meixner Mine but is better-known among geoscientists and rock hounds as Hagendorf-North being located north of its larger brethren Hagendorf-South (Figs. 1.1b and 1.4e). At the floor of the quarry, the miners began driving by underground means several adits which could, however, not extend the lifetime very long. Before World War II broke out, the mine had to be closed in 1937 for economic reasons. A total of five shafts and one raise were put down to as much 45 m depth where they reached the contact between the pegmatite and the gneissic country rocks. All attempts made during trial mining operation in 1948/1949 were not greeted with success. The output at Hagendorf-North amounted to a total of 220,000 t of feldspar. Considering the total volume of this zoned pegmatite, measuring 100 m in length and 50 m in thickness, yields about 800,000 t of quartz-feldspar. The amount of phosphate minerals stands at 300 t. Only recently the remaining parts of the open pit have been backfilled to avoid damage to the country road and the housing area so that even for the mineral collectors this site is not more than anything of the past. Technically this mine will still be standing out from the overall mining activities in the region by its ropeway carriages that transported the broken feldspar to the railway station and the processing plant after the horse-drawn buggies had turned out to be no longer economic over a distance of 3.5 km and had to be put out of service.

Waidhaus-Silbergrube

In 1938, the entrepreneur Max Schmidt established a plant processing feldspar at Waidhaus. To feed the mill, he sank a 30-m-deep shaft into an aplite (rather call aplite granite) called Silbergrube. The aplitic rock consists of 50–70 %

Table 1.3 Run-off mine feldspar from various pegmatites and aplite granites of the HPPP

Element	Brünst	Silbergrube	Silbergrube	Silbergrube	Pleystein-Trutzhofmühle	Pleystein Trutzhofmühle	Hagendorf 49 m-level	Hagendorf 109 m-level	Hagendorf 109 m-level
SiO_2	71.82	72.69	76.31	72.99	77.81	74.24	64.69	63.89	69.11
Al_2O_3	16.30	16.06	12.57	14.92	13.01	13.85	19.62	19.01	18.52
Fe_2O_3	0.95	0.30	0.55	0.50	0.76	1.04	0.31	0.04	0.04
CaO	0.50	0.20	0.41	0.09	0.36	0.97	0.16	0.07	0.43
MgO	0.25	0.08	0.12	0.01	0.17	0.11	<0.01	<0.01	<0.01
MnO	0.92	0.05	0.00	0.07	0.03	0.25	0.02	0.00	0.01
K_2O	4.62	4.80	3.71	3.71	2.90	4.40	8.83	13.49	0.33
Na_2O	4.38	5.40	4.14	5.02	3.38	1.60	4.56	1.39	9.95
P_2O_5	0.34	0.60	0.20	0.43	0.60	1.52	0.68	0.90	0.90

Teuscher and Weinelt (1972)
Novák et al. (1996)
Breiter and Siebel (1995)

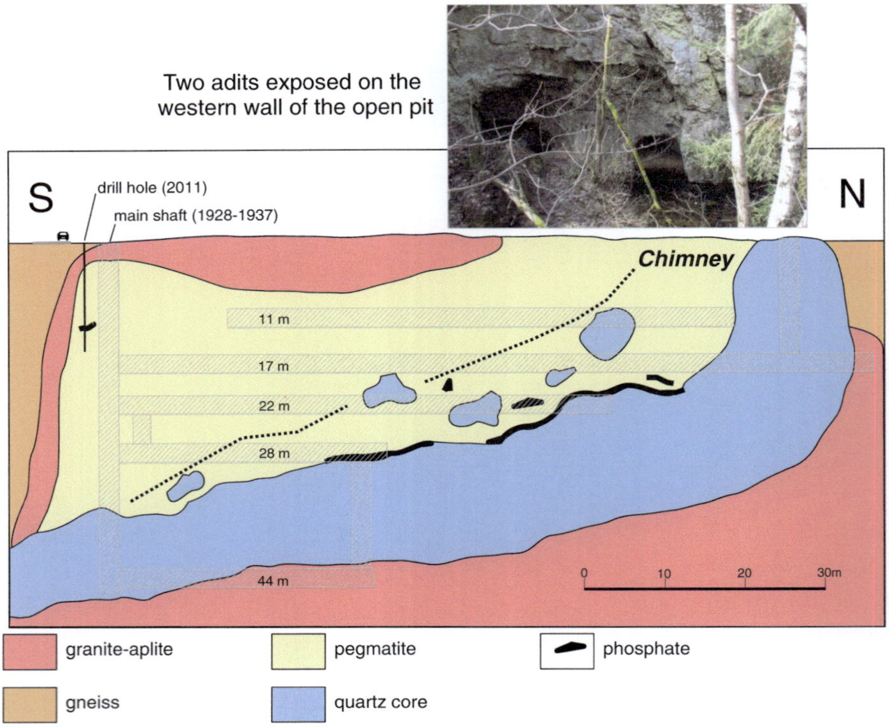

Fig. 1.4e A cross section through the Hagendorf-North deposit with underground workings of the Meixner Mine partly flooded. The "chimney" structure is discussed in Sect. 6.4.2. For comparison see also Fig. 5

albite-oligoclase and K feldspar. Another shaft was sunken down to the 40-m level after the destruction of the head frame of its predecessor by a blaze in 1954 (Fig. 1.4f). During the initial mining period handling and transport was carried on-track underground. To depict the miners' labors and work sites during these initial phase of mining pegmatites in the Hagendorf-Pleystein Region a sequence of photographs taken on site has been compiled to show the extraction and the transition from the on-track to the off-track haulage technique finally ending up in an opencast mine-only operation which went into production during the 1960s (Figs. 1.4g, 1.4h, 1.4i, 1.4j, 1.4k and 1.4l). The opencast mine has become the only source of raw material to feed the mill (Fig. 1.4f).

Today the run-off mine feldspar is no longer the staple product of the company when a pilot plant set up in 1998 to process ceramic raw materials and glassware turned into full swing. Natural raw materials and recycled materials from household and the industry are going to hold an ever increasing share in the mill feed. The current processing plant has been designed to produce basic raw materials for the ceramic industry and decor auxiliaries, offering a wide grain size range from micron to millimeter particle size on demand for different ceramic finished goods.

Fig. 1.4f The Silbergrube opencast mine west of Waidhaus from the bird's-eye-view (Courtesy of Max Schmidt Corporation). The different lithologies exposed within and adjacent to the open pit are given and their contact zones marked by stippled lines. See also the abandoned galleries opened up during the underground mining period and later cut when the open cut operation went deeper into the aplitic body

Fig. 1.4g Exploitation of feldspar in the Waidhaus Silbergrube Mine, using techniques from rail-bound to load-haul- and dump method (LHD) (Sources of images : Max Schmidt Corporation and J. Forster Heimatkundlicher Arbeitskreis Waidhaus e.V.). Miner drilling the blast holes in the aplite using a special drill stand near the roof of the adit

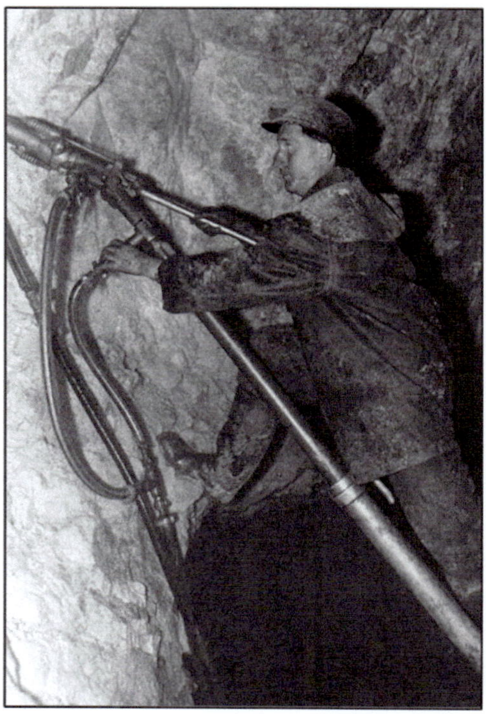

Fig. 1.4h Filling the blast holes with the cartridges and installing the detonators with the ignition cables

Fig. 1.4i Hauling the broken aplite in a rail-bound wagon by hand from the stope to the shaft

Fig. 1.4j The operator on the controls of the hoist

Fig. 1.4k Making the ball mill near the open pit operational. In the background the head frame of the old shaft is visible

Fig. 1.4l A front-end-loader manned by one driver in the underground extraction site of the aplite mine-see the white rock on the roof and the blasted feldspar "ore" piled up in the foreground

Brünst

Since 1959 Buchtal A.G. at Schwarzenfeld had been mapping and trenching the environs of Brünst for feldspar. In 1960, this survey succeeded in finding a pegmatite about 500 m SSW of Lesslohe which was exploited by underground and opencast mining operations (Fig. 1.1b). In 1999 the Deutsche Steinzeug Cremer & Breuer AG filed an application to the regional mining authority in order to shut down all its mining operations in this area. The run-off –mine material is a rather homogeneous intergrowth of quartz, K feldspar, Na-enriched plagioclase and biotite. No phosphate minerals were spotted in this material. The run-off mine output was in 1961 5400 t of feldspar, in 1963 10,360 t.

Pleystein-Kreuzberg

The quartz reef of the Kreuzberg in the city of Pleystein can hardly called a quartz pegmatite but rather downgraded to a "pegmatite ruin"; only the quartz core of a larger zoned feldspar-quartz pegmatite resisted hydrothermal alteration and subsequent chemical weathering. Those parts that braved the elements in the course of shaping the landscape now stand out from the hillock (Figs. 1.1b and 1.3, Sect. 5.1.2). Measuring roughly 200 m in N-S direction with a height of 30 m above ground, the size of this cone-shaped quartz body amounts to 1,500,000 t of quartz which is equal to the Hagendorf-South quartz core when it was hit by the underground drilling and mining operations (see also Table 4.4,). An inclined gallery bears witness of trial mining operations in footwall of the Kreuzberg Pegmatite. Despite of the size of the quartz core, the attempts failed and it was not out to commercial use. By and large, the failed mining operations in the past turned out to be good for the present generations. The quartz reef is now a protected natural monument or, in terms of economic geology, an "aesthetic deposit" (Fig. 1.3).

Numerous feldspar-quartz pegmatites and aplites occur in the Oberpfälzer Wald and in the Bayerischer-Böhmer Wald, many of them have been mined by local companies almost exclusively for ceramic purposes and, in case of inferior qualities of quartz and feldspar, also for aggregates. The rock strength of the quartz renders it suitable to renew the road surfaces. These local mining operations centered around pegmatites were compiled by Schmid and Weinelt (1978), who traced these activities back into history, making use of the excellent works of geologists from the Bayerisches Oberbergamt (1924) and the "Old Master" of Bavarian Geology, C.W. von Gümbel (1868, 1879), whose studies cannot be overstated in any current study covering NE Bavaria.

1.3.2.2 The Extractive Geology of Pegmatoids

Along the southeastern endocontact of the Münchberg Gneiss Complex, numerous Na-enriched feldspar pegmatoids and aploids evolved in the "Hangend-Serie" (hanging-wall series) (Bauberger 1957) (Fig. 1.5a). The pegmatoids were trenched by local mining companies supported by a state-run organization the GAB (Company for the Exploitation of Mineral Resources in Bavaria Ltd.). In addition a great deal of the license area was mapped and documented by the former Geological Survey of Bavaria, the geological work of which contributed to increasing mining activities until the shutdown of the last mine near Friedmannsdorf on October 15, 1978 (Bauberger 1957; Stettner 1960, 1964; Emmert et al. 1960; Emmert and Stettner 1968).

About 30 underground and opencast mines exploited the feldspar pegmatoids of the Münchberg Gneiss Complex from 1937 through 1978. Too often a considerable thickness and striking length of the feldspar layers and schlieren, as a precondition for a successful mining operation was not fulfilled in all of these pegmatoid deposits. Therefore smaller equivalents of these pegmatoids are occasionally exposed in road cuts or during re-building the highway (Dill 1979). The average thickness of the pegmatoid layers stands at 4 m; the thickness was increased when pegmatoid layers underwent deformation and feldspar crystallized in the trough or hinge zones of fold structure where sufficient accommodation space was created by the preceding structural deformation (Figs. 1.5b and 1.5c). Along the limbs, the pegmatoid layers saw a diminishing of their thickness whereas in the synclines the feldspar, in places, increased in thickness to as much as 15 m, as it was the case, e.g., at Friedmannsdorf, where more than 100,000 t of albite-oligoclase were mined and at Lübnitz with a total amount of 60,000 t. The feldspar schlieren and layers were worked down to a cut-off thickness of 0.8 m.

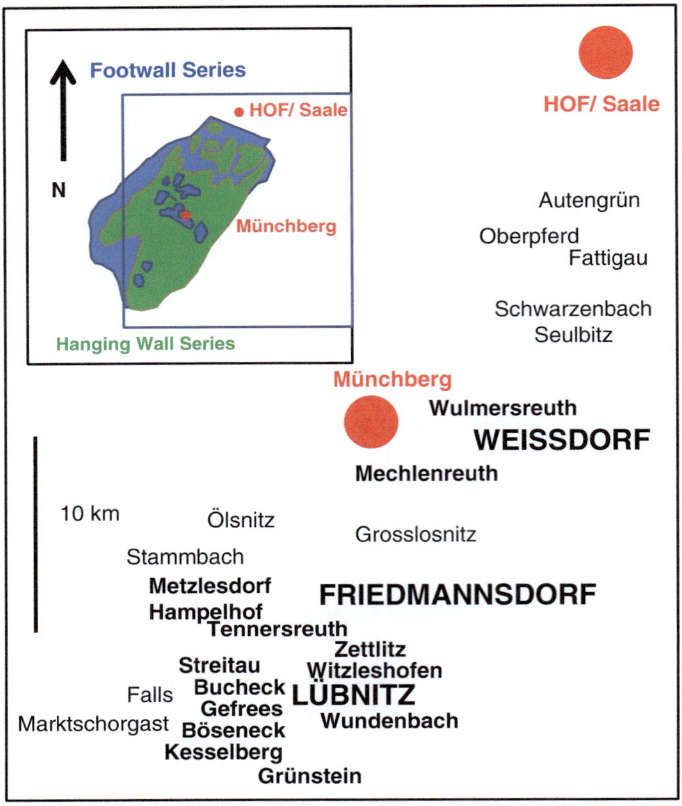

Fig. 1.5a Index map to show the position of the Na-enriched pegmatites (pegmatoids) in the southeastern part of the Münchberg Gneiss Complex (Bauberger 1957; Teuscher and Weinelt 1972; Dill 1979). The inset shows the hanging wall and footwall series of the Münchberg Gneiss Complex (See also Fig. 2.1d). The size of the font in each inscription corresponds to the size of the pegmatoid deposit. Friedmannsdorf was the largest deposit totaling 100,000 tons of feldspar

Mining of these feldspar pegmatoid deposit was feasible even if the total amount fell below 10,000 t per operational site because all feldspar layers are located close to one another which made the management of the blue-collar staff very economic and the feldspar lenses are arranged in a way that geological made predictions rather easily as to dip and strike with high chance of success making new discoveries (Figs. 1.5a and 1.5c). A central processing plant at Grohenhammer and a quick exchange of machinery between the various working sites rendered mining possible, even of those deposits valued at marginal or close to subeconomic. The feldspar crops either out at the surface or was found near-surface so that sinking shafts was only necessary at few mining sites such as at Zettlitz (10 m), Lübnitz (27 m), Böseneck (40 m) and Friedmannsdorf (90 m). After trenching down to a depth of 2 m, feldspar was exploited by opencast mines down to a maximum depth of 8 m,

Fig. 1.5b The Friedmannsdorf Pegmatoid Mine in the Münchberg Gneiss Complex. Haulage installation with the head frame and installation to load the feldspar raw material on the lorries, just after shut down of the mine

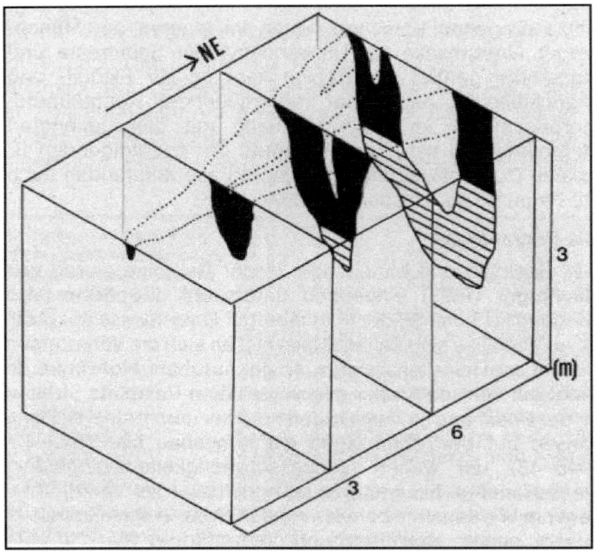

Fig. 1.5c The Friedmannsdorf Pegmatoid Mine and the abandoned open pit in the stage of backfill with the two major lithologies exposed in the open pit. The cartoon of the inset illustrates where the pegmatoids were preferably concentrated in the synclinal and anticlinal structures (See also Fig. 5.5d. Redrawn from Bauberger (1957) who used the Seulbitz structure to exemplify the mode of concentration (Dill 1979))

e.g., at Bösneck. In case of promising drill results or positive indications of the feldspar to extend to greater depth a ramp was built at an angle of 30–40°. Underground mining strongly depended upon the thickness of the feldspar layers. In those parts of the mine, where the feldspar layers reached the average thickness normal and inclined cut-and-fill stoping was applied. In those section, where the thickness increased the cross-working technique proved to be the most successful way to safely extract the feldspar. All opencast operations were backfilled and reclaimed so that agricultural use in the wake of mining was not impeded. The landlords, mainly farmers, were entitled to mining royalties and paid compensation for losses or damages incurred as a result of opencast or underground mining operations. Of the 30 deposits only two to three were permanently mined, while new deposits were explored ahead of these extractive measures and abandoned ones subject to rehabilitation, so that today no spoiled landscape has been returned to the people living in this region. During the heyday of mining about 30 miners were active in the area working in three shifts. Ongoing improvement of the machinery and exhaustion of the feldspar deposits caused a reduction of the workforce down to 10. During 1937 the former mining authorities of the "Third Reich" ordered the Mandt Company in Stettin, in northern Germany, to exploit these deposits of the Münchberg Gneiss Complex reducing the dependency on feldspar imported from Scandinavia. They bundled the activities of artisanal mining activities and could

Fig. 1.5d Lower Triassic arkoses (Buntsandstein) enriched in kaolin and feldspar are currently more competitive than the pegmatitic ore bodies. Two industrial minerals can be extracted from the large open pits in the Hirschau-Schnaittenbach area, while quartz is dumped and form the "Monte Kaolino" which is the basis for a recreation center and leisure site to practice summer skiing and operate a toboggan run (See inset)

supply the domestic market with raw materials to produce abrasives and ceramic goods until the late 1970 when cheaper imports from Sweden, Portugal, South Africa, India, Norway and the former CSSR put a halt on the mining operations in these metamorphic rocks. Moreover these feldspar deposits were no longer competitive to the feldspar deposits in the Triassic arkoses which are mined by large opencast mining operations in the area of Hirschau-Schnaittenbach, only 100 km south-east of the Münchberg Gneiss Complex in Germany (Dill et al. 2016) (Fig. 1.5d).

1.3.2.3 The Extractive Geology of Metapegmatites

Metapegmatites are another source of feldspar, quartz and mica in the NE Bavarian Basement. Many of these metapegmatites are located along the western edge of the Bohemian Massif and but only a few of them attained economic grade near Menzlhof and Döltsch where mining was under way at Wilma and Getrude Mines until the late 1980s (Fig. 1.6a). At Püllersreuth another underground mining operation named Maier Mine was located not far away from the afore-mentioned ones. It was well

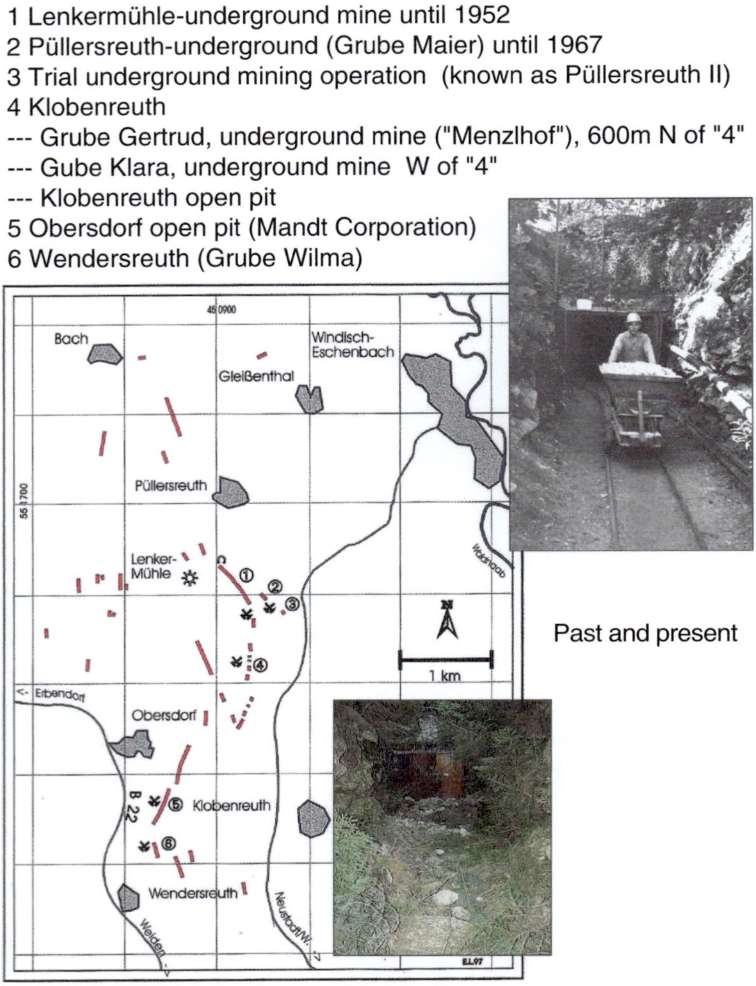

Fig. 1.6a Map to show the mines within the Zone of Erbendorf –Vohenstrauß (ZEV) exploiting metapegmatite for feldspar and quartz. The inset at the bottom right of the map shows the adit of the mine "Gertrude" near Menzlhof. Vegetation has reclaimed what once was the adit to the underground galleries. The image *-top right-* will give the reader an impression of the haulage of feldspar, applying a rail-mounted transport system from the underground workings to the dumps at surface. The mine-car is designed for two-way dumping (Courtesy of Gottfried Feldspat GmbH)

known for its large packages of muscovite crystals, locally, exceeding the size of a palm, yet of lesser economic importance with as to the feldspar content than the Wilma and Getrude Mines, the major reason why this mine was forced to abandon mining already in 1967. These two deposits and the more complex one near Püllersreuth show a good match with the feldspar pegmatoids in the Münchberg Gneiss Complex as far as their mineralogical composition is concerned but they differ greatly from the pegmatoids when viewed with respect to their textural inventories and timing relative to the structural deformation (Weber 1978a). The latter attests to a prekinematic emplacement and gives these felsic rocks the outward appearance of an orthogneiss. Their tabular and layered shape demonstrates full compliance with the structural system and great similarity to the afore-mentioned pegmatoids. The striking length of the Gertude Pegmatite measured up to 80 m in N-S direction attaining thickness of 20 m (Kopp 1987). The pegmatitic layer composed of K- and Na feldspar dips towards the west at an angle of 30°. Getting closer to the contact between metapegmatite and gneiss, the quartz content increases considerably in the endocontact zone. Metapegmatites and pegmatoids, both similar in shape and mineralogical composition are conducive to similar underground and opencast mining technique and ended up in applications resembling those of the pegmatoids (Figs. 1.6b and 1.6c).

The Püllersreuth metapegmatite deserves a special treatment, as its outward appearance and mineralogical composition is akin to that of the pegmatites of the HPPP as far as the zonation into a quartz core at depth, a feldspar rim and graphic intergrowth and the mineralogy are concerned, which is more varied than the

Fig. 1.6b Open pit mining of metapegmatites for feldspar near Wendersreuth during the initial phases of operation as the metapegmatite was opened up. The white feldspar body is exposed at the deepest level of the open pit. The upper working level exposed an off-shoot (v) of the metapegmatite into the gneissic country rocks. The feldspar is transported out of the open pit by means of track-mounted mine cars of two-ways dumping type to the loading bridge, where the feldspar is handed over on to lorries (Courtesy of Gottfried Feldspat GmbH)

Fig. 1.6c At an advanced level the feldspar has been transported with a front-end loader from the working site in the open pit to the crusher. The pre-treated raw material is transported out of the mine by a staircase-like set of conveyor belts (Courtesy of Gottfried Feldspat GmbH)

overall lithology of the metapegmatites containing columbite and beryl (Linhardt 2000). It strikes NW-SE and dips towards the SW at an angle of 50°. The mining technique applied does not differ from what is applied to the exploitation of the metapegmatites.

1.3.2.4 Overview of Small-Scale Mining Operations on Pegmatitic Deposits in the Bavarian Basement and Neighboring Areas

Mining of feldspar pegmatites in the Bayerischer and Böhmer Wald was mainly sparked by the numerous glassworks in this region of the NE Bavarian Basement, which were in desperate need of quartz to feed their kilns. All of the pegmatite listed were prevalently opened up for that purposes (Table 1.2). Inferior qualities need not be dumped but could be used for improving the infrastructure and the feldspar used for abrasives. A full-discussion of all pegmatitic deposits would go far beyond the scope of the book and will be reserved for a more regional representation.

Chapter 2
Pegmatitic Rocks and Their Geodynamic Setting in the Central European Variscides

Abstract Pegmatitic rocks are not randomly distributed across the Variscan/Hercynian basement in Central Europe. The evolution of pegmatites s.l. in the course of a complex orogeny of Meso-Europe took rather long, from the Devonian (419 Ma) through the Permian (252 Ma). In terms of structural geology and geodynamics, pegmatitic deposits primarily occur in ensialic Variscan-type orogens (calc-alkaline) with a thickened crust and a preponderance of thrusting and nappe stacking. In Rift-type settings (alkaline) a strong subcrustal impact is evident and as reactivated/reworked pseudopegmatites in Alpine-type orogens (calc-alkaline) these deposits developed during the initial stages when the crustal section was still rather thick. Both types pertain to the marginal ensimatic settings. They left their hallmarks to some extent also within the Central European Variscides and at its southern edge in the Alpine-Carpathian Orogen. The geodynamic units subjected to very-low-grade- to low-grade stage metamorphism at the margin of the Central European Variscides are barren with regard to pegmatites and aplites. Pegmatoids with minor B-(Li)-P-REE-U-Be mineralization occur along a suture zone extending across the present-day continents. It resulted from the late Variscan closure of the Rheic Ocean between Gondwana and Laurussia with remnants of an arc-related plutonism. Within allochthonous metamorphic complexes and nappes barren feldspar-quartz pegmatoids plus metapegmatites developed. Further south another part of this former coherent nappe also contains a small Be-Nb-P mineralization. Within the Subfluence zone, marked by continent-continent collision and thickening of the crust pegmatite, granite- pegmatite (miarolitic), pegmatite-aplite and pegmatoid abundant in B, Be, F, Li, Sn, U, P and As are encountered. Heading further to the core zone of the Variscan orogen, strong diapthoresis and shearing in the contact zone between the Saxothuringian and Moldanubian zones *sensu lato* favored the emplacement of pegmatite and aplite enriched in B, P, Be, Nb, As, Zr and F. High grade metamorphic rocks in an autochthonous position with a protolith mainly of Proterozoic age exist in the core zone. At the margin they are overthrusted onto adjacent geodynamic units and penetrated by multiple intrusions. The Hagendorf-Pleystein Pegmatite Province is located near the root zone for the nappe complexes thrusted onto the north-western geodynamic realms. Pegmatites and aplites with minor pegmatoids of the Hagendorf-Pleystein Pegmatite Province show the most varied concentration of rare elements in pegmatitic and aplitic rocks in this crustal section (B-P-REE-Nb/Ta-Li-Sc-Zn-Be). In some parts in core zone pegmatites can

H.G. Dill, *The Hagendorf-Pleystein Province: the Center of Pegmatites in an Ensialic Orogen*, Modern Approaches in Solid Earth Sciences 15,
DOI 10.1007/978-3-319-18806-5_2

also be observed associated with skarns. Variscan lithologies were incorporated into the Alpine orogen and reactivated during the Alpine orogeny at the southern edge of the Meso-Europe. They contain granitic pegmatites, meta-pegmatites, pegmatoids and pseudo-pegmatites (B-Be-P-Nb-U-F-As-Li-Sn-REE-U). By quality this element assemblage is not very much different from that of the neighboring Variscan parent rocks. The suite of pegmatitic and aplitic mineral deposits is associated with mineral deposits of non-pegmatitic origin. They include thrustbound deposits (Au-As-Sb-(Hg)-Fe-Cu-Pb-Zn), plutonic/granite-related deposits (Sn-W-Mo-Pb-Ag-Zn-(In)-Cu-U), and unconformity-related (U-Pb-Zn-F-Ba). While the deposits can at least in parts structurally and compositionally related to the various types of pegmatites and aplites, stratabound deposits are mainly marker deposits for geodynamic units prone to aplitic or pegmatitic rocks in an ensialic orogen (SMS >>VM FeS-Cu-Zn, SEDEX Fe deposits, black-shale –hosted U-Cu-Mo-Sb-Zn-REE (low-grade-large-tonnage) and graphite). As an exception from this rule, the two last-mentioned mineralization with organic compounds can be considered (see geophysical surveys).

Pegmatitic rocks are not randomly distributed across the Variscan basement in Central Europe. Everybody may find out by himself that they reveal a compositional and textural variation as a function of the geodynamic position within an ensialic orogen. Therefore it makes sense to look at these basement units in more detail shifting our view from the small-scale geodynamics to the large scale regional (economic) geology in order to facilitate a correlation between the study area in Germany and areas outside Europe hosting also pegmatites. Even if Precambrian Shields endowed with pegmatitic rocks are not exposed in Central Europe, which went through the orogenic phase during the Paleozoic, the results obtained in the present study might also contribute to a better understanding of the pseudopegmatites as they went through their incipient stages of emplacement.

2.1 The Geological and Metallogenetic Evolution of the Central European Variscides with Special Reference to Pegmatites

The Central European Variscides that gave host to the pegmatites of the HPPP and their adjacent pegmatoids/aploids and metapegmatites saw mining of metallic and non-metallic commodities during more than 2000 years. Compared to the mining period of 2000 years in the Central Europe, the evolution of pegmatites s.l. in the course of complex orogeny of Meso-Europe took rather long, from the Devonian (419 Ma) through the Permian (252 Ma), while some scientists even extend these orogenic processes into the Triassic (201 Ma) (Suess 1888). In the Anglo-Saxon literature the term Hercynian is a synonym for the German word “variszisch” or Variscan which is used in preference to the term Hercynian throughout this presentation.

To give scientists also interested in pegmatites in Northern America a chance for a comparative study and to geodynamically tie up the Variscan pegmatite provinces on both sides of the present Atlantic Ocean it is inevitable to have a closer look at the western prolongation of the European Variscides. It is the Alleghanian or Appalachian orogeny that formed the Appalachian and Allegheny Mountains (Bartholomew and Whitaker 2010). During the afore-mentioned mountain-building processes, North America which was part of the Euramerica super-continent collided with Gondwana resulting in the newly formed super-continent Pangaea. Taking a closer look at Central Europe, we find the African plate in the south which formed part of the afore-mentioned Gondwana continent and in the North Laurussia (Ziegler et al. 1977; Matte 2001) (Fig. 2.1a). Whatever name given above you might prefer for the northern continent, there is only one geodynamic event responsible for the emplacement of this supercontinent Laurussia or Euramerica. During the Silurian two continental blocks named Laurentia and Baltica amalgamated, closing the existing Iapetus Ocean in between the two during what is called the Caledonian Orogeny. Seafloor spreading and the closure of the resultant ocean was not anything but another of the same during the Variscan Orogeny as the Rheic Ocean between Laurussia and Gondwana was closed during the Carboniferous leading to the Variscides. It is imperative to keep an eye on this Late Paleozoic period of time

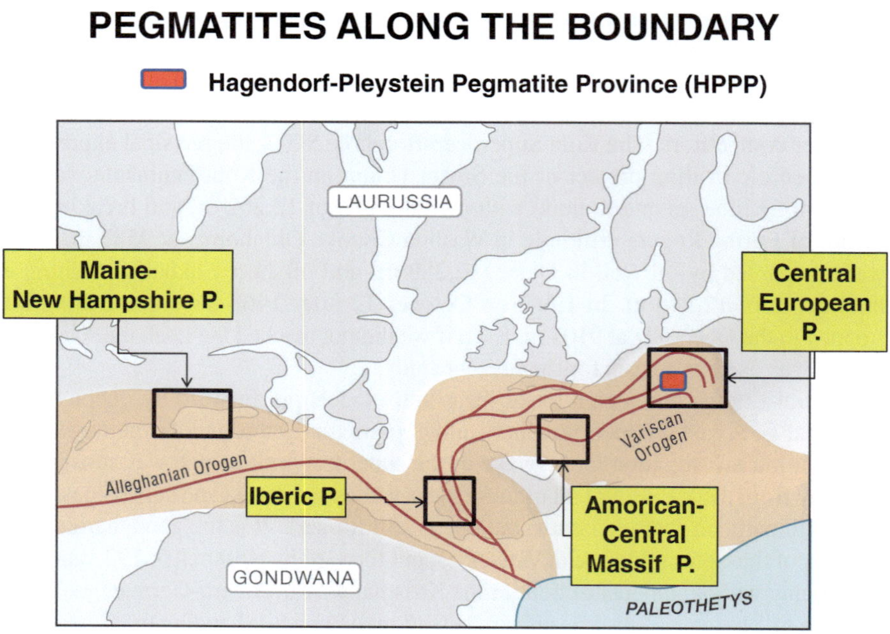

Fig. 2.1a Reconstruction of the mountain ridges produced in the course of the Alleghanian/ Appalachian and Variscan/Hercynian orogenies during the Late Paleozoic on both sides of the present-day Atlantic Ocean amalgamating Laurussia and Gondwana and giving host to four pegmatite provinces

because it is the moment when the granites, pegmatites and aplites came into existence in the Central European Variscides. Pangea was the final result of this Variscan collision and originated from amalgamating of another large continent Siberia to the existing landmass during the Late Permian. Within this geodynamic framework our felsic intrusive rocks originated in what is called today Central Europe but was then the easternmost end of the Variscan mountain ridge (Fig. 2.1a).

The Central European Variscides rank among the most intensively studied crustal sections on Earth with a plethora of papers being available and part of them now stored in the various databases. These investigation starting off more than a century ago when the term Variscides was first coined by Suess (1888). The region experienced a tremendous increase in publications as the Continental Deep Drill Program of the Federal Republic of Germany (KTB) was in full swing and supplementary data derived from a deep-geophysical project, abbreviated to DEKORP were processes and interpreted to the full extent. These data sets were retrieved and made available for the deep-geology study of pegmatitic rocks. A brief retrospective of the data collection and goals of these ambitious programs is necessary.

At the very beginning, prior to the site selection, some held a continental drill hole down to 15,000 m in the midst of Europe feasible. During the late 1970s and early 1980 the motto was "Nothing but the very best will do". The Russian engineers and scientists had raised the bar, or to be more precise, sunken the drill bit very deep into the upper crust, slowly but steadily. Whatever delicate excuses may have been put forward (I was a staff member of KTB Management Group, in charge of economic geology, mineralogy and geochemistry) and how many attempts made to adjust the goals, everything comes to light at the end (Emmermann and Lauterjung 1997). We were only fifth in the row, trailing by some considerable margin behind the other competitors. The Kola Superdeep Borehole SG-3, the physical expression of a scientific drilling project of the Soviet Union on the Kola Peninsula was for quite a long time second to none with a final depth of 12,262 m, and breaking the record of Bertha Rogers drill hole in Washita County, Oklahoma, at 9583 m. They were surpassed by oil well in Qatar (12,289 m) and offshore Sakhalin reaching a final depth of 12,345 m. In 1994, on October 12 after 1468 days the drilling bit stopped in the Oberpfalz at 9101 m. Even if we did not reached the goal, the scientific fallout was used as a basis for the present study.

Central Europe has seen a lot of ideas and models put forward by geophysists, structural geologist and petrographers, alike, particularly throughout the heyday of geoscientific investigation in the wake of the superdeep hole which was sunken not far away from the HPPP. Not all of these ideas were solid-based and sometimes build at least in parts on shaky grounds, but one is still for sure. It is the geodynamic subdivision of the Central European Variscides, put forward by Kossmat in 1927 and still valid today without any restriction. Franz Kossmat was an Austro-German geologist but not a full-blown or full-time university staff member while heading the Geological Survey of Saxony in Germany for almost 20 years. In the succeeding paragraphs this classical subdivision is used as a basis to describe the various geodynamic units, placing particular emphasis on those units of interest for pegmatites.

Beforehand some of the major publications providing a comprehensive overview of this crustal section in Central Europe need to be cited so as to give those a chance

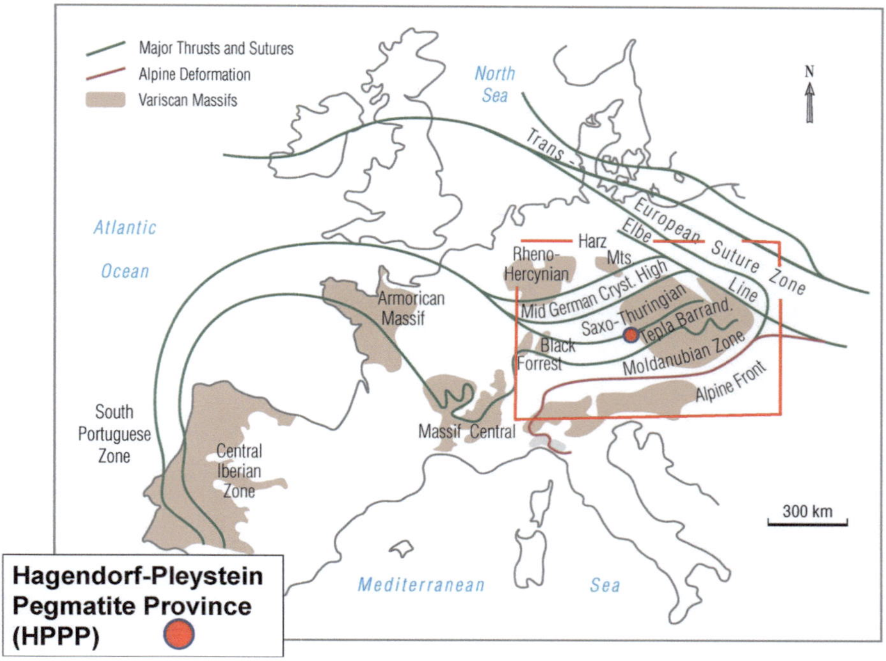

Fig. 2.1b Variscan massifs in Europe with sutures and lineamentary faults bounding the geodynamic units (Matte et al. 1990; Franke et al. 1995; McKerrow et al. 2000). The area framed by the *red line* demarcates the study area of the Central European pegmatites in this book (see Fig. 2.1d)

who want to immerse themselves in the geology of the Central European Variscides apart from the study of pegmatites (Walter 1992; Dallmeyer et al. 1995a, b; Winchester et al. 2002; Raumer von et al. 2003; McCann 2008a, b). In Fig. 2.1b the reader gets an idea what the geodynamic situation looks like in the Central and Western European Variscides while the diagrammatic N-S cross-section through the Central European Belt in Fig. 2.1c makes the reader familiar with the ideas of Behr et al. (1984), how to interpret the geodynamic evolution from the Moldanubian core zone through the northern passive margin along the western edge of the Bohemian Massif. The geodynamic units in Central Europe were discussed in the succeeding sections based upon the geodynamic map drafted by Dallmeyer et al. (1995a, b) (Fig. 2.1d).

2.1.1 The Subvariscan Foredeep

2.1.1.1 Lithology and Structural Geology

During the Upper Carboniferous, the Subvariscan Foredeep subsided along the northwestern boundary of the Central European Variscides leading to a basin extending from the Ardennes, Belgium, in the West, through the Ruhr District, Germany, to

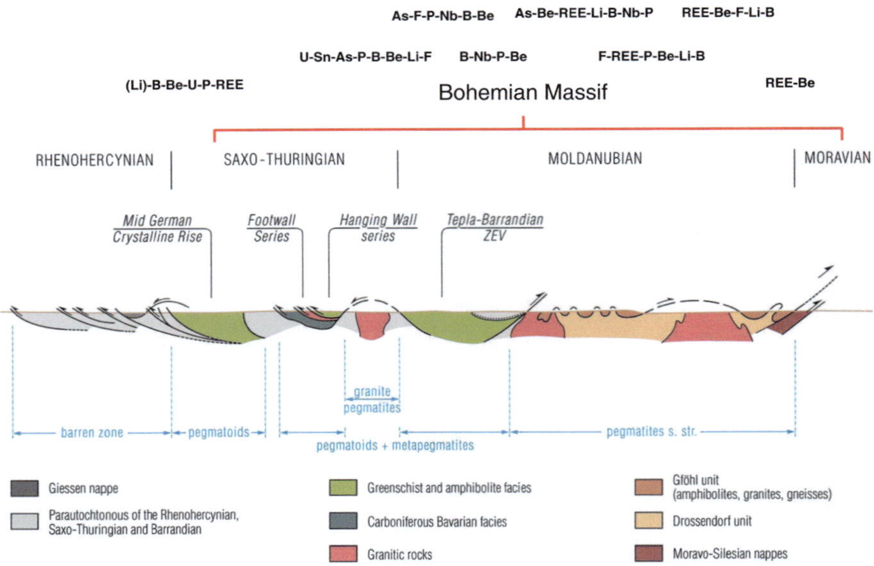

Fig. 2.1c Diagrammatic cross section through the allochthonous units and autochthonous units of the Central European Variscides (Behr et al. 1984). The types of pegmatitic rocks are given for each unit and their various element associations are shown on *top*

the Upper Silesian Coal Basin, Poland. Basin fill began in a flysch-type style and eventually faded out in what might be called a molasse-type sediment now being covered under the Mesozoic series of the North German Basin on the passive margin of Laurussia. The most striking feature of the basin is its cyclothems with more than 100 interbedded paralic coal seams that are still mined today in the Ruhr and in the Upper Silesian Basin for high volatile bituminous to anthracite-rank coal, but no longer in the Wallonian and Campine basins (Drozdzewski 1993; Langenaeler 2000; Gaschnitz 2001). The basin fill diminishes from approximately 5000 m in the South to 3000 in the North. The thickness of the Carboniferous molasse deposits in the eastern foreland of the Moravo-Silesian fold belt attains a much higher thickness of up to almost 9000 m (Jura et al. 2000). Igneous rocks are absent, excluding some tonstein horizons, in the Subvariscan Foredeep (Fig. 2.1d).

2.1.1.2 Mineral Deposits

Several Pb-Zn vein-type deposits perpendicular to the fold axes of the large anticlines crosscut the coal seams and are terminated by Cretaceous platform sediments resting upon an unconformity which truncates the Late Paleozoic sediments. Neither granitic mobilizates, nor pegmatites or aplites can be expected in this very-low grade metamorphosed to unmetamorphosed lithologies at the northwestern margin of the Central European Variscides.

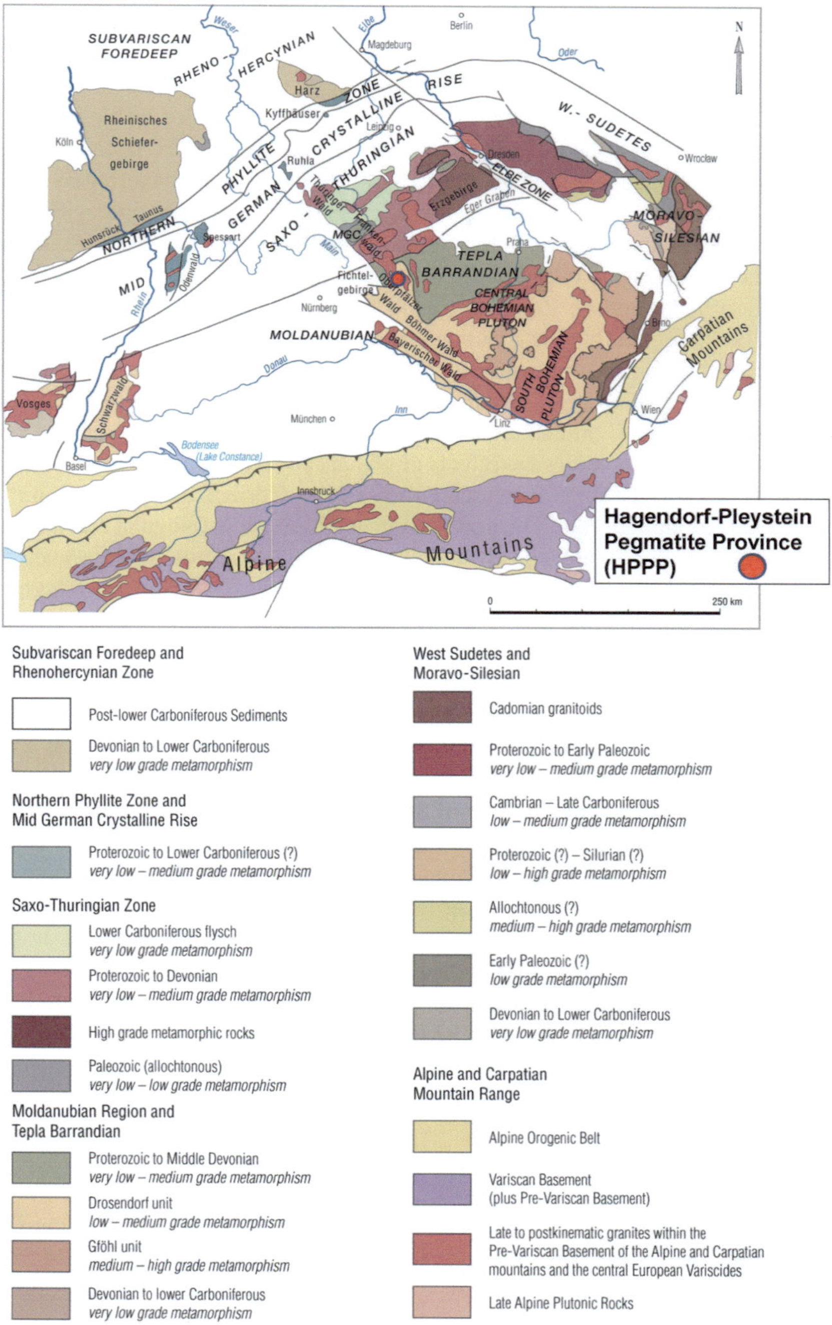

Fig. 2.1d The Hagendorf-Pleystein pegmatite Province the center of pegmatites among the geodynamic units and lithology of the Central European Variscides (Modified after Dallmeyer et al. 1995) and the Alpine-Carpathian Mountain ridges (Redrawn from Dill et al. 2008b) *MGC* Münchberg Gneiss Complex

2.1.2 *Rhenohercynian Zone*

2.1.2.1 Lithology and Structural Geology

The Rhenohercynian Zone follows immediately south of the Subvariscan Foredeep and forms the German low mountain ranges of the Rheinisches Schiefergebirge and the Harz, both of which are the source for the composite name Rhenohercynian used for this geodynamic zone (Fig. 2.1d). Towards the West, the clastic shelf sediments predominantly of Devonian age in the Ardennes, Belgium, Cornwall, Great Britain and the South Portuguese Zone, Portugal, pertain to the Rhenohercynian Zone (Franke 2000) (Fig. 2.1b). Basic submarine volcanic activity is widespread across this tectonostratigraphic unit, whereas felsic plutonic rocks are scarce. The seismic results of the DEKORP (Deutsches Kontinentales Reflexionsseismisches Programm = German Continental Seismic Reflection Program) indicate the presence of NW-vergent tectonics the effects of which can be traced down to the deepest parts of the crust (DEKORP Research Group 1990). Horizontal compression must have played a dominant role. The geodynamic position and geological results of the Rhenohercynian zones are controversially discussed and not unequivocal. The Rheic Ocean located between Avalonia and the Amorican Terrane Assemblage was said to be closed, whereas the Saxo-Thuringian ocean was subducted to the S beneath the Teplá-Barrandian Zone during Early through Mid Devonian time (Winchester et al. 2002). This geodynamic zone is held to be a foreland fold-and thrust belt on a Devonian-Carboniferous passive margin (Oncken 1997). It developed from southward subduction and underthrusting underneath the Saxo-Thuringian Zone. The Northern Phyllite Zone stretching immediately south of the Rhenohercynian Zone contains relic of the Rheic suture zone of the former Rheic Ocean which persisted from the Early Ordovician through the Lower Devonian W. (McKerrow and Ziegler 1972; Linnemann et al. 2007). Weber (1978b) proposed a subfluence model which may be regarded as a plate tectonic model including the special features of an ensialic orogen. The superimposed uplift and tectonic shortening leads to horizontal overthrusts and nappe formation in places. Subfluence in the Rhenohercynian zone is interpreted in connection with the movement of larger lithospheric plates which transgress the limits of the Rhenohercynian zone and the Subvariscan Foredeep.

2.1.2.2 Mineral Deposits in the Rhenohercynian Zone

Many metallogenetic studies have been published by economic geologists on this geodynamic zone of the Variscides, highlighting prevalently the SEDEX or SMS-type mineralizations ("Rammelsberg-type") during the syn-rift phase of the Rhenohercynian Basin, as metalliferous basin-dewatering brines conduced to these Pb-Zn-Cu-pyrite-barite deposits at Meggen, at Eisen, Lohrheim, in the Rohberg Mine near Wiesbaden and in the Hartz Mts. at Goslar and Elbingerode (more VMS)

(Werner and Walther 1995). The Lahn-Dill Fe deposits are another stratabound type of ore deposits closely related to the Devonian basic and keratophyric volcanic rocks (Bottke 1963). Apart from these stratabound deposits a variegated group of vein-type deposits developed synkinematically during the Variscan orogeny and postkinematically in its aftermaths, containing mainly Pb, Zn, Cu, Sb, Fe, Ag, and barite. The geodynamic and lithological settings were not favorable neither for the development of pegmatitic nor aplitic rocks, only deep-seated quartz veins formed along its southern margin and are still operated today near Usingen, Germany (Fig. 2.1e, f, g).

Quartz veins at Usingen are surrounded by Devonian slates and were emplaced during the Permian as a result of deep-seated lineamentary fault zones governing the circulation of hydrothermal fluids but without the mediating effect of felsic intrusive rocks which played a significant part at the same time further south within the Saxo-Thuringian and Moldanubian zones. The quartz veins formed near the SSE boundary of the Rheinisches Schiefergebirge and act as an intermediary between the ore-bearing vein-type deposits in the basement and the unconformity fault bound mineral veins in the platform sediments (Fig. 2.1d). To see how these vein-type deposits are related to the quartz bearing pegmatitic deposits geodynamically, the reader is referred to Fig. 2.2a. Among the domestic deposits relevant for the supply with industrial minerals, these high-purity quartz veins still play an important part. They are selectively worked so as to attain a grade of almost 100 wt% SiO_2.

Fig. 2.1e Cross section through a massive quartz vein surrounded by Devonian slates and operated in an open pit near Usingen, Germany, for high-purity quartz

Fig. 2.1f Palisade quartz form the Usingen Quartz vein

2.1.3 Mid-German Crystalline Rise

2.1.3.1 Lithology and Structural Geology

The Mid German Crystalline High (Rise) does not show up geomorphologically as a coherent ridge, forming the central highland in Germany but is exposed only in isolated uplifted basement blocks, e.g., Odenwald, Spessart, Ruhla and Kyffhäuser massifs that are lined up in NE-SW direction and composed of metamorphic rocks very much different in their metamorphic grade which hits its peak with a series high-pressure eclogites and orthogneisses (Will and Schmädicke 2001) (Fig. 2.1d). It is often considered as part of the Saxo-Thuringian Zone, discussed subsequently, but in this study devoted to pegmatitic rocks it is treated as a separate entity for its rather different lithology and physical-chemical regime (Weber 1995). Lithologically the zone is characterized by a strong arc-related plutonism and Proterozoic (?) through Devonian medium to high grade metamorphic rocks (Anthes and Reischmann 1996). It is mainly late to post-kinematic felsic rocks of the granitic to dioritic suite that may be found in these isolated basement blocks. Detailed

Fig. 2.1g Open cavity fillings of quartz crystals (Bergkristall = rock crystals) in the Usingen quartz vein. See biro for scale

investigations about the lithology and geodynamic position have been performed by Will and Schmädicke (2001), Zeh et al. (2005), Zeh and Will (2010). The Cambrian to Ordovician rocks are metamorphosed attaining P-T conditions of the granulite-facies. Pre-Variscan granulites occur in the western Odenwald crystalline basement (Will et al. 2010). Medium-pressure metamorphic rocks formed during the Ordovician. The Silurian through early Devonian orthogneisses have calc-alkaline affinities. The Northern Phyllite Zone acting as the transition between the Rhenohercynian zone *sensu stricto* and the Mid German Crystalline High consists of low-grade stage regionally metamorphosed rocks of Ordovician through Devonian age.

The Mid-German Crystalline Rise forms part of a suture zone extending from Mexico to Turkey, resulting from the late Variscan closure of the Rheic Ocean opening up between Gondwana and Laurussia and taking a wide range of Paleozoic sediments sourced from Baltica and from Gondwana.

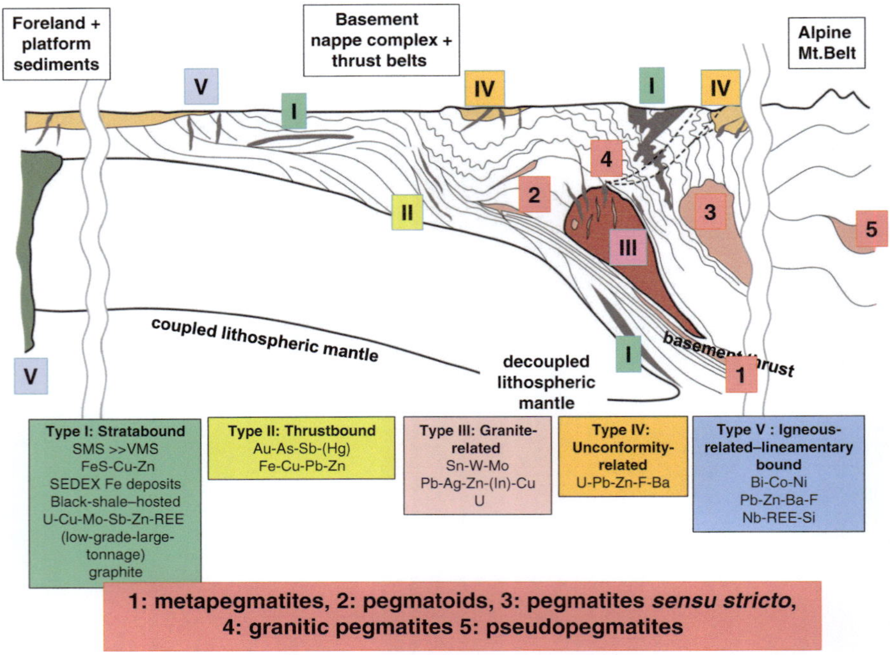

Fig. 2.2a An overview of mineral deposits typical of an ensialic orogen associated with pegmatitic rocks in time and space. As a consequence, non-pegmatitic mineral deposits are cast as marker deposits for pegmatite-prone or barren crustal sections and vice versa. *SMS* sedimentary massive sulfides deposits, *VMS* volcanic massive sulfide deposits, *SEDEX* sedimentary exhalative deposits

2.1.3.2 Mineral Deposits in the Mid-German Crystalline Rise

At the southern margin of the Rhenohercynian Zone, in the strongly metamorphosed Mid-German Crystalline Rise, pegmatitic mobilizates were mapped in the pre-Variscan gneisses of the Spessart Mts. as well as further SW, within the Odenwald Mts., where pegmatitic schlieren (pegmatoids) infiltrated magmatic rocks of granitic through gabbroic composition (Nickel and Fettel 1985). Compared with the pegmatitic rocks of the Bohemian Massif that form the gist of the matter in this book, the mineral assemblage of the Odenwald pegmatitic rocks is rather poor with quartz, feldspar and muscovite as the diagnostic rock-forming minerals and tourmaline and garnet forming only accessory minerals. Neither a temporal nor a spatial relationship can be established between felsic intrusive rocks and these alkaline mobilizates along the Mid-German Crystalline Rise. Pegmatoids developed in the environs of Aschaffenburg where a more elevated temperature provoked an intensive mobilization. The resultant pegmatitic schlieren contain quartz, K feldspar, plagioclase (albite-oligoclase), muscovite and biotite, arranged in order of decreasing abundance. Graphic intergrowth of feldspar and quartz is locally common, but rare elements typical of highly fractionated granitic pegmatites are missing. Spessartite, magnetite and titaniferous magnetite formed instead. From the northern Kyffhäuser Crystalline Complex, only pegmatitic granites are known, yielding an K/Ar cooling age of 333 Ma (Neuroth 1997).

To get an impression, what a physical-chemical regime these pegmatites were faced with during their emplacement, the metamorphic conditions for the Odenwald Mts. are briefly depicted. In the Heppenheim Complex the metamorphic rocks experienced a P-T history of 8–9 kb at a temperature of around 580 °C, yielding maximum conditions of 650 ° C at a more moderate pressure between 4 and 5 kb. (Willner et al. 1995). The retrograde pathway passed through the muscovite-quartz breakdown at 630 °C and about 3 kb. To allow for a better chronological correlation the minimum and maximum ages are reported for units I through III, yielding cooling ages for hornblende at 360 Ma (Kirsch et al. 1988) and concordant zircon ages of 335 Ma in migmatitic rocks (Todt 1979). An exceptionally high temperature of formation achieving 550 °C through 800 °C between 8 and 9 kb has reported by Willner et al. (1995). The age of formation span the range 380 Ma to 325 Ma (Kreuzer and Harre 1975; Todt 1979; Lippolt 1986). While the physical-chemical conditions are within the limits of what has been given for the Odenwald Mts., the age data of the Spessart are significantly younger with K/Ar and Ar/Ar mineral data ranging from 326 to 318 Ma (Lippolt 1986; Dombrowksi et al. 1994).

With regard to the pegmatitic rocks, the Mid-German Crystalline Rise marks the passage from the barren zone into the zone of pegmatoids (Fig. 2.1c).

2.1.4 *Saxo-Thuringian Zone*

Lithology and structure of the Saxo-Thuringian Zone are representative of a Cambro-Ordovician rift basin separated from the afore-mentioned Mid-German Crystalline Rise by a south-facing reverse fault, which some geologists also call a strike-slip fault (Franke et al. 1995; Franke and Stein 2000; Linnemann 2003; Kroner and Hahn 2004). A great variety of sedimentary and volcanic rocks formed in this basin from the Precambrian through the Lower Carboniferous, when the Visean tectonic disturbances once and for all put an end to the basin development. In the south this geodynamic unit is bordered by Pre-Variscan basement blocks, stretching from the Vosges Mts., France, into the Bohemian Massif covered to a large extent by Czech Republic, which is eponymous to the Bohemian Massif (Fig. 2.1d). Separated by a pronounced NW-SE-striking Lineament named after the river Elbe (in Czech Labe) the Lugicum and the West Sudetes form the eastern prolongation of the Saxo-Thuringian Zone (Sect. 2.1.6.2). The crustal section of Saxo-Thuringian Zone in the Central European Variscides was named after Thüringen (Thuringia) and Sachsen (Saxonia), two German states which have a large share in this geodynamic real, the first one through the Thüringer Wald and the second one through the Erzgebirge, straddling the border with the Czech Republic where the Erzgebirge is called Krušne Hory in the Czech language. In the Free State of Bavaria the Frankenwald, which is the southern prolongation of the Thüringer Wald is also attributed to this zone. Those who are accustomed to work in Precambrian terrains in ancient cratons where the lithology does not significantly vary over hundreds of kilometers may feel bewildered because of the variegated lithology in Central Europe and the great number of facies and site names, that sometimes change within a distance measuring only a few tens of kilometers.

2.1.4.1 Lithology and Structural Geology of the Fränkisch-Thüringisches Schiefergebirge

The Thuringian Facies in the Franconian-Thuringian Slate Mountains, or in German Fränkisch-ThüringischesSchiefergebirge, is composed of neritic clastic sediments of Cambrian through Ordovician age resting upon greywacke-dominated Proterozoic units (Falk et al. 1995) (Fig. 2.1c – see parautochtonous of the Saxo-Thuringian). The Silurian succession is dominated by black-shales and cherts and the younger sediments up to the Early Carboniferous encompass shales, calcareous rocks and turbidities (Wignall 1991). Volcanic rocks of bimodal character are known to occur during the Ordovician and late Devonian at a large extent in this geodynamic zone.

The Bavarian Facies is closely related tectonic units which have an allochthonous character at least in parts and may be found also further south near the heartland where the pegmatites crop out. Olistholiths and gravitational nappes are unique in this region made up of a wide range of rocks from shales, greywackes and tuffs of Ordovician and Cambrian age (Fig. 2.1c -see Bavarian Facies). The Silurian and Devonian units lithologically resemble largely those of the Thuringian Facies with cherts, black shales, calcareous and volcanic rocks. During the Tournasian and Visean wild-flysch and mass flow deposits are common. The sedimentary rocks of the Bavarian Facies are a deeper-marine equivalent of the Thuringian Facies which was laid down originally further to the South-East and subsequently displaced towards the North-West by thrustal movements (Fig. 2.1c). The most striking example of tectonic mass movement in the Saxo-Thuringian Zone, the Münchberg Gneiss Complex can also make a significant contribution to the understanding of these feldspar-quartz mobilizates but prior to their discussion it needs a more detailed description as to the lithology and geodynamic setting of this tectonic klippen (Fig. 2.1c)

2.1.4.2 Lithology and Structural Geology of the Münchberg Gneiss Complex

Passing through the Münchberg Gneiss Complex from North to South by car along the "Autobahn" (motor highway) may provide the driver with an outline of the geomorphology that is anything but exciting. Neither eye-catching mountain ridges nor prominent bluffs or rock exposures are lined up along the route in this morphological depression between the deeply dissected plateau of the Frankenwald in the

North and the high-rising granitic domes of the Fichtelgebirge Anticline at its southern border (Fig. 2.1d). By contrast, the lithology and the structural geology of the Münchberg Gneiss Complex together with the other klippens at Wildenfels and Frankenberg both being located North-East of the Münchberg Gneiss Complex are far from dull and rather extraordinary from the petrographic point of view.

Recalling the regional metamorphism of the central Saxo-Thuringian Zone, the Early Carboniferous flysch sediments of the Bavarian Facies in the Frankenwald on top of the Berga Anticline are of low-grade stage to very low grade stage which is the case with most rocks of the Thuringian Facies of the in the remaining parts of the Saxo-Thuringian Zone. Heading further South-East and approaching the Münchberg Gneiss Complex, we will enter a succession with an increasingly higher metamorphic grade that has been overthrusted by the Münchberg Gneiss Complex *sensu stricto* (Fig. 2.1c). The afore-mentioned succession is subdivided into the "Phyllit-Prasinit-Series" (phyllite-prasinite series [prasinite=fine-grained metabasalt bearing actinolite, chlorite, albite, epidote subjected green schist facies conditions]) and the "Rand-Amphibolit" (rim amphibolite) which contains minerals of the epidote-amphibolite facies and has been derived from a basic magmatic protolith (Fig. 2.1d – Paleozoic allochthonous of the Saxo-Thuringian Zone). Schüssler et al. (1986) found out that the mafic volcanic rocks exhibit a calc-alkaline affinity. Passing into the Münchberg Gneiss Complex *sensu stricto*, we are faced within this allochthonous synform with another dual subdivision into the so-called "Liegend-Serie" (footwall series) and the "Hangend-Serie" (hanging-wall series) (Fig. 2.1c). The "Liegend-Serie" is composed of meta-pelites and meta-greywackes, intercalated with orthogneisses/augengneisses and lenses of metagabbro and metagranodiorites. The orthogneiss yielded a Rb/Sr whole rock age of 499 ± 20 Ma (Söllner et al. 1981). U/Pb age dating using zircon and monazite from metagabbros and metagranites gave and age of intrusion around 500 Ma for both intrusive rocks (Gebauer and Grünenfelder 1979). Other than in the "Liegend-Serie", in the "Hangend-Serie", amphibolites, hornblende gneisses and eclogites prevail over metasedimentary rocks. The eclogites are of MORB affinity and the original basalt which they derived from were vented at 525 Ma based upon U/Pb zircon ages (Gebauer and Grünenfelder 1979). They derived from gabbros and tholeiites which underwent metamorphic pressure of 13 kb. The "Hangend-Serie" gave host to the albite-bearing pegmatoids which almost all are confined to the south-eastern part of the this allochthonous gneiss complex and bound to the hornblende gneisses (Figs. 1.5a and 2.1c). A rather strange type of pegmatoids, called zoisite pegmatite on account of its diagnostic accessory mineral, is closely-related to the ecologite and the eclogite amphibolite. Their emplacement and origin are discussed later in the book in context with the remaining felsic mobilizates from the Moldanubian Zone (Sect. 6.3)

Deformation and metamorphism of the Münchberg Gneiss Complex are of polyphase type. Both are treated in this paper only to the extend to understand the development of the pegmatoids and aploids in this allochthonous nappe complex.

An initial high- to medium-pressure regime was succeeded by a reequilibration when the entire nappe was thrusted towards the Northwest. Metapelites formed under a physical-chemical regime of 8.2 kb and 607 ± 50 °C, while the eclogites saw a pressure of 13 kb and a temperature of 600 °C (Blümel 1986). The HP metamorphism occurred during the Silurian. Cooling ages of 394 ± 14 Ma derived from a Rb/Sr isochrone and 395 Ma from a Sm/Nd isochrone refer to the Early Devonian (Stochs and Lugmair 1986, 1987). Kreuzer et al. (1989) provided K/Ar mineral ages for hornblende and mica in the range between 340 and 410 Ma. The large number of age date reported here for the various minerals point to a thrust movement between the Early Devonian and Early Carboniferous. As early as 1912, Suess postulated a nappe tectonic based on field geology which was only 70 years later proven by a painstaking investigation, involving geophysics, petrology, structural geology and geochemistry.

2.1.4.3 Lithology and Structural Geology of the Fichtelgebirge-Erzgebirge Anticline

The Fichtelgebirge-Erzgebirge Anticline strikes ENE–WSW and is characterized in the geophysical maps by a significant gravity minimum of less than 100 mgal that may be traced southward into the Moldanubian Zone, discussed in the following section (Behr et al. 1989). Several granitic complexes are accountable for this gravity low typical of the Fichtelgebirge-Erzgebirge Anticline. Two main intrusive complexes have been distinguished in the eastern part of this Variscan anticline, in the Erzgebirge (Tischendorf and Förster 1990; Sebastian 2013). The entire intrusive suite began with an older series, coded OIC and aged 330 through 310 Ma, that is of monzogranitic composition and of a mixed I/S type affiliation. Separated by a period of igneous quiescence, a younger series coded YIC followed with ages of intrusion in the range 305–290 Ma. It is mostly of I-type with some A-type affinities. These monzo- to syenogranites are strongly fractionated, in parts autometasomatically altered and enriched in Li and F (Tischendorf et al. 1987). Förster et al. (1999) put forward a chemical subdivision of the granitic intrusive rocks into (1) medium-F and low-P biotite granites (A-type), (2) high-F- and low -P lithium mica granites (A-type), (3) high-F- and high -P lithium mica granites (S-type), (4) low-F- two-mica granites (S/I-type). A peculiarity as to the chemical specialization in the Mid-European Variscides warrants mentioning; it is the abundance of Li mica- and topaz-bearing granites. Although very much different with respect to their chemical composition, the granites were emplaced in a small time window immediately after the collisional phase when the crust attained its maximum thickness of more than 60 km. A high-heat-producing process as a result of the radioactive decay within the time span of roughly 20 million years has been held accountable for the evolution of the late-collisional granites (Förster et al. 1999). The same authors envisioned as an additional heat source to generate these granites an intrusion or underplating of mafic magmas from the mantle (see also Sects. 5.1.1, 5.1.2, and 5.1.3). However, the first-mentioned process on its own is already sufficient to provide the heat necessary to create these granites.

In the western part of the anticline, geomorphologically termed Fichtelgebirge, the felsic intrusive rocks have some basic to intermediate predecessors, ranging in their chemical composition from gabbroic, through dioritic into granodioritic which were given the collective term "redwitzite" after the town of Marktredwitz located in the region (Richter and Stettner 1979). Their origin and emplacement is rather complex, and supposed to involve differentiation of a basaltic magma and/or mixing with a granitic melt (see also Sects. 5.1.1, 5.1.2, and 5.1.3).

Because of their outstanding part in terms of the origin, of what was denominated as "granitic pegmatites" in Table 1.1, the role they play among the magmatic evolution and their importance for the metallogenesis in the Saxo-Thuringian Zone, these granitic rocks deserve a particular treatment in a study like that. This is true especially in the western part of the anticline where their emplacement overlaps with that of pegmatites and aplites and on account of the cross border extension of these granites into the southern Moldanubian Zone which suggests a common process to have been responsible for their evolution in both geodynamic zones. Chemical composition and petrological description of these granites are based upon the study conducted by Richter and Stettner (1979) (Table 2.1). Similar to the equivalent intrusive rocks further NE, in the Fichtelgebirge also two intrusive suites can be distinguished. An older group of granites is coded G 1 and occurs together with its differentiates around Weißenstadt, Markleuthen, Reuth and Selb. An account of its central position within the Fichtelgebirge Anticline it was named Central Granite. The Central Granite complex is prevalently made up of monzogranites grading towards the edge into what might be categorized as a granodiorite or taken to the extreme in its most basic facies as a diorite, when its chemically data points overlap with the felsic end members of the redwitzites. The older granites which are correlative to the OIC granites of the Erzgebirge formed throughout multiple intrusions between 330 Ma and 315 Ma (Lenz 1986; Carl and Wendt 1993). The oldest granite shows a porphyritic texture in a coarse groundmass and takes an overall tabular shape, dipping gently towards the south. Based upon its almost identical cooling ages of mica its magma consolidation was very fast around 319 ± 3 Ma.

The remaining granites G 2 (Marginal Granite), G 3 (Core Granite) and G 4 (Tin Granite) have been placed in order of their intrusion on geological grounds. Using radiometric age dating, however, does not allow a precise age of formation to be attributed to the individual granite bodies and proves these granite intrusions to be more or less coeval in the range 301–295 Ma. The G-2 granite is a fine-grained porphyritic monzogranite. Its texture attests to an intrusion of the granitic magma in an environment unbalanced as to the temperature. Approximately 20 million years after the intrusion of the G 1, the temperature had significantly dropped.

The coarse-grained G-3 granite is a syenogranite, whereas the youngest member of this series of granite straddles the boundary into monzogranite. The latter is a fine-grained granite that already by its name reveals a strong chemical fractionation leading to U and Sn mineral assemblages. While the granite-hosted uranium mineralization was only investigated by underground trial mining operations, the

Table 2.1 Chemical composition of the four major granite types in the Fichtelgebirge

Site	Unit	Granite G 1	Granite G 2	Granite G 3	Granite G 4
SiO_2	wt%	67.9	74.2	75.5	74.80
TiO_2	wt%	0.68	0.23	0.17	0.05
Al_2O_3	wt%	15.00	13.40	13.00	14.10
Fe_2O_3	wt%	0.74	0.32	0.23	0.10
FeO	wt%	2.90	1.60	1.50	1.20
MnO	wt%	0.06	0.04	0.03	0.03
MgO	wt%	1.22	0.30	0.21	0.06
CaO	wt%	2.30	0.75	0.57	0.37
Na_2O	wt%	3.30	3.00	3.00	3.70
K_2O	wt%	4.52	5.25	5.00	4.74
P_2O_5	wt%	0.32	0.21	0.19	0.24
H_2O/LOI	wt%	0.80	0.60	0.50	0.30
Li	ppm	60	129	153	306
Be	ppm	4.6	7.7	9.9	17.6
B	ppm	2	2	2	17
F	ppm	602	1665	1243	2315
S	ppm	166	69	74	50≤
Cl	ppm	76	96	88	67
Sc	ppm	13	5	4	3
V	ppm	61	12	7	≤3
Cr	ppm	11	9	6	6
Ni	ppm	4	5	4	4
Cu	ppm	7	3	6	4
Zn	ppm	62	44	43	43
Ga	ppm	20	22	21	41
Rb	ppm	211	404	427	805
Sr	ppm	251	44	29	5
Y	ppm	18	26	25	14
Zr	ppm	267	112	80	25
Nb	ppm	15	12	11	15
Mo	ppm	≤2	2	2	2
Sn	ppm	8	12	15	24
Cs	ppm	9	24	33	59
Ba	ppm	930	283	129	13
La	ppm	73	46	33	39
Ce	ppm	99	56	29	22
Pb	ppm	35	31	34	19
Th	ppm	34	21	12	7
U	ppm	4	6	10	17

Fig. 2.3a Tin Granite (Granite No 4), the most strongly fractionated granite shown in its normal or un-mineralized facies. Kugler Quarry near Tröstau, Fichtelgebirge-Germany. Lens-protection for scale

abundance of cassiterite, present mainly as greisen-type or in alluvial-fluvial placer deposits in the clastic apron around the G 4, also sparked some mining operation in the past (Fig. 2.3a, b, c).

The peraluminous granites are of intracrustal and comagmatic origin based upon their trace element differentiation (Richter and Stettner 1979) and $^{87}Sr/^{86}Sr$ isotope ratios (Besang et al. 1976; Wendt et al. 1986). The granite suite reveals a textbook-style fractionation with a steady increase of elements such as Li, Be, Sn, Cs, and U with time (Table 2.1). The younger granites G 2 through G 4 can thus be grouped among the so-called fertile granites as to the elements Li, Be, Sn, W, U, and Nb. As to the P/T regime, the G 1 granite in the area of Weißenstadt and Marktleuthen formed around 660 °C at a pressure of 7 kb to be equal to a depth of between 20 and 25 km (Richter and Stettner 1979). The granite at Selb, belonging to the same intrusive stage formed at a depth of 7 km, equivalent to 2 kb and at a temperature of between 700 and 720 °C. The older granite has been derived from a migmatitic shear zone penetrating the upper crust at the afore-mentioned depth.

Applying the same parameters of the albite-anorthite ratios to the younger suite of granites yielded a pressure of between 0.5 and 2 kb which equals to a depth of formation of between 2 and 7 km at a temperature of 650–700 °C. The temperature can only be approximated owing to the abundance of fluorine in the range 1665–2370 ppm F which has a lowering affect on the temperature of formation. The younger granites in the Saxo-Thuringian Zone came into existence when the Asturian Orogeny deformed the sediments in the westernmost part of the Rhenohercynian Zone (Fig. 2.1c, d). The

Fig. 2.3b Tin Granite (Granite No 4). Specimen has been taken from the mineralized facies or greisen facies with cassiterite (*black*) and corroded arsenopyrite. Rudolfstein, Fichtelgebirge-Germany

Fig. 2.3c Tailings dumped in the softwood-tree forests around the Tin Granite. The dumps are relics of the mining operations targeting upon the cassiterite placers in the clastic apron around the Granite No 4

abnormally high rare element contents of, e.g., Sn and U may have derived from remobilization of Carboniferous and Devonian sediments. A preferred orientation of phenocrysts frequently observed in the younger granites may be interpreted as representative or a relic of the pre-existing cleavage and stratification in the surrounding country rocks. Metasomatic processes are more likely to have been caused these texture than a mechanically intrusive emplacement which might lead to flow banding in the magma similar in its outward appearance to such alignments of minerals. Willner et al. (1995) made an attempt to account for the complex geodynamic evolution of the Fichtelgebirge-Erzgebirge Anticline and put forward a four-step model involving (1) oceanic crust to be subducted and converted under eclogite-grade conditions, (2) continent-continent collision and thickening of the crust, (3) attenuation of the thickened crust with simultaneous and continuous under thrusting of continental crust, (4) exhumation of the existing stacked crustal section in the course of another thinning.

The Paleozoic country rocks of the Fichtelgebirge granites not only form part of the Saxo-Thuringian but also an integral part of the LP-metamorphic zone characteristic for the western edge of the Bohemian Massif (Stein 1988). It is the low-grade equivalent of the LP unit being present also further south.

2.1.4.4 Mineral Deposits

From the Cambrian through the upper Devonian a wide range of stratabound ore deposits evolved in the autochthonous part of the Saxo-Thuringian Zone (Baumann 1979; Baumann et al. 2000; Bernard 1980; Dill 1989; Pertold et al. 1994; Tischendorf et al. 1995b)- see also Fig. 2.2a. During the early Paleozoic several massive sulfide deposits evolved and were mined on both sides of the Czech-German border. They are abundant in pyrite and pyrrhotite with little Zn and Cu. During the early phases of rifting submarine hydrothermal fluids gave rise to these polymetallic deposits such as at Waldsassen in Germany and Tisova in the Czech Republic. The ironstone deposits of the Thuringite-/Wabana-Type have been reported from different parts of the world throughout the Ordovician, among others from the Saxo-Thuringian Zone of the Middle European Variscides (Van Houten and Hou 1990). The oolithitic Fe ores, containing siderite, thuringite, chamosite and magnetite were formed during transgressions in a shallow-marine environment (Thuringian Facies). Similar to the afore-mentioned ironstones of worldwide occurrence, the Silurian and Lower Devonian alum shales of the Lower and Upper Graptolite Shales have also a share in the Saxo-Thuringian metallogenesis, yet only as low-grade-large tonnage deposits enriched in U, Au, Mo, Cu, Sb, Zn and REE. Passing vertically upward in the stratigraphic column, the exhalative magnetite-hematite iron ore deposits in the Saxo-Thuringian Zone show mineralogically and geological similarities to the Lahn-Dill deposits from the Rhenohercynian Zone where these iron deposits are closely linked in time and space to the metabasalts (diabase) and keratophyres vented on the sea floor.

Another group of deposits common to both geodynamic zones, the Rhenohercynian and the Saxo-Thuringian Zones, are denominated as metamorphogenic, thrust bound and fold-related cleavage veins that are common to during the Devonian and Early Carboniferous. Activation of the continental margin of the Mid-German Crystalline Rise resulted in the initiation of southward subduction during which Rhenohercynian and Saxo-Thuringian crust was consumed. Without any doubt, these synkinematic fault-related deposits are more widespread in the Rhenohercynian than the Saxo-Thuringian zone, where only some mesothermal gold, mercury and antimony vein-type deposits along the Schwarzburg-, Berga and Fichtelgebirge-Erzgebirge Anticlines may be attributed to this stage of the Variscan metallogenesis. In the northwestern geodynamic zone, structurally equivalent vein-type deposits can be subdivided into two major groups, Ag-enriched Pb-Zn vein-type deposits and siderite-Cu-Pb-Zn vein-type deposits named as Siegerland-Type (Werner and Walther 1995; Wagner and Boyce 2001, 2003). Hein (1993) suggested that the Siegerland siderite veins, which extend down to a depth of 1000 m, formed from low salinity CO_2-undersaturated fluids at temperatures between 180 and 320 °C following the peak of metamorphism and prior to the postkinematic magmatism.

Postkinematic magmatism which is intensive in the Saxo-Thuringian zone marks the late stages of Variscan convergence in mid-to late Carboniferous times resulting in the emplacement of abundant synorogenic granites and genetically related Sn-, W-, Pb-,Zn-, and U deposits (Seltmann and Faragher 1994; Štemprok and Seltmann 1994; Henk et al. 2000). The highly fractionated S- and I-type granites are the most likely source for the fluids to create these mineralizations, present as vein-type, greisen and skarn deposits in the Erzgebirge Mts. Particular attention should be drawn to the Poehla-Haemmerlein deposit at Gottesberg, Germany, which is a large, low-grade, refractory tin skarn with its reserves standing at 12.3 Mt @ 0.42 % Sn (Buder et al. 1993). Wolframite occurs in quartz veins (Krásno, Czech Republic, Ehrenfriedersdorf, Germany, Altenberg, Germany, Zinnwald-Cínovec, Germany/Czech Republic), Pechtelsgrün, Germany. In the Altenberg Sn deposit, Germany, the missing link between the true pegmatite bodies and the granites exists. The afore-mentioned Sn deposit is characterized as "greisen" and "stockscheider", both technical terms coined by the ancient miners in search of Sn. Greisen may be described as a pervasively altered lithium-albite granite in which feldspar and biotite are converted to a disseminated assemblage of quartz, topaz, muscovite, zinnwaldite and protolithionite (both Li-micas), cassiterite, sericite, fluorite, dickite, kaolinite, wolframite and scheelite. Stockscheider is an altered granite of pegmatitic texture abundant in topaz (pyknite), zinnwaldite, and quartz with cassiterite, wolframite, and molybdenite as the main ore minerals (Baumann et al. 1986). Both terms describe alteration zones within the host granite caused by the fluids that were concentrated in the apical parts of the granites during their consolidation. In the classical German literature these peculiar alteration zones were classified as "pegmatitic-pneumatolitic" alteration, a term which has become a bit obsolete in present-day literature and rather replaced by supercritical with some description of the mineralizing processes in more detail. Although occurring in an environment strongly enriched in highly volatile chemical compounds no individual pegmatite bodies resembling those from the core zone of the Central European Variscides have been encountered in the Erzgebirge. Host structures are mainly druses and cavities in the felsic intrusive rocks filled with pegmatite-type minerals, present on a collector's rather than on a miner's scale. Quite the same scenario may be reported from the Fichtelgebirge in the western part of the anticline. The so-called "pegmatitic-pneumatolitic" mineral assemblages mainly with schorl as diagnostic mineral may be observed in the apical parts of the G-3- and G-4 granite of the younger suite. They did neither spawn large individual pegmatitic bodies that were self –intrusive into the surrounding country rocks nor did they give rise to aplitic veinlets as it is the case with the older granite G-1, the "Central Granite". The south-eastward dip of the tabular granite is marked by swarm of aplitic and pegmatitic veins in the roof of the Central Granite. Richter and Stettner (1979) reported aplite granites and aplitic through pegmatitic veins prevalently from those areas where meta-carbonates and calcsilicate rocks make up a great deal of the metamorphic country rocks. Aplitic veins prevailing over pegmatitic veins intersect not only meta-carbonate and calcsilicate rocks but also mica schists. Far away from the variegated mineral assemblage which we will see later in the pegmatites located farther south, these fault-related aplites and pegmatites in the vicinity of the G-1 granite have only a poor assemblage of schorl and beryl together with some apatite and sphene.

The majority of minerals which are listed in Table 1.2 for this zone is bound to druses and miaroles. Although well endowed with ample supplies and a variety of metallic, non-metallic and energy deposits, mining has peaked in the Erzgebirge-Fichtelgebirge Anticline on account of a long-lasting mining activity mainly focused on Sn, W, U, Pb, Zn and Ag. Only those commodities which our ancestors exempted from their mining operations, because they had no use for them, such as fluorite and barite are still worked from vein-type deposits at Niederschlag near Bärenstein. When the European countries realized their dependency on a few supply countries outside Europe, particularly in raw materials for future technologies such as lithium and indium, but also more traditional ones, e.g., tin and tungsten, mining companies mainly from outside Germany revisited this crustal section of the Central European Variscides and started drill on sites well known for ages for its Sn-W accumulations such as Geyer-Ehrenfriedersdorf and Gottesberg, in the western Erzgebirge (Dill et al. 2008a). As these deposits do no longer play in the champions league of extractive geology or called deposits by world standards neither as to size nor as to grade, their fate hangs by a silk thread in terms of exploration and exploitation. As being viewed realistically, a glimpse of hope is there but anything else.

2.1.5 *The Moldanubian Zone*

2.1.5.1 Lithology and Structural Geology

The rivers Moldau (Vltava in Czech) and Donau (Danube) drain the southern part of the Bohemian Massif and, logically sound and comprehensible for the audience, they lend their names to designate the central geodynamic or core zone of the Mid-European Variscides, the Moldanubian Zone. The southern half of the Bohemian Massif, which is identical to this geodynamic zone, includes two tectonostratigraphic units, one called the Moldanubicum *sensu stricto*, the other allochthonous unit is named the Teplá-Barrandian zone or Bohemicum (Malkovsky 1979; Tollmann 1982; Weber and Behr 1983). An overview of this geodynamic zone has been given by Franke (2000), Matte (2001) and Raumer et al. (2003). The Moldanubian zone represents a stacked pattern of nappes which were superimposed onto each other during the late Variscan (Fig. 2.1c, d).

The southern part of the Oberpfälzer Wald, Bayerischer and Böhmer Wald (Český les in Czech) are underlain by clastic and magmatic rocks in the Moldanubicum s.s., which were converted into high-grade gneisses granulites and orthogneisses, summarized under the lithostratigraphic term "Monotonous Group" or Ostrong unit. It is the lowermost part of the Moldanubicum *sensu stricto*.

In contrast to these lithologies, the "Varied Group" or Drosendorf unit consists of Proterozoic and early Paleozoic meta-greywackes, metapelites, marbles, calcsilicates, quartzites and graphite schists which reflect rather mobile dynamic processes in the original environment of deposition, an impression strengthened by the great variety of metabasic igneous rocks intercalated into these sedimentary rocks. In the map presented by Mazur et al. (2005), the majority of the Drosendorf encompass the Moldanubian zone.

The Drosendorf units was trusted over onto the Gföhl unit which was metamorphosed under amphibolite and granulite facies conditions. Its HP granulites are accompanied by pyrope- and spinel-bearing peridotites, pyroxenites, eclogites, migmatites, as well as orthogneisses and paragneisses (Fuchs and Matura 1976; Tollmann 1982). The source rocks were both sedimentary and igneous in origin. The protolith of the Gföhl Gneiss was an Ordovician granitic rock.

First records of detrital material in paragneisses on zircons showed an event around 2600–2400 Ma (Hansen et al. 1989). Metafelsic and metabasic igneous rocks yielded SHRIMP age of 555 ± 12 Ma, 549 ± 7 Ma and 549 ± 6 Ma (Teipel et al. 2002). Another event that brought about felsic metamagmatic rocks occurred at the Cambrian-Ordovician boundary, based on SHRIMP ages of 486 ± 7 Ma and 480 ± 6 Ma (Teipel et al. 2002). There is an indistinct sign referring to a medium-pressure metamorphic event at about 380 Ma (Hansen et al. 1989). A strong low-pressure-high-temperature regional metamorphic event about 320 Ma overprinted inherited metamorphic material (Hansen et al. 1989). A lower intercept age of 530 Ma can be recognized. The major metamorphic events include HT-LP (periplutonic) regional metamorphism with cordierite and andalusite ensued by the late Variscan felsic intrusions.

The Teplá-Barrandian zone is composed of volcano-sedimentary sequences affected by synsedimentary faulting and continental clastic rocks laid down in grabens subsequently to the Cadomian subduction and collision (Dallmeyer and Urban 1998; Zulauf et al. 1999; Dörr et al. 2002). Typical of the Teplá-Barrandian zone and its analogue in NE Bavarian Basement, the Zone of Erbendorf-Vohenstrauß (ZEV), their metamorphosed mafic igneous rocks revealed chemical affinities to suboceanic mantle, ocean-ridge, calc-alkaline and within-plate settings (Kastl and Tonika 1984; Schüssler et al. 1986; Jakes and Waldhauserova 1987). Similar to the Moldanubian s.s. the first zircon ages can be recorded from the interval 2500 Ma through 2400 Ma. Rubidium-strontium age dating of paragneisses from the ZEV were dated as Lower Cambrian (530 Ma) (Weger et al. 1998). A more detailed picture has been drawn by Drost et al. (2004) who claimed the Teplá-Barrandian unit to be part of the Avalonian-Cadomian belt at the northern margin of Gondwana during Proterozoic and Early Cambrian times- see also Fig. 2.1a. Its volcano-sedimentary series developed in a back-arc basin. Around 2500 m of Lower Cambrian continental siliciclastics were deposited in a basin-and-range-type setting accompanied by magmatism, which shows within-plate features in a few cases, but is predominantly derived from anatectic melts. The geochemistry of clastic sediments suggests a deposition in a rift or strike-slip-related basin, respectively. Upper Cambrian magmatism is represented by 1500 m of subaerial andesites and rhyolites demonstrating geochemical characteristics of an intra-plate setting. Zircons from a rhyolite give a U-Pb-SHRIMP age of 499 ± 4 Ma (Drost et al. 2004).

Metamorphic monazite from the Teplá-Barrandian zone indicated a Barrovian-type isograde in the Domažlice Crystalline Complex around 551–540 Ma (Zulauf et al. 1999; Timmermann et al. 2002). A magmatic event can be identified in the rock record between 530 and 460 Ma. The climax of metamorphic overprinting was reached around 390 Ma under medium-pressure conditions, with a rapid cooling until 360 Ma (Hansen et al. 1989). The final nappe emplacement took place around 330 Ma.

According to Holub et al. (1995) magmatic rocks of granitic composition in the Moldanubian zone may be attributed to the South Bohemian and Central Bohemian Plutonic Complexes with some of their satellite batholiths being also exposed along the western edge of the Bohemian Massif, in the NE Bavarian Crystalline Basement. As in this part the interrelationship between granites and pegmatitic rocks is crucial for the understanding of the origin of the latter felsic rocks they will be discussed later in more detail and only a general overview will given in this chapter. The granitic rocks are subdivided into three successive suites each being directly correlated with the evolution of the Variscan Orogen in this geodynamic unit. During the early Carboniferous, synorogenic granitic rocks evolved as the Variscan orogeny went through its thermal climax around 350 through 335 Ma. The older suite of the synorogenic granitic rocks of the

South Bohemian Pluton developed during this period of time, starting off with minor gabbros, diorites and quartz monzonites and ending with the large coarse-grained Weinsberg Granite (Koller 1996). Today this suite of granitic rocks has a representative in the "Kristallgranite I", as it is denominated in SE Germany (crystal granite I) by regional geologist. It is a coarse-grained biotite granite to granodiorite of I and I/S affiliation rife with mega crystals of K feldspar, that account for its regional name "Kristallgranite I" (crystal granite I). The subsequent suite of granite formed between 329 ± 7 Ma and 303 ± 6 Ma. The oldest intrusions of this suite of the South Bohemian Pluton in Germany may be traced into neighboring Austria where among others the Eisgarn (318 ± 7) and Mauthausen Granites are held to be contemporaneous with them (Scharbert and Vesela 1990). They are classified as slightly deformed S- and I-type granites. The youngest representatives of this group yielded Ar/Ar muscovite ages of between 312 and 308 Ma (Dallmeyer et al. 1995b). It is a suite of very shallow intrusions which very rapidly cooled down to less than 400 °C.

The third suite was emplaced between 300 ± 41 Ma and 295 ± 5 Ma (Holub et al. 1995). It is a group of anorogenic granites related to an extensional regime. Pegmatites and aplites are very rare, excluding the extraordinary phosphorus Homolka granite (Breiter 1998c).

More prominent than the South Bohemian Pluton in size and outstanding for its chemical variation, the Central Bohemian Pluton covers about 3200 km^2 and extends along the Central Bohemian Suture in NE-SW direction along the contact between the Tepla Barrandian and the Moldanubian *sensu stricto* (Kodym 1966). It shows the full blown succession of plutonic rocks with gabbros at the beginning and granites at the end. A peculiar type, in many places related to pegmatite is called durbachite, what is synonymous with potassium-enriched melanocratic syenites (Holub 1997). Numerous age data obtained by Pb/Pb single-grain dating of zircon and field evidence attest to a late Devonian to early Carboniferous age of formation of the Central Bohemian Pluton (Janoušek et al. 1995; Holub et al. 1997): Sázava (349 ± 12 Ma), Požáry (351 ± 11 Ma), Blatná (346 ± 10 Ma), Čertovo břemeno (343 ± 6 Ma). According to Janoušek et al. (1995) the Sázava unit is the most primitive one. It is supposed to be generated either by melting of metabasic igneous rocks as they were found as roof pendants on the Central Bohemian Pluton, by partial melting of mantle material or magma mixing, involving both source mentioned before. The remaining intrusive units give a similar very divers picture as far as the source and magma generating processes are concerned, invoking mantle material being contaminated with crustal material from paragneisses (Blatná unit), or leucogranites, e.g. at Čertovo břemeno.

A comparison of the characteristic features of the Teplá-Barrandian and Moldanubian terranes, as they were called by Mazur et al. (2005) is given below since both terranes are host to a wide varied of pegmatitic rocks, yet very unevenly distributed across the Moldanubian Zone.

	Teplá-Barrandian terrane	Moldanubian terrane
sedimentation	Neoproterozoic and Early Palaeozoic to Middle Devonian	Proterozoic and poorly constrained Early Palaeozoic; in the Orlica-Śnieżnik unit — pre-Ordovician
plutonism	Cambrian; Carboniferous at the contact with Moldanubian terrane	Ordovician, Carboniferous
metamorphic age	close to the Proterozoic/Cambrian boundary, Early Carboniferous overprint at the contact with the Moldanubian terrane	intense Early Carboniferous HT/M-LP overprint
HP metamorphism	lacking	bodies of eclogites and granulites ranging in age from 360 to 330 Ma
metamorphic grade	low-grade to unmetamorphosed	medium- to high-grade
uplift/exhumation	Late Devonian	Early Carboniferous

2.1.5.2 Mineral Deposits

Evidently, not only the lithology but also the metallogenetic evolution in the Moldanubian Zone is very much diverse, giving rise to a large number of stratabound and vein-type ore deposits besides the common granite-related and pegmatite deposits (Dill et al. 2008a). Volcanic-hosted massive sulfides, sedimentary massive sulfides and kieslager-type ore deposits were mined in the past and investigated in great detail by Czech and German geoscientists on both sides of the border.

Silver-, lead-, zinc and copper accumulations in the Příbram ore district, Czech Republic, were also attributed to the thrust-bound and fold-related metamorphogenic ore mineralization (Dill et al. 2008a). Several base metal veins were emplaced around the Central Bohemian Pluton where underground mines reached an operational depth of more than 1500 m. The ore veins are feather structures accompanied by diabase dykes of similar shape (Pouba and Ilavsky 1986). Important Ag carriers besides galena in this deposit are pyrargyrite, stephanite and diaphtorite.

Several studies have been published by Boiron et al. (2001), Zachariáš and Stein (2001), Zachariáš and Pudilová (2002) about the gold deposits located in paragneisses and migmatites with intercalated quartzites, calc-silicate rocks, felsic volcanic rocks, amphibolites, and marbles of the Central Bohemian Metallogenetic Zone. The majority of these gold deposits is low in sulfur, enriched in siliceous gangue and shows a prevalence of metamorphic aqueous low-salinity fluids. The fluids belong to a fluid system enriched in C, N, O and H and probably resulted from fluid–rock interactions within the metamorphic series at high P–T conditions ($T \approx 450$–550 °C and $P \approx 250$–400 MPa) (Boiron et al. 2001). The Mokrsko deposit, Czech Republic, is one of the largest Au resources in Central Europe.

For the central part of the Bohemian Massif, Mrázek (1986) compiled the mineral deposits pertaining to the Late Proterozoic metallogeny and related them to volcanic and post-volcanic thermal activity. Pyrite and pyrrhotite mineralization with subordinate amounts of Cu and Zn sulfides occur in low-grade regionally metamorphosed basic and intermediate metavolcanic rocks. Representatives of this type of ore deposits are located near Struhadlo/Klatovy, Czech Republic, at the SW edge of the Teplá-Barrandian Zone, where Fe sulfide-Cu-Zn deposits are held equivalent to the modern Cyprus-type ore deposits, and in the Jílové Belt, where Cu-Zn sulfide mineralization occurs in basic and acidic metavolcanites near the boundary between the Teplá-Barrandian and Moldanubian *sensu stricto*, closely resembling those mineralizations known from the Achaean greenstone belts (Morávek and Pouba 1990). A unique lithological series silicites bearing uranium and vanadium in amounts of up to 0.2 wt% and with Fe contents attaining as much as 35 wt% cover large areas in the southeastern and western parts of the Teplá-Barrandian Zone (Mrázek and Pouba 1995). In terms of geochemistry and geodynamic setting, this stratabound mineralization is similar to the Siluro-Devonian black shale mineralization in the Graptolite Shales elsewhere in Central Europe. The role of organic material in the formation of these metal-rich shales in the late Precambrian Bohemian Massif has been noted by Pašava et al. (1996).

Along the south-western part of the Moldanubian *sensu stricto*, the Lam-Bodenmais Kieslager Belt extends from the Northwest through the Southeast, parallel to the shear zone of the "Pfahl" ("Great Bavarian Quartz Lode") and runs through the middle of the area hosting the majority of pegmatites of the Bayerischer-Böhmer Wald, which were mined for feldspar and quartz for several decades (Dill 1985a). The afore-mentioned sediment-hosted Fe-Zn-Cu-Pb sulfides located in the Drosendorf Unit show a zonation into a proximal Fe-Zn-Cu association with Fe-enriched sphalerite, argentiferous galena, pyrite and pyrrhotite and a distal Pb-Ba mineralization. Elevated barium contents were found genetically associated with these massive sulfide deposits and their Pb and S isotopes are well in accordance with those sediment-hosted deposits classified as Sullivan-type/Meggen-type deposits *sensu* Jiang et al. (1998) and Taylor and Beaudoin (2000).

Sedimentation, diagenesis and thermal activity contributed very much to the concentration of metals in this stratabound ore deposits throughout the Late Proterozoic, but these processes tell us little about the evolution of physical-chemical regime in the immediate surroundings and as such they are of minor relevance for the emplacement of adjacent pegmatites. Argentiferous galena and the pyrite-pyrrhotite association agree well with the maximum temperature of 700 °C achieved at the climax of low-pressure regional metamorphism (Dill 1990). The variation of the activity a_{FeS} in pyrite coexisting with sphalerite was determined to be between 15.8 and 16.5 mol% and used to calculate log a_{s2} as 1.27 and log a_{O2} as 14.2 for the metamorphic country rocks pierced by the granites and pegmatites in the period of time from 340 through 280 Ma. After the heyday of metamorphism was hit the sulfur fugacity decreased whereas the salinity of mineralizing fluids increased. Silver-bearing tetrahedrite, zincian spinel and the study of fluid inclusion are appropriated tools to the temperature path of the retrograde metamorphism. Quartz mobilization became very widespread during temperature drop. Along with desulfurization, at about 410 °C zincian spinel appeared and around 390 °C quartz mobilization began. The physical-chemical regime observed in the retrograde pathway displays many similarities with that reported from the vein-type gold deposits of the Central Bohemian Metallogenetic zone mentioned earlier in this chapter.

The Czech part of the Moldanubian zone, that is synonymous with the core of the Moldanubian Bohemian Massif and dominated by the two large igneous complexes of the Central and South Bohemian Plutons, deserves a special treatment on account of its numerous pegmatites which have been mined for feldspar for decades but now almost all have been closed. Although of subeconomic grade, considerable amounts of rare elements were determined in these mineralogically and structurally very much different pegmatites resultant in a wealth of uncommon minerals which render these felsic rocks highly attractive for mineral collectors and mineralogists and economic geologists to conduct genetic investigations (Novák 2005). Given the large outcrops of the Central and South Bohemian Plutons in the Moldanubian Zone, the pegmatitic rocks' preference of metamorphic host rocks to granitic host rocks comes as an unexpected surprise: Třebíč Pluton at Oslavice near Velké Meziříčí resides in syenogranite "durbachite" (shoshonitic association) and gneisses (Škoda et al. 2006; Škoda and Novák 2007; Novák et al. 1999), the Horní Bory pegmatite near Velké Meziříčí in granulites, cordierite migmatites, biotite-sillimanite migmatitic gneisses (Povondra et al. 1992; Novák et al. 1992), the Přibyslavice pegmatite near Čáslav in orthogneiss and two-mica and biotite paragneisses (Němec 1973, 1978; Čech et al. 1978; Prachař et al. 1983; Povondra et al. 1987, 1998), The pegmatite at Vlastějovice near Zruč nad Sázavou has Fe-skarns as host rocks besides the common migmatized biotite-sillimanite gneisses amphibolite, pyroxene gneiss, quartzite, marbles, two- mica and tourmaline-bearing orthogneisses (Novák and Hyršl 1992; Žáček et al. 2003; Ackerman et al. 2007). Some of them are also emplaced within serpentinized lherzolite such as the Věžná I pegmatite and the pegmatite at Ruda nad Moravou (Novák and Gadas 2010; Dosbaba and Novák 2012). The pegmatite dikes Bližná I cut through calcite–dolomite marble, while its neighbor Bližná II intersects graphite-bearing biotite (Novák et al. 1999b, 2012).

Almost each district mineralized with pegmatites shows its peculiar host rock lithology. Metapegmatites have been recorded from Maršíkov I and III by Černý et al. (1992).

Lithium-bearing pegmatites make up a great deal of the pegmatites, mainly scattered in the southern part of the Moldanubian zone *sensu stricto* (Novák 2005). It is prevalently lepidolite-bearing pegmatites which prevail over elbaite-bearing ones in that region (Novák and Povondra 1995). Moreover other Li hosts such as petalite or Li phosphates are also present in the pegmatites of Moldanubian region, all of which are situated in metamorphic country rocks omitting the large batholiths of the Bohemian plutons and their offshoots (Novák 2005).

As a reference type of these lithium-bearing pegmatites, the Rožná pegmatite, where rossmanite [$LiAl_8Si_6O_{18}(BO_3)_3(OH)_4$] and lepidolite [$KLi_2AlSi_4O_{10}F(OH)$] have been found for the first time, is described in more detail because of its intensive mineralogical studies through time by numerous geologists (Němec 1998; Selway et al. 1998, 1999; Novák and Černý 2001; Cempírek and Novák 2006a). The pegmatitic dike extends over a length of about 1000 m and is about 35 m wide. It is oriented parallel to the NWN-trending strike of the foliation and runs along the boundary between the Strážek Moldanubicum and the Svratka Unit (Novák 1992) (Fig. 2.4a, b). The country rocks consist of biotite gneisses with minor hornblende gneisses, serpentinites, lenses of amphibolites and leucocratic gneisses. The various minerals at Rožna are listed below (Pezzotta and Guastoni 1998; Selway et al. 1998):

Albite, amblygonite, apatite, bertrandite, beryl, brazilianite, columbite-(Mn), cookeite, dravite, elbaite ("rubellite, indigolite, verdelite"), feldspar, foitite, mica, hydroxylherderite, cassiterite, lepidolite, montebrasite, muscovite, quartz, rossmanite, schorl, topaz, triplite, zircon

The mineralogical and chemical setting observed in this part of the Moldanubian Zone or in other words along the SE boundary of the Bohemian Massif looks like a mirror image of what we have already observed along its NW boundary near the

Fig. 2.4a Slender crystals of albitic feldspar in a matrix of lepidolite. Rožna pegmatite, Czech Republic

Fig. 2.4b Massive milky quartz with nests filled with cassiterite (*black*) and amblygonite (*pinkish*). Rožna pegmatite, Czech Republic

Fig. 2.4c Porous episyenite at the rim of the Rožna pegmatite. The mineral replacing quartz is a minerals of the smectite group (pers. com. M. Novák, Brno)

Saxo-Thuringian subfluence zone. A closer look at the cross section of Fig. 2.1c provides us with an explanation. Both boundaries are marked by deep-seated fault and trust zones dipping towards the core zone of this uplifted basement block (Fig. 2.1c, d).

In a later stage episyenitization has affected this pegmatite along fault zones causing a porous desilicified rock with smectitic phyllosilicates (pers. com. M. Novák) (Fig. 2.4c). The most conspicuous group of minerals identified in this pegmatite vein is mica, being composed of biotite, muscovite, lepidolite (trilithionite to polylithionite) and illite (Černý et al. 1995). They are replaced by phyllosilicates such as kaolinite and chlorite. The tourmaline-bearing mineral assemblages made up of schorl-foitite, elbaite and rossmanite were investigated by Novák and Selway (1997) and by Novák et al. (1998). Further diagnostic minerals are cassiterite and Nb-Ta-oxide minerals, e.g., columbite-(Fe) and columbite-(Mn). Lithium also entered together with aluminum the structure of phosphates resulting in the formation of the amblygonite–montebrasite s.s.s. There are a few Li-free phosphates such as lacroixite, brazilianite, goyazite, eosphorite, and fluorapatite in this pegmatite (Němec 1998). One of the few pegmatites hosted by granites is the euxenite-type pegmatite of Kožichovice II attributed to the NYF series (Novák and Filip 2010). It stands out predominantly by its beryllium contents causing the precipitation of a great variety of primary and secondary Be minerals disseminated among the rock-forming minerals of the ultrapotassic orogenic Třebíč syenogranite (beryl, bavenite, bazzite). A representative of the peraluminous P-rich tourmaline-bearing pegmatite is situated at Přibyslavice near Čáslav within a host rock series of orthogneisses and two-mica and biotite paragneisses (Čech et al. 1978; Němec 1973, 1978; Prachař et al. 1983; Povondra et al. 1987, 1998). The pegmatitic schlieren and dikes are abundant in phosphates and Al-enriched silicates. An overview of the central parts of the Moldanubian Zone as to the pegmatites and their mineralogy is given in Table 2.2.

2.1.6 Geodynamic Zones Along the Northeastern and the Southeastern Margin of the Bohemian Massif

Dealing with the pegmatitic rocks in central Europe from a geodynamic point of view while claiming that the HPPP is the center of pegmatites and to ignore the Sudetes along the northeastern margin and the Moravo-Silesicum along the southeastern margin of the Bohemian Massif would simply be only half the story (Fig. 2.1d). Therefore a brief overview of the lithologies, structural features and the mineral assemblages of both units is presented to thwart any possible accusations that these Variscan pegmatites and their geodynamic setting had been cast aside in this pegmatitic regional overview for whatever reasons.

2.1.6.1 The Sudetes

Lithology and Structural Geology

The Sudetes (West Sudetes = Lugicum, East Sudetes = Silesicum), was once considered by Kossmat (1927), the father behind the idea of geodynamic subdivision in the Central European Variscides, as the eastern prolongation of the Saxo-Thuringian Zone, and while doing so made an object for a lively debate among geoscientists lasting until today (Żelaźniewicz 1995). Its geotectonic position as part of one of the

Table 2.2 Host rocks, mineral assemblage and morphology of pegmatitic rocks from the central zone of the Bohemian Massif (Czech Republic)

Localitity	Host rock		Morphology	Reference
Oslavice near Velké Meziříčí Třebíč Pluton	Syenogranite shoshonitic association	Quartz, oligoclase, phlogopite ± amphibole, microcline, quartz, tourmaline, ilmenite, titanite, allanite-(Ce)	Dikes	Škoda et al. (2006) and Škoda and Novák (2007)
Oslavice near Velké Meziříčí Třebíč Pluton	Syenogranite shoshonitic association	K feldspar,quartz, oligoclase, phlogopite ± amphibole, albite, titanite, phenakite, beryl, euxenite, green amazonite, Mg-rich biotite,allanite-(Ce), aeschynite, ilmenite, niobian rutile, actinolite, zircon, monazite-(Ce), bavenite, bazzite, milarite, bertrandite, chlorite, pyrochlore pseudorutile	Dikes, pockets	Škoda et al. (2006) and Škoda and Novák (2007)
Oslavice near Velké Meziříčí Třebíč Pluton	Gneiss	Quartz, albite, K-feldspar (locally amazonite), cleavelandite, muscovite, zinnwaldite, lepidolite to masutomilite, schorl (elbaite), topaz, spessartine, F-rich hambergite, monazite-(Ce), xenotime-(Y), zircon, columbite, wolframoixiolite, cassiterite, fergusonite-(Y), samarskite and pyrochlore group, beryl, bertrandite, bavenite, fluorite	Dikes	Novák et al. (1999)
Horní Bory near Velké Meziříčí	Granulites, cordierite migmatites, biotite- sillimanite migmatitic gneisses	Cordierite, schorl, oxy-schorl, grandidierite–ominelite, boralsilite, dumortierite (locally Sb-enriched), ferberite, rutile – niobian, W-rutile, wolframoixiolite, ilmenite, monazite-(Ce)	Pockets, veinlets, dikes	Povondra et al. (1992) and Novák et al. (1992)
Starkoč near Čáslav, Kutná Hora Unit –	Granulites, garnet peridotites, eclogites, garnet-biotite gneiss to migmatite biotite gneiss	Tourmaline, dumortierite, garnet (Alm 80–62 Sps 30–10 Prp 8–3 Grs 3–2), chrysoberyl, plagioclase, K-feldspar, kyanite, staurolite, fluorapatite, monazite-(Ce), xenotime-(Y), löllingite	Discordant pegmatite veins, zoned pegmatite	Cempírek and Novák (2006a, b)

(continued)

Table 2.2 (continued)

Localitity	Host rock		Morphology	Reference
Přibyslavice near Čáslav	Orthogneiss in two-mica and biotite paragneisses	Albite, K-feldspar, muscovite, biotite (annite), trilithionite, oxy-schorl, schorl, elbaite, garnet, staurolite, dumortierite, sillimanite, nigerite, fluorapatite, triphylite, sarcopside,graftonite, ferrisicklerite, heterosite, ferroalluaudite, lipscombite, ludlamite, melonjosephite, messelite, mitridatite, phosphophyllite, rockbridgeite, strunzite, vivianite, niobian rutile, cassiterite, ferrocolumbite, manganocolumbite, tungstenite, tantalian rutile, zircon	Lenticular schlieren-like pegmatite, dike	Čech et al. (1978), Němec (1973, 1978), Prachař et al. (1983), and Povondra et al. (1987, 1998)
Vlastějovice near Zruč nad Sázavou	Migmatized biotite- sillimanite gneisses amphibolite, pyroxene gneiss, quartzite, marbles, two- mica tourmaline-bearing orthogneisses, Fe-skarns	Amphibole (hastingsite + edenite), fluorite, biotite, hedenbergite, andradite–grossular, epidote, titanite, calcite, magnetite, bastnaesite, fluorapatite, zircon, rutile, monazite-(Ce), xenotime-(Y), allanite-(Ce), arsenopyrite, pyrite, uraninite, cassiterite, niobian rutile, Sn-rich titanite, gadolinite–hingannite minasgeraisite, Y-rich milarite, pyrochlore-group minerals, manganocolumbite, datolite, bavenite, tourmaline (schorl, dravite, elbaite), garnet	Veins, layers	Novák and Hyršl, (1992), Žáček et al. (2003), and Ackerman et al. (2007)
Myšenec near Protivín, Písek region	Amphibole-biotite syenite (durbachite)	Tourmaline, inclusions of chromite in tourmaline, beryl, phenakite, danalite, biotite, ilmenite, muscovite, schorl–dravite, dravite	Veins, zoned	Novák et al. (1997)
Podlesi Stock (see also Lázni Kynžvart and HPPP for gradual changes along a N-S lineamentary fault zone)	Iotite granite, phyllites,	Alkali feldspar, zinnwaldite, protolithionites, quartz, topaz, apatite, childrenite–eosphorite, zwieselite, triphylite, monazite, Nb–Ta rutile, columbite, cassiterite, ilmenorutile, U-tantalite, U-microlite, ixiolite, wolframite, huebnerite, scheelite, rutile, ilmenite, haematite, pyrite, bismuthinite, powellite, roosweltite	Stock + dyke granite apical veins	Breiter et al. (1997)

Lázni Kynžvart		K feldspar, albite, muscovite, biotite, quartz, opaline, tridymite, topaz, schorl, chlorite, beryl, euclas, betrandite, zircon, helvine, coffinite, thorite, triplite, montebrasite, lacroixite, brazilianite, fluorapatite, greifensteinite, rockbridgeite, lipscomite, crandallite, perhamite, monazite, brabamtite, fluellite, hurlbutite, hydroxylherderite, herderite, chernikovite, torbernite, autunite, xenotime, vivianite, wolfrmaite, fluorite, hematite, ilmenite, ixiolite, columbite, magnetite, pseudorutile, rutile, uraninite, arsenopyrite, bismutite, emplectite, galena, greenockite-hawleyite, galenobismutite (?), chalcopyrite, cassiterite, madiltite, molybdenite, pyrite, sphalerite, stannite, tennantite, wittichenite, wurtzite		Veselovský et al. (2007)
Rožná, western Moravia	Granulitic to migmatitic biotite hornblende and biotite gneisses	Quartz, albite, muscovite, biotite, lepidolite, cookeite, tourmaline (dravite, elbaite, var. rubellite, var. indigolite, schorl, foitite), bertrandite, beryl, hydroxylherderite, amblygonite, montebrasite, brazilianite, triplite, apatite, rossmanite, topaz, columbite-(Mn), cassiterite, hornblende, zircon		
Bližná I	Calcite–dolomite marble	Microcline, quartz, albite, plagioclase, tourmaline (schorl, elbaite–liddicoatite, uvite), axinite, datolite, dumortierite, diopside, bastnäsite-(Ce), allanite-(Ce), parisite-(Ce), monazite-(Ce), titanite, epidote, calcite, zircon, apatite, pyrochlore, scheelite	Dike	Novák et al. (1999b)
Bližná II	Graphite-bearing biotite gneiss	Microcline, quartz, albite, plagioclase, muscovite, spessartine, tourmaline (dravite, schorl, elbaite, olenite, thortveitite, monazite-(Ce), zircon, apatite, columbite-(Mn), rutile, niobian rutile, cassiterite, pyrite	Dike	Novák et al. (2012)

(continued)

Table 2.2 (continued)

Localitity	Host rock		Morphology	Reference
Scheibengraben pegmatite 1.5 km E of Maršíkov	Medium-grained hornblende gneiss	Garnet, beryl, columbite-tantalite, fluorapatite, schorl, zircon, garnet, triplite, topaz, native bismuth, albite, K-feldspar, quartz, muscovite, euclase, bertrandite, uranmicrolite, microlite, rynersonite, gahnite	Lenticular body	Novák et al. (2003)
Maršíkov I and III		Quartz, muscovite, K-feldspar, sillimanite,beryl chrysoberyl, bavenite, chlorite, epidote, biotite, zircon, gahnite, garnet, fersmite, columbite-(Mn), tantalite, pyrochlore, microlite	Metapegmatite	Černý et al. (1992)
Ruda nad Moravou	Serpentinized lherzolite	Quartz, plagioclase, K-feldspar, grossularite, diopside, epidote, clinozoisite, dissakisite, allanite, titanite, zircon, baddeleyite, zirconolite, gittinsite, uraninite, thorite, fluorapatite, monazite-(Ce), fersmite, pyrochlore-group minerals, rutile, niobian rutile, biotite, Ti-rich magnetite, ilmenite, chromite, magnesiohornblende, actinolite, pargasite, tremolite	Dike	Novák and Gadas (2010)
Věžná I pegmatite	Serpentinized lherzolite	K-feldspar, oligoclase, quartz, biotite, smoky quartz, "cleavelandite", muscovite, fluorapatite, pollucite, "fluor-elbaite", albite, polylithionite, trilithionite, muscovite, niobian rutile, monazite-(Ce), xenotime-(Y), zircon, lepidolite, triplite, Cs-rich beryl, cheralite, hübnerite, native Bi, anthophyllite, Cr-enriched actinolite, phlogopite, vermiculite, chlorite, phlogopite, cordierite, celadonite, beryl, bertrandite, epididymite, milarite, Cs-rich analcime, chabasite-K, harmotome, "kerolite" (talc, serpentine, saponite)	Dike	Dosbaba and Novák (2012)

G 1 Central Granite, *G 2* Rim Granite, *G 3* Core Granite, *G 4* Tin Granite (Richter and Stettner 1979)

above mentioned geodynamic zones or a unit of its own undergoing Caledonian and Variscan deformations is still controversially debated (Bederke 1924; Collins et al. 2000; Mazur et al. 2006).

Here I only provide an abstract of that complex history and concentrate on those geological facts that are closely related to the emplacement of pegmatites only.

As a basic principle, the most modern models on these very divers lithologically units, favor an eastward extension of the Variscan tectonostratigraphic units into the Sudetes (Aleksandrowski and Mazur 2002; Kryza et al. 2004; Mazur et al. 2006). Interrupted by periods of extensional tectonic, the amalgamation of the Sudetes took place from the Silurian through the early Carboniferous. Flysch-type sediments are interpreted as a sign of active subduction of oceanic crust during the Late Devonian, suggesting that ophiolite obduction and significant overthrusting in the Sudetes occurred as an integral part of the Variscan orogeny (Collins et al. 2000).

The geodynamic unit is cut through by some prominent lineamentary fault zones such as the Elbe-Fracture zone, forming its southwestern edge and the Moravo-Silesian Thrust Zone, dipping toward the northwest. There are also internal ductile faults, e.g., the Main Intra-Sudetic Fault, which according to Don (1991) represents a deep-seated crustal feature separating Caledonian and Variscan parts of the Sudetes. The pre-Permian fault movements are compressional, extensional and of strike-slip type, bounding tectonic units with metamorphic rocks attaining, in places, high-pressure conditions of blue schist, eclogite and granulite facies conditions (Aleksandrowski et al. 1997; Aleksandrowski and Mazur 2002). Such fracture reflecting the internal subdivision of basement block have been active over a long period of time even if the change the deformational style and direction of movement through time. They play a role, not only for the geodynamic evolution of the Bohemian Massif but also fostered the heat flow from the mantle and the emplacement of minerals deposits, *inter alia* of pegmatitic deposits.

During the late Variscan orogeny, as in many other parts of the Central European Variscides, felsic intrusions penetrated the crust during the Carboniferous and bimodal volcanic volcanism was active by the beginning of the Permian (Kryza 1995a). Although the plutonic activity started very early, at approximately 350 Ma, there are two distinct phases alternating with periods during which granitic activity was low. One was around 325–330 Ma the other around 280 Ma (Kryza 1995a). SHRIMP zircon dating of granitic rocks conduced to a fine-tuning of the granitic activity with magmatic events around 340, 328, 312, 305–300 and 295–280 Ma (Oberc-Dziedzic et al. 2010; Kryza et al. 2012; Oberc-Dziedzic and Kryza 2012).

Also not uncommon to other geodynamic zones of the Central European Variscides, this granitic activity can be correlated with a late Variscan metamorphism (Kryza 1995b; Awdankiewicz et al. 2013). The Góry Sowie Block, in the West Sudetes, SW Poland, consists of a gneiss–migmatite complex with subordinate amounts of amphibolites, calc-silicate rocks, ultrabasic rocks and granulites. U/Pb dating of monazite and xenotime yielded ages of c. 380 Ma and constrain the timing of the last metamorphic–migmatitic event, a Devonian high-temperature metamorphism in the Sudetes. Rb–Sr mica–whole-rock dating for samples with D2, D3 and D5 deformation characteristics provide ages between 362 and 375 Ma (Bröcker et al. 1998).

Mineral Deposits

A wide range of mineral deposits, were mined in the past but today no longer considered as economic in the Sudetes on the Czech, German and Polish territories (Mochnacka et al. 1995). According to these authors the five principal categories of minerals deposits may be established for this north-easternmost part of the Bohemian Massif: (1) Metamorphosed deposits related to submarine volcanism and sedimentation (Fe- oxide- and Fe-sulfide deposits), (2) deposits related to pre-Variscan and Variscan magmatism (Cr-, Ti-Fe-, Cu-Ni-, U-, Th- deposits), (3) Vein-type and stratabound hydrothermal deposits (U-, Cu-, Pb-, Zn-, Ag-, As-, Fe-, Sn-, Au deposits), (4) Stratabound deposits of sedimentary affiliation (Cu-Ag-, U-, Au-Ti deposits), (5) Deposits and occurrences related to weathering (Ni-Mg-Fe-, Al deposits). For the current topic, only those mineral associations related to granites, pegmatites, skarns and contact-metasomatic processes are mentioned in this chapter and were referred to in the next paragraph.

In the Karkonosze Granite veinlets host pitchblende and uraninite, mostly related to a strong episyenitization (Lis and Sylwestrzak 1979). In the same region at Markocice near Bogatynia, in the metamorphic wall rocks of the afore-mentioned granite, Th-bearing pegmatites, called metasomatic syenites occur (Jęczmyk and Juskowiakowa 1989; Mochnacka and Banaś 2000). They contain a rather exotic mineral assemblage with monazite, thorite, cheralite, grayite, huttonite, ningyoite, voglite, thorogummite accompanied by various sulfides. Kucha (1980) reported ThO_2 contents of 56.4–69.9 wt% from the huttonitic monazite-(Ce), which lies between $ThSiO_4$ and $CePO_4$. In the Fore-Sudetic Block rare metal-bearing pegmatite veins are known from the Szklary serpentinite massif containing chrysoberyl, spessartite, columbite-(Mn) and manganotantalite, stibiocolumbite, holtite, pyrochlore, beusite, paradocrasite, stibarsen, and manganiferous apatite (Pieczka 2000). Another lens-shaped zoned pegmatite called Skalna Brama pegmatite is located near Szklarska Poręba within the Karkonosze Granite. It is a REE-pegmatite containing zirconolite, gadolinite, fergusonite–formanite, aeschynite, arsenopyrite, uraninite, monazite, zircon, and xenotime (Gajda 1960 a, b; Kozłowski and Sachanbiński 2007; Szełęg and Škoda 2008). In view of the age of intrusion of the Karkonosze granite, the pegmatite is younger than 329 ± 17 Ma (Duthou et al. 1991).

In the Strzegom-Sobótka Massif, miarolitic pegmatites hosting more than 90 different minerals were encountered in the two-mica monzogranite whose of age of formation was chronological constrained to 324 ± 7 Ma by Pin et al. (1989). The various minerals allow for an attribution of this pegmatitic mineralization to a REE-Nb-Ta-Be-B-Sc-F-W granitic pegmatites (Janeczek and Sachanbiński 1989; Ciesielczuk et al. 2008). The Michałkowa pegmatite complex made up of lenses and veins with its mineralized structures cutting through gneisses and amphibolites of the Góry Sowie Mts. Its lenses run subparallel to the foliation. Van Breemen et al. (1988) reported an age of 370 ± 4 Ma of this pegmatite. Apart from the typical minerals of pegmatites, the phosphate sarcopside which is pseudomorphosed by vivianite was described for the first time from this locality by Websky (1868). According to its diagnostic minerals, the pegmatite has been classified as tabular to vein-type B-P pegmatite

Proximal to the Karkonosze Granite at Kowary magnetite and hematite-bearing skarns occur. Beyond the border, in the Czech part of the Sudetes, at Obří Důl scheelite mineralization is related to the granitic influence on marble layers (Chrt 1959). This idea did not remain unchallenged as Pertold (1978) advocated a syngenetic origin of these tungsten deposits.

2.1.6.2 The Moravo-Silesicum

Lithology and Structural Geology

The Moravo-Silesian zone extends from Austria, through the Czech Republic into Poland, bounding the Bohemian Massif, or in other words the Moldanubian Zone towards the SE (Misař and Urban 1995). It consists of the autochthonous Cadomian basement, called the Brunovistulicum overlain by Devonian to Carboniferous sedimentary rocks and the allochthonous Variscan units of the Moravicum and the Silesicum. Several tectonic thrust planes mark the overriding Moldanubian and Lusatian nappes onto the Moravo-Silesian zone. Towards the East, the Moravo-Silesian zone submerges underneath the Carpathian foredeep as far east as the Peripeninic Lineament which is the concealed boundary of the Bohemian Massif (Máška and Zoubek 1960). Like its western parts in the Alpine Mountain range, these Variscan rocks of the Carpathians also became reactivated during the Alpine orogeny and reappeared as intra-Alpine massifs (Grecula and Roth 1978) (section 2.2.4). Paleofacial comparisons with other geodynamic units, which have been discussed previously in this book, stress the striking similarities between the coal-bearing Upper Silesian Basin and the Subvariscan Foredeep (Sect. 2.1.1) while the late Variscan flysch facies of the Moravo-Silesian zone finds its match in the Rhenohercynian Zone (Sect. 2.1.2). As a logical consequence, the basement in the eastern parts of the Silesian zone was correlated with the Mid-German Crystalline Rise (Misař et al. 1983) (Sect. 2.1.3).

Mineral Deposits

After emphasizing the many similarities to exist between the Rhenohercynian and the Moravo-Silesian zones in terms of paleofacies and geodynamic evolution both of which are characterized by extensional processes during the Devonian and subsequent compressional tectonic processes, it is more than a tempting idea to see also striking similarities between the mineral deposits in both geodynamic realms, located so far away from each other in the Central European Variscides (Aichler et al. 1995). In the Moravo-Silesian Zone, submarine Lahn-Dill iron ore are encountered closely related to basic volcanic rocks. Fe-bearing base metal deposits with subordinate amounts of Au formed in a geodynamic setting similar to that known from Rhenohercynian Basin. The Devonian volcano-sedimentary series (Vrbno Group), however, underwent regional metamorphism up to green schist facies conditions (Patočka and Vrba 1989; Kalenda and Vaněček 1989). Extensive mining was focused on deposits in the N part of the Moravo-Silesian Zone at Zlaté Hory, Horní Město, Oskava, and Horní Benešov in the Czech Republic. The total ore exploited from these deposits amounts to 100 Mt of mostly low grade ore (Aichler et al. 1995). The sulfur isotope ratios obtained from barites closely resemble those from Meggen and Rammelsberg, whereas the sulfide sulfur is isotopically much lighter (Hladíková et al. 1992). High radiogenic Pb contents of the galena-enriched ore shows that upper crustal rocks have contributed much to the metal-bearing hydrothermal or exhalative solutions creating the stratabound mineralization in the Moravo-Silesian Zone (Vaněček et al. 1985).

There are also conspicuous similarities between the pegmatites in the Spessart, being part of the Mid-German Crystalline Rise and its most likely counterpart at the easternmost part of the Bohemian Massif, where barren feldspar or primitive pegmatites exist, according to Novák (2005). A few REE-bearing pegmatites, hosting allanite as it is the case in the Spessart, occur in the Moravo Silesian zone. At Maršíkov I and III Be-REE metapegmatites with sillimanite crop out (Černý et al. 1992) (Table 2.2). Boron and beryllium is more widespread than lithium. It has to be noted that there are also differences such as the Scheibengraben pegmatite 1.5 km E of Maršíkov, tabular Be-Nb pegmatite in medium-grained hornblende gneisses (Novák et al. 2003) (Table 2.2). While pre-Variscan granites are common, granitic pegmatites are rare, a scenario also known from the northwestern edge of the Central European Variscides. According to Novák (2005), the following relations can be quoted – B>>P+F, Be>>Li-for the Moravo Silesian zone, which is not only with respect to the paleofacies a match to the Mid-German Crystalline Rise.

2.2 The Geological and Metallogenetic Evolution of the Variscides Within the Alpine Mountain Range with Special Reference to Pegmatites

2.2.1 The Variscan Massifs in the Alpine-Carpathian Mountain Range

Less than 50 km south of the highland boundary fault which runs along the River Donau (Danube) and terminates the uplifted block of the Bohemian Massif, the geomorphological expression of the Moldanubian Zone, the spectacular mountains of the Alpine Orogen rise from hilly landscape of the molasse basin (Fig. 2.1d) (Froitzheim et al. 2008; Rasser et al. 2008; Reichert et al. 2008). The present investigations reveal that there were two major tectonic activities, one during the Cretaceous and another during the Paleogene and Neogene provoking that the late Paleozoic and Mesozoic rocks that were under the sea within the Neo-Tethys elevated over the sea level and were transformed into the present-day high-altitude mountain ridge, stretching from Grenoble, France, to Vienna, Austria. Immediately East of Vienna, Austria, this modern fold belt of the Alpine Mountain Range changed its direction of strike from ENE towards NNE, and running along the southeastern border of the afore-mentioned Bohemian Massif, forms another branch of the Alpine-Himalayan Fold Belt, called the Western Carpathian Mountains. The late tectonic phases of the Alpine Orogeny resulted from the northward plate movement of Africa, getting it closer to the Eurasian continent, whose geodynamic history was related in the previous sections with special reference to the emplacement of pegmatitic rocks in the various geodynamic units along a NW-SE transect from the Subvariscan Foredeep through the Moldanubian Zone (Sect. 2.1). This transect does not end in the southernmost tip near Vienna, where the Monotonous Series, the Varied Series and the Gföhl Unit represent the crystalline basement of the Moldanubian Zone in Austria nearest to the Alpine Mountain Range. Pegmatitic rocks have also been encountered further south in a geodynamic setting closely related to the one known from the Moldanubian Zone of the Central European Variscides (Petrakakis 1997).

In the western Alps crystalline basement rocks are exposed in the Helvetic realm, among others in the Gotthard and Aare Massifs, while immediately south of it Paleozoic rocks are at outcrop in the Penninic realm (Trümpy 1980). Heading east, in the Tauern Window and the "Altkristallin" (Old Crystalline Rocks) Paleozoic rock are of more widespread occurrence (Tollmann 1977).

Another series of rocks next to the Moldanubian Zone is located within the Western Carpathian Mountains which are traditionally divided into the Outer, Central and Inner Western Carpathians. The Central Western Carpathians comprise three superunits, the Tatricum, Veporicum and Gemericum (Hovorka et al. 1992; Vozárova and Vozár 1988, 1996; Ludhová and Janák 1999). Of those superunits, the two last ones are of particular interest as to the presence of Paleozoic rocks and pertinent mineral deposits.

2.2.2 The Variscan Massifs and Their Associated Pegmatites in the Swiss Alpine Mountain Range and the External Moldanubian Zone

Among the most well-studied and -dated Variscan massifs in the Swiss Alps, the Aar Massif reveals several similarities with its extra-Alpine Variscan counterparts to the north (Schaltegger and Corfu 1995). The Tödi Granite evolved around 333 Ma contemporaneously with the older granites from the Bayerischer-Böhmer Wald (318–338 Ma) and the granite at Triberg (333 ± 20 Ma) and Münsterhalden

(333 ± 5 Ma), in the Schwarzwald, all of which belong to the Moldanubian Zone (Walther 1992). The Central Aar Granite (289 Ma) and its microgranite (299 Ma) are considerably younger than the older granite. Aplitic veins have been reported from the Aar Massif, but mineral assemblages typical of pegmatites are missing (Amacher und Schüpbach 2011). Schneiderhöhn (1961) has already reported in his comprehensive study on pegmatites, that pegmatites and aplites are poorly represented in the Paleozoic rocks of the Swiss Alps. He cited the Aar Massif as an example, where in the amphibolites of the schistose envelope pegmatitic and aplitic rocks are scarcely exposed whereas in the Mont Blanc Massif, these felsic rocks are absent. Why, in spite of their close chronological and geodynamic resemblance to the most prominent pegmatite terrane in the Central European Variscides, are these Variscan massifs in the Western Alps devoid of these rocks ? To answer this question needs a closer look at the two uplifted basement blocks of the Moldanubian Zone in southwestern Germany and France next to the internal massifs of the Alpine realm. One of them forms the eastern (Schwarzwald Mts.) and the other the western flanks of the Rhein Graben Rift (Vosges Mts.) (Fig. 2.1d).

The crystalline basement of the Schwarzwald has been subdivided into four tectonometamorphic complexes. At the northernmost tip of the basement uplift the Baden-Baden Zone is assigned to the Saxo-Thuringian zone by Franke (1989) and was intruded by a suite of Early Carboniferous high-K, calc-alkaline I-type plutonic rocks composed of diorites, granodiorites and granites which are also exposed in the Vosges Mts. (Altherr et al. 2000). In the Central Schwarzwald Gneiss Complex metapsammitic gneisses were brought about by a HT-LP regional metamorphic event of approximately 730–780 °C/0.40–0.45 GPa (Kalt 1995). All intrusions into this gneiss complex are homogeneous two-mica or only muscovite-bearing S-type granites, the youngest of which were dated at 325 ± 7 Ma and developed from granitic magmas of crustal origin (Kalt et al. 2000). Further south, in the Badenweiler-Lenzkirch Zone, arc relics were identified in the metamorphic rocks. At the southern extremities of the Schwarzwald, the Southern Schwarzwald Gneiss Complex is located. Metaaluminous to slightly peraluminous biotite granites and peraluminous two-mica granites were intruded into this metamorphic complex, spanning an age of intrusion from 334 ± 2 Ma, determined on monazite, through 328 ± 2 Ma, obtained from a dike of granite porphyry (Schaltegger 2000). The peak of HT-LP metamorphism was reached almost contemporaneously with or immediately before the intrusion of the granitic magmas into the metamorphic rocks. According to Altherr et al. (2000), the granitic suite of the Schwarzwald and the Vosges Mountains resulted from a series of intrusion of crustal-derived felsic and mantle-derived more basic dioritic magmas. The Zone of Baden-Baden and its prolongation to the West, the Lalaye-Lubine Zone dissecting the Vosges Mts. in WNW-ENE direction, represents a deep-seated suture zone between two different terranes, the Saxo-Thuringian and Moldanubian zones.

The Schwarzwald is one of the oldest mining regions in Germany, mainly based upon vein-type Pb-Zn deposits, e.g., in the Schauinsland and in the Münstertal-Wiesental, Ag-Bi-Co-Cu-U vein deposits around Wittichen, Sb-As veins in the environs of St. Ulrich and Sulzburg and fault bound uranium deposits at Menzenschwand, the latter yielded a U/Pb age for its pitchblende mineralization of 310±3.5 Ma (Gehlen von, 1989). In stark contrast to the regional presence of base metal, precious metal and nuclear fuel deposits is the Schwarzwald's notoriously poor presence of elements known to be genetically related to granitic intrusions, such as Sn, Be, Nb or Li which is totally absent, even though almost half of the crystalline rocks exposed in the uplifted block of the Schwarzwald is made up of Variscan granites. The only granite-hosted mineralization bearing Sn and Be and hence has been mineralogically revisited again and again for its peculiar position, is located in the Triberg Granite Complex (Osann 1927; Fettel 1971; Oppelt 1988; Markl 1995; Markl and Schumacher 1996; Achstetter 2007).

The afore-mentioned mineralization within the composite intrusion of the Triberg Granite is a greisen-type Sn occurrence rather than a pegmatitic mineralization present in narrow veins, in miarolitic cavities and disseminated in parts of the granite (Schleicher 1994; Markl and Schumacher 1996). A small beryl-bearing pegmatite is hosted by the same granite, a two-mica leucogranite unit, and it was altered by the postdating greisen-forming fluids resulting in secondary beryl, albite, phenakite, bertrandite, and kaolinite. Pressure was estimated by the authors to be 1500 bars and the measured salinity is said to be 4–5 wt% NaCl equiv., while the temperature decreased from 550 to about 250 °C along a transect from the internal to the marginal parts.

A list of minerals in miarolitic cavities of the Triberg Granite at Hornberg is given below (Fettel 1971; Schorr 1984).

Albite, bertrandite, beryl, cassiterite, fluorite, hematite, kaolinite, metazeunerite, muscovite orthoclase, phenakite, quartz, zeunerite, zinnwaldite.

Corresponding to Schleicher (1994), the host granite derived from partial melting of a mid to lower crustal source. Analogous to this mineralization in the eastern Schwarzwald referred to above, in the Vosges Mts., on the western banks of the River Rhein, a similar mineral association also containing beryllium and tin was found in veinlets near Rothau-Alsace, France, (Hohl 1994).

List of minerals in a granitic pegmatite at Rothau-Alsace, France (Hohl 1994)

Bertrandite, beryl, fluorite, hematite, orthoclase (var. adularia), phenakite.

With respect to the chemical composition and textural type these Be-B-Sn pegmatites are more akin to the pegmatites straddling the boundary between the Saxo-Thuringian- Moldanubian boundary at the western edge of the Bohemian Massif see the pie chart diagram of Fig. 2.5a (for statistical reasons no equivalent pie chart diagram has been calculated for the Schwarzwald and the Vosges Mts., where only one occurrence is known in each basement uplift). Both pegmatites from the Vosges and Schwarzwald Mountains are lithologically alien elements, demarcating different suture zones between different geodynamic settings. Niobium, commonly accommodated in columbite s.s.s in Central European pegmatitic rocks, went quite a different way, as it was accommodated in the lattice of perovskite of the calcite carbonatite of the Kaiserstuhl in the Upper Rhine Graben (Chakhmouradian and Mitchell 1997). The extra- and intra-Alpine parts of the Moldanubian Zone to the West of the Bohemian Massif are strongly depleted both in pegmatites and aplites and, if present, the rare-element members of this group of felsic rocks

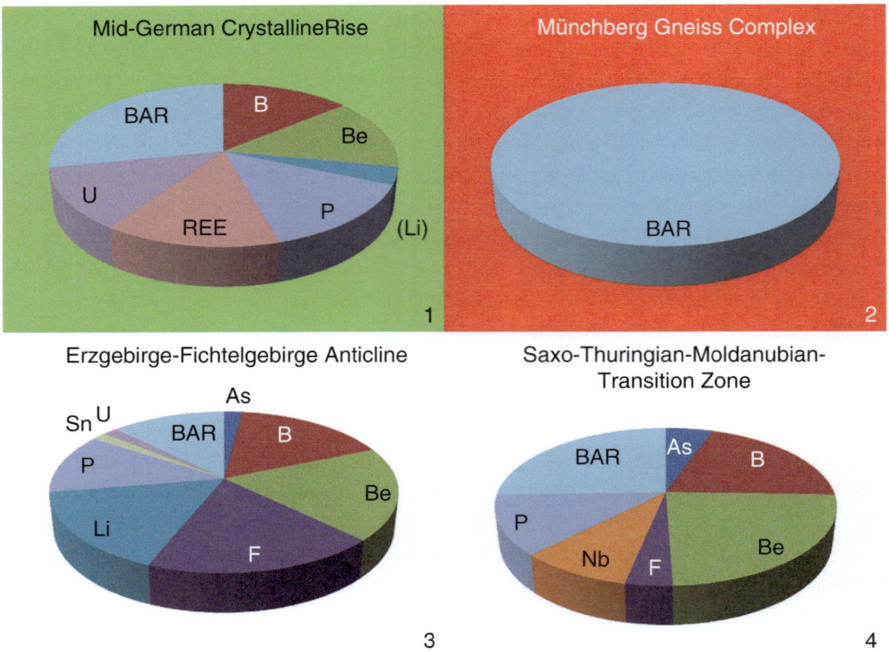

Fig. 2.5a Pie chart diagrams showing the chemical composition of pegmatitic rocks of the Central European Variscides and the intra-Alpine/Carpathian massifs. The numbers and colors in Fig. 2.5a, b, c, d refer to the areas on display in Fig. 2.5e the geological basis of which is given in Fig. 2.1d

designate a concealed suture zone between two ophiolites in Central Europe, the Saxo-Thuringian towards the North and the Ligurian-Massif-Central-Moldanubian ophiolites to the South (Matte et al. 1990; Franke et al. 1995; McKerrow et al. 2000).

2.2.3 *The Variscan Massifs and Their Associated Pegmatites in the Austrian Alpine Mountain Range*

In contrast to the westernmost Swiss part of the Alpine Mountain range, which is poor in pegmatites in the Austrian, more central part of the Alps, where the Moldanubian extra-Alpine Bohemian Massif comes as closely as possible to the Alpine fold belt, the number of pegmatitic rocks significantly increased, and forces to some explanation from the lithological and geodynamic point of view.

The most prominent intra-Alpine massif, made up of pre-Variscan crystalline rocks, subsequently being intruded by Carboniferous granitic melts, is exposed in the Tauern Window. Arc, fore arc and ensialic back arc environment are juxtaposed (Eichorn et al. 1999, 2000). As a consequence of the Late Devonian amalgamation of Gondwana, represented by the Tauern Window, and Laurussia-Avalonia an Early

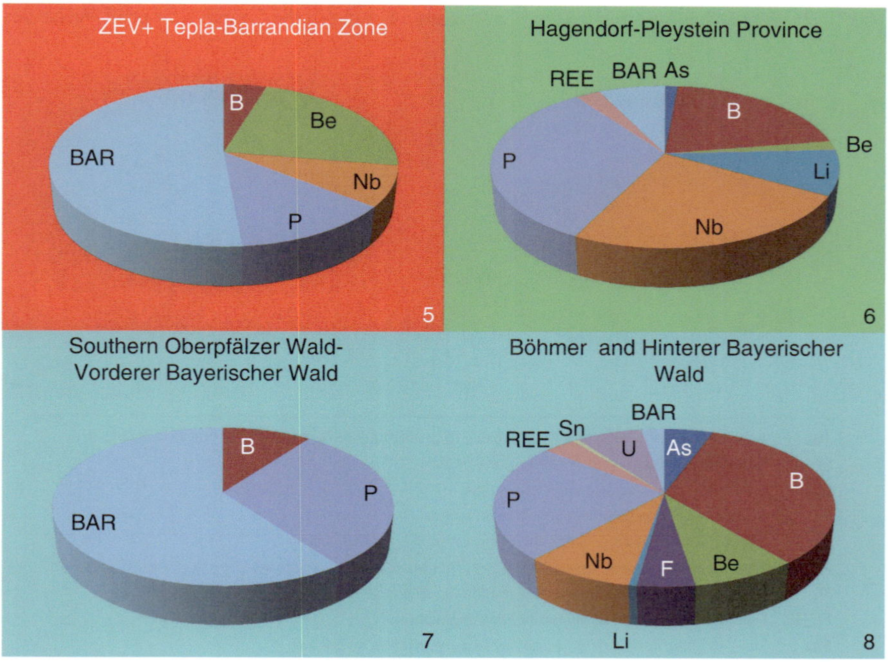

Fig. 2.5b Pie chart diagrams showing the chemical composition of pegmatitic rocks (see also Fig. 2.5a)

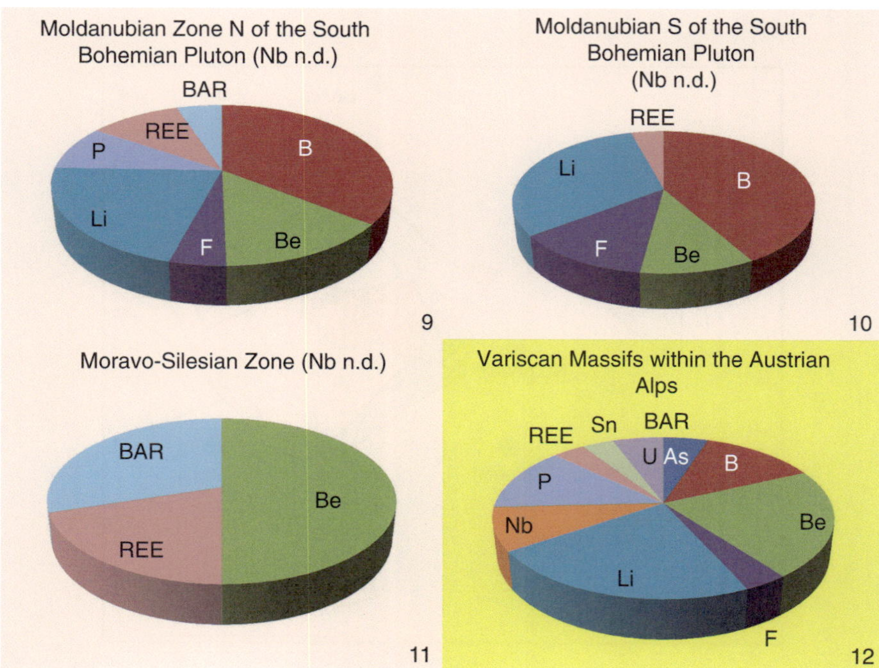

Fig. 2.5c Pie chart diagrams showing the chemical composition of pegmatitic rocks (see also Fig. 2.5a)

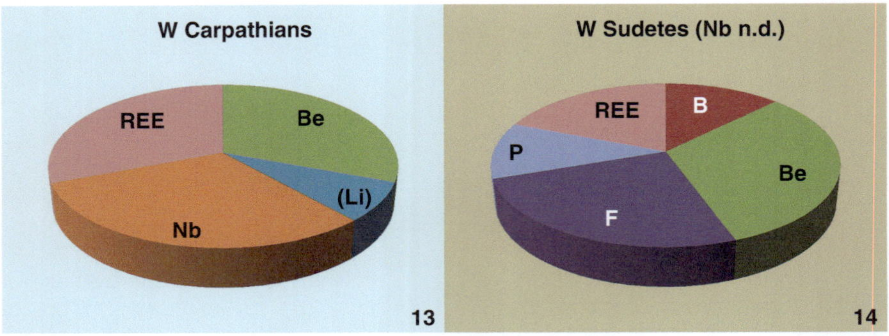

As	Arsenic-, bismuth-, zinc- bearing pegmatite (e.g. arsenopyrite, bismutinite, Fe-enriched sphalerite)
B	Boron pegmatites (e.g. tourmaline, dumortierite)
Be	Beryllium pegmatites (e.g. beryl, phenakite)
F	Fluorine-bearing pegmatites (e.g. fluorite, topaz)
Li	Lithium pegmatites (e.g. Li-bearing phosphates, -tourmaline, mica)
Nb	Niobium and tantalum-bearing pegmatites, plus scandium (e.g. columbite-(Fe)), n.d. = not determined
P	Phosphate pegmatites (e.g. apatite, Fe-Mn-Mg-Al phosphates, excluding Li, Be phosphates)
REE	Rare earth elements and yttrium-bearing pegmatites
Sn	Tin – and tungsten-bearing pegmatites (e.g., cassiterite, wolframite)
U	Primary and secondary uranium and thorium minerals in pegmatites
BAR	Barren as to rare elements, only feldspar, quartz and mica-bearing pegmatites

Fig. 2.5d Pie chart diagrams showing the chemical composition of pegmatitic rocks (see also Fig. 2.5a)

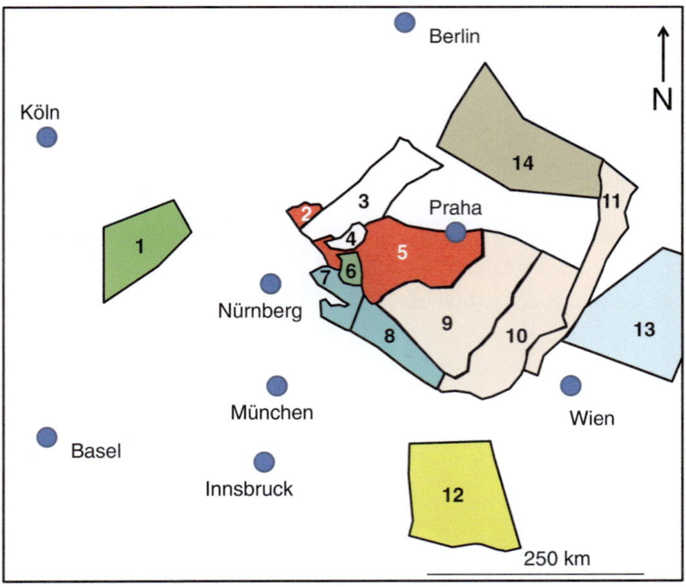

Fig. 2.5e Overview of the topographic position of the pegmatitic provinces (see also Figs. 2.1d and 2.5a)

magmatism was initiated in this crustal section, now present as the Central Gneiss Complex (Lammerer and Weger 1998). I-type granitic rocks lithologically reflect subduction process (Finger et al. 1997). The afore-mentioned magmatic activity in the Austrian Alps closely resembles that from the Swiss and French Alps in the Aar Massif, Aiguilles Rouges Massif and Mont Blanc Massif (Fig. 2.1d).

Featuring commodity groups, a variegated group of ore deposits and of pegmatitic deposits, the intra-Alpine Paleozoic massifs performs metallogenetically almost as well as their northern extra-Alpine analogue, the Bohemian Massif. A comprehensive overview of the mineral deposits has been provided in Dill et al. (2008a, b). Rating the metallogenetic evolution along the Alpine Mountain Range in terms of the abundance of mineral deposits, simply by a visual inspection of the metallogenetic map 1: 2,500,000 unravels, where the frequency of mineral occurrences and deposits reaches its maximum (Dill et al. 2008b). It is the area between the Tauern Window and the Vienna Basin.

In the Eastern Alps, S of Innsbruck, the Monteneve/Schneeberg deposit, Italy, forms part of a horizon mineralized with Zn-Pb minerals that extend over about 20 km, within a paragneiss formation of pre-Silurian age (Frizzo et al. 1982). The ore bodies contain Cd- and Mn-rich sphalerite and Ag-rich galena, with minor pyrrhotite, chalcopyrite, pyrite and stibnite. Another mineralization in the Eastern Alps of Paleozoic age occurs in the Speik Terrane at Kraubath, Austria, with Cr and PGE minerals in a highly dismembered back-arc ophiolite (Malitch et al. 2003). Amphibolites and gneisses in the Ötztal and Kreuzeck Mountains contain stratiform polymetallic Fe-Cu-Zn-Pb deposits in back-arc settings at Raggabach, Austria (Ebner et al. 2000). There are several more mostly rather small deposits in the Paleozoic rocks of the Eastern Alps.

Without any doubt the most outstanding island-arc related metallotect of the Habach Terrane (Frisch and Neubauer 1989) is exposed in the Tauern Window in the Eastern Alps, where 1967 the two ore bodies of the Mittersill scheelite deposit were discovered (Höll 1975). These deposits bridge the gap between the wealth of base metal deposits and the pegmatitic deposits in this region of the Alps. According to Eichhorn et al. (1999) the evolution of the Mittersill scheelite deposit commenced with the development of a volcanic arc at ca. 550 Ma as indicated by the emplacement of volcanic-arc basalts. Approximately coeval crustal thinning occurred in a back-arc region, accompanied by the emplacement of tholeiitic basalts and the intrusion of minor diorites. Subsequently, gabbroic and ultramafic melts intruded into the arc and back-arc region followed by normal I-type granitoid melts with mantle signature until 530 Ma. Subsequently, highly differentiated, yet still mantle-dominated granitic melts were locally intruded between 530 Ma (EOZ) and 520 Ma (K2) in the Mittersill ore deposit. Variscan-age magmatism around 340 Ma is likely to have brought about a second phase of scheelite mineralization which was superimposed on the primary Early Paleozoic phase in the Mittersill deposits.

In the Middle Eastern Alpine unit *sensu* Tollmann (1986) more than ten spodumene occurrences have been discovered since 1876 (Seeland 1876), the larger ones are listed in Table 1.2. Pioneer studies during this case history on spodumene have been performed by Meixner (1952, 1966) and by Höller (1959, 1964) who mineralogically confirmed the early discoveries and increased the number of new finds in Steiermark and Kärnten. Many of these felsic rocks attracted the attention of mining engineers, mainly for the considerable amount of Na- and K-enriched feldspar concentrated in lenses with up to 100,000 t, sufficient to supply the domestic ceramic industry but too low in their rare element contents to spark any exploration for Nb, Li or Be (Ucik 2005). Only one area stands out among these pegmatites in Austria and, not surprisingly, was scouted by many mineralogists. It is the spodumene pegmatite of the Koralpe that eventually turned into trial mining operation and after suspending this preparatory work in the run up of exploitation is currently subject of a new exploration campaign (Postl and Golob 1979; Göd 1978, 1989; Niedermayr 1990; Taucher et al. 1992, 1994). The area can be designated the type locality of the rare Ca-Be Phosphate Weinebeneite [$CaBe_3(PO_4)_2(OH)_2$ 4 H_2O] (Walter et al. 1990). Representative of the great number of Li-bearing pegmatites in the Middle Eastern Alpine unit, the mineralogically most renown one and the only subjected to underground operations, the spodumene deposit from the Koralpe, is given a more detailed treatment by summarizing the wealth of data collected by the above authors.

The lithology of the crystalline basement of the Koralpe is rather varied, being composed of kyanite- and garnet-bearing mica schists, paragneisses, amphibolites, eclogites, metagabbro and marbles. Kleinschmidt et al. (1975) unraveled the complex lithology and claimed Ordovician and Silurian rocks to contribute to the built-up of the "Phyllitgruppe" (phyllite group) at the southern border of the Saualpe. Habler and Thöni (2001) dealt with the metapelites and metapegmatites intercalated into the crystalline basement of the Austroalpine nappe complex in the Eastern Alps. Their geothermobarometric investigations on gneisses of this basement section yielded temperatures around 600 °C at a pressure of 0.4 GPA. According to these authors the pegmatite formation is correlated with the low-pressure metamorphism in the metapelites, which based upon Sm–Nd-dating of magmatic garnet from the pegmatite gneiss is placed at 249 ± 3 Ma. This high temperature metamorphism provoked the mobilization of felsic mobilizates with quartz, feldspar and mica, termed as pegmatites by regional geologists. The geothermobarometric studies above well agree with the phase diagram, showing the various stability fields of Li-bearing silicates in a P-T plot (London 2008). Under the existing physical-chemical conditions spodumene is the stable phase. During a more recent study, including garnet, xenotime, apatite, monazite and feldspar and using Sm-Nd mineral isochrones, Thöni et al. (2008) were able to fine-tune their previous petrological and chronological investigations and furnished clear evidence for multiple emplacement of pegmatitic melts between 273 ± 2 and 258 ± 3 Ma, in some sites even younger with age down to 251 ± 7 to c. 230 Ma. Ensuing overprinting processes under eclogite-facies conditions with peak temperatures around 700 °C and a pressure at 2.2 GPa accompanied by intense deformation during Cretaceous time, were unable to obliterate previous isotopic signals and blur the magmatic nature of the afore-mentioned rocks.

Spodumene has been concentrated in layers conformably intercalated among the varied metamorphic lithologies. According to Göd (1989) two different types of spodumene ore can be distinguished from each other, the amphibolite-hosted AH pegmatite and the mica schist-hosted MH pegmatite. In the AH – type spodumene crystals are aligned subparallel to each other in the central pegmatite. The marginal part of the felsic rock is aplitic in texture and the contact zone to the barren amphibolite saw the growth of biotite, holmquisite, in places, associated with garnet, beryl, tourmaline and apatite.

The pegmatitic rocks of the MH-type are intensively deformed and foliated. As the orientation of their minerals is taken to the extreme it is hardly to be distinguished from the surrounding kyanite-bearing mica schists. Any aplitic margin so often seen in these felsic intrusive rocks is completely absent from the MH pegmatite. Similar to many other pegmatites in central Europe there is no parent granite close by. Different from many central European, this Alpine Li pegmatite was overprinted by an early Alpine regional metamorphism under amphibolite facies conditions and consequently the result has to be referred to as unzoned pseudopegmatite (MH type) while the AH type has to be called a zoned pseudopegmatite (Frank et al. 1987). Even if these Alpine pegmatitic rocks differ by is mineral association from the pegmatoids and metapegmatitic rocks of the Zone of Erbendorf-Vohenstrauß along the western edge of the Bohemian Massif, there are also striking textural similarities which may be accounted for by the horizontal thrustal movement. Disregarding the presence of spodumene, there are many similarities between the primary pegmatitic minerals of the Koralpe pseudopegmatites and those pegmatites scattered along the western edge of the Bohemian Massif. Some of these minerals are listed here: Quartz, albite, microcline, muscovite, apatite, beryl, ferrisicklerite, columbite-(Fe), heterosite, triphylite, lithiophyllite, tourmaline, cassiterite, Nb rutile. Holmquisite, on the other hand, seems to be one of the diagnostic minerals typical of the Alpine regional metamorphism and as such its absence from the pegmatites in the NE Bavarian Basement is not a surprise.

The various spodumene-bearing pegmatitic rocks mentioned here in this section and exemplified by the pegmatite-hosted Koralpe lithium deposit are not the only evidence for the emplacement of pegmatites in the central part of the Alpine mountain range (Thöni and Miller 2004). Older meta-pegmatites were recorded by both authors from three localities in the Ötztal Basement, in Tyrol (Eastern Alps). Garnet-whole rock or garnet-feldspar Sm–Nd isochrone ages span the interval 445 ± 3 and 473 ± 3 Ma, indicating a Middle to Late Ordovician heat event which lead to the emplacement of pegmatites, now called metapegmatites owing to its subsequent overprinting (Thöni and Miller 2004)- see also Fig. 3.1a. The Alpine Mountain Range has metapegmatites similar to those from the western edge of the Bohemian Massif – see allochthonous ZEV and Tepla Barrandian Zones – and pseudopegmatites, which were subjected to a high-T remobilization during the early Alpine orogeny in its bounds. The latter are interpreted as Late Variscan rare-metal pegmatites s.str. of the Bohemian Massif which underwent Mesozoic remobilization.

2.2.4 The Variscan Massifs and Their Associated Pegmatites in the Slovak Carpathian Mountain Range

Evidence for intensive Silurian rifting and volcanism is not only found in the Eastern Alps but was also recorded from the Western Carpathians, in Slovakia (Grecula 1982; Grecula et al. 1995). The oldest known stratabound mineral deposits in the Carpathian Mountains are subeconomic black shale-hosted sulfide mineralizations with disseminated pyrite and a varied spectrum of rare elements such as V, Sb, Pb, Zn, Cu, As, Ag and Ni, particularly widespread in the Lesser Carpathians (Tatricum) and in the Gemericum (Chovan et al. 1992). During the Devonian, in the Gemericum base metal deposits developed closely associated with hematite and magnetite deposits at Jalovičí vrch and Hutná dolina both of which are genetically linked to basalts and keratophyres (Grecula 1982; Grecula et al. 1995). Rb/Sr isotopic ages of granites in the Gemericum indicate a Permian age of intrusion for these felsic rocks (290 ± 40 to 220 ± 32 Ma; Kovách et al. 1986). They are held by some as the heat source of siderite-sulfide veins in the Carpathian Mts., while others discard this idea and put forward a metamorphic-hydrothermal model for these siderite-sulfide mineralizations in the Gemericum (Radvanec et al. 2004). A large Sb-Au-As province extends across the Western Carpathians.

As far as the pegmatites are concerned, not unexpectedly, an intra-Alpine counterpart to the Bohemian Massif can also be found within the northern Carpathians. This mountain ridge is not only the NE prolongation of the eastern Alpine Mountain Range but it is closer to the Bohemian Massif than the Easter Alps, although the direct boundary between the two is concealed by a thick pile of sediments laid down in the younger foreland basin of the Carpathian Mts.. With regard to the mineralogical composition there are some similarities, but as to the economic potential, the Carpathian pegmatitic province trails behind equivalent rocks in neighboring Austria where lithium pegmatites are widespread and currently under exploration in southeastern Austria (Sect. 2.2.3). By comparison, the Slovak part of this modern fold belt in Central Europe surpasses by some orders of magnitude the Paleozoic massifs of the western Alps which are located far off the central European uplifted basement blocks of the Variscides as to the potential of pegmatitic occurrences and the number of the mineralized sites (Uher and Broska 1995; Uher and Černý 1998; Uher et al. 1998, 2012).

It is mainly Be-Nb-Ta pegmatites in granites, one group bearing Ti- and Mg-poor minerals and another group of pegmatites carrying Ti- and Mg-enriched phases (Nb-Ta oxide minerals, garnet, beryl) that is characteristic for this mountain range. Uher et al. (1998a) attributed the first group of pegmatites to monazite-bearing orogenic granites and the second to allanite-bearing orogenic granites.

Sphene, fersmite, pyrochlore-group minerals and romeite occur in small dikes of relatively poorly fractionated pegmatite in the Variscan Prasiva biotite granodiorite-granite, in central Slovakia: They underwent strong hydrothermal alteration, involving concentration and depletion. Percolating fluids mobilized Nb, Ta, Ti, U, Fe, Si, and probably also Ca and Na from the primary minerals of the pegmatite while introducing Sb and Pb from an external source. Uher et al. (1998b) held metamorphic-magmatic solutions responsible for this hydrothermal overprinting. As shown above, there are many Sb-, Au-, Pb-, Zn- and Sb sulfide deposits, although many of them no longer economic by international standards, in the Slovak branch of the Carpathian Mts. that are worth to be considered as a potential source for these fluids accountable for the alteration of the pegmatites. Uher et al. (2001) reported black tourmaline of the Adolf adit pegmatite near the Magurka gold-antimony deposits to have been penetrated by pyrite and stibnite in the wake of a late-stage hydrothermal overprint of the pegmatite.

Ferrotapiolite as a dominant product of alteration pseudomorphosing primary stibiotantalite was described by Novák et al. (2004) from the lepidolite pegmatite at Laštovičky, western Moravia, Czech Republic, which is geodynamically "round the corner". Two compositionally distinct varieties of ferrotapiolite were recognized together with a Sb-rich phase. A rare mineral-bearing pegmatite from the Szklary serpentinite massif, in the Fore-Sudetic Block, SW Poland, also gave host to columbite-(Mn), tantalite-(Mn) and stibiocolumbite as the latest member among these Nb- and Ta oxides (Pieczka 2000). Uher and Černý (1998) calculated the temperature of zircon, present in barren and rare-element pegmatites to 700–580 °C. Beryl is the characteristic mineral of the Variscan granitic pegmatites which have a rather high age relative to the many pegmatites located in the adjacent Bohemian Massif, of 350 Ma and which are associated with S- and I- type granites-granodiorites of in the Malé Karpaty (Bratislava Massif), Považský Inovec, and Nízke Tatry Mountains. Beryl is the only rare mineral while Nb, Ta and Sn accessory minerals only appear at a more advanced level of pegmatite evolution,e.g., Moravany nad Váhom and Jezuitské Lesy, Slovakia (Uher et al. 2012). According to the authors, the average lithium contents fall in the range 120–830 ppm Li, with maximum values of 1400–1800 ppm (Švábsky Hill and Kamzík II). The highest Li contents are in beryl from the Moravany nad Váhom pegmatite with up to 5600 ppm. Lithium minerals of their own were not reported by the authors, a situation also known from the Spessart along the Mid-German Crystalline Rise, where Li is present as trace element in micas. It is not the only chemical congruence that can be reported from the pegmatitic rocks of these region being located at the periphery of the core zone of the Variscan orogen. Beryl plays a significant part among the pegmatitic suite of rocks, not only in the NW part of the Variscan Orogen and eastern part of the Alpine Orogen but also at the north-easternmost border of the Bohemian Massif, in the western and eastern Sudetes, which belong to the eastern extremities of the Saxo-Thuringian and Moravo-Silesian zones, respectively.

The geodynamic zonation based on pegmatites finds further support by the presence of REE-bearing pegmatites in all three peripheral geodynamic zones.

In conclusion, the Austrian Li-Be pegmatites, and the Carpathian pegmatite province with its Be-REE-Nb-Ta pegmatites is an example for the persistence of Variscan or Mesoeuropean pegmatites into Alpine or Neoeuropean pegmatites, being incorporated into ancient stable massifs with or without subsequent reactivation. Metapegmatites of early Variscan age suffered least during their incorporation into Neoeurope. Chemical similarities between pegmatitic rocks found at the exo- and endocontact of the Bohemian Massif and pegmatitic rocks within the Alpine-Carpathian Mountain Ridge support this idea.

2.3 Pegmatites and Geodynamics-a Synopsis and Exploration Strategies

This section is not designed to only repeat and concentrate the most significant issues in form of a synoptical overview but to amalgamate these facts and forge a first-order exploration model for the selection of target areas and secondly to pinpoint the most fertile zones within such an exploration area with respect to pegmatite-related mineral deposits. Pegmatites are part of the metallogenetic evolution in an ensialic fold belt, the characteristic deposits of which were treated in the previous sections. Thereby some of these reference types of mineral deposits illustrated in the cartoon of Fig. 2.2a are cast into the role of marker mineralization for pegmatite-prone crustal sections while other ore deposits are held to be diagnostic for crustal section barren as to pegmatites.

2.3.1 First Order – Model Pegmatites Like Ensialic Mobile Belts

In the previous Sects. 2.1 and 2.2 two data sets, one summarizing the litho-tectonic results gathered during the most recent past and the other manifesting visibly the metallogenetic inventory were discussed as the geodynamic setting of pegmatitic rocks. Both data sets can hardly be brought in line with a full-blown Wilson cycle that might be held accountable for the evolution of the mid-European Variscides.

Many of the complex models describe rifting and ocean spreading, active continental margin settings with island arc magmatic activity caused by subduction, nappe stacking and last but not least collision-related felsic intrusions for the various uplifted basement blocks in central Europe (Hann 2003; Konopásek and Schulmann 2005). Similar processes were also claimed by students of the Cadomian orogeny a predecessor of the Variscan orogeny, to take place in the period of time from more than 550 Ma to 510 Ma (Willmer et al. 1995; Zulauf et al. 1999; Dörr et al. 2002). Subduction-related volcanic activity in the Gondwana continental crust accompanied off-shore by a volcanic arc similar to its modern counterpart in the western Pacific Ocean has been claimed for the core zone of the Variscides in the Bohemian Massif, where the Tepla-Barrandian is supposed to be representative of this geodynamic setting.

While in the western Pacific ophiolites and their genetically related deposits with Cu, Cr, Ni and PGE are abundant, similar deposits are almost absent in Central Europe, as has been demonstrated in the individual sections dealing with the economic geology of each geodynamic unit. Porphyry Cu-Au and their "smoking guns" the epithermal Cu-Au deposits which render this Pacific area one of the most strongly exploited parts in terms of these types of deposits did not show up in central Europe either. This is not a matter of age and the erosive level but the different style of orogenies.

Going further east, in the central Asian interior, where gold, copper, molybdenum, rare earths elements and tungsten deposits were concentrated in the Kazakhstan-Kyrgyzstan border region during the Middle to Early Carboniferous age. The ore deposits in this region are situated within the Urals-Mongol Fold Belt and developed along an active plate margin. This compressional regime, including plate subduction, partial melting and igneous activity, was in full swing from the Devonian through the Carboniferous. Subsequent extension during Permian and early Triassic times may already herald the break-up of the Variscan/Hercynian continent and the incipient stages of the newly formed Tethys (Abdakhmanov et al. 1997; Rafailovich 1997). While granite-related Au, Cu, Mo, W, REE deposits, porphyry-type and epithermal gold deposits are widespread during the late Paleozoic, pegmatite deposits are a rarity.

The inferred metallogenetic evolution throughout the Variscan orogeny is typical of an ensialic mobile belt leading from an initial stage of rifting into a collisional stage giving rise to a complex stacking of crustal sections, involving subhorizontal thrustal movements and the intrusion of felsic magmatic bodies, with the pegmatites and aplites being part of this widespread mobilization of feldspar-quartz magmas. Partial melting of hot lower crust and the interaction with the mantle may account for the complete suite of intrusives from the early Carboniferous more basic syntectonic magmatic rocks to the posttectonic true granites. This sequence has a parallel from the metapegmatites, through the pegmatoids to the true pegmatites at the very end of the orogeny. Crustal thickening, an abnormally high radioactivity in this huge pile of continental crust and shear-zone related friction, in places added up by heat from the mantle are the "kitchen" or "incubator" for the granitic and pegmatitic rocks.

Pegmatites like ensialic orogens but hate ensimatic orogens (Fig. 2.2b). The idea of a metallogenetic evolution in an ensialic mobile belt, exemplified by the Central European Variscides is not a new model because it has already been published in detail by Dill (1985a), streamlined by Dill (1989) and corroborated in the aftermaths by the data of the numerous publications cited in this paper. The ideas were stimulated by the classification scheme put forward by Tankard et al. (1982) and Pitcher (1979, 1982, 1987). The ideas of these authors have been modified and adapted so as to create a geodynamic-metallogenetic basis to build upon succeeding discussion of pegmatites in an ensialic orogen. The characteristic features of the five geological environments are listed in the cartoon of Fig. 2.2b. The pegmatite-prone Variscan type is devoid of the classical subduction of oceanic crustal slaps, a characteristic feature shared also by the Alpine type orogen where pegmatitic rocks, although strongly modified in places re-appear next to the uplifted Variscan blocks. It is not really a surprise to find Namibia among the countries rife with pegmatites, a country where Tankard et al. (1982) shaped their model in the Proterozoic Damara Province.

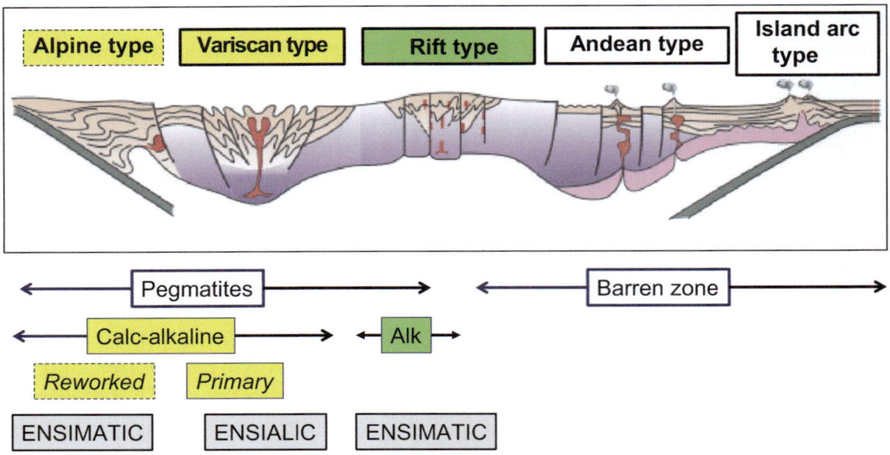

Fig. 2.2b Cartoon to show those environments favorable for the generation of pegmatites in relation to geodynamic settings detrimental to the evolution of pegmatitic rocks. *Alk* alkaline

Granites and pegmatites are "old ironsides" in geological terms, or in other words, they survive even under harsh physical-chemical conditions and may be tracked down to their predecessors or source area using the rare-element spectrum, e.g.,. Li, Nb-Ta, F and Be. The pseudopegmatites, pegmatoids and metapegmatites, predominantly widespread in the Austrian Alps are striking evidence for the incorporation into this modern fold belt and reactivation in the course of the Alpine metamorphic and kinematic process (reactivated Variscan pegmatites) (Sect. 2.2.3). It is not by chance that the Bohemian Massif is closest to the Alpine Mountain Range, the Li pseudopegmatites are of economic grade and where the Variscan uplifted blocks are depleted in pegmatites, as it is the case with the Schwarzwald-Vosges massifs the correlative Alpine massifs are devoid of productive pegmatites (Sects. 2.2.1 and 2.2.2). In the Western Carpathians, Slovakia, reworked slices of the crystalline basement of the Variscides are of widespread occurrence in the Mesozoic and Cenozoic sediments (Kohút 2013). Granitic rocks and their more basic predecessors originated from melting processes from the Devonian through the Carboniferous. Ensimatic mobile belts may not be the first choice for pegmatites to show up but if it comes to a reactivation of ensialic orogens at their margin, they cannot only stand the new geodynamic process within ensimatic fold belts, but even unfold their whole potential and upgrade.

Coupling the Variscan type orogen, the Alpine type orogen, and the primary *loci* of pegmatitization provide the basis to create a role model that not necessarily is confined to the Paleozoic-Mesozoic orogen. The processes inherent to this model may also be expect in older geological series, although their features may be less well documented there than in the type examples from Meso- and Neoeurope. Vice versa, the various types of pegmatitic rocks held to be typical of the various geodynamic realms can also be used as marker mineralization to disentangle Precambrian geodynamic systems in ancient cratons. These pegmatitic rocks may guide us during exploration also to crustal sections prone to mineral deposits others than pegmatites which may be attributed to the five types of deposits illustrated in the cartoon of Fig. 2.2a, where a crustal wedge with coupled and decoupled lithospheric mantle and basement thrusts is shown.

2.3.2 Second Order – Model: Pegmatite Like It Hot and Need Friction

Giving just a glimpse at the geodynamic map of central Europe reveals that the pegmatites are distributed not randomly across the uplifted basement blocks of the Variscides. They tend to be linked to particular architectural elements and they are lithologically-controlled as demonstrated by the pie-chart diagrams (Fig. 2.5a, b, c, d, e).

Even though granites are exposed in the north-westernmost part of the Rhenohercynian Zone (Brocken Granite in the Harz Mts.) neither there nor in the immediate foreland pegmatitic mobilizates developed. The country rocks are unmetamorphosed and only, in places, attain very low grade regional metamorphism (Onken et al. 1995). Only near the southeastern margin of the Rhenohercynian Zone, along the Mid-German Crystalline Rise, pegmatites and pegmatoids formed in a suture zone which comparing to the adjacent Rhenohercynian shows a sudden increase in the metamorphic grade (Sect. 2.1.3) (Oncken 1997). You can trace this geodynamic zone through the external zone of the Sudetes in SW Poland (Sect. 2.1.6.1), whose boundary towards the East European Craton is marked by the prominent Tornquist Suture, into the Moravo-Silesian Thrust Belt (Sect. 2.1.6.2). The horseshoe-like belt bounding the Bohemian Massif is marked by deep-seated shear and thrust zones as far as the structural inventory is concerned and its pegmatites are dominated by REE-specialized pegmatites. The reactivated pendant to these REE-bearing pegmatites in the Moravo-Silesian Zone in the Alpine Massifs is situated in the Western Carpathians (Sect. 2.1.4). A second more central belt is confined to the Moldanubian Zone, where some REE-bearing pegmatites are lined up like pearls on the string along the thrust plane separating the medium to high grade Gföhl nappe from the low to medium grade Drosendorf unit (Sect. 2.1.5.1). De Jonge et al. (1997) recorded alteration and brecciation accompanied by saline solution activity at temperatures of greater than 450 °C in shear-zones of the Mount Isa Inlier, Australia. In the course of this retrograde metamorphism Y, Nb and LREE were enriched by 15 times as much as in the country rocks. A similar setting was investigated by Rolland et al. (2003) focusing on mid-crustal shear zones of the granite complex of the Mont Blanc Massif in the western Alps. The physical conditions determined to be at ~ 0.5 GPa, 400 °C lead to a selective enrichments of LREE or HREE together with Y, Ta, and Hf. The authors claim that REE mobility is unrelated to the deformation style and the intensity of strain and ascribe this REE enrichment of up to 5:1 compared to the initial granite whole-rock REE budget to a synkinematic alteration of pre-existing magmatic REE-bearing minerals during deformation-related fluid-rock interaction. Fine-grained accessory minerals smaller than 20 µm in the host rock present in amounts of less than 2 wt% are sufficient to build up this REE concentration.

According to the current investigations, fluorine is an element diagnostic for the Saxo-Thuringian zone and its eastern prolongation the, W Sudetes. Its enrichment in granites and pegmatites and marks the zone of continent-continent collision with off-shoots towards the core zone of the ensialic orogen. Deep-seated structure zones striking perpendicularly to the NE-SW-trending collision front of the Saxo-Thuringian Zone formed the conduits for the element to get enriched as south as the Saxo-Thuringian-Moldanubian contact zones. Yet this process did not influence the mineral assemblage of the HPPP to the full extent.

The allochthonous units of the Tepla-Barrandian, once forming a coherent nappe with the Münchberg Gneiss Complex and the Zone of Erbendorf-Vohenstrauß are barren and not considered as prospective for rare-element-bearing pegmatitic rocks. Only in those parts of the ensialic orogen where it overrides the Moldanubian characterized by its HT event, its element association comes close to that of the varied assemblage known from the HPPP, showing similar amounts in Nb, B and P.

Lithium plays an important part in the primary ensialic orogen, the Central European Variscides and in its successor where the crustal section was reactivated under Alpine-type geodynamic conditions. At the boundary of the massif, along the shear and collision zones, Li is mainly accommodated into phyllosilicates such as lepidolite and zinnwaldite with an incipient stage of Li-bearing biotite already present in the suture zone of the Mid-German Crystalline Rise in front of the collision zone. Moving southward into the Moldanubian core zone, pegmatite contain lithium to be accommodated into the structure of phosphates and, as the boron contents rise, into tourmaline s.s.s., such as rossmanite and elbaite in the SE part of the Bohemian Massif. In the pseudopegmatites of the Austrian part of the Alpine mountain range, the re-activated lithium re-appears in form of spodumene (Sect. 2.2.3) and in the western Carpathians as trace element in beryl (Sect. 2.2.4). Lithium minerals are a direct response to the geodynamic variation in the ensialic orogen, reflecting changes in the physical-chemical regime as well as in the fracturing of the Bohemian Massif.

Chapter 3
Pegmatites and Their Country Rocks in the Central European Variscides

Abstract Aplites and pegmatites can be correlated with their country rocks using the temporal, the compositional/chemical and physical relationships. The emplacement of pegmatitic rocks began during the early Paleozoic (470–440 Ma) which are encountered as metapegmatites in nappe units in the Bohemian Massif, proper, and in the Ötztal Massif within the Alpine Mts. Range. At the end of the MP-HT metamorphism around 370 Ma pegmatoids came into existence in the allochthonous units. From the core to the margin near the collision zone the granites and the pegmatites get younger. The belong to the Variscan heat event, but a close-up view of some of the pegmatites bear witness of an older cooling age of the muscovite from the HPPP pegmatite than the whole-rock age of the nearby Flossenbürg granite. The pegmatite show different ages as the major-rock-forming silicates and rare-element minerals, e.g. columbite are considered. Considering the element assemblage of the pegmatites and aplites in the Hagendorf-Pleystein-Pegmatite Province reveals that the element has not been derived from one source only. Intracrustal sources may be claimed for Li, F, U, Sn, B, P, As and Mn. Niobium, beryllium and bismuth (?) are of subcrustal derivation. In addition, there are subcrustal - intermediate repositories as it is the case with the Zn, REE and Y. Apart from the granitic intrusive rocks and metamorphic rocks, another group of subcrustal magmatic rocks, (meta)-lamprophyres, has not been drawn the attention to, which they deserve. If the fractionation and zonation in a pegmatite field or province is investigated and an attempt is made to compare these individual pegmatites and aplites with a nearby granite some critical points have to be considered.

1. the direction of fractionation in and emplacement of the granite
2. the polyphase zonation within the pegmatite field or province.

The Flossenbürg granite shows a E-W direction of emplacement and an opposite trend of fractionation. The HPPP, which is situated to the south of the aforementioned granite, shows an emplacement from the SW towards the NE and a polyphase zonation as a function of the mineral association. With this in mind the granites and pegmatites are supposed to be part of the Variscan thermal event but unrelated with regard to the structural and fractionational processes. They are sisters but not parents and children. Geophysical surveys lend support to this structural-compositional plan. The Flossenbürg Granite dips away from the ZEV metamorphics towards the E, similar to its northern counterpart, the Tirschenreuth Granite. South

H.G. Dill, *The Hagendorf-Pleystein Province: the Center of Pegmatites in an Ensialic Orogen*, Modern Approaches in Solid Earth Sciences 15,
DOI 10.1007/978-3-319-18806-5_3

of the Oberpfalz granites, an ENE to NE trending deep structure, called Luhe Line exists. It is marked by ultrabasic to basic rocks of mantle affiliation. At the point of intersection, where it is crossed by another deep structure, this strong linear magnetic anomaly is perforated by a couple of negative rounded to oval-shaped magnetic anomalies like a "Swiss Cheese". The zone of intersection coincides with the cluster pegmatites and aplites of the HPPP.A low-resistivity zone known from the foreland dips down under the basement. The low resistivity layer at shallow depth was interpreted as an accumulation of graphitic matter or highly saline waters. Graphitic material accounts for the deep low-resistivity zone. Seismic methods were extensively used during DEKORP. They show a series of southward dipping reflectors which formed the guide-lines for the differentiation within the pegmatite-aplite fields located between the core zone and collision front and mark the nape stacking at the western edge of the Bohemian Massif. The intersection of these nappe piles and deep-seated lineamentary fault zone are of ore control for the HPPP.

3.1 Geochronology

3.1.1 Geochronology of Granites

Granites bear the closest resemblance to pegmatites and aplites among all intrusive rocks, in terms of their rock-forming minerals feldspar, quartz and mica as well as their texture. Considering their size of minerals, granites are often in stark contrast to that of pegmatites and aplites. The first ones are much coarser-grained whereas the aplites may be similar to granites or finer-grained, as exemplified by the aplitic granite at Silbergrube at the western outskirts of Waidhaus. Understandably, the temporal relation between pegmatites and neighboring granites has been a key element during studies of pegmatites and this issue has to be resolved first when it comes to a discussion of the origin of pegmatites in general and along the western edge of the Bohemian Massif in particular (Table 3.1, Fig. 3.1a).

In the following paragraphs, the chronological relationship between these intrusive rocks is looked at on regional and local scale. In an overview, the emplacement of granites is documented for each geodynamic zone of the Variscan Orogen in Central Europe, placing emphasis upon those geodynamic zones known for their abundance in pegmatites. On a local scale, the nitty-gritty of the evolution of granites is debated along a NW-SE transect along the western edge of the Bohemian Massif.

In a comprehensive study, Finger et al. (1997) outlined the spatial distribution of Variscan granitoids in Central Europe and subdivided them into five groups based upon their chemical signature and age.

Group 1 is of late Devonian through Early Carboniferous age (370–340 Ma). These intrusive rocks are tonalitic, dioritic, gabbroic and granodioritic composition. The Odenwald granitoids, the Central Bohemian Pluton, Czech Republic and the

Table 3.1 Radiometric age dating of granitic and pegmatitic rocks along the western edge of the Bohemian Massif

Rock type and site	Rb/Sr whole rock (Ma)	Mineral ages (Ma)						
		Biotite	Muscovite	K feldspar	Columbite	Monazite-Zircon	Horn-blende	Garnet
Oberkotzau pegmatoid			372.5–377.0					
Eclogite pegmatite			370–380					
G 4 Tin Granite	290 ± 4							
G 3 Core Granite	298 ± 4							
G 2 Marginal Granite	301 ± 8							
G 1 Central Granite	321 ± 14							
G 1 Central Granite	319 ± 3	315.8 ± 0.2						
Friedenfels Granite (Steinwald)	314.8 ± 2.3	304	312					
		302	309					
Mitterteich Granite monzogranite	310 ± 3							
Falkenberg Granite monzogranite	312 ± 3							
Falkenberg Granite	311 ± 4	300 ± 1	307 ± 1 310 ± 1					
Liebenstein monzogranite	320 ± 40							
Granite dikes intersecting the Falkenberg Granite	309.5 ± 2.2	299	308					
		292	298					
Leuchtenberg granite	321 ± 8							
Leuchtenberg monzogranite-granodiorite	326 ± 2							
Flossenbürg Granite	304 ± 11							
Flossenbürg Granite	311.9 ± 2.7	292						
Flossenbürg Granite leucogranite	311 ± 5							

(continued)

Table 3.1 (continued)

Rock type and site	Rb/Sr whole rock (Ma)	Mineral ages (Ma)						
		Biotite	Muscovite	K feldspar	Columbite	Monazite-Zircon	Horn-blende	Garnet
Bärnau Granite	313.2 ± 2.0	297	306					
		289						
		(Chlorite)						
Bärnau Granite leucogranite	313 ± 2							
Georgenberg- Brünst pegmatite			316 ± 3					
Plössberg pegmatite			315 ± 3					
Hagendorf-South pegmatite			317 ± 3					
Menzelhof metapegmatite			mega crystal		475			
			474 ± 5 normal					
			crystal 464 ± 5					
Wildenreuth metapegmatite			mega crystal 476 ± 5 Ma					
			normal crystal					
			352 ± 5 Ma					
Störnstein metapegmatite			mega crystal 440 ± 4 355 ± 3					
Wendersreuth metapegmatite			mega crystal			480 + 7/–9		
			473 ± 5					
			446 ± 5 Ma					
Püllersreuth metapegmatite		363 ± 4	mega crystal					
			481 ± 5					

Irchenrieth metapegmatite			mega crystal					
			477 ± 5 normal crystal					
			468 ± 5					
Oedenthal			mega crystal					
			479 ± 5					
Domazlice metapegmatite					482.2 ± 13			
KTB Pilot hole <3455 m		319	372 Ma					
		316	363 Ma HP/MP Meta.					
		LP Meta.						
KTB Pilot hole >3455 m		302	348					
		293	304 HP/MP Meta.					
		LP Meta						
KTB Pilot hole 2847 m			<2 µm					
			260					
			220					
KTB Pilot hole 2847 m			<2 µm					
			202					
			170					
KTB Pilot hole 433 m pegmatoid		375						
KTB Pilot hole 3297 m pegmatite		370						
KTB Pilot hole 2231 m lamprophyre		306 ± 4						
		295.1 ± 3						

(continued)

Table 3.1 (continued)

Rock type and site	Rb/Sr whole rock (Ma)	Mineral ages (Ma)						
		Biotite	Muscovite	K feldspar	Columbite	Monazite-Zircon	Horn-blende	Garnet
Trutzhofmühle Aplite					302.1 ± 3.3			
					376 ± 14			
Silbergrube aplite granite					302.8 ± 1.9			
Hagendorf South pegmatite					299.6 ± 1.9			
Hagendorf North pegmatite		314.0 ± 5.5	305.3 ± 3.4					
		266.9 ± 10.6						
Blötz pegmatite		256	317	281				
Blötz pegmatite		261 ± 16	309	326				
Blötz pegmatite			306 ± 14					
Blötz pegmatite			329					
Blötz pegmatite			307.8 ± 4					
Großaulüß pegmatite							390 ± 4	
Hörlberg pegmatite		332	313	290				
Hörlberg pegmatite			280					
Hörlberg pegmatite			309					
Hühnerkobel granite	318	291	301					
Hühnerkobel granite	313	316						
Hühnerkobel granite	317.5 ± 4.8	289	300					
Hühnerkobel granite	325				318.4 ± 0.5	3073 ± 845		
Hühnerkobel granite	315							
Hühnerkobel bi-pl-gneiss	494			481				
Hühnerkobel bi-pl-gneiss	481							

Matzersdorf				295	316.4 ± 8.7 gadolinite			305 ± 7
Matzersdorf (Tittling Granite)	337 ± 15							
Pochermühle pegmatite			312	335	313.1 ± 8.5	3547 ± 1400		
Sattelbrunnen pegmatite				296				
Sattelbrunnen pegmatite				325				
Sattelbrunnen pegmatite		280	312					
Schwarzeck pegmatite		309	296	264				
Schwarzeck			314	372				
Schwarzeck			329					
Kirchberg (Erzgebirge) granite	309.4 ± 3.8							
Kirchberg (Erzgebirge)		317–325						
Bergen granite	312.8 ± 7							
Bergen			318.2 ± 3.1					
Bergen		323.6 ± 2.6						
Ehrenfriedersdorf, granite	316 ± 11							
Ehrenfriedersdorf, granite	317.2 ± 4.2							
Ehrenfriedersdorf, aplite	291.3 ± 3.2							
Eibenstock 1 granite	305 ± 4							
Eibenstock 2 granite	299 ± 6							
Altenberg granite porphyry Ar/Ar	307–309							
Podlesi stock granite	312 ± 1.8							
Podlesi dyke granite	311.2 ± 1.8							
Podlesi pegmatite	309.7 ± 1.7							

(continued)

Table 3.1 (continued)

Rock type and site	Rb/Sr whole rock (Ma)	Mineral ages (Ma)						
		Biotite	Muscovite	K feldspar	Columbite	Monazite-Zircon	Horn-blende	Garnet
Horni Blatná granite	315.4 ± 1.8							
Horni Blatná granite	311.9 ± 2.1							
Horni Blatná granite	313.2 ± 2.3							
Křížový Kámen/Kreuzstein Granite	297 ± 2							
Stará Knížecí Hut fine-medium grained tonalite granite	296 ± 9							
Rozvadov Granite leucogranite	304 ± 19							
Bor granite monzogranite	337 ± 7							

Kreuzer et al. 1993; Besang et al. 1976; Lenz 1986; Wendt et al. 1986, 1992, 1994; Siebel et al. 1995a, b, 1997; Müller et al. 1998; Köhler et al. 1974; Köhler and Müller-Sohnius 1976; Glodny et al. 1995b, 1998; Wemmer and Ahrendt 1993; Dill et al. 2008b, 2009a, b, 2013a, b, c; Schaaf et al. 2008; Gerstenberger et al. 1995; Velichkien et al. 1994; Schabert cited by Breiter 1998a; Seltmann and Schilka 1995; Velichkien et al. 1994

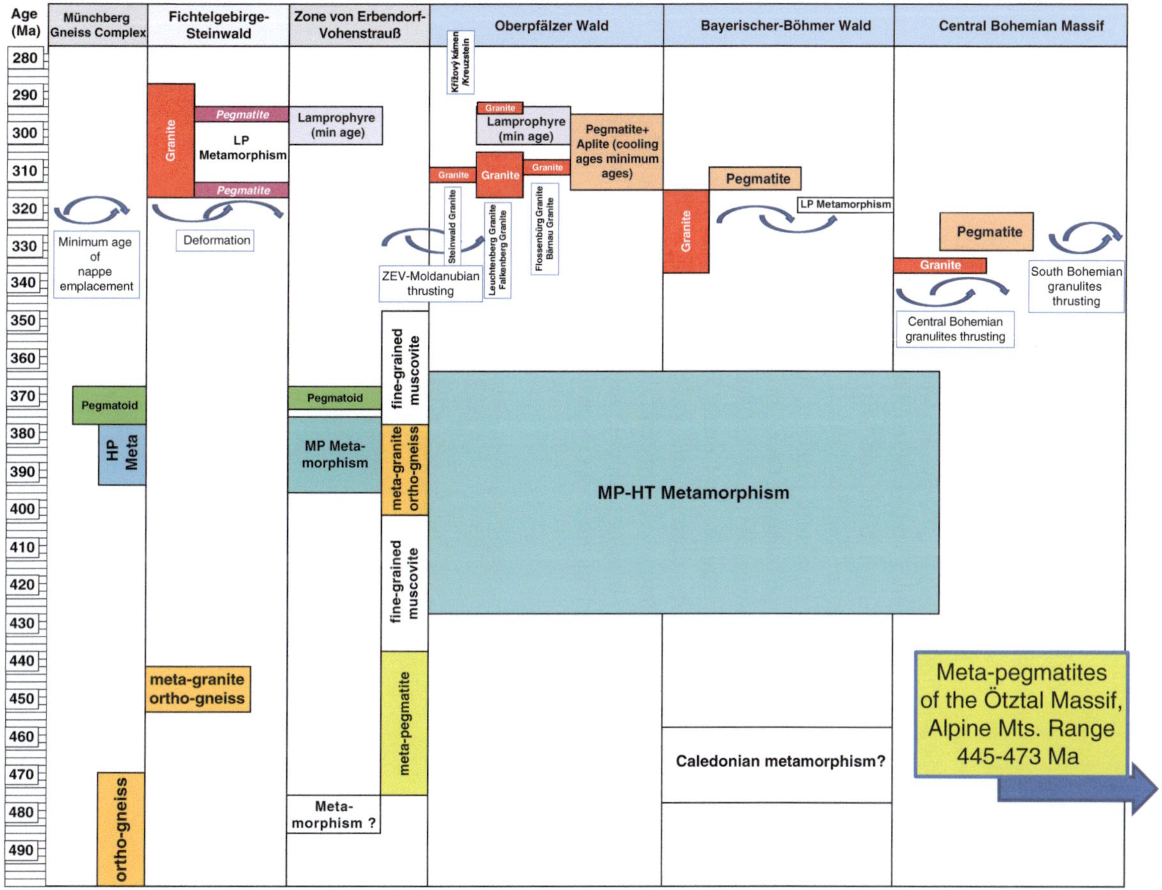

Fig. 3.1a Geochronology of magmatic activity and metamorphism in the various geodynamic units at the western edge of the Bohemian Massif (for reference see Table 3.1). The metapegmatites in the Ötztal Massif of the Alpine Mts. Range are discussed in Sect. 2.2.3

Sudetic Pluton at the northeastern edge of the Bohemian Massif belong to these I-type Cordilleran granitoids. They are held to be the only intrusive rocks that are coeval with the Variscan subduction undergoing a strong fractionation and assimilation on their ascent. Excluding the pegmatitic rocks of the central Bohemian Massif, most pegmatites trail far behind these granitic rocks (Fig. 3.1a).

Group 2 is representative of S-type granitoids of Early Carboniferous age, around 340 Ma. They developed syncollisionally in the course of crustal melting and nappe stacking. These felsic rocks are exclusive to the south-eastern edge of the Bohemian Massif, near the Alpine Front (Figs. 2.1d and 3.1b).

The majority of all Variscan granitoids formed between 340 and 310 Ma in Central Europe and has been categorized as group 3. These late Visean and early Namurian S-type and high-K I-type granites are also relevant for the granites along the western edge of the Bohemian Massif, extending from the granites of the Saxo-Thuringian Fichtelgebirge through to the southern granites of the Bayerischer and Böhmer Wald in the Moldanubian Region (Table 3.1)

Post-collisional as to their structural position, the group-4 granitoids comprise granodioritic and tonalitic I-type intrusive rocks along the "Pfahl" ("Great Bavarian Quartz Lode") and formed between 310 and 290 Ma (Fig. 1.1a).

The youngest group of leucogranites in Central Europe formed between 300 and 250 Ma, during the late Carboniferous and Permian, when highly-fractionated S-type granites and A-type granites were concentrated across the Fichtelgebirge and the Erzgebirge (Fig. 2.1d). They have no temporal equivalent rocks in the Moldanubian zone where pegmatites and aplites are of widespread occurrence.

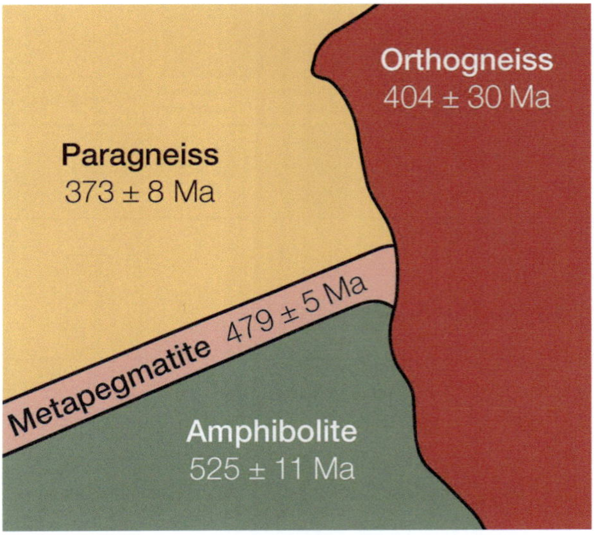

Fig. 3.1b Chronology of metabasic and metafelsic igneous rocks emplaced within paragneisses (Modified from Glodny et al. 1995b)

Viewed on a larger scale and starting off with the northwestern-most granitic suites, the granitic rocks of the Fichtelgebirge in the south-western the Saxo-Thuringian Zone are brought to the area of attention first. Three granites abundant in rare elements were dated by Besang et al. (1976) as Granite G 4 (290 ± 4 Ma), G 3 (298 ± 4 Ma) and G 2 (301 ± 8 Ma). Although the timeframe set by Finger et al. (1997) on a regional scale has to be slightly modified, correlating these Fichtelgebirge granites with their group- 5 granites would be the best fit. The older granite G 1 in the Fichtelgebirge is significantly older than the afore-mentioned granites (Table 3.1) (Lenz 1986). On comparison of the whole rock analyses of the granite (319 ± 3 Ma) and the pertinent mineral ages of biotite of 315.8 ± 0.2 Ma, a rapid cooling of these felsic rocks is evident, even if there is still some dispute on the blocking temperature of biotite, spanning the T range from 280 °C to more than 400 °C (Dickin 2005). The Fichtelgebirge granites are not associated with large bodies of highly-fractionated granitic pegmatites or aplites although their rare metal contents are abnormally high (Table 2.1). The older granites formed in a rather unbalanced environment as far as the temperature is concerned. The conclusion can be drawn from the texture of the Granite G 2, that it is made up of a few isolated feldspar phenocrysts floating in an aplite granitic matrix. Only the smaller felsic intrusions of the granites G 3 and G 4 were emplaced in a well-balanced environment. They are accompanied by some granitic pegmatites of comparatively minor relevance in view of what we will learn later, as we discuss the HPPP.

Heading south, the Friedenfels Granite is situated astride the Saxo-Thuringian-Moldanubian boundary (Fig. 2.1d). Based on a Rb-Sr-whole rock isochrone, this granite was dated as 314.8 ± 2.3 Ma (Wendt et al. 1992). Its K-Ar data of muscovite scatter between 312 and 309 Ma, while its biotite ages run from 304 to 302 Ma. There are two options to interpret these isotope data, either by a slower cooling rate than in the northwestern part or by a stage of subsequent reheating. Not far from this granite, the small Hopfau pegmatite forms the northernmost representative of the phosphate-bearing pegmatites in the study area (Table 1.2). It can be taken as a missing link between the granitic pegmatites of the Saxo-Thuringian and the pegmatites s.st. of the Moldanubian because it contains minerals diagnostic of both pegmatite provinces.

Just south of the Saxo-Thuringian-Moldanubian boundary pegmatites and aplites show up on both sides of the Falkenberg Granite and also within the granite itself, e.g., the Pilmersreuth Pegmatite (Table 1.2). The granite suite in the northern Oberpfalz bears striking similarities in their chronological evolution to the granites recorded from the Fichtelgebirge. There is an older granite, named Leuchtenberg Granite and a group of younger granites, comprising the Flossenbürg, Bärnau and Falkenberg Granites, the latter is endowed with a set of granite dykes. The Leuchtenberg Granite dated as 321 ± 8 Ma has all the lithological hallmarks of the G1 granite from the Fichtelgebirge (Köhler et al. 1974; Köhler and Müller-Sohnius 1976). Unlike its northern equivalent, it is accompanied by numerous small pegmatites along its western edge (Figs. 2.5b and 2.5e – area 5). The Falkenberg Granite is significantly younger and yielded a Rb-Sr whole-rock isochrone interpreted as an age of formation around 311 ± 4 Ma (Wendt et al. 1986). The mineral ages of biotite and muscovite are significantly different, trailing behind the whole rock ages by about 10 Ma. Quite similar to the Friedenfels Granites, these differences may be accounted for either by a slow cooling rate or by a subsequent reheating. The interpretation is fraught with difficulties because of the presence of a swarm of granite dykes intersecting the Falkenberg Granite around 309.5 ± 2.2 Ma (Wendt et al. 1992). The widely scattering of K/Ar dates of their muscovites (308–298 Ma) and biotites (299–292 Ma) interpreted conventionally as cooling ages, resulted from a younger hydrothermal activity which, in places, gave rise here a subeconomic fault bound U mineralization (Dill 1986). As illustrated in the map of Fig. 2.5e – area 6 – especially the eastern endo – and exocontact zones of the Falkenberg Granite are rife with pegmatites around Pilmersreuth; the area is also the center of granite dyke swarms. The remaining two granite in the region, the granite complexes of Flossenbürg and Bärnau come closest to the HPPP and so deserve a more detailed consideration as to the age relationship to the neighboring pegmatites. The Flossenbürg Granite with a Rb-Sr whole rock age of 311.9 ± 2.7 Ma and the Bärnau Granite with an equivalent age of 313.2 ± 2.0 Ma do not differ from each other within the 1σ range (Wendt et al. 1994). The muscovite and biotite age data recorded by the above authors reflect a rather slow cooling. Significantly younger age of biotite are due some chloritization (Table 3.1).

In the Bayerischer and Böhmer Wald, even though very widespread between Bad Kötzting and Zwiesel, pegmatites seldom are associated in space and time with granites (Fig. 2.1d). Only the most well known Hühnerkobel Pegmatite is located within a fine-grained two-mica granite which was also subjected to radiometric age dating (Madel 1967; Strunz 1971, 1974; Tennyson, 1981). The host granite was dated by Schaaf et al. (2008) who calculated a Rb/Sr isochrone age of 317.5 ± 4.8 Ma with a $^{87}Sr/^{86}Sr$ initial ratio of 0.70591 ± 0.00996 which is the lowest ratio encountered in the granites along the NW-SE transect through the pegmatite provinces of the western edge of the Bohemian Massif (Table 3.1) The authors published a cooling age of 300 Ma for muscovite and 289 Ma for biotite. Using the closure of the minerals at temperature of 500 °C for muscovite and 320 °C for biotite (Harrison and McDougall 1980). Provided the temperature was around 700 °C during its emplacement, a slow cooling rate of 11.8 °C existed on the way down to 500 °C and a cooling rate of 16.4 °C/Ma may be calculated during further cooling of the granite down to 300 °C. Another pegmatite resides in the Tittling Granite/Granodiorite near Matzersdorf, the age of which is 337 ± 15 Ma (Schaaf et al. 2008; Köhler et al. 2008). Köhler et al. (2008) concluded from a whole rock isochrone of 10 granite samples taken from various sites in the Bayerischer and Böhmer Wald that a magmatic event extended from 350 Ma through 300 Ma in the Moldanubian Zone. The postkinematic granites intruded into the Moldanubian Zone getting gradually younger in the NE Bavarian Basement from SE to NW leading eventually to the Tin Granite G 4 in the Fichtelgebirge at 290 ± 4 Ma (Table 3.1).

3.1.2 Geochronology of Pegmatitic and Aplitic Rocks

A wealth of isotope data has been generated particularly during the last 20 years for the gneisses and metabasic rocks in the Saxo-Thuringian and Moldanubian basement in NE Bavaria. Some of these age data are also mentioned within this chapter, on providing an overview of both geodynamic zones, the Saxo-Thuringian and Moldanubian. To draw a full-blown picture for the NE Bavarian Basement of the temporal relationship of the various crystalline rocks among each other, would go far beyond the scope of this study about pegmatites and aplites. Therefore the non-granitic crystalline rocks are dealt with on a case-by-case basis in context with the chronology of the pegmatites and aplites and are only discussed as to the extent to which these variable country rocks contributed to the formation of the felsic intrusive rocks under consideration.

Pegmatites create a lot of problems as they were targeted upon during radiometric age dating, making use of the Rb-Sr method for whole rock samples and mineral specimens. This is prevalently due to their peculiar texture and crystal size of rock-forming minerals, but also caused by the common interaction of their melt and rocks with magmatic, hydrothermal and meteoric fluids in the course of their evolution (Marakushev and Gramenitskiy 1983; Küster 1995; Lowenstern 1995).

Uranium-lead dating of accessory minerals bearing considerable amounts of uranium in their structure is a reliable way out of this unsatisfactory situation. Aldrich et al. (1956) addressed this problem identified above and used for the first time columbite a common rare-element mineral in pegmatites. Romer and Smeds (1994) successfully dated lithium-cesium-tantalum- and niobium-yttrium-fluorine-type pegmatites from Sweden, using columbite. The same authors applied in 1996 this method to post-kinematic and post-metamorphic rare-metal pegmatites from the Sveconorvegian province where they were able to demonstrate that these pegmatites are more than 50 Ma older than the granites spatially associated with these pegmatites (Romer and Smeds 1996). In another study on columbite age dating from Swedish pegmatites they found a wide range of ages matched by migmatic gneisses (Romer and Smeds 1997). Even if the ages obtained by their studies on columbite are much higher than those of equivalent LCT- and NYF-type pegmatites from the study area, they also show many similarities and the results encouraged geochronologists to use also columbite for constraining the age of formation of pegmatites and aplites in the Bohemian Massif (Glodny et al. 1998; Dill et al. 2008a, b, 2009a; Schaaf et al. 2008; Melleton et al. 2012). Unlike their counterpart's muscovite and biotite, commonly used in radiometric age dating, columbites seem to have much higher closure temperatures at around 700 °C and thus are less vulnerable to postmagmatic alteration (Glodny et al. 1998). By all accounts, the columbite age dating is certainly not a jack of all trades in pegmatite chronology and a lot of questions arise, particularly as these age data are compared with those obtained by different methods such as K/Ar or Rb/Sr using silicates in the same rocks.

The southernmost pegmatite in the Bayerischer-Böhmer Wald which successfully was subjected to U/Pb dating, using columbite, is located at Pochermühle. It yielded a U/Pb age of 313.1 ± 8.5 Ma with an upper intercept at 3547 ± 1400 Ma characteristic for an Achaean age (Schaaf et al. 2008). The neighboring Hühnerkobel pegmatite give a similar age of 318.4 ± 0.5 Ma in the U-Pb concordia diagram with a similar upper intercept corresponding to an Achaean age. The radiometric data overlap with the granite age of 317.5 ± 4.8 Ma so that they are indistinguishable from each other on the basis of U/Pb dating and hence a close kinship existed between granitic and pegmatitic rocks stressing that both belong to the same genetic process. This assumption is corroborated by the U/Pb data of another mineral, gadolinite. Although a rather rare mineral among the pegmatite-related minerals in this central European crustal section, its age 316.4 ± 8.7 Ma is another clue to the 320 Ma – year heat event in the Bavarian sector of the Bohemian Massif. An independent method using K/Ar of muscovite was applied to six samples from different pegmatites by Schaaf et al. (2008) with the same result of 305.8 ± 14.1 Ma, albeit the error was rather big. The cooling of the pegmatites was rather swift compared with that of the granites which they are related in space and share the same heat event in this region.

Some pegmatites in this region also contain as minor constituents hornblende and garnet, whose Rb/Sr- and Sm/Nd isotope ratios point to an early Devonian event (e.g. 390 ± 4 Ma) or even a much older process at 414 ± 7.7 Ma which is equivalent to the late Silurian period of time. These isotope data remained unaffected even when the heat exceeded temperatures of more than 650 °C as it was the case during the magmatic event around 320 Ma known from almost all pegmatites in the Bayerischer-Böhmer Wald. Analyses of monazite from one Moldanubian cordierite-sillimanite gneisses gave a concordant age of 460 Ma, while Rb/Sr whole rock analyses of cordierite-sillimanite gneiss samples scattered around 390 Ma (Hansen et al. 1989). This early Devonian regional medium-pressure/high-temperature metamorphic event left its imprint in the melt leading to the pegmatites. The succeeding low-pressure metamorphic event between 320 and 325 Ma, however, was of no impact on the evolution of pegmatites in the south-western part of the Bohemian Massif.

All remaining pegmatites in the southeastern most part of the study area are hosted by non-granitic rock such as cordierite-sillimanite gneisses (e.g., pegmatite at Blötz) or metabasites (e.g., pegmatite at Großaulüß) (Figs. 2.5a, 2.5b, 2.5c, 2.5d, and 2.5e). The latter rock type yielded a U-Pb zircon age of 511 ± 3 Ma (Gebauer and Grünenfelder 1982; Gebauer 1993) and Sm/Nd four-point whole rock isochrone of 639 ± 154 Ma (Miethig et al. 2008). These ages determined in the host rock of the pegmatite reflect the age of the tholeiitic protolith of the metabasites.

In the Oberpfälzer Wald the lithological diversity among the feldspar-quartz mobilizates is much greater than in the Bayerischer-Böhmer Wald and the existing rocks therefore need a more detailed treatment considering their geochronological evolution. Not only postkinematic pegmatites and aplites but also synkinematic pegmatoids and even prekinematic meta-pegmatites can be distinguished from each other within an area measuring a few tens of kilometers across.

The potassium-argon and argon-argon methods were applied to the phyllosilicates biotite and muscovite of the Hagendorf-North pegmatite as well as to the pegmatoids of the Zone of Erbendorf-Vohenstrauß providing information on the cooling history of stocklike zoned pegmatites and the metamorphic feldspar-quartz-mica mobilizates (Kreuzer et al. 1993; Dill et al. 2013a). For the recrystallized muscovite and feldspar of the strongly foliated sections of metapegmatites and the large mega crystals of the undeformed pegmatites, both the Rb-Sr technique has proved to be a successful tool (Glodny et al. 1995a, b, 1998). Finally an analysis of the U/Pb ratios obtained from columbite-(Fe) in these felsic rocks, proved this Nb-Ta oxide to be second to none in the metapegmatites, stocklike pegmatites and tabular aplites as the measurements are carried out by means of laser ablation inductive coupled plasma-mass spectrometry (LA-ICP-MS) (Glodny et al. 1998; Dill et al. 2008b, 2009a, b).

Age dating of pegmatites and their country rocks is a key element to answer the question how these felsic intrusive rocks have formed. In the following papers the various radiometric techniques used for the different types of pegmatites and aplites either in the course of a whole rock analysis or as applied to individual minerals are briefly described and can be taken as reference for dating in pegmatites, as they are embedded in a couple of geological and mineralogical data (Glodny et al. 1998; Dill et al. 2008b, 2009a, b, 2013a).

The northern Oberpfälzer Wald could in a certain sense be considered as the most complicated part of the study area as far as the tectonic setting is concerned and in terms of diversity of felsic intrusive rocks, which can be classified in three different groups (1) metapegmatites, (2) pegmatoids and (3) pegmatites *sensu stricto*, including their fine-grained analogues of the aplitic mobilizates. From whatever angle you may look at the pegmatitic and aplitic rocks the strong relationship between the felsic intrusive rocks and the litho-tectonic units is obvious. There are two major units, the Moldanubian metamorphic rocks well-known from the Bayerischer-Böhmer Wald where these gneisses and their granites are the sole host of the pegmatites. In addition to the above mentioned autochthonous gneissic unit of the Moldanubian Region another allochthonous unit called "Zone von Erbendorf-Vohenstrauß" (ZEV) in Germany and Tepla-Barrandian in the western part of the Czech Republic reside on top of the Moldanubian crystalline rocks. What was previously a coherent nappe, now has been broken apart into klippen and windows by erosion in a textbook-like way so as to match the nappes known from the modern fold belt of the Alpine Mountain range.

The ZEV in Germany and the Tepla-Barrandian unit in the western part of the Czech Republic are both host of the oldest metapegmatites in the study area (Fig. 3.1a, Table 3.1). In the ZEV several of these pegmatitic rocks composed of alkaline feldspar, quartz, biotite, muscovite, garnet, tourmaline and beryl were dated by Glodny et al. (1995b). The metapegmatites at Wildenreuth, Störnstein, Wendersreuth, Menzelhof, Püllersreuth, Irchenrieth and Oedenthal which all passed through the Variscan deformation and also subjected to younger medium-pressure and high-temperature regional metamorphism formed between 480 and 440 Ma, as it is documented by the cooling ages of mega-crystals ($4 \times 4 \times 0.5$ cm) of muscovite

(Fig. 3.1a). Their maximum age of formation coincides with the age data obtained during U/Pb dating of zircon in metabasic rocks (Hölzl et al. 1993). The opinions are split on this isotope data, as some assumed the age of 476 ± 4 Ma as the age of emplacement whereas argument against it and favor an age of metamorphism. The cartoon of figure (Fig. 3.1b) illustrates the interrelationship of the various metamorphic rocks and the metapegmatites. The metapegmatites (479 ± 5 Ma) are intercalated between amphibolites (525 ± 11 Ma) and paragneisses (373 ± 8 Ma). The pegmatites were emplaced within crystalline country rocks that had already reached a high-grade regional metamorphism. The orthogneisses were emplaced in the interval 404 ± 30 Ma and 379 ± 15 Ma and thus cannot be held as the parent material from which the pegmatites evolved but have to be held liable for the thermal overprinting of the pegmatites. The Devonian high-temperature regional metamorphism with its temperature increasing to as much as 600 °C and the structural disturbance dealt another physical blow to the pegmatites but without doing any harm the mica that preserved their Ordovician signature of the Rb/Sr system. According to the authors' assumption, the fine-grained muscovite resulted from a neomorphism of micas during a later stage of deformation. The Püllersreuth metapegmatite stands out in the group of pegmatites not only by its presence of beryl but also by its occurrence of columbite (Linhardt 2000). The metapegmatite from the Domažlice crystalline complex, Czech Republic, a part of the Teplá-Barrandian unit, shows a mineralogical development analogous to the that of Püllersreuth by its presence of columbite (Glodny et al. 1998). Although there were different minerals and methods used for age dating in both pegmatites, muscovite in Püllersreuth (Rb/Sr) and columbite at Domažlice (U/Pb), both host rocks have almost the same age 481 ± 5 Ma and 482,2 ± 13 Ma, respectively. In addition to what has already been concluded from the HT metamorphic overprinting of the metapegmatites, this is another incontrovertible proof that the cooling temperature of muscovite is much higher than expected and obviously comes close to that of columbite. One can draw the conclusion that the metapegmatites in these allochthonous units unambiguously formed without any parent granites, independent from any felsic predecessor. Moreover, the oldest members of this suite of metapegmatite are the only giving host to columbite.

Pegmatoids in the ZEV are eclipsed by the afore-mentioned metapegmatites, as they neither bear rare minerals nor did they attain a size as big as to draw the attention to these synkinematic feldspar-quartz-mica mobilizates. According to Kreuzer et al. (1993) these pegmatoids developed between 374 and 359 Ma in the ZEV. They were hit at 433 m by the pilot hole of the Continental Deep Drilling Program of the Federal Republic at Windischeschenbach drill site. Albeit the drill bit on its way down hit almost the same lithological units, alternating with each other like onions, peppers, meet and sausages on a skewer of a barbecue grill party, there was no other layer or vein similar to this pegmatoid near the surface. They are confined to special zones of high shearing stress and resultant of metamorphic and kinematic processes around 370 Ma.

The only magmatic rock left unaffected by the thrusting of this nappe towards the Northwest, a lamprophyre dike was chronologically constrained by the biotite cooling ages to 306 ± 4 Ma and 295,1 ± 3 Ma (Kreuzer et al. 1993). This fine-grained magmatic rock type is not uncommon to the HPPP either, where dark magmatic rocks show up in the form of swarms of dikes, measuring as much as 0.1 m in width, in the cordierite-sillimanite gneisses and the aplites of Pleystein.

The Moldanubicum of the Oberpfälzer Wald is host to the HPPP and was intruded by the Leuchtenberg, Falkenberg and Flossenbürg, the latter is the granite closest to the HPPP. The pegmatites and aplites are postkinematic to the youngest Variscan metamorphic and structural event. Structural adjustments in the Oberpfalz are constrained to the period 450–330 Ma by Weber and Vollbrecht (1989) terminating the age of formation of granites and pegmatites towards higher ages. The metamorphic processes preceding in the area under study the emplacement of intrusive rocks is typical of the cordierite-K-feldspar zone described by Vrána et al. (1995), for which Lee and Holdaway (1977) obtained temperatures of formation of approx. 650 °C and pressure data in the range 3–4 kbar. In the geologist's mind working in that region and supporting the traditional ideas of all pegmatites to have been derived from a large granitic body underneath, the melt creating the Flossenbürg Granite was the parental magma for all pegmatites and aplites south of this granite –see summary of ideas in Strunz (1974) and Strunz et al. (1975). A simple comparison of the intrusive age of the Flossenbürg Granite dated at 311,9 ± 2,7 Ma and the blocking ages of its muscovite of 300 Ma and biotite of 292 Ma with muscovite ages from various pegmatites would discard this hypothesis (Wendt et al. 1994). Glodny et al. (1995b) published for the pegmatite at Hagendorf-South a Rb/Sr age for its muscovite of 317 ± 3 Ma, for the pegmatite near Georgenberg (Brünst) of 316 ± 3 Ma and for Plössberg of 315 ± 3 Ma. The real age of emplacement is certainly higher than these cooling ages. Considering the U/Pb ages for columbite obtained from ICPMS analyses of Hagendorf-South, Silbergrube and Trutzhofmühle sites renders the situation even more complex than simplifying it (Dill et al. 2008b, 2009a, b). The largest pegmatite stock in Central Europe at Hagendorf-South yielded a U/Pb age for columbite-(Fe) of 299.6 ± 1.9 Ma (Fig. 3.1c, Table 3.2), the tabular Silbergrube aplite/aplite granite a U/Pb age of 302.8 ± 1.9 Ma (Fig. 3.1d, Table 3.3), and the Table 3.2), Upon closer inspection of the isochrone calculated for the Trutzhofmühle aplite a relic age of 376 ± 14 Ma can be confirmed. This age has already been recorded from the shear zone-related pegmatoids, it may be recognized within the newly grown mica and coincides with the waning stages of the HT-MP metamorphism to be seen all across the Moldanubian Zone in Bavaria (Fig. 3.1a). Considering the cooling ages around 314.0 ± 5.5 Ma of Hagendorf – North pegmatite for which no columbite age is available due to the scarcity of this Nb-Ta oxide there, the interrelationship to its larger brethren Hagendorf-South cannot be overlooked while a derivation from the nearby Flossenbürg Granite is as unlikely as for any other pegmatite or aplite in the HPPP (Strunz 1952; Forster et al. 1967). The Rozvadov Granite beyond the Czech-German border has also to be eliminated from the list of potential parent granites for chronological reasons. A small granitic intrusion north of the HPPF on Czech territory, the Křížový kámen (German: Kreuzstein) Granite, is known as a strongly fractionated phosphorus-enriched intrusion. It has been held accountable by Siebel et al. (1995b) for the emplacement of the HPPF pegmatites (Breiter and Siebel 1995). Its age of 297 ± 2 Ma is younger than those of the HPPF pegmatites and aplites (Hagendorf-South Pegmatite/muscovite: 317 ± 3 Ma and columbite 299.6 ± 1.9 Ma,

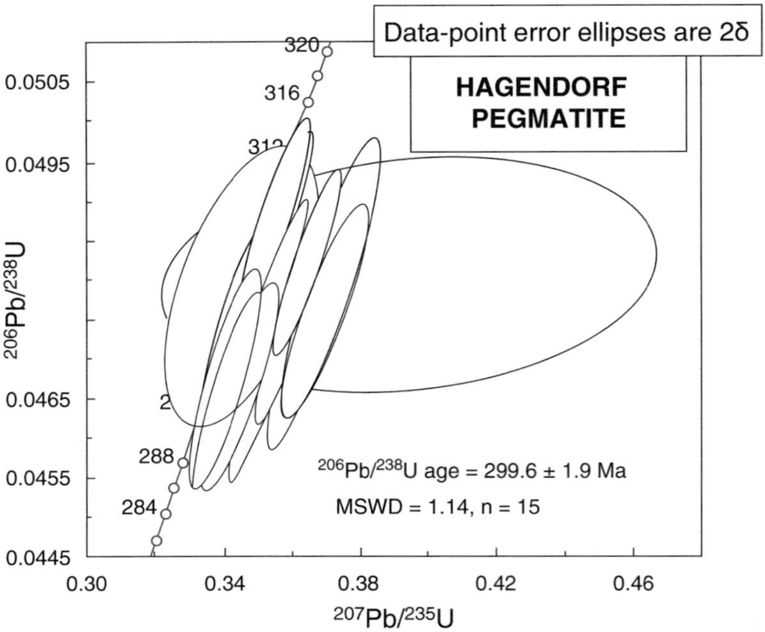

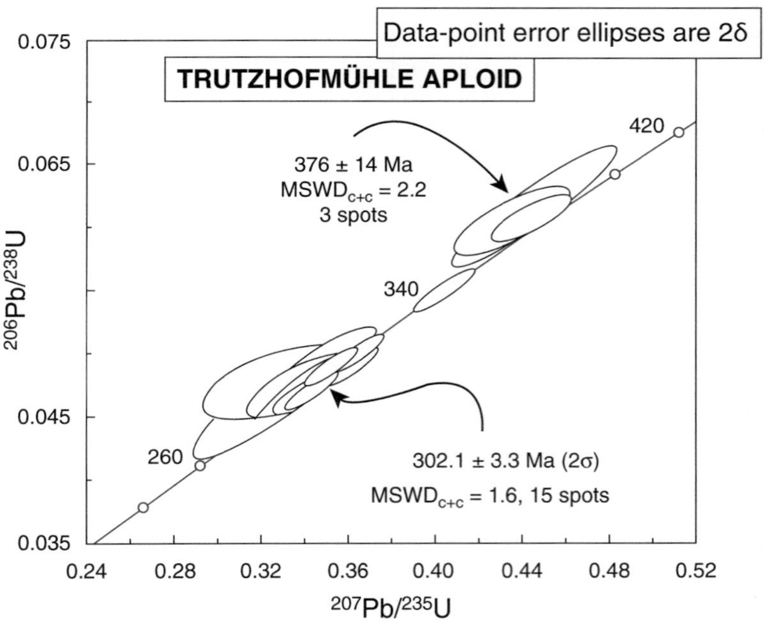

Fig. 3.1c Age dating of hypogene minerals from the Hagendorf Pegmatite Province (Dill et al. 2008b) *Top*: Ferrocolumbite from Hagendorf *Bottom*: Ferrocolumbite from Trutzhofmühle

Table 3.2 Age dating of ferrocolumbite from Hagendorf and the Trutzhofmühle (Dill et al. 2008b)

														Age (Ma)					
	Spot (μm)	^{207}Pb[a] (cps)	U[b] (ppm)	Pb[b] (ppm)	$\frac{^{206}Pb}{^{204}Pb}$	$\frac{Th}{U}$[b]	$\frac{^{206}Pb}{^{238}U}$[c]	±2σ (%)	$\frac{^{207}Pb}{^{235}U}$[c]	±2σ (%)	Rho[e]	$\frac{^{207}Pb}{^{206}Pb}$[c]	±2σ (%)	$\frac{^{207}Pb}{^{235}U}$	±2σ	$\frac{^{206}Pb}{^{238}U}$	±2σ	$\frac{^{207}Pb}{^{206}Pb}$	±2σ
Hagendorf Pegmatite																			
Ch2	60	55691	1346	59	104500	0.002	0.04762	2.4	0.3536	2.6	0.95	0.05385	0.8	307	7	300	7	365	19
Ch3	60	67248	1355	62	102213	0.003	0.04825	2.0	0.3647	2.2	0.92	0.05482	0.9	316	6	304	6	405	20
Ch4	60	71629	1305	60	20171	0.002	0.04760	2.4	0.3613	2.6	0.92	0.05505	1.0	313	7	300	7	414	23
Ch5	60	67946	1275	57	6666	0.002	0.04683	2.4	0.3405	2.6	0.90	0.05275	1.1	298	7	295	7	318	26
Ch6	60	63771	1276	57	3449	0.003	0.04727	3.1	0.3561	3.3	0.95	0.05463	1.1	309	9	298	9	397	24
Ch7	60	50019	1186	51	88477	0.002	0.04658	2.2	0.3410	2.5	0.89	0.05310	1.1	298	6	293	6	333	25
Ch8	60	65200	1546	69	76851	0.003	0.04849	2.7	0.3531	2.8	0.94	0.05281	0.9	307	8	305	8	321	22
Ch9	60	48634	1094	47	5644	0.002	0.04664	2.3	0.3454	2.6	0.87	0.05372	1.3	301	7	294	7	359	30
Ch10	60	85010	1321	63	2759	0.005	0.04811	3.0	0.3529	3.1	0.96	0.05320	0.9	307	8	303	9	337	20
Ch11	30	7671	1358	60	16347	0.003	0.04750	3.0	0.3479	3.6	0.83	0.05265	2.0	303	10	299	9	314	46
Ch12	80	66504	1331	60	135899	0.003	0.04737	2.6	0.3653	2.9	0.92	0.05471	1.1	316	8	298	8	400	26
Ch13	80	152436	1536	99	3959	0.002	0.04793	2.6	0.3434	2.7	0.94	0.05196	0.9	300	7	302	8	284	21
Ch14	80	45275	1299	58	85934	0.003	0.04833	2.5	0.3732	2.9	0.86	0.05533	1.5	322	8	304	7	426	32
Ch15	80	45145	1297	57	87552	0.003	0.04760	2.4	0.3699	2.7	0.89	0.05590	1.2	320	7	300	7	448	27
Ch16	80	56473	1449	66	3867	0.003	0.04809	2.5	0.3945	15.0	0.17	0.05429	3.9	338	44	303	8	383	89
Trutzhofmühle (THM) Aploid																			
Pl2	60	190906	2569	123	1154	0.008	0.04834	4.7	0.3517	4.9	0.96	0.05277	1.4	306	13	304	14	319	32
Pl3	30	22564	1483	66	2652	0.005	0.04675	3.3	0.3406	3.8	0.87	0.05284	1.8	298	10	295	9	322	42
Pl4	20	24314	1476	77	323	0.006	0.04576	7.5	0.3232	8.6	0.87	0.05122	4.2	284	22	288	21	251	98
Pl5	20	9415	1345	59	1859	0.005	0.04719	4.4	0.3369	5.6	0.80	0.05178	3.4	295	14	297	13	276	77
Pl7	20	9005	1024	49	1009	0.005	0.04897	4.7	0.3506	5.5	0.85	0.05193	2.9	305	15	308	14	282	67
Pl8	20	2436	298	14	1473	0.002	0.04752	5.2	0.3283	8.6	0.60	0.05010	6.9	288	22	299	15	200	160

(continued)

Table 3.2 (continued)

	Spot (μm)	^{207}Pb[a] (cps)	U[b] (ppm)	Pb[b] (ppm)	$\frac{^{206}Pb}{^{204}Pb}$	$\frac{Th}{U}$[b]	$\frac{^{206}Pb}{^{238}U}$[c]	±2σ (%)	$\frac{^{207}Pb}{^{235}U}$[c]	±2σ (%)	Rho[e]	$\frac{^{207}Pb}{^{206}Pb}$[c]	±2σ (%)	Age (Ma) $\frac{^{207}Pb}{^{235}U}$	±2σ	$\frac{^{206}Pb}{^{238}U}$	±2σ	$\frac{^{207}Pb}{^{206}Pb}$	±2σ
Pl11	30	15591	1225	55	28264	0.004	0.04858	4.0	0.3571	4.2	0.94	0.05332	1.5	310	11	306	12	342	33
Pl12	30	46689	3418	151	1348	0.006	0.04690	2.8	0.3422	3.2	0.87	0.05292	1.6	299	8	295	8	325	36
Pl13	40	66019	1860	88	698	0.008	0.04742	4.8	0.3449	5.0	0.97	0.05275	1.1	301	13	299	14	318	26
Pl14	40	140222	5395	246	596	0.009	0.04666	3.9	0.3360	4.2	0.94	0.05222	1.4	294	11	294	11	295	33
Pl15	40	31006	1554	69	57894	0.005	0.04861	2.5	0.3534	2.8	0.90	0.05273	1.2	307	7	306	8	317	27
Pl16	40	48492	1831	86	1608	0.006	0.04953	3.0	0.3596	3.2	0.95	0.05266	1.0	312	9	312	9	314	23
Pl17	40	55772	2696	119	1746	0.008	0.04793	3.9	0.3490	4.0	0.97	0.05282	0.9	304	11	302	12	321	20
Pl18	60	209117	3514	158	888	0.009	0.04667	3.1	0.3414	3.6	0.87	0.05306	1.8	298	9	294	9	331	41
Pl20	60	54222	697	34	447	0.004	0.04885	2.5	0.3588	3.7	0.67	0.05327	2.7	311	10	307	7	340	62
Pl1	60	150848	2063	110	955	0.006	0.05473	2.5	0.4058	2.8	0.90	0.05378	1.2	346	8	343	8	362	27
Pl9	20	13627	1009	63	659	0.008	0.06020	3.8	0.4390	5.3	0.71	0.05289	3.7	370	16	377	14	324	84
Pl10	20	25454	1943	121	556	0.010	0.06144	6.5	0.4555	7.1	0.92	0.05378	2.8	381	23	384	24	362	62
Pl19	60	74299	751	47	455	0.003	0.06055	2.6	0.4479	3.3	0.78	0.05365	2.1	376	10	379	9	356	47
Reference zircon[f]																			
Plesov. n=	60	28528	551	28	19017	0.11	0.05401	1.1	0.3962	1.2	0.72	0.0532	0.7	339	4	339	3	337	16
GJ1, n=7	60	27266	296	26	57443	0.032	0.09764	1.5	0.8070	2.0	0.00	0.05994	0.8	601	9	601	9	601	17

[a]Within run background-corrected mean ^{207}Pb signal in counts per second
[b]U and Pb content and Th/U ratio were calculated relative to GJ-1 reference (LA-ICP-MS values, Gerdes, unpublished)
[c]Corrected for background, within-run Pb/U fractionation and common Pb using Stacy and Kramers (1975) model Pb composition and subsequently normalised to GJ-1 and Plesovice reference (ID-TIMS value/measured value); $^{207}Pb/^{235}U$ calculated using $^{207}Pb/^{206}Pb/(^{238}U/^{206}Pb*1/137.88)$
[d]Mass bias corrected by normalising to GJ-1 (c. 0.5 % per amu)
[e]Rho is the error correlation defined as $err^{206}Pb/^{238}U/err^{207}Pb/^{235}U$
[f]Mean and standard deviation (2σ) of reference zircon GJ1 and Plesovice

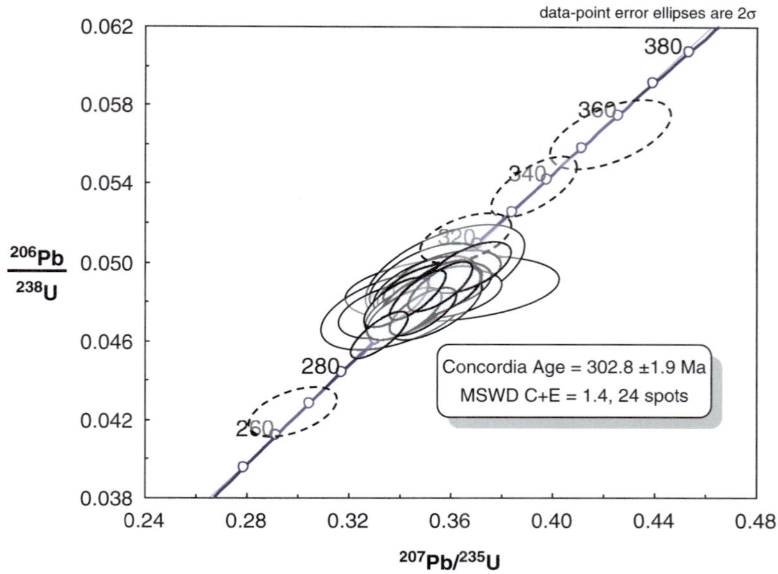

Fig. 3.1d LA-ICP-MS U/Pb age dating of ferrocolumbite from Silbergrube Aplite (Dill et al. 2009a)

Trutzhofmühle Aplite/columbite: 302.1 ± 3.3 Ma and 376 ± 14 Ma, Silbergrube Aplite/columbite: 302.8 ± 1.9 and forces us to discard this hypothesis of a derivation of the pegmatites and aplites in the HPPF from this granite. Younger cooling ages of 305.3 ± 3.4 Ma well agree with the U/Pb ages of columbite and may be interpreted as a thermal adjustment throughout the formation of columbite-(Fe). In conclusion, the temporal relation between granites and pegmatites as well as aplites rule out any simple derivation of the pegmatites and aplites from one of the granites exposed in the northern Oberpfälzer Wald. The bottom line is that: The granites and the pegmatites as well as the aplites belong to a thermal event at the end of the Variscan orogeny and that granites and pegmatites are two different sides of the same coin.

Heading further north, granites are still widespread in the Fichtelgebirge-Erzgebirge Anticline. They were comprehensively investigated particularly with regard to their age of formation and subdivided into two different phases, on older one around 319 ± 3 Ma and a younger active between 301 ± 8 Ma and 290 ± 4 Ma (Table 3.1). No large pegmatitic or aplitic bodies like in the southern area are exposed here. It is a miarolitic type of granite pegmatites closely connected in time and space with the granites, so that the ages reported for the various granites are also applicable to the pegmatitic and aplitic mineral assemblages within or in the wall rocks of the granites.

The story of the chronology of pegmatites and granites ends in the Münchberg Gneiss Complex where Na-enriched pegmatoids exceeded by far their counterpart from the Zone of Erbendorf-Vohenstrauß not only by the number but also in size, so that they could be targeted upon by opencast and underground mining operations

Table 3.3 Age dating of ferrocolumbite from Silbergrube near Waidhaus (Dill et al. 2009a)

Number	$^{207}Pb^{a}$ (cps)	U^{b} (ppm)	Pb^{b} (ppm)	$\frac{Th}{U}^{b}$	$\frac{^{206}Pb}{^{204}Pb}$	$\frac{^{206}Pb}{^{238}U}^{c}$	2 σ %	$\frac{^{207}Pb}{^{235}U}^{c}$	2 σ %	$\frac{^{207}Pb}{^{206}Pb}^{c}$	2 σ %	rho	$\frac{^{206}Pb}{^{238}U}$	2 σ (Ma)	$\frac{^{207}Pb}{^{235}U}$	2 σ (Ma)	$\frac{^{207}Pb}{^{206}Pb}$	2 σ (Ma)
A1	1596	28	1.3	0.002	886	0.04946	4.0	0.3581	7.1	0.05251	5.9	0.56	311	12	311	19	308	134
A2	2835	56	2.6	0.001	4523	0.04854	2.5	0.3535	3.7	0.05282	2.7	0.69	306	8	307	10	321	61
A4	1563	20	0.9	0.004	1409	0.04913	3.0	0.3503	5.1	0.05171	4.1	0.58	309	9	305	13	273	94
A5	4200	44	2.1	0.001	5610	0.04851	2.3	0.3558	3.9	0.05320	3.2	0.58	305	7	309	11	337	72
A6	5161	45	2.2	0.014	1156	0.04862	2.8	0.3609	9.5	0.05384	9.1	0.30	306	8	313	26	364	204
A8	2853	39	2.0	0.002	3066	0.05379	2.4	0.3935	3.6	0.05305	2.7	0.66	338	8	337	10	331	61
A10	1905	30	1.6	0.001	3518	0.05645	2.6	0.4230	4.6	0.05434	3.8	0.56	354	9	358	14	385	85
A14	3070	75	3.3	0.001	5839	0.04807	2.3	0.3497	3.9	0.05276	3.2	0.59	303	7	304	10	318	72
A15	3576	85	3.8	0.001	6269	0.04842	2.2	0.3553	3.8	0.05322	3.1	0.59	305	7	309	10	338	70
A16	6434	84	4.2	0.001	3136	0.04749	2.5	0.3388	5.4	0.05174	4.8	0.46	299	7	296	14	274	109
A17	3446	75	3.4	0.001	1309	0.04748	3.4	0.3383	7.0	0.05167	6.1	0.49	299	10	296	18	271	141
A19	2184	71	3.1	0.001	4061	0.04760	2.8	0.3544	4.3	0.05400	3.2	0.65	300	8	308	11	371	73
A20	1898	68	3.0	0.001	3621	0.04853	2.3	0.3528	4.9	0.05272	4.3	0.47	305	7	307	13	317	99
A21	1212	47	2.1	0.002	2321	0.04809	3.2	0.3474	5.0	0.05239	3.8	0.64	303	9	303	13	303	87
B1	3881	48	2.3	0.001	5517	0.04868	2.2	0.3550	4.9	0.05288	4.4	0.44	306	6	308	13	324	99
B2	8408	61	3.1	0.003	6751	0.04753	2.4	0.3434	4.0	0.05240	3.3	0.58	299	7	300	11	303	75
B3	2873	33	1.5	0.001	1199	0.04692	2.3	0.3418	4.8	0.05284	4.2	0.47	296	7	299	13	322	96
B4	4474	44	1.7	0.001	7435	0.04234	2.4	0.2984	4.8	0.05111	4.1	0.50	267	6	265	11	246	95
0	8347	77	3.4	0.001	3316	0.04828	2.1	0.3474	2.6	0.05219	1.6	0.80	304	6	303	7	294	36
B6	7883	76	3.2	0.001	4259	0.04626	2.1	0.3321	2.8	0.05207	1.8	0.76	292	6	291	7	289	41
B7	4516	26	1.2	0.002	5360	0.04832	2.1	0.3591	4.9	0.05390	4.4	0.43	304	6	312	13	367	100

B8	6150	56	2.5	0.001	10719	0.04873	2.1	0.3564	2.9	0.05304	2.0	0.73	307	6	310	8	331	45
B9	4427	37	1.6	0.001	7398	0.04733	2.3	0.3423	3.5	0.05245	2.7	0.64	298	7	299	9	305	62
B10	6243	57	2.5	0.001	10380	0.04743	2.1	0.3493	3.1	0.05342	2.2	0.69	299	6	304	8	347	50
B11	5197	51	2.3	0.001	9562	0.04967	2.2	0.3681	3.6	0.05375	2.9	0.60	312	7	318	10	360	65
B12	3536	34	1.6	0.002	6372	0.05118	2.0	0.3664	4.0	0.05192	3.5	0.50	322	6	317	11	282	80

[a]Within run background-corrected mean ^{207}Pb signal

[b]U and Pb content and Th/U ratio were calculated relative to GJ-1 reference (LA-ICP-MS values, Gerdes, unpublished)

[c]Corrected for background, within-run Pb/U fractionation and common Pb using Stacy and Kramers (1975) model Pb composition and subsequently normalised to GJ-1 (ID-TIMS value/measured value); $^{207}Pb/^{235}U$ calculated using $^{207}Pb/^{206}Pb/(^{238}U/^{206}Pb \times 1/137.88)$

[d]Rho is the error correlation defined as $err^{206}Pb/^{238}U/err^{207}Pb/^{235}U$

[e]Degree of concordance = ($^{238}U/^{206}Pb$ age × 100/$^{207}Pb/^{206}Pb$ age)

(Bauberger 1957; Dill 1979) (Figs. 1.5a, 1.5b, and 1.5c). The pegmatoid at Oberkotzau was successfully investigated in the course of a K/Ar- and Ar/Ar dating project which lead to a muscovite age of 372.5–377.0 Ma, almost identical to the age range of 370–380 Ma reported for the eclogite pegmatoids (Kreuzer et al. 1993). They reflect retrograde metamorphic process (Fig. 3.1a).

3.1.3 Synopsis – Chronology of Metapegmatites, Pegmatoids and Pegmatites

Four of the five different types of pegmatitic rocks listed in the classification scheme of Table 1.1 are exposed along the western edge of the Bohemian Massif and were shown as to their temporal relationships to the granitic rocks in the various geodynamic units (Fig. 3.1a). The metapegmatites are unrelated to the metagranites or orthogneisses in the Zone of Erbendorf-Vohenstrauß (Fig. 3.1b). Probably, they are contemporaneous with a Caledonian regional metamorphism which is poorly preserved in the geodynamic units under study. It does not come as a surprise that neither the orthogneisses nor the metapegmatites are temporally related to the geodynamic evolution in the Central European Variscides whose central zone is on display in the transect of Fig. 3.1a. Because of their allochthonous position, as a consequence of the Variscan nappe tectonics, they tend to get older towards the NW and go against the common trend in the Variscan geodynamic units.

Pegmatoids behave like their predecessors, the metapegmatites, and show neither a relation in time nor in space to meta-granites or orthogneisses. Their position within the succession of Fig. 3.1a is only controlled by the rock-forming processes during the MP-and HP metamorphism. They are almost contemporaneous with the allochthonous klippen of the Münchberg Gneiss Complex and the Zone of Erbendorf-Vohenstrauß (see Sects. 2.1.4.2 and 2.1.5.1).

According to our expectations based on the data collected in section 2, granites, pegmatites *sensu stricto* and granitic pegmatites get younger along the transect from the core zone of the ensialic orogen towards its margin in the NW. Contrary to the image quite popular with some mineralogists, who see the granite as the "father" of the pegmatites, their kinsman like relation to the granites is more complicated than the afore-mentioned general trend (Fig. 3.1a). The temporal gap between the age of formation of the large granites batholiths and the pegmatites is rather wide in the SE and no longer exists in the NW, where the term "granitic pegmatites" is filled with contents, although only in a few places.

In the Central Bohemian Massif the age of pegmatites is significantly younger than that of the large granites. Tectonic events were operative within that interval and can bridge this gap between granites and pegmatites. The temporal gap narrows down and deformational phases leading to the formation of the "Bayerischer Pfahl" shear zone system and to the deformation by dextral shear of the migmatites are located in the surroundings of the Fürstenstein Pluton (Galadí-Enríquez et al. 2010). The author's ideas are based on the subsolidus Variscan evolution of several small granitic bodies from the southern Bayerischer Wald. They constrain the age of this deformation to 315 Ma, while one of the reference granitic rocks, the Saunstein dyke was emplaced pre-kinematically at 324.4±0.8 Ma. The pegmatite-granite relation even reverses in the Oberpfälzer Wald, where some granites, exemplified by the small granitic intrusion north of the HPPP, the Křížový kámen Granite, is younger than the pegmatites and lamprophyres. Contrary to the common view of an undeformed dyke, these lamprophyres have been folded and faulted (Fig. 3.1a). Based upon the geochronological data, young granites, pegmatites and aplites meet at equals. In the Fichtelgebirge and in the northern Oberpfälzer Wald seven granite massifs were investigated for their $^{87}Sr/^{86}Sr$ initial ratios by Wendt et al. (1986) and displayed in an x-y plot (Fig. 3.1e). All granites, straddling the contact between the Saxo-Thuringian and the Moldanubian geodynamic zones were interpreted by the authors as having resulted from a fractionation of one parental magma leading eventually to the highly-specialized granite No 4, called "Tin Granite". The Hühnerkobel Granite stands out from the overall granites in this cross plot. It is not only host of the well-known pegmatite but also has the lowest initial strontium ratio.

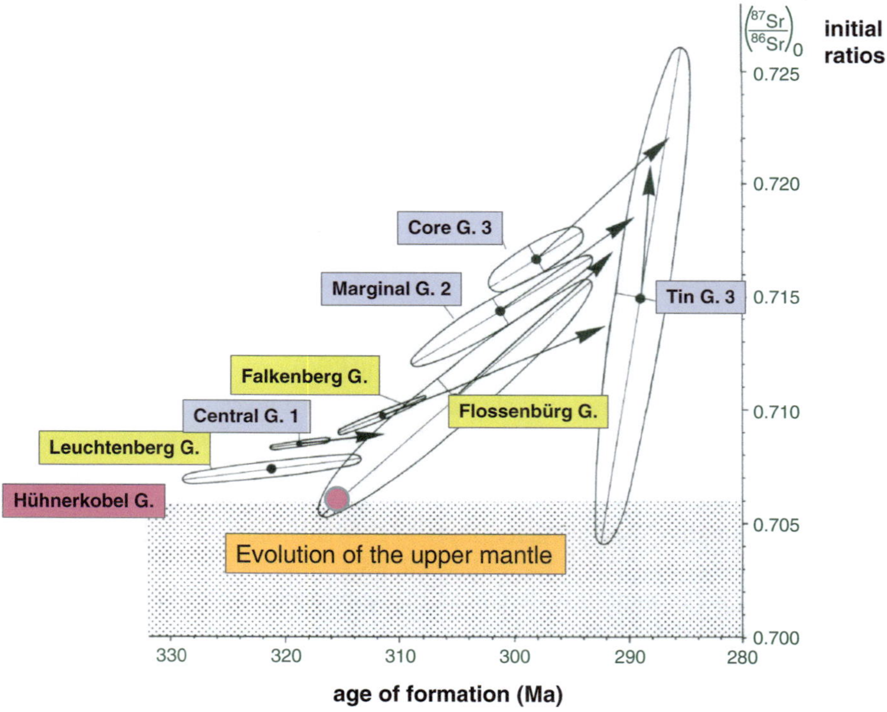

Fig. 3.1e Diagram to show the evolution of the Fichtelgebirge and Oberpfalz granites by plotting the $^{87}Sr/^{86}Sr$ ratios versus the age of formation (Wendt et al. 1986). The x-y plot also shows the error by an ellipse for each granite and denotes by the *arrowheads* the direction of the $^{87}Rb/^{86}Sr$ ratios. For comparison, the ratios of the Hühnerkobel Granite were shown in the diagram, which plot near the upper mantle evolution of the Sr isotopes

The geological and geochronological relation between granites and pegmatites was taken to the extreme in the Fichtelgebirge-Erzgebirge Anticline. The rare-metal bearing granites take over the role of the rare-metal pegmatites and aplites further south and became the target of mining operations for Sn, W and U. These granites outnumber the pegmatites, justified to be called granitic pegmatites in this region. The entire evolution from pegmatites *sensu stricto* to rare metal-bearing granites plus some affiliated granitic pegmatites is a function of the geodynamic evolution of the Central European Variscides, closely related to a window between 290 and 320 Ma when the youngest deformational processes were intertonguing with magmatic processes. Pegmatites, granitic pegmatites and granitites resulted from lithogenic processes which are referred to as late syn-to postkinematic.

3.2 Geochemical Survey of Rare Elements in Magmatic and Metamorphic Rocks of the NE Bavarian Basement

3.2.1 Lithium

Lithium is the only rare element that attracted the interest of mining engineers in the area under study and was extracted from the Hagendorf-South pegmatite in addition to the common raw materials for ceramic purposes. Moreover lithium plays a prominent part as a marker for the integration of the pegmatites into the geodynamic evolution and regional economic geology along the western margin of the Bohemian Massif. Lithium was also one of the elements on the list of elements during the geochemical survey conducted by the Federal Institute for Geosciences and Natural Resources (Fauth et al. 1985) (Fig. 3.2a). Even if the sample spacing was very wide for the Oberpfälzer Wald, the area where the HPPP is located, does not attract the attention by abnormally high values. Compared with the adjacent Fichtelgebirge-Steinwald area and the Bayerischer Böhmer Wald, the area stands out in the chemical data array as a relative low from the lithochemical point of view. There is a moderate high in the Fichtelgebirge-Steinwald area and in the south, where, in places, the maximum Li values ever measured in stream sediments were achieved. The southern part, which is synonymous with the Böhmer Wald and the Southern Oberpfälzer Wald-Vorderer Bayerischer Wald is not well endowed with Li pegmatites (Figs. 2.5a, 2.5b, 2.5c, 2.5d, and 2.5e).

The region in the south is underlain by metapsammopelitic rocks whose age of metamorphism is older than 340 Ma and which were intruded by granites of the older generation (Fig. 3.1a). Lithium and Al_2O_3 contents correlate positively for most sediments and the average Li contents in shales are known to be 49 ppm Li and in greywackes 31 ppm Li (Teng et al. 2004). According to Marschall et al. (2009) lithium can be used as a tracer for metasomatic processes in metamorphic rocks.

In anticipating any chemical assessment, the granitic rocks under study are abnormally high in lithium by international chemical standards of felsic igneous rocks, that lie in the range 34–43 ppm Li (Teng et al. 2004). Only some granodioritic

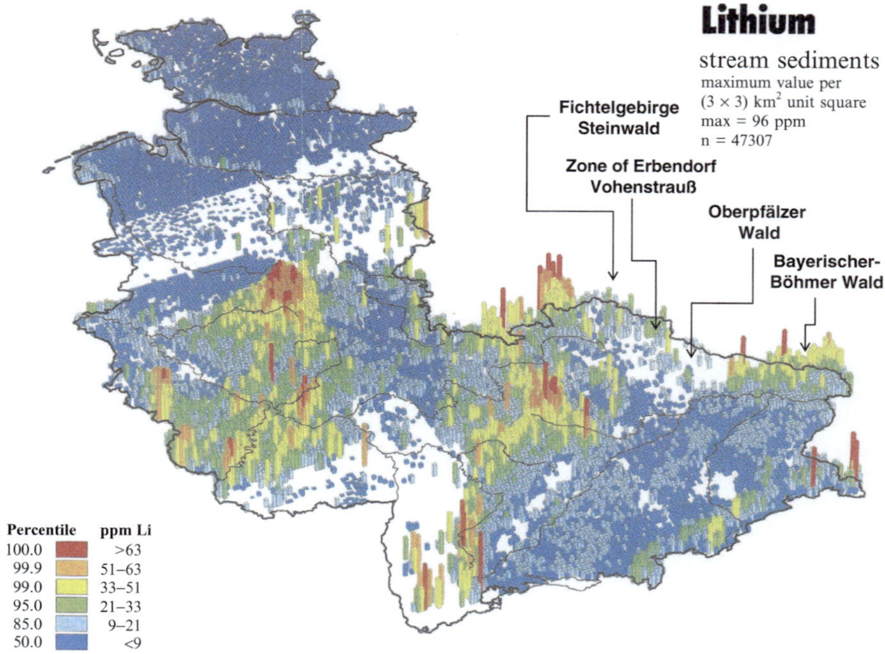

Fig. 3.2a Map of the Federal Republic of Germany representing the results of the official geochemical survey conducted by the Federal Institute for Geosciences and Natural Resources (Fauth et al. 1985). The map visualizes the lithium contents in view of histograms with the highest lithium contents given by the length of the columns and by a color code with *red* indicative of the maximum content

early intrusions and the granites in the southernmost part of the study area are moderate in lithium.

The coarse-grained granites from the granites west of Neunburg vorm Wald have mean values of 47 ppm Li, their finer-grained equivalents only 33 ppm Li. (Saarberg-Interplan open file report/CEA excursion 1982). Using the Rb/K ratio as a criterion for the different ion of the magma yields a coefficient of $R_{\text{Li-Rb/K}} = +0{,}77$ for the coarse-grained variety and $R_{\text{Li-Rb/K}} = +0{,}40$ for the fine-grained granites

The Leuchtenberg Granite, the oldest intrusion among the Oberpfalz granites has a mean Li content of 65 ppm, its younger neighbor the Falkenberg Granite of 105 ppm Li and the Flossenbürg Granite grades 159 ppm Li.

The higher lithium contents in the north of the study area result from an enrichment of this alkaline element in the course of granite differentiation granite G 1 through G 4 – G1/60 ppm Li ⇒ G 2/129 ppm Li ⇒ G 3/153 ppm Li ⇒ G 4/306 ppm (Richter and Stettner 1979) (Fig. 3.1a) (Table 2.1). Heading further east towards the Erzgebirge/Krušne Hory Mts. shows a moderate increase in the Nejdek Pluton/Eibenstock and in a similar way in the Slavkovský les/Kaiserwald (Breiter 1998a) (Table 3.4). The minimum contents there are roughly the same as those found near

Table 3.4 Lithological units at the western margin of the Bohemian Massif and their lithium contents (P_2O_5 in wt. %, Li in ppm Li)

Locality	Lithology	P_2O_5	Li	Reference
Dračí Skála near Pernink/Nejdek Pluton = Eibenstock 958 (OIC)	Porphyr medium-gr. Bt.granite	0.22	107	Breiter (1998a)
Hradecká/Nejdek Pluton = Eibenstock 973 (OIC)	Medium-gr. Bt.(+msc) granite	0.22	98	Breiter (1998a)
Rotava/Nejdek Pluton = Eibenstock 981 (OIC)	Fine-gr. msc granite	0.31	205	Breiter (1998a)
Bilá Skála/Nejdek Pluton = Eibenstock 961 (YIC/MG)	Porphyry fine-gr. Bt.(+msc) granite	0.20	191	Breiter (1998a)
Merklin/Nejdek Pluton = Eibenstock 955 (YIC)	Porphyr medium-gr. Bt. granite + topaz	0.21	350	Breiter (1998a)
Vidlice/Nejdek Pluton = Eibenstock 265 (YIC)	Medium-gr. Bt. granite + topaz	0.27	410	Breiter (1998a)
DDH Český Mlýn/Nejdek Pluton = Eibenstock 968 (YIC)	Coarse-gr. Bt.granite + topaz	0.40	438	Breiter (1998a)
Nové Hamry/Nejdek Pluton = Eibenstock 913 (YIC)	Porphyry coarse-gr. Bt. granite + topaz	0.32	368	Breiter (1998a)
Slavkovský les Loket 999 OIC	Porphyry coarse-gr. Bt.granite	0.31	65	Breiter (1998a)
Slavkovský les Kfely 1000 TG	Medium-gr. 2 mica granite	0.22	154	Breiter (1998a)
Slavkovský les Dolni- Žandov 1006 TG	Coarse-gr. 2 mica granite	0.16	61	Breiter (1998a)
Slavkovský les Třidomi 1001 TG	Porphyry fine-gr.granite	0.23	144	Breiter (1998a)
Slavkovský les Miliře 1002 YIC	Medium-coarse-gr. 2 mica granite	0.18	270	Breiter (1998a)
Slavkovský les Čistá 1003 YIC	Medium-gr. albite-Li-mica-granite + topaz	0.30	932	Breiter (1998a)
Slavkovský les Lysina 1004 YIC	Coarse-gr. albite-Li-mica-granite + topaz	0.29	932	Breiter (1998a)
Krušne Hory Hora Svaté Kateřiny 1390	Fine-medium-gr. albite-Li-mica-granite	0.01	652	Breiter (1998a)
Krušne Hory Preiselberg DDH E-6,242 m 2014	Biotite-granite	0.03	56	Breiter (1998a)
Krušne Hory Cinovec DDH E-6,492 m 1017	Albite-zinnwaldite-granite	0.01	979	Breiter (1998a)
Krušne Hory Cinovec DDH E-6,540 m 1019	Albite-zinnwaldite-granite	0.01	746	Breiter (1998a)
Krušne Hory Cinovec DDH E-6,708 m 2022	Albite-protholithionite-granite	0.02	284	Breiter (1998a)
Krušne Hory Markersbach	Biotite-granite	0.01	326	Förster et al. (1995)
DDH Zinnwald 1	Albite granite zinnwaldite (lepdidolite) fine-gr.	0.01	341	Seltmann et al. (1998)
DDH Zinnwald 2	Albite granite zinnwaldite (lepdidolite) fine-gr.	0.01	912	Seltmann et al. (1998)
DDH Zinnwald 3	Albite granite zinnwaldite medium-gr.	0.01	1,057	Seltmann et al. (1998)

(continued)

Table 3.4 (continued)

Locality	Lithology	P_2O_5	Li	Reference
DDH Zinnwald 4	Albite granite zinnwaldite medium-gr.	0.01	710	Seltmann et al. (1998)
DDH Zinnwald 5	Albite granite protolithionite medium-gr.	0.01	1,078	Seltmann et al. (1998)
Podlesi 2666	Granite stock	0.53	1,375	Breiter (1998b)
Podlesi 2670	Granite stock	0.64	1,230	Breiter (1998b)
Podlesi 2360	Dike	1.64	1,650	Breiter (1998b)
Podlesi 2669	Dike	0.99	2,130	Breiter (1998b)
Podlesi 2674	Dike	0.55	1,314	Breiter (1998b)
Podlesi 2361	Pegmatite	0.91	1,650	Breiter (1998b)
Podlesi 2650	Pegmatite	0.82	1,477	Breiter (1998b)
Podlesi 2661	Pegmatite	0.65	904	Breiter (1998b)

the top level of the Fichtelgebirge. Only the albite-Li-mica-granites with topaz are significantly enriched, a fact that is mineralogically proved by the presence of the Li ore mineral zinnwaldite. The ultimate Li level is reached in the easternmost parts near Zinnwald where medium-grained albite granite with zinnwaldite and lepidolite are exposed. The plunge of the axis of the Erzgebirge-Fichtelgebirge anticline towards the SW is accompanied by a decrease in the lithium contents determined in the granitic rocks. This regional trend goes unnoticed when it comes to the Li variation during the differentiation of each granitic suite along the Erzgebirge-Fichtelgebirge anticline. The fluorine- and phosphorus-enriched Podlesi stock granite is a small granite in the southern part of the Nejdek Pluton/Eibenstock pluton (Breiter et al. 1997, Breiter 1998b). In the Podlesi stock granite, these differentiation are certainly present, but does not lead to the expected results. It is not the highly differentiated pegmatite where the most elevated lithium contents have to be reported from but the dike rocks with as much as 2130 ppm Li (Table 3.4). The lithium contents in the various lithologies exposed in the border zone between the Saxo-Thuringian and the Moldanubian geodynamic units have been collected by Richter and Stettner (1986) and listed in a final activity report for the German Science Foundation. The highest Li contents of 75 ppm Li were reported from aplitic gneisses of the Zone of Erbendorf- Vohenstrauss. These metasedimentary rocks are host to the metapegmatites which were operated for feldspar in the Gertrude and Wilma mines where Li contents lie in the range 24–50 ppm Li. In order of decreasing average Li contents, the country rocks can be arranged as follows: redwitzites (granodioritic to gabbroic magmatic rocks): 59 ppm Li, aplitic gneisses: 46 ppm Li, metaultrabasic magmatic rocks: 29 ppm Li, metagabbro-amphibolite-eclogite-chlorite schists: 20 ppm Li, chert: 16 ppm Li.

In conclusion, with this chemical data set in Figs. 3.2a and 3.2b, we can track down the accumulation of lithium in this ensialic orogen in Central Europe to Precambrian and Early Paleozoic rocks. At the beginning of the tripartite pathway lies the preconcentration of this alkaline element in the basement of the Moldanubian zone. Particular attention for the preconcentration of lithium is drawn to the abundant metapelites at outcrop in the Bayerischer-Böhmer Wald as paragneisses in the "Monotonous" and "Varied Group". While these metalliferous rocks or low-metal concentrations of lithium are exhumed in the south (Fig. 3.2a) so as to be targeted upon by geochemists during the chemical survey, they are covered at depth, absorbed to completeness and recycled in the north, at the boundary between the Münchberg Gneiss Complex and the Fichtelgebirge-Steinwald area, where the active margin of the ensialic orogen is located (Fig. 2.1c). For the description of the process operative along this active margin the reader is recommended to consult Fig. 2.2a.

Along the branch No 1 on lithium's pathway of crustal recycling we even touch the uppermost parts of the mantle, as deduced from the evolution of the initial strontium isotope ratios which stress a continuous process of the felsic magmas of the granitoids when passing from the Saxo-Thuringian into the Moldanubian zones (Fig. 3.1e). On the anatectic conversion of crustal material into felsic magmas at depth there are sporadic magmatic processes of subcrustal origin, e.g., the formation of the redwitzites and the remobilization of metapegmatites at Püllersreuth, Germany and in the Domažlice Crystalline Complex, Czech Republic along the boundary between the autochthonous and allochthonous units of the Moldanubian region, in the Zone of Erbendorf-Vohenstrauß and the corresponding Tepla Barrandian zone (Fig. 3.2b). These subcrustal basic rocks attest to a subcrustal heat source to be involved in the mobilization of the this element from the metasediments. Bierlein et al. (2009) speculate on a large scale mantle-crust interaction in the lower crust.

The pegmatoids emplaced during the waning stages of northward-thrusting of the nappes had no impact on the accumulation of lithium (Fig. 3.1a). From the lower crust to the upper crust the lithium contents got gradually upgraded reaching its climax in the large highly differentiated granitic batholiths along the Fichtelgebirge-Erzgebirge Anticline. It cannot be described how the release of Li took place at the brink of metamorphism to anatexis in this region for lack of exposures. It would be a matter of conjecture to speculate on this.

Cuney and Barbey (2014) made the granulite facies metamorphism in the lower crust accountable for the release of large amounts of incompatible elements such as Li and Rb, resulting from the breakdown of mica and amphiboles. These theories on the processes involved are mentioned here as it is the case with the mantle-crust interaction of Bierlein et al. (2009) for the sake of completeness only, but left undiscussed as it would shift us too much into the territory of speculation in the area under study and does not significantly add to the built-up of an exploration model.

Lithium was concentrated intragranitically, in their apical parts, in cupolas and dikes, and to a lesser degree in granitic pegmatites *sensu stricto*, locally called "stockscheider". The granites did not spawn individual bodies of lithium pegmatites as it is the case in the "heartland" of the ensialic orogen. Pegmatites containing Li as exemplified by the HPPP are equivalent to the granites in time and chemical composition but not in space and structure. The issue of accommodation space for

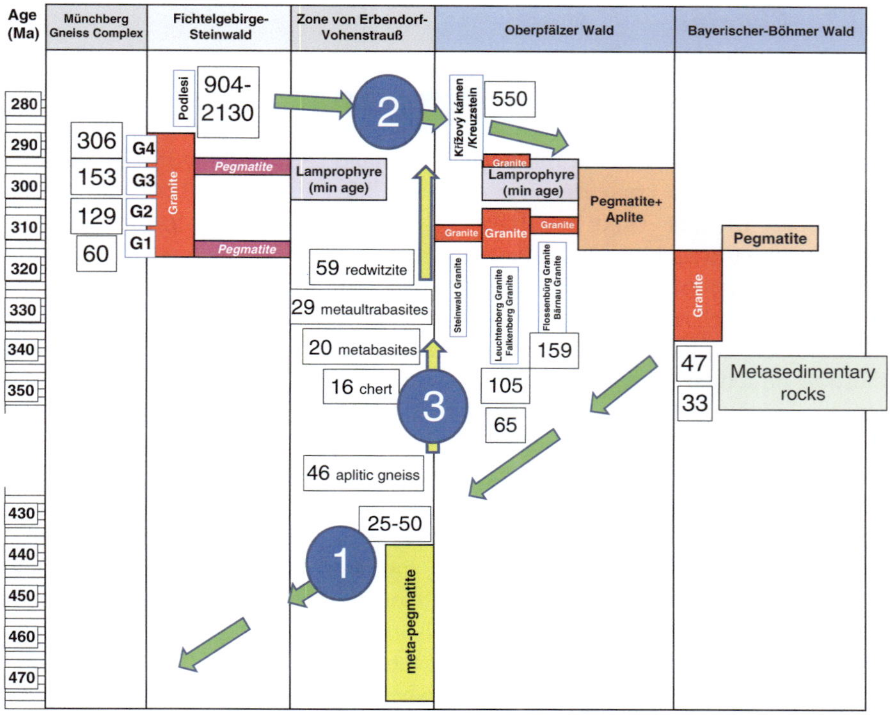

Fig. 3.2b The Li concentration in the Moldanubian autochthonous (*blue*), the allochthonous (*gray*) and Saxo-Thuringian zones (*white*). The mean Li contents are set in *boxes*. As a chronological base see also Fig. 3.1a (1) Remobilisation of Moldanubian rocks as a result of subluence (autochthonous units) (2) Concentration by lineamentary-bound deep-seated mobilization (autochthonous units) (3) Reactivation in the allochthonous units Li-bearing granites in the Saxo-Thuringian zone located near the subluence zone are chemically equivalent to Li pegmatites in the central parts of the ensialic orogen (HPPP)

the pegmatites cannot be tackled by looking at it from a chemical point of view, it is the principal subject of structural geology which can only explain size and shape of it (Sect. 5.1.4). However, the model on lithium recycling can provide an active contribution to the structural differences between granites plus granitic pegmatites on one side and pegmatites *sensu stricto* on the other side. Off the active geodynamic zone towards the core of the ensialic orogen the pathway enters zones of geodynamic tranquility, fostering the growth of larger crystals and the development of pegmatitic textures. The model of concentrating lithium which was elaborated for the northwestern edge of the Bohemian Massif experiences a repetition in the reverse order at its southeastern margin along the Moravian Thrust (Fig. 2.1c) (Novák et al. 1999; Dosbaba and Novák 2012). More than along the northwestern edge of the Bohemian Massif, the prominent role of structural geology is particular apparent along this active margin.

3.2.2 Fluorine

The map of the Federal Republic of Germany displaying the aerial distribution of fluoride in stream sediments resembles that chemical map visualizing the data of lithium which were gathered throughout the same chemical survey (Fauth et al. 1985). The fluorine contents are abnormally high in the southern part of the Bayerischer-Böhmer Wald and have a second maximum in the area of the Fichtelgebirge-Erzgebirge Anticline. Although being completely covered during the sampling campaign, the Oberpfälzer Wald and the HPPP only stand out with values of as much as 120 ppm F. Due to the widespread occurrence of paragneisses in this region, the large chemical anomalies of fluorine which are more striking than those of lithium are not unexpected. The second-order anomaly is caused by the granites of the Erzgebirge-Fichtelgebirge Anticline, where in addition to the widespread occurrence of micaceous phyllosilicates, topaz is responsible for the elevated level of fluorine in the strongly fractionated granites. The succession of granites in the Fichtelgebirge G 1 (602 ppm F), G 2 (1665 ppm F), G 3 (1243 ppm F) and G 4(2315 ppm F) do not show a similar textbook-like increase upon differentiation as reported for lithium. Nevertheless, the youngest members of this granite suite tend to be most strongly enriched in fluorine and contain the highest amount of topaz. The Nedjek Pluton which was also referred to on recording the lithium contents in this region has more elevated F contents of as much as 6400 ppm F in the porphyritic coarse-grained biotite granite where topaz is present (Breiter 1998a). From 1200 ppm F onward topaz is so common in the mineral assemblage to render this phase a key mineral for the classification of the Nejdek/Eibenstock Pluton. In the Podlesi granite the differentiation swept its way through the felsic complex in an ideal way (Breiter 1998b). The stock granite, parasitic to the Nejdek/Eibenstock Pluton, has a mean of 8000 ppm F, the dike granites attain a more advance level with 9900 ppm F and the granitic pegmatites reach the top level of 14700 ppm F. Not all granite complex at the western edge of the Bohemian Massif behave in a similar way, as exemplified by the Homolka and Josefsthal granitic system for which chemical data were recorded by Breiter (1998c). It is a muscovite granitic stock with no significant mineralization despite elevated F contents in the range 2400–4790 ppm F. Dyke granites have contents between 3880 and 5300 ppm F, whereas the pegmatite takes an intermediate position with 4270 ppm F, below those of the stock granite.

In the Slavkovský les, Czech Republic, south of the afore-mentioned site the F contents in the granites drops to 3800 ppm F, a sustained downward trend to be backed up by the contents in the fine-grained biotite granite near Rozvadov (300 ppm F) and Mraveniště Hill (1700 ppm F). Crossing into the HPPP the trend is upheld by the fine-grained leucocratic granite underneath the Kreuzberg Pegmatite in Pleystein (1600 ppm F) and the mean value of the Silbergrube aplite granite (2500 ppm F) (Müller et al. 1998). Neither site is enriched in fluorine minerals of their own. The medium-grained albite-zinnwaldite granite stock at Křížový Kámen is the exception from the rule with 9600 ppm F.

The fluorine data show a similar aerial distribution as lithium and hence are interpreted in a similar way as those of lithium. The elevated F values determined from the basement rocks of the Bayerischer and Böhmer Wald in the southern part of the study area supports the idea of a fluorine low-element concentration in the S which was later reworked in the course of the compressional stage of this ensialic orogen in the northwestern parts giving rise to the F concentration along the Erzgebirge-Fichtelgebirge anticline (Fig. 3.3). Granites and granitic pegmatites along the Erzgebirge-Fichtelgebirge Anticline are the only representatives of the late Variscan fluorine remobilisation and enrichment. There are offshoots along deep-seated lineaments like the F-enriched granite at Křížový Kámen "invading" the Moldanubian Zone. Yet they were of no effect on the mineralizations in the HPPP mainly for chronological reasons. The Silbergrube aplite granite is older than the Křížový Kámen granite and topaz has no meaning for the HPPP mineral assemblages. In geodynamic and geochemical terms, fluorine is an element enriched to anomalously high quantities close to the active margin of the ensialic orogen while its low-metal concentration or protore lies away from this zone towards the center of the Bohemian Massif. Considering the ensialic orogen in Central Europe,

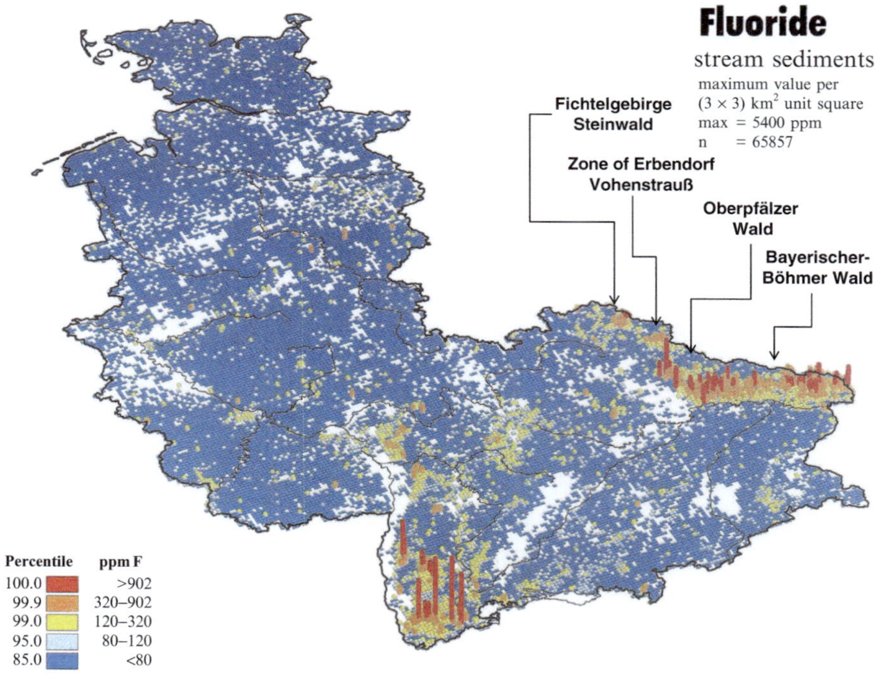

Fig. 3.3 Map of the Federal Republic of Germany representing the results of the official geochemical survey conducted by the Federal Institute for Geosciences and Natural Resources (Fauth et al. 1985). The map visualizes the fluoride contents in view of histograms with the highest fluoride contents given by the length of the columns and by a color code with *red* indicative of the maximum content

fluorine is concentrated in the frontal parts of the nappes and close to collision zone, whereas lithium saw its concentration at a more distal position closer to the root zone. A different approach has to be taken in areas, where the mantle impact is highlighted by the presence of mantle-derived felsic intrusive rocks.

3.2.3 Tin

Tin was one of the elements analyzed during the country-wide geochemical survey but its sampling campaign did not cover the former Federal Republic of Germany to completeness (Fauth et al. 1985). Figure 3.4 is a striking examples for a sustained downsizing trend of element remobilization during the waning stages of the Variscan orogeny. Lithium had an almost equal share in the formation of rare-element granites and granitic pegmatites in the Saxo-Thuringian as well as Moldanubian zones, whereas fluorine is more or less confined to the margin of the basement block with little effect on the HPPP located closer to the center. Tin which is a rare element in

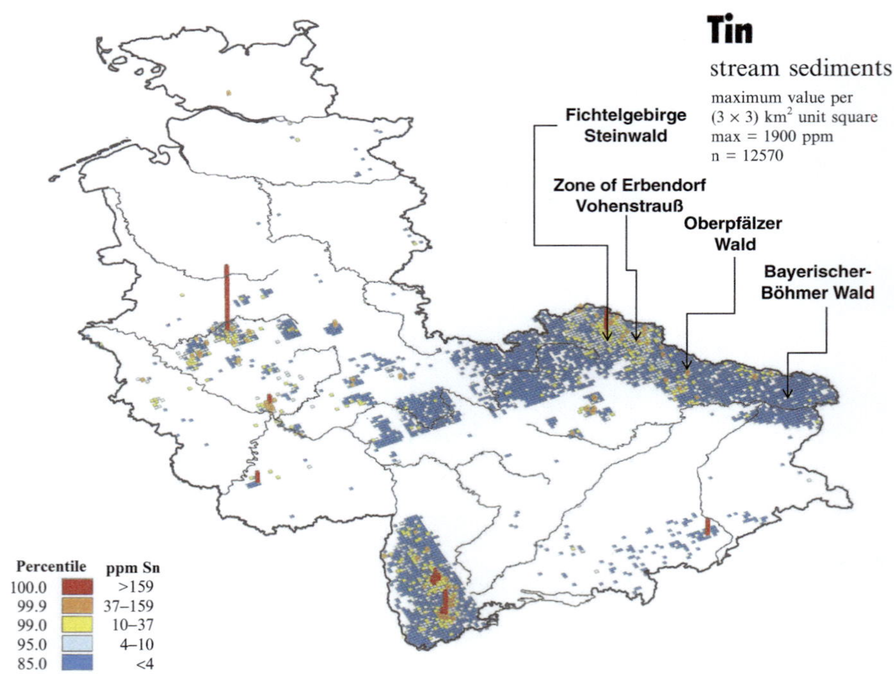

Fig. 3.4 Map of the Federal Republic of Germany representing the results of the official geochemical survey conducted by the Federal Institute for Geosciences and Natural Resources (Fauth et al. 1985). The map visualizes the tin contents in view of histograms with the highest lithium contents given by the length of the columns and by a color code with *red* indicative of the maximum content

the central parts of the ensialic orogen is the marker element of the Saxo-Thuringian zone, where is was exploited for centuries and currently again the target of exploration activities. Concentration of Sn is exclusive to the active margin zone where it is disseminated *inter alia* in greisen zones and stockscheiders in the apical parts of the granites but fails to form Sn-bearing granitic pegmatites of its own of economic significance similar to what we encounter further south in the Oberpfalz. The protore of Sn, if it exists at all, cannot be delineated by the chemical survey.

3.2.4 Uranium

Uranium is quite a common element in the NE Bavarian Basement and one of the most important elements in economic term, accountable for several uranium deposits along the western edge of the Bohemian Massif which mainly belong to the vein-type and vein-like mineralizations (Bültemann 1954; Ziehr 1957; Strunz 1962; Dill 1981, 1982, 1983a, b, 1986; Carl and Dill 1983, 1984, 1985; Dill and Kolb 1986; Dill et al. 2010a).

Structurally controlled epigenetic U black ore mineralizations (U oxides, U titanates, and U silicates) are found in the geotectonic units of the Saxo-Thuringian (Frankenwald, Fichtelgebirge) as well as in the Moldanubian of the Oberpfälzer and Bayerischer-Böhmer Wald of the Central European Variscides. The vein-type occurrences may be classified into three principal types of U deposits with respect to their structural properties: fault zones with stockwork-like ore shoots, mineralized structure zones and veins *sensu stricto*. The relationship between the host and the source rocks is best described by a twofold division. The element content of the Paleozoic vein-type deposits hosted in black shales is derived from the enclosing wall rocks (host rocks and source rocks are identical), whereas in the rocks of the crystalline basement the protore or the low metal concentration is assumed to have been metabiolites of the Upper Proterozoic "Varied Group". From which uranium, carbon and phosphorus were expelled during metamorphism and anatectic mobilization. Concentration and transport took place via convectively circulating fluids. The heat necessary for the initial phase of circulation is suggested to have been generated by the Late Variscan to Early Alpine igneous activity (granites, subvolcanoes and magmatic dikes) owing to the results obtained from radioactive age dating. Uranium mineralization is accompanied, in places, with intensive hydrothermal wall rock alteration, leading to episyenites-see also Fig. 2.4c. They are in time and space related to silicification and the emplacement of quartz veins. Furthermore argillitization, encompassing chloritization, sericitization, smectitisation and kaolinization favored the thick fault gouches along the mineralized structures zones. Carbonaceous matter, present as epi-, meso- and kataimpsonite (metamorphosed bitumen) has derived from the metabiolites of the Proterozoic "Varied Group" as may be concluded from trace elements and from isotope data. Two different characteristic uranium parageneses were determined within ore-bearing structure zones. The so-called "monotonous uranium paragenesis" is characterized by its simplicity of element composition, including U, Si, Fe and Ti, but variable bonding of uranium in brannerite, "uranium leucoxene", neouraninite, pitchblende, "sooty pitchblende", coffinite and U-Ti silicates. Contrary, the "polymetallic uranium paragenesis" displays a great variety of elements such as U, As, Au, Bi, Co, Ni, Sb, Se, Cu, Fe, Pb, and Zn which are responsible for different sulfides, selenides, arsenides and native elements. Uranium is present in this mineral association only as pitchblende, its highly oxidized decomposition products, and to a lesser extent as coffinite. Selenides are considered to be diagnostic for the "polymetallic uranium paragenesis".

The pegmatite-hosted uranium mineralization in NE Bavaria has not been mentioned in the previous paragraph about uranium deposits, as it will be discussed in the section on the mineralogy of pegmatites and was never the focus of any uranium exploration compaign until the Germans began declaring this radioactive element as an "economic and chemical no-go", due to the limited size of uranium mineral assemblages in the NE Bavarian pegmatites. There are several pegmatitic granites of the Fichtelgebirge, e.g., Epprechtstein, Waldstein, Gregnitzgrund and Brand, hosting U minerals. In the Oberpfalz, uranium is known from the pegmatites at Sägmühl, Beidl, Schönthan, Wildenau-Plößberg, Pleystein and Hagendorf. Even the rather sterile metapegmatites near Klobenreuth contain uranium minerals. In the southeastern part of the study area, the pegmatites at Katzberg near Cham, Hühnerkobel, Oberfrauenau, Frath and Blötz, both of which are located near Bodenmais, gave host to a minor U mineralization.

The geochemical survey in Fig. 3.5 shows a rather homogeneous U distribution in the range 2.8–9 ppm U. Out of this greenish yellow background colors coverin vast areas along the Czech-German border, some prominent red columns stand out well from the background and evidence U anomalies accounted for by the vein-type and vein-like deposits briefly described in the previous paragraph. Only the Pleystein and the Hagendorf areas of the HPPP form two prominent anomalies with values in the range 9–22 ppm U but they are not match to the abnormally high U contents in and around the Falkenberg and Flossenbürg Granites. The uranium deposits Höhenstein and Mähring close to the Czech-German border cannot be overlooked in the geochemical survey. Upon an assessment of the effectiveness of the different elements Li, Sn, F, and U for the localization of these pegmatites, it is U that can most effectively applied during a geochemical survey to localize these pegmatite bodies, despite their subordinate U mineral assemblages.

3.2.5 Barium-Rubidium-Zirconium

Barium is a rare element in the Trutzhofmühle Aploid, where it is accountable for a hitherto unnamed K–Ba–Sc–Zr phosphate and also known from the Kreuzberg Pegmatite, the Miesbrunn Pegmatite Swarm and the Plößberg Pegmatite (Dill et al. 2008c, 2011a, c, 2012a). Hagendorf-South has also among its wide range of phosphates a Ba-bearing member, named perloffite [$Ba(Mn,Fe)_2Fe_2(PO_4)_3(OH)_3$] (Mücke 1987; Kastning and Schlüter 1994). All of these mineral shows are located

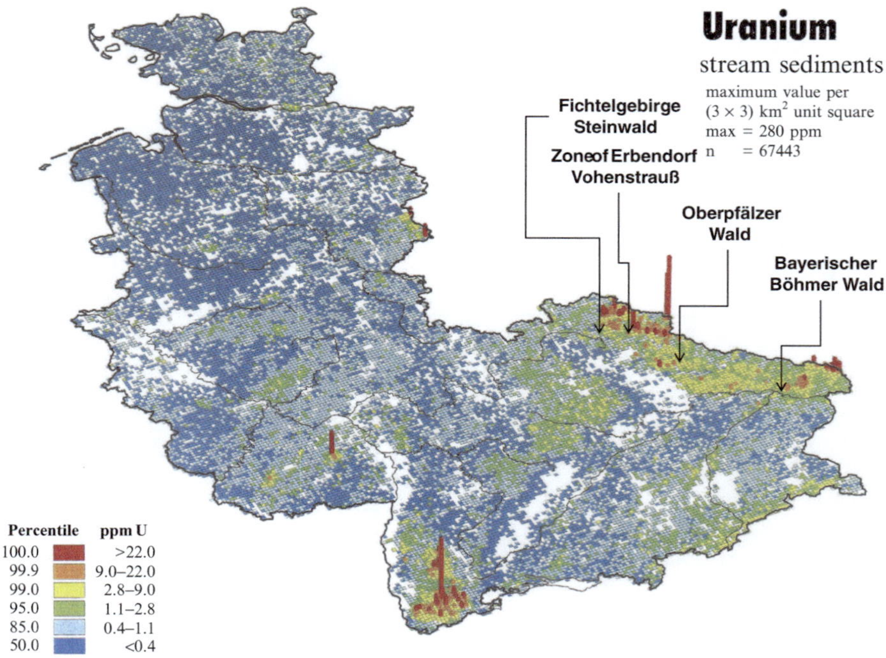

Fig. 3.5 Map of the Federal Republic of Germany representing the results of the official geochemical survey conducted by the Federal Institute for Geosciences and Natural Resources (Fauth et al. 1985). The map visualizes the uranium contents in view of histograms with the highest uranium contents given by the length of the columns and by a color code with *red* indicative of the maximum content

along the western and southern perimeter of the Flossenbürg Granite which is the largest granite close to them (Fig. 3.6a). The Ba distribution in the Flossenbürg and Bärnau granites was reported by Wendt et al. (1986) who demonstrated in their colored contour plot a NE-SW trend, with the maximum of Ba along the southwestern edge of the Flossenbürg Granite (Fig. 3.6a). The neighboring Bärnau Granite, straddling the Czech-German border has only been covered on the German side of the border by the geochemical survey so as to make any far-reaching and reliable predictions as to the Ba variation fraught with difficulties. Similar to its western neighbor a SW-NE trend from high Ba contents to low Ba contents may be deduced from the map presented by Wendt et al. (1986). Using the granites G 1 through G 4 in the Fichtelgebirge as a model type for granitic fractionation in this region, provides us with a clear decreasing-upward trend for Ba: G 1: 930 ppm Ba, G 2: 283 ppm Ba, G 3: 129 ppm Ba, G 4: 13 ppm Ba (Richter and Stettner 1979). The granite sequence of the Fichtelgebirge is going to be depleted in barium along different ion. Applying these results obtained from the Fichtelgebirge granites to the Flossenbürg and Bärnau granites implies a trend of differentiation in SW-NE direction in the Flossenbürg Granite, and in the Bärnau Granite too. If both granites

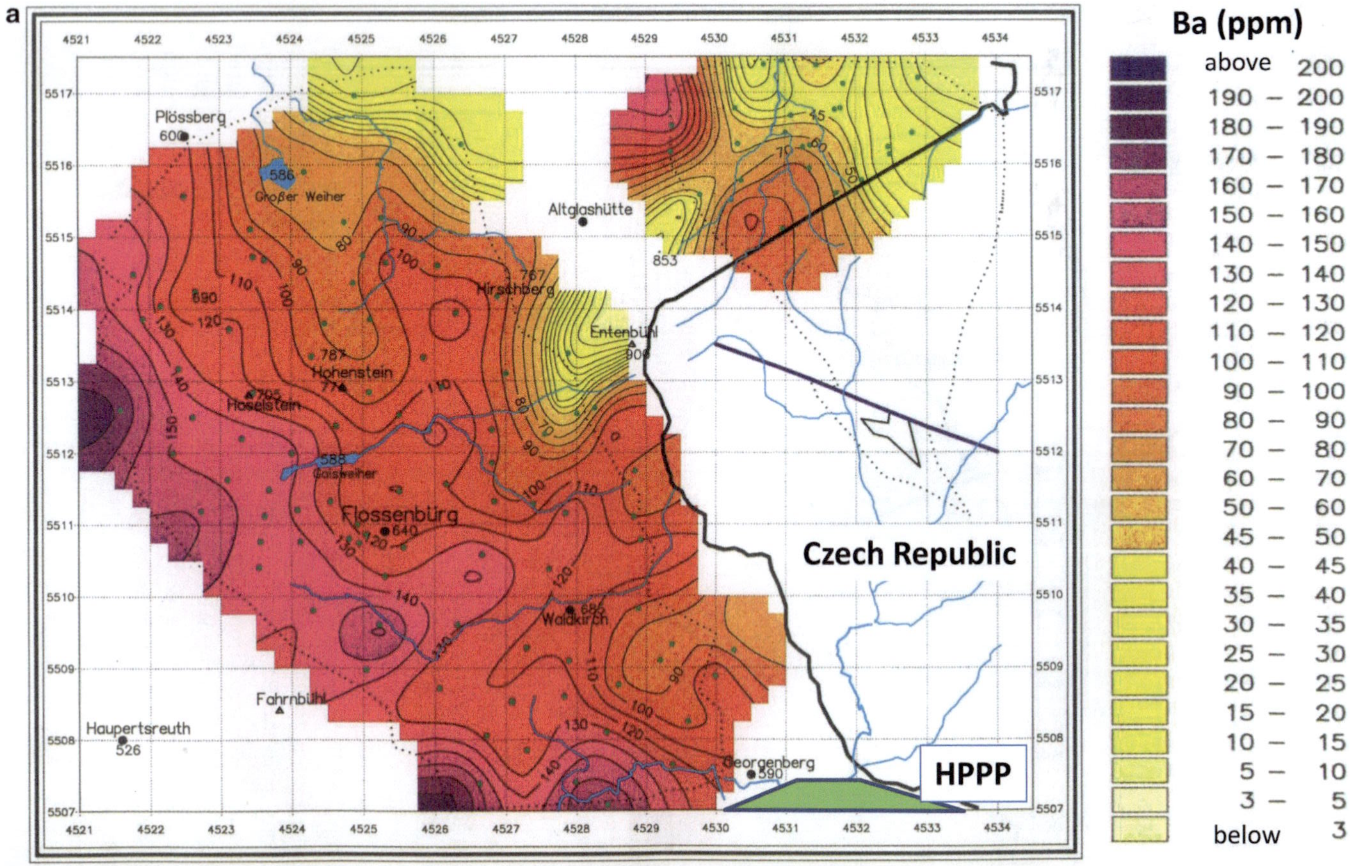

Fig. 3.6a Barium distribution in the Flossenbürg (west) and Bärnau Granites (east). For reasons of clarity different colors are used for the graphical representation of contour lines. The color key on the right-hand side denotes the content of Ba in ppm. *Small dots* mark the sampling sites in both granites, modified from Wendt et al. (1986). The position of the HPPP is shown *down right*

formed part of homogeneous granite complex with the apical part in the center, the trend in the Bärnau Granite should have been reversed. That is not the case, with far-reaching consequences for the origin of the HPPP. The conclusion is evident, that there is no south-eastward differentiation trend and the HPPP pegmatites and aplites cannot be the most strongly differentiated apical system of the Flossenbürg Granite simply for chemical reasons. Comparing the degree of fractionation, the Flossenbürg granite system is less strongly fractionated (difference 180 ppm Ba) than the system G 2 to G4 (difference 270 ppm Ba).

Considering the previous trends in Fig. 3.6a), the trend in Fig. 3.6b) is simply the mirror image. Rubidium underlines in a more pronounced way the SW-NE trend of different ion and marks conspicuously the sharp contact between the eastern Bärnau and the western Flossenbürg Granites (Fig. 3.6b). There is no hint as to a affiliation of the HPPP to the both granite either. Using again the northern granites for reference leads to the following sequence: G 1: 211 ppm Rb, G 2: 404 ppm Rb, G 3: 427 ppm Rb, G 4: 805 ppm Rb (Richter and Stettner 1979). The degree of differentiation in the Fichtelgebirge (G 2 to G 4) (difference 401 ppm Rb) is much higher than in the Oberpfalz (difference 180 ppm Rb). It provides further evidence that a derivation of the HPPP pegmatites and aplites from the northern granites is highly unlikely, not to say to be excluded. Rubidium is contained in the different types of mica in the HPPP pegmatites and aplites, mainly in the Li-enriched varieties as know from elsewhere (Knanna 1977). The behavior of rubidium in pegmatites was studied in detail by Černý (1982).

The distribution of zirconium in the four granites shows the same trend know from barium: G 1: 267 ppm Zr, G 2: 112 ppm Zr, G 3: 80 ppm Zr, G 4: 25 ppm Zr (Richter and Stettner 1979). The degree of differentiation in the Fichtelgebirge (G 2 to G 4) (difference 87 ppm Zr) is only moderately higher than in the Oberpfalz (difference 80 ppm Zr) and the distinction not so pronounced as for the elements discussed previously. Zircon is present in the major pegmatites Hagendorf North and South, Kreuzberg and the Miesbrunn pegmatite-aplite dike swarm, yet zirconium only plays a minor part as rock-forming mineral in the HPPP. It has to be noted that in the Trutzhofmühle Aploid, where zircon is absent, an unnamed K–Ba–Sc–Zr phosphate and an unnamed Zr–Sc phosphate–silicate developed instead (Dill et al. 2008a, b, c).

Of potential importance for the HPPP, the sharp contrast between Bärnau and Falkenberg Granites in the chemical distribution pattern stresses the presence of a NW- to NNW- deep-seated structure zone. It disconnected chemical trends in both granites. The HPPP resides at the south-eastern most tip of this predicted structure zone. A structural and magmatic event post-dating the emplacement of Flossenbürg Granite is supposed to have impacted on the mineralization in the HPPP- see also sections 5.1.4 and 6.

b

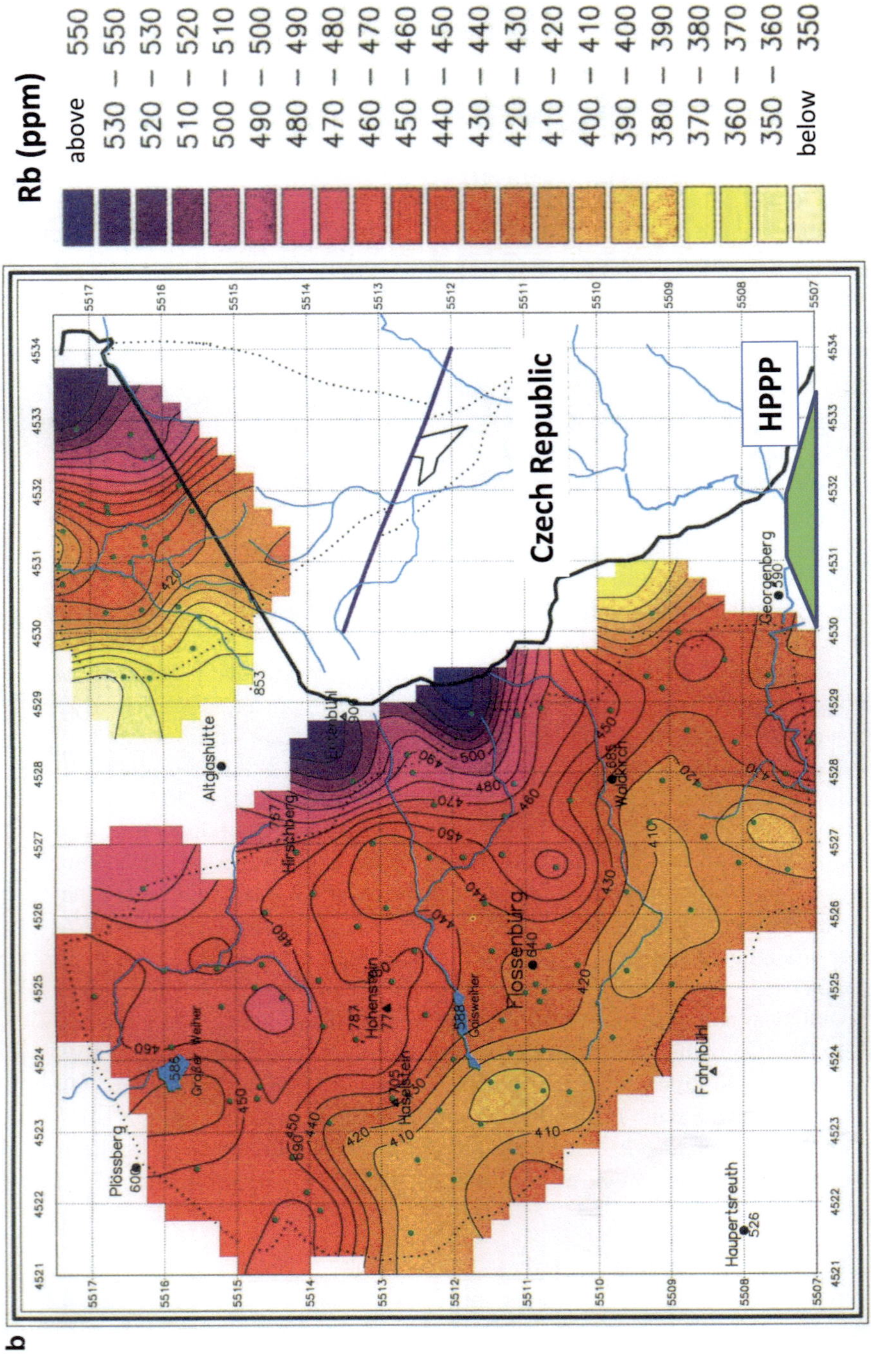

Fig. 3.6b Rubidium distribution in the Flossenbürg (west) and Bärnau Granites (east). For reasons of clarity different colors are used for the graphical representation of contour lines. The color key on the *right-hand side* denotes the content of Rb in ppm. *Small dots* mark the sampling sites in both granites, modified from Wendt et al. (1986). The position of the HPPP is shown *down right*

3.2.6 Niobium-Tantalum

Niobium is accompanied in almost all sites from NE Bavaria, hosting Nb-Ta oxides by tantalum. However, Nb always exceeds the contents of Ta. In contrast to Ba, Rb and Zr which show clear unidirectional trends upon fractionation of granites, niobium does not: G 1: 15 ppm Nb, G 2: 12 ppm Nb, G 3: 11 ppm Nb, G 4: 15 ppm Nb (Richter and Stettner 1979) (Figs. 3.6a, 3.6b, and 3.6c). The chemical composition of metamorphic country rocks in the surroundings of the HPPP shows rater moderate average contents of niobium: chert: 5 ppm Nb, aplitic gneiss: 14 ppm Nb, amphibolite: 11 ppm Nb. The granites north of the HPPP have only slightly higher average contents of niobium: Bärnau Granite: 14 ppm Nb, Flossenbürg Granite: 19 ppm Nb, Falkenberg Granite: 18 ppm Nb (Wendt et al. 1986, 1994). There is only one rock type that falls out of the series of magmatic and metamorphic rocks by its significantly higher Nb values, the metalamprophyres. Basic dikes crossing into the HPPP average 39 ppm Nb. Rohrmüller (1998) encountered these rocks during his mapping campaign for the geological map 1: 25,000 Flossenbürg (No 6240) which is located immediately north of the HPPP. Four specimens listed in his report yielded a mean value of 24.75 ppm Nb. Their maximum value stands at 29 ppm Nb, their minimum value at 20 ppm. Niobium in the pegmatitic rocks of the HPPP can be traced back to these basic dikes which have been either sidelined during geological mapping or were totally overlooked in the wall rocks of the pegmatites, as exemplified by the metamorphic wall rocks of the Kreuzberg Quartz Pegmatite in the city center of Pleystein- see also Sect. 5.1.3. A closer look at the chemical composition of various metamorphic and magmatic rocks reveals an intimate chemical relationship between niobium and titanium in the metamorphic country rocks ($r_{\text{calcsilicate}}=0.96$, $r_{\text{gneiss aplitic}}=0.80$, $r_{\text{lamprophyre}}=0.88$, $r_{\text{amphibolite}}=0.74$, $r_{\text{gneiss (biotite-sillimanite)}}=0.58$). Among the granitic rocks in the vicinity of the HPPP only the Falkenberg Granite shows a positive correlation between niobium and titanium ($r_{\text{granite}}=0.63$), whereas the Flossenbürg Granite at a more proximal position to the HPPP shows no correlation between titanium and niobium. Its niobium contents are prevalently correlated with tin ($R_{\text{Sn-Nb}}=0.92$, $R_{\text{Rb-Nb}}=0.84$, $R_{\text{P-Nb}}=0.78$). The aplitic and pegmatitic rocks in the HPPP do not show an correlation with titanium but take up different positions. The aplitic rocks not unexpectedly are correlated to elements typical of these highly differentiated felsic rocks ($R_{\text{B-Nb}}=0.99$, $R_{\text{Be-Nb}}=0.96$, $R_{\text{Ta-Nb}}=0.79$, $R_{\text{Rb-Nb}}=0.72$). The correlation coefficients for the pegmatitic rocks are not very much different ($R_{\text{K-Nb}}=0.99$, $R_{\text{Zn-Nb}}=0.96$, $R_{\text{Rb-Nb}}=0.96$, $R_{\text{Al-Nb}}=0.95$).

Country rocks and wall rocks with a highly positive correlation between niobium and titanium reflect the pathway of niobium into the HPPP, a transport process mediated by Ti. The granites, pegmatites and aplites show the last leg of this migration just before concentration of Nb and Ta took place in the columbite s.s.s. of the HPPP. Along the first leg of transport from the source to the site a decisive role has to be assigned to titanium, while throughout the second leg of transport and concentration processes, Sn, Be, B, K and B took over the position previously played by Ti.

Among the potential source rocks, a key role is played by the lamprophyres, a group of hypabyssal porphyritic rocks with phenocrysts of dark-colored minerals,

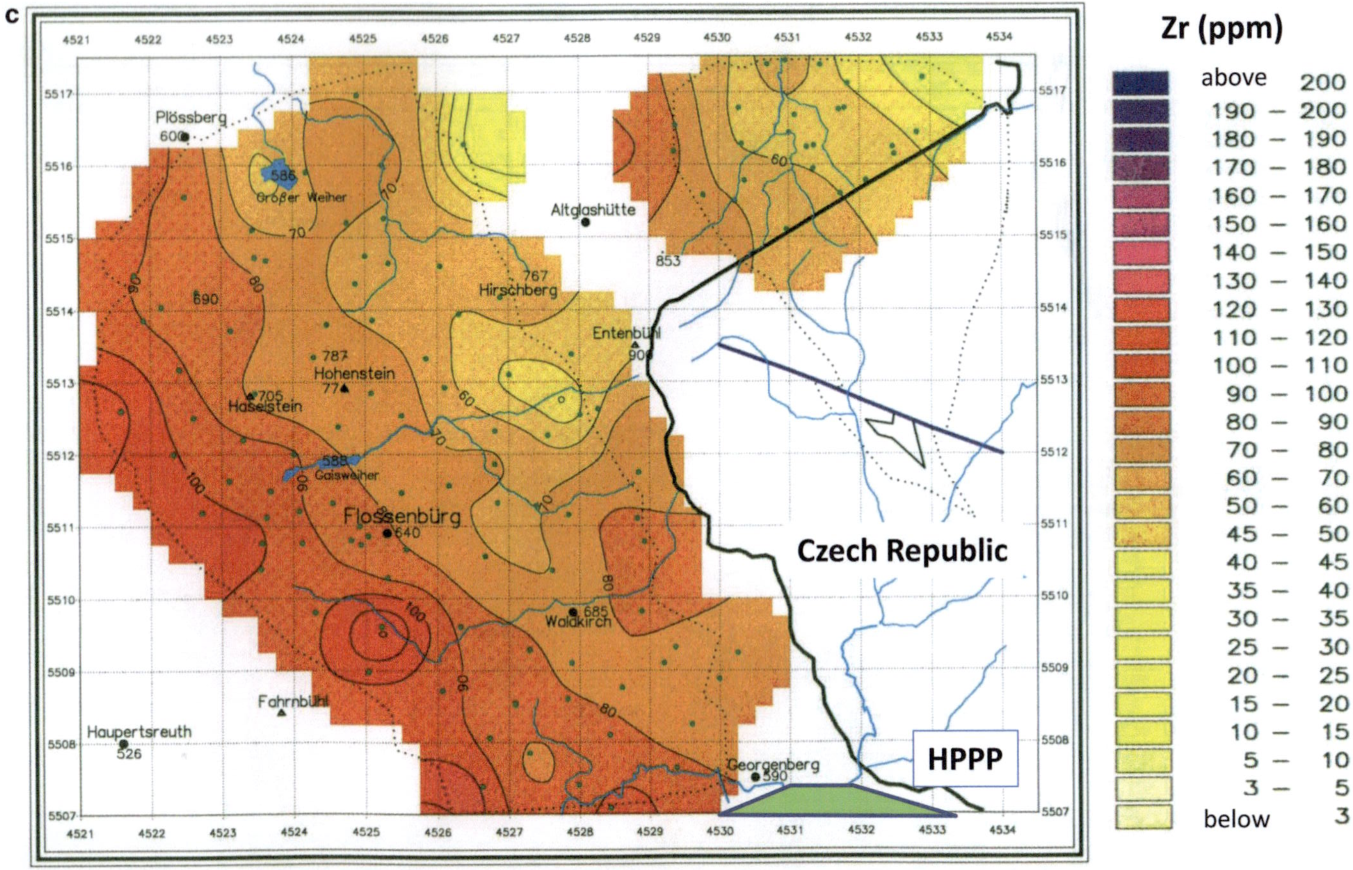

Fig. 3.6c Zirconium distribution in the Flossenbürg (west) and Bärnau Granites (east). For reasons of clarity different colors are used for the graphical representation of contour lines. The color key on the right-hand side denotes the content of Zr in ppm. *Small dots* mark the sampling sites in both granites, modified from Wendt et al. (1986). The position of the HPPP is shown *down right*

whose classification scheme has not yet found common consensus and whose origin is still under debate (Métais and Chayes 1963, 1964; Rock 1991; Wooley et al. 1996; Le Bas 2007; Seifert 2008; Vasyukova et al. 2011).

Among the students of these basic to ultrabasic rocks a subcrustal source is held to be likely. On their ascent to shallow crustal levels as in the HPPP they react with the country rocks along the deep-seated fault zones. Concluding from the aforementioned correlation coefficients metacarbonates are the most likely wall rock for such chemical reactions. Calcsilicate rocks and skarn are quite common at the edge of the HPPP. They were discussed in more detail from the petrographic and lithochemical point of view as it is the case with the lamprophyres in a later section (Sect. 5.1.3). To come right to the point, these magmatic dike rocks in and around the HPPP have the highest ever reported Ti contents of all lithologies in the NE Bavarian Basement and render accountable the widespread occurrence of "nigrine", an intergrowth of ilmenite and rutile, in the bedload of the creeks draining the HPPP (Sect. 4.9).

3.2.7 Beryllium

Beryl is a mineral common to some pegmatites of the NE Bavarian Basement. The mineralogical facets of beryl and the Be minerals associated with it, is discussed later (Sect. 4.23). The element beryllium is a rare element in the continental crust averaging only 3 ppm and it comes close to the mean value of uranium another granitophile metal (Taylor and McLennan 1995). In felsic intrusive rocks, pegmatites and aplites, the amount of beryllium is increased by several orders of magnitude by the combined action of anatexis and crystal fractionation and commonly precipitates as beryl as stated above for the NE Bavarian Basement. The northernmost S-type granites in the Saxo-Thuringian zone with mean values exceeding the average content of 3 ppm Be are located proximal to the collision zone. Fractionation of beryllium takes place in these granites in a textbook- style as documented by the succession of granites, arranged in order of decreasing age of granite formation: G 1: 4.6 ppm Be, G 2: 7.7 ppm Be, G 3: 9.9 ppm Be, G 4: 17.6 ppm Be (Richter and Stettner 1979). These contents do not differ very much from the Be contents recorded from the granites further East in the Erzgebirge, where strongly mineralized albite granite and zinnwaldite–lepidolite-protolithionite granites reside in the same geodynamic realm (Seltmann et al. 1998). Their beryllium contents fall in the range 5.1–11.0 ppm Be. Aplites in the HPPP have Be contents below crustal average 0.7–1.1 ppm Be. Pegmatites, however may reach up to as much as 42.1 ppm Be, as at Plößberg, where beryllium formed minerals of its own and beryl developed. Not a host of beryl but also standing out from the overall igneous rocks, lamprophyres have considerable Be values with 10.6 ppm Be. But the sample number is too scarce to claim these rocks to be a potential source of beryllium. As far as the metamorphic rocks are concerned, the majority of cherts, gneisses, metaultrabasic rocks, amphibolites and metalimestones are undernourished with respect to the discussed element and average beryllium of less than 3 ppm Be.

To come to grips with the issue where the beryllium for the Be minerals in the NE Bavarian pegmatites came from, the laboratory work of Evensen and London (2002) was tested for its applicability in the current study area in and around the HPPP to track the distribution of Be between aluminous quartzofeldspathic source rocks and an anatectic melts. The most common country rock in the area is biotite-cordierite-sillimanite gneiss. According to the authors, if cordierite is present, most of the available Be in the source will be retained in restite, leaving derivative melts highly depleted in Be. Cordierite-bearing protoliths will not achieve beryl saturation at any point in their evolution, even after strong fractionation. Beryllian cordierite may be found in this case, as exemplified by Černý and Povondra (1966) from Věžná, Czech Republic. Beryl is common to many pegmatites in the region and having a closer look at a sandwich texture of gneiss and aplite/pegmatite at a meter spacing, in what might be called the initial phase of mobilization, the following Be distribution has been found in the field (Dill et al. 2012a). Is it a support of the laboratory trials above or may we have to discard this theory for the field under study ? The values obtained from sampling a dyke swarm and listed below speak for themselves.

No	Lithology	Be (ppm)
1	Gneiss	1.1
2	Contact zone	1.1
3	Aplite	1.1
4	Gneiss	0.8
5	Aplite	0.7
6	Gneiss	0.9
7	Contact zone	0.9
8	Aplite	0.9
9	Gneiss	1.6

Due to its high reactivation potential enabling beryllium to show up also in pseudopegmatites, this element is locally rather widespread and so masking its ultimate derivation from a subcrustal source. The way of beryllium from its source to the pegmatite is not yet fully tracked down for the area under study.

3.2.8 *Boron*

Tourmaline is a jack-of-all-trades in the crystalline basement rocks along the western edge of the Bohemian Massif and abundant even in the arenaceous platform sediments, where it share the upper ranks in terms of resistant to weathering with rutile and zircon (Fig. 3.7). In granites, pegmatites and aplites the black schorl crystals are common. Unfortunately, this element does neither rank among the elements

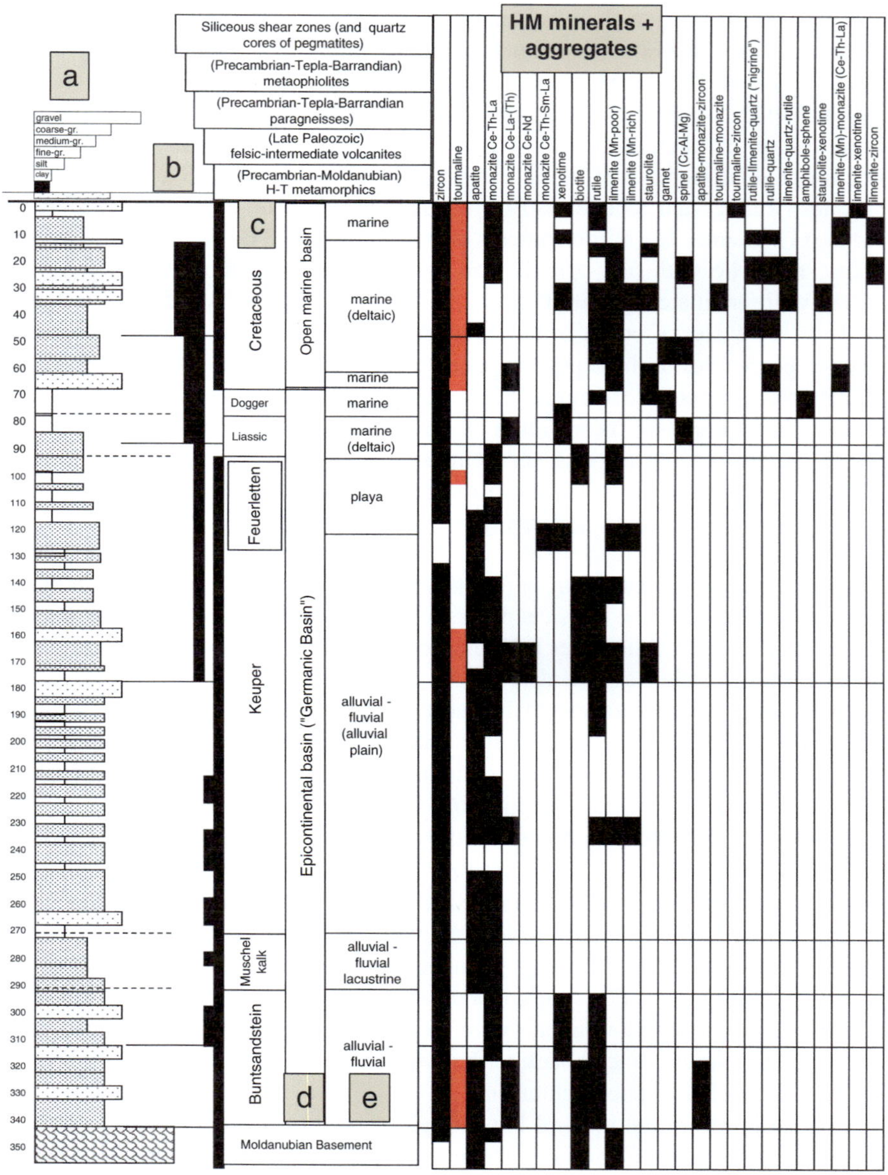

Fig. 3.7 Variation of heavy minerals and heavy mineral aggregates as a function of lithology of the platform sediments in the epicontinental basin immediately to the west of the NE Bavarian Basement (see key to lithology in the header) (**a**) Unroofing of the source area (*black vertical bars* are hooked up with description of the provenance area in the *rectangular box* in the header) (**b**). Stratigraphy of the Mesozoic platform sediments (**c**) Geodynamic position of the epicontinental basin (**d**) Environment of deposition of the stratigraphic units (**e**) -The drill depth is given in meter. The distribution of tourmaline is highlithened in *red* and shown to be most strongly enriched in the Late Cretaceous modern sediments (Modified from Dill and Klosa 2011)

analyzed during routine chemical analyses nor is found among the elements being commonly used for stream sediment analyses of a regional survey. Only a few lithologies have been analyzed and therefore can be used to deal with boron on a regional level. The granites from the Fichtelgebirge frequently used in previous sections to show the chemical trends in the NE Bavarian granites fail to reveal any trend and no gradual increase of boron can be claimed for the G 1, G 2, G 3 and G 4 granites (Richter and Stettner 1979). A maximum value of 400 ppm was reported for the G 2 granite. The boron contents known from the NE Bavarian pegmatitic and aplitic rocks lie with 672 ppm B well above the average reported for the most strongly differentiated granite G 4 called also "Tin Granite" yielding 17 ± 8 ppm B. Thus, a transfer of boron from one of the granites even into the most proximal granitic pegmatites or those pegmatites bodies at a more distal position is unlikely. As the average grades of pegmatitic and aplitic rocks are precisely be monitored and checked against the mean values of the gneissic country rocks with 682 ppm B the paragneisses seem to be closer to the source of boron than to the felsic magmatic rocks. The boron contents in the gneissic country rocks under consideration in the HPPP show strongly positive correlation with three elements ($R_{B\text{-}Al} = 0.78$, $R_{B\text{-}K} = 0.80$, $R_{B\text{-}P} = 0.89$), that stress a close genetic link between boron and the marine phosphate-bearing (meta)pelites. The data are in line with the findings and published in the classical paper by Harder (1970), that clays contain more boron than sands or limestones and micaceous phyllosilicates contain more boron than smectite-enriched sediments.

Who can give you an answer to the question as to the individual boron concentration in the different geodynamic units of the NE Bavarian basement ? Where was the concentration the strongest one so that boron might have taken accommodation in the B-bearing pegmatites and aplites of the HPPP ? The key to this lock lies in the foreland of the basement as documented by one of the deepest and most proximal deep drill holes through the Mesozoic platform sediments. The litholog of Wackersdorf drill hole is plotted alongside the heavy mineral log providing a full coverage of all heavy mineral species, including tourmaline, which is accumulated in peculiar layers, mainly during younger Late Cretaceous sediments (Dill and Klosa 2011) (Fig. 3.7). The quantification of the boron content is accomplished by counting the tourmaline mineral grains in the heavy mineral association of each stratigraphic unit and traced back to the provenance area in the basement. It is an indirect but economic approach taken.

Concluding from the mineral log in Fig. 3.7, single crystals of tourmaline are accompanied by zircon and monazite in some reservoir rocks of upper Triassic to Upper Cretaceous age. These aggregates and mineral assemblages going through an intermediate repository known as the lower Cretaceous weathering mantle, have two principal source rock areas: (1) Moldanubian H-T metamorphics (2) gneisses of the Tepla-Barrandian unit- for location see Fig. 2.1d. The Moldanubian unit also gives host to the felsic mobilizates of the HPPP. The Tepla-Barrandian unit only acts as a source of boron for the metapegmatites and to a lesser degree for the pegmatoid mobilizates. The boron is supposed to be directly incorporated into these felsic mobilizates, by-passing the granites, whose boron trend is not corroborating the granites'role as an intermediate repository for this element.

Boron is widespread in parametamorphic rocks and a geodynamic proximity marker for the core zone of ensialic, orogens where a deep erosional level has been exposed.

3.2.9 Phosphorus

Unlike beryllium or boron, which were dealt with in the previous sections together with many other rare elements, phosphorus is one of the major elements in the earth crust, contributing with approx. 1010 ppm P to the chemical composition of the lithosphere. Phosphorus is among the top-ten elements in the earth's crust, although at the bottom of this "league" and it is responsible for a wide range of phosphate minerals as well as contained in many rocks, particularly phosphorites and their metamorphic equivalents (Nriagu and Moore 1984). Graphite-quartzite, graphite schists, and biotite-andalusite fels with minor graphite are widespread around the HPPP, but do not occur in the HPPP, itself (Dill 1983b). The afore-mentioned metasedimentary rocks have remarkable contents of tourmaline and the amount of phosphate increases to as much as 2.49 wt% P_2O_5, particularly as their lithology changes from pure metabiolites into metaphosphorites. These metasedimetary rocks are Silurian in age and as such form the early Paleozoic roof on top of the Moldanubian basement in NE Bavaria, where they are exposed along the contact of the allochthonous unit of the Tepla-Barrandian and the underlying autochthonous Moldanubian basement rocks (Stettner 1990). The phosphate-enriched metabiolites are located immediately at the western and northern margin of the HPPP.

Amphibolites are more largely exposed around the granites and pegmatites and found within the allochthonous unit of the Tepla-Barrandian as well as the underlying autochthonous Moldanubian basement rocks. Their P_2O_5 contents are as follows min: 0.03 wt%,mean: 0.26 wt%, max: 0.75 wt% P_2O_5. Similar to the afore-mentioned metabiolites and metaphosphorites, the HPPP is also undernourished with regard to amphibolites when compared with the overall Moldanubian or the Tepla-Barrandian geodynamic units. Rarely these metabasic igneous rocks are intercalated into cordierite- and biotite-bearing gneisses.

Apart from apatite which is the most common source in the metamorphic basement rocks, there are some other rock-forming minerals that warrant mentioning when the question is raised where the phosphorus in pegmatites and granites has been derived from. Phosphorus, although a minor element, is accommodated in the lattice of the most widespread minerals in the earth crust. The feldspar group, with K feldspar being more likely a source of P than albite, has to be envisaged as a potential source of P too. This agrees well with the petrographic studies performed by Breiter and Siebel (1995) and Frýda and Breiter (1995), who reported from the nearby Czech granites at Podlesi (0.83 wt% P_2O_5) and Homolka (0.77 wt% P_2O_5)

elevated amounts of phosphorus contained in the K feldspar of these granites. The phosphorus variation in granitic rocks has been among others studied by Broska et al. (2004) and treated in experiments by Tollari et al. (2006). The influence of the iron content and oxidation state on the saturation of phosphate minerals in magmatic systems have been studied by the latter authors in the temperature range from 1030 to 1070 °C and oxygen fugacity from 1.5 log units below to 1.5 log units above the fayalite–magnetite–quartz buffer. The P content of garnet-group minerals could also achieve rather high values of 1.21 wt% P_2O_5, but to quote the authors themselves, "the partitioning of P among garnet and its associated minerals in granitic systems remains still unclear" (Breiter et al. 2005).

Considering the granite suites in the immediate vicinity of the HPPP on a case-by-case basis reveals a clear trend. The granites of the Fichtelgebirge, being located at a distal position relative to the HPPP, average 0.24 wt% P_2O_5. The Falkenberg Granite Complex, including all of its varieties has a more elevated mean value of 0.33 wt% P_2O_5, while mean value of the Bärnau Granite is only moderately higher at 0.34 wt% P_2O_5, and the Flossenbürg Granite, lying next to the HPPP attains the highest mean value among the granites straddling the Saxo-Thuringian – Moldanubian boundary with 0.37 wt% $P_2O_{5.}$ An analysis published by Müller et al. (1998) for the albite-zinnwaldite granite at Křížový Kámen/Kreuzstein Granite north of the HPPP, close to the border on Czech territory, stands out by a much higher value of 0.43 wt% $P_2O_{5.}$ Only the Podlesi Granite exceeds this average value reported from the Křížový Kámen/Kreuzstein Granite by a considerable margin with a mean content of 0.84 wt% $P_2O_{5.}$

Multiple chemical relations between phosphate and minor as well as major elements can be discussed for the different lithologies in the NE Bavarian Basement. In metabasic igneous rocks phosphate does not correlated with other elements yielding a correlation coefficient greater than 0.50, although considerable amounts of P_2O_5 were reported from these metamorphic rock. The results from the correlation of chemical elements in gneissic country rocks hint a more nuanced relationship, in that aplitic gneisses, particularly those within the HPPP, are strongly positive correlated with Mn (common aplitic gneisses $R_{\text{Mn-P}}=0.80$, aplitic gneisses within the HPPP $R_{\text{Mn-P}}=0.90$). Aplites, aploids and aplite granites are identical to the common aplitic gneisses, yielding the same correlation coefficient of $R_{\text{Mn-P}}=0.80$). Granites in the immediate vicinity, like the Falkenberg and Flossenbürg Granites have a correlation coefficient of $R_{\text{Mn-P}}=0.56$ and $R_{\text{Mn-P}}=0.50$, respectively. Manganoapatite is not out of the ordinary in granitic pegmatites and has been described form different sites so that gradual improvement of the P-Mn correlation is not unusual (Cruft 1966; Keller and von Knorring 1989; Pieczka 2007). Two principal conclusions may be drawn from the correlation of P and Mn. The conspicuously low correlation between P and Mn in neighboring granites relative to the correlation of P and Mn in paragneisses of the HPPP and its felsic mobilizates, in particular, does not strengthen the theory of pegmatites and aplites to have derived from the Oberpfalz granites by magma differentiation. The trend of Mn and P in gneissic country rocks shows more and more positive values of the correlation coefficients as heading for the pegmatites and aplites of the HPPP, a fact that can be explained through a mobilization of phosphate from the country rocks in the course of a thermal process discussed later in the book (Sects. 4.14 and 4.15). Viewed from the applied angle of economic geology, the P-Mn ratio can be called a lithochemical ore guide.

There are two distinct phosphate systems in the area, one represented by the succession from the Fichtelgebirge through the Flossenbürg Granite Complex. The southward trend towards more elevated P contents cannot be ignored. It is the chemical expression of an increased phosphate content in the crustal parts of the basement where the granites'kitchen is located. It is driven geodynamically by the compressional processes in this ensialic orogen. Some granites, rather isolated stock-like granites vented along deep-seated lineamentary fractures zones depart markedly from the general trend and even may reverse it. Prefiguring the tectonic influence upon the emplacement of pegmatites, these highly-differentiated granites are bound to the afore-mentioned deep-seated lineamentary fault zones postdating the larger granites complexes. They do not reflect this lithochemical zonation visualized by the N-S trend of these larger granites in the Fichtelgebirge and the northern Oberpfälzer Wald. A closer look at the various granites in the Fichtelgebirge reveals that there does not exist any increase of phosphorus upon fractionation (Granite G 1: 0.32 wt% P_2O_5, Granite G 2: 0.21 wt% P_2O_5, Granite G 3: 0.19 wt% P_2O_5, Granite G 4: 0.24 wt% P_2O_5). The same holds true for the Podlesi granite complex, whose dikes have a higher mean value of 1.06 wt% P_2O_5 than the pegmatites attaining only 0.79 wt% P_2O_5 starting off from a stock granite with 0.59 wt% P_2O_5. The chemical trends in granites support an increase of phosphate towards the south but do not substantiate the idea of granites being the immediate source of P in the HPPP.Moreover this hypothesis has to be discarded on chemical grounds as the internal trends, e.g. in the Flossenbürg and Bärnau Granites are considered (Figs. 3.6a, 3.6b, and 3.6c). In context with the distribution of P-enriched country rocks around the granites and the HPPP, alike, a specialization of the metamorphic country rocks are likely to have been responsible for the presence or absence of phosphate-bearing pegmatites. The preponderance of P-enriched granites and aplitic and pegmatitic derivates on both sides of the Czech-German border in this part of the NE Bavarian Crystalline Basement may be genetically related to phosphate bearing metasediments (metabiolites, metaphosphorites, phosphate-bearing paragneisses) and amphibolites whose protolith has also to be looked for among the sedimentary realm and not only among igneous parent rocks. This is especially valid for the series of amphibolites alternating with calcsilicate rocks. Abnormally high values of phosphate in the aplitic mobilizates and in the pegmatite cupolas are convincing evidence that P specialization of the geodynamic setting of this part of the Moldanubian zone was crucial for the overall presence of phosphate in these felsic rocks, be it granitic or pegmatitic. Both lithologies were supplied by the same geodynamic units but accumulated P to a different extent during their emplacement (for processes sections 5 and 6).

3.2.10 Rare-Earth Elements (REE) and Thorium

Minerals accommodating rare-earth elements into their lattice, such as monazite and xenotime, are common to the aplites and pegmatites in the NE Bavarian Basement but they do not take such a prominent position as to become an identifier, relevant for the classification of the pegmatites of the HPPP. Both minerals are of widespread occurrence in the crystalline basement rocks of the Oberpfälzer and Bayerischer/Böhmer Wald where they are frequently associated with zircon and to a lesser extent with orthite. Each of the minerals represents a source of thorium and REE, mainly of the LREE group (monazite, zircon, orthite) and to a minor degree of HREE (xenotime). Since REE minerals have not systematically been investigated so far in the basement areas hosting pegmatites and aplites in NE Bavaria, their distribution cannot be treated on a regional basis. The only area with stratabound REE-, Th- and U minerals being side-by-side with pegmatites (Hühnerkobel, Frauenau, Blötz, Dachselsried) extends along the belt of "Kieslager-type deposits" from Bodenmais to Roter Koth in the environs of Bodenmais-Zwiesel in the Bayerischer Wald (Strunz 1962; Dill 1990). In some places, these REE-, Th- and U-bearing heavy minerals take the shape of a placer-type accumulation, intercalated into the high-grade paragneisses where they are accountable for the radiometric anomalies parallel to the strike of the metamorphic country rocks. Against all expectations, the pegmatites around are undernourished as to REE minerals when compared with their counterparts in the Oberpfälzer Wald, being located further north (Figs. 2.5b and 2.5e). Only the pegmatite at Blötz, about 2 km ENE of Bodenmais is host to samarskite-(Y) that may contain considerable amounts of HREE (Pfaffl 1966). Another site in the Bayerischer Wald mineralized with REE minerals is located in the granodiorite at Tittling, which is also host to pegmatite veins at the contact to a younger granite. It is again samarskite-(Y) that attests to the preponderance of HREE in the southern Moldanubian zone. Claus (1936) has already stressed the important role that xenotime plays among the REE-bearing accessory minerals in the metamorphic rocks of southern Moldanubian region.

More recent studies of the Gföhl Unit, one of lithological units of the Moldanubian Zone made up of anatectic gneisses, HP-HT granulites peridotites, pyroxenites and eclogites, revealed variably high contents of REE (Janousek and Finger 2003). According to both authors, U and Th indicate higher mobility during the high-pressure metamorphism and got depleted, while HREE (+Y) seem to be in the upper crustal range. The Gföhl unit underlies the afore-mentioned Drosendorf unit. To provoke a REE enrichment, obviously shearing and friction are more relevant than simple regional metamorphism (De Jonge et al. 1997; Rolland et al. 2003) (see Sect. 2.3.2). REE-bearing pegmatites along the thrust plane between the medium to high grade Gföhl nappe and the low to medium grade Drosendorf unit support this theory (Sect. 2.1.5.1).

Skimming the various country rocks around the HPPP does not conduce to the discovery of another REE-Th concentration similar to that from the southern Moldanubian Zone. Amphibolites have mean values of 44 ppm REE_{tot}, cherts 149 ppm REE_{tot}, and aplitic gneisses 73 ppm REE_{tot}. More elevated mean values are known from the gneisses and calcsilicate rocks, averaging 153 ppm REE_{tot} and 131 ppm REE_{tot}, respectively.

A totally different picture exists for the pegmatites and aplites in the HPPP. Aplitic rocks interbedded sandwich-like with these REE-bearing gneisses and calcsilicate rocks mentioned in the previous paragraph average 8 ppm REE_{tot}, tabular aplites like that at Trutzhofmühle gain 36 ppm REE_{tot} and the stocks like Hagendorf North and Pleystein have 62 ppm REE_{tot} on average. The Fichtelgebirge granites show a reverse trend to the known fractionation: G 1: 172 ppm REE_{tot}, G 2: 102 ppm $REE_{tot,}$ G 3: 62 ppm REE_{tot}, G 4: 61 ppm REE_{tot}. The same reversal of the chemical trend during differentiation can also be observed in the Falkenberg Granite 141 ppm REE_{tot} and the Flossenbürg Granite, proximal to the HPPP with an average of 10 ppm $REE_{tot.}$

Magma differentiation during the emplacement of postkinematic granites caused a depletion in REE_{tot} providing another good argument against a genetic link between the Flossenbürg Granite and the HPPP felsic mobilizates which have monazite, xenotime-(Y) and allanite among the rock-forming minerals similar to their enclosing gneisses. Considering the variation of REE, no granitic-pegmatitic trend exists in the sites under study in the northern Oberpfälzer Wald, which is documented at best by the HPPP. The variation of REE in pegmatites is controlled by a dynamo-metamorphic processes leading to an incorporation of accessory minerals into the felsic mobilizates. This well agrees with the finds of De Jonge et al. (1997) and Rolland et al. (2003) (see Sect. 2.3.2). In the southern Moldanubian Zone, in NE Bavaria, a similar dynamo-metamorphic processes cannot be identified. Therefore the pegmatites lack any REE enrichment, although being surrounded by country rocks abundant in REE. The minor presence of samarskite-(Y) in some pegmatites is driven by the protolith with is relatively enriched in HREE and/or the depth of emplacement of the pegmatites (?).

The REE and Th contents are to low and, hence, the database too weak so as to pinpoint a source for the pegmatites under study. Looking at the REE and Th contents on more global scale, the subcrustal level is likely the most plausible source for these elements.

3.2.11 Arsenic, Bismuth and Zinc

Arsenic and gold tend to be closely associated with each other in many gold deposits worldwide, a chemical affinity which also by recognized in several gold occurrences in the NE Bavarian Basement, although being subeconomic with regard to grade and size by modern standards (Lehrberger et al. 1990; Herzog et al. 1997; Morávek and Lehrberger 1997). Gold-arsenic-bismuth mineralization evolved in the cordierite-sillimanite gneisses at different places and is held to be a potential source of gold in the placer deposits in the environs of Hagendorf-Pleystein (Lehrberger et al. 1990). Each heavy mineral survey focused on the stream sediments of the drainage systems in the northern Oberpfälzer Wald will end up with some findings of milled gold or minute nuggets in the washing pan.

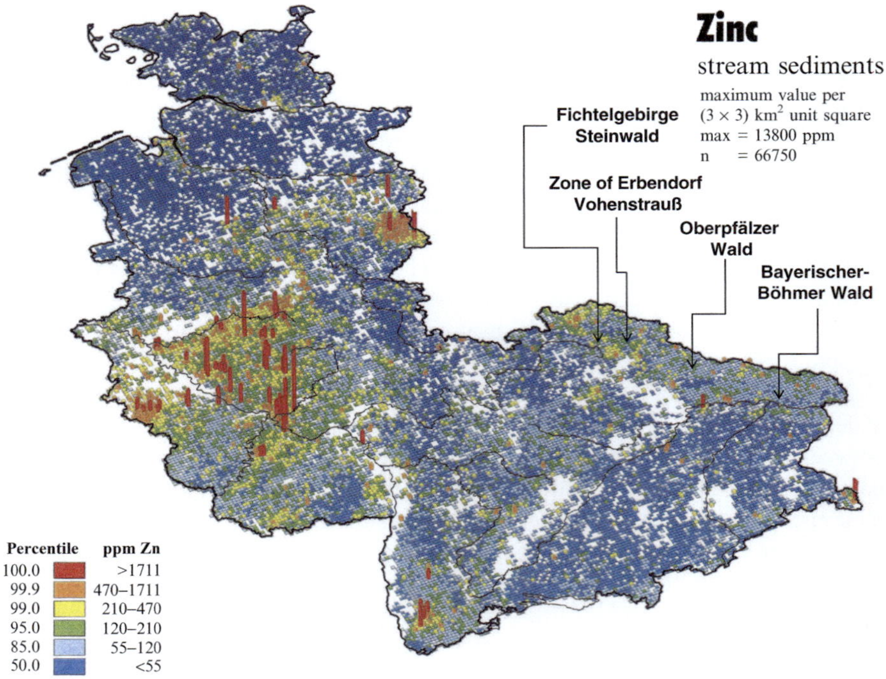

Fig. 3.8 Map of the Federal Republic of Germany representing the results of the official geochemical survey conducted by the Federal Institute for Geosciences and Natural Resources (Fauth et al. 1985). The map visualizes the zinc contents in view of histograms with the highest zinc contents given by the length of the columns and by a color code with *red* indicative of the maximum content

On the other hand, zinc does not accompany any of these elements mentioned above in the various structure bound and stratiform mineralization in NE Bavaria but used to be coupled with Pb as in many deposits worldwide. All three elements are encountered side-by-side with each other in the pegmatites at Hagendorf and Pleystein and, in places, also form minerals in some aplites of the HPPP, e.g., the Miesbrunn aplite-pegmatite dike swarm (Dill et al. 2012a). While for zinc the chemical background values of the crystalline basement rocks are quite well known, and the number of chemical data available for arsenic is only slightly less lower, the bismuth contents are often too low as to attract attention during routine chemical analyses and consequently no chemical survey was done including this element.

The map produced in the course of the geochemical survey by the Federal Institute for Geosciences and Natural Resources visualizes the zinc contents for the NE Bavarian Basement (Fig. 3.8) (Fauth et al. 1985). It has to be noted that the regional Zn contents increase towards the NW in the NE Bavarian Basement and achieve extensive anomalies in the Saxo-Thuringian low to very low grade regionally metamorphosed rocks. The Oberpfälzer Wald Region is not very much different from the Bayerischer Wald Region with regard to the Zn contents. In both areas the blue and green colors dominate attesting to background values between 55 and 210 ppm Zn. Only in the small area in the Oberpfälzer Wald where the HPPP is located, is marked by an anomaly striking roughly N-S and attaining maximum contents of 1711 ppm Zn.

Considering the chemical composition of the gneissic country rocks, reveals Zn to be coupled with Al (R=0.70), K (R=0,83) and Ba (R=0.93). The results obtained for the aplitic gneisses attest to an accommodation of Zn in the lattice of micaceous phyllosilicates and to a lesser extent to zincian spinel and staurolite. Vrana et al. (1995) reported from the Moldanubian Zone of SE Bavaria (Bayerischer Wald) zinc present in amounts between 320 and 550 ppm Zn. The authors assume that this element has a stabilizing effect on the Fe staurolite and on spinel – see also Sect. 4.27- in the biotite-garnet-chlorite zone. Staurolite was mentioned by the above authors to contain 3.2 wt% ZnO. It has to be noted that these data refer to metamorphic rocks of the Moldanubian Zone being located to the SE of the HPPP and as such can be considered as a potential source rock underneath it.

In addition to the elements Al, K and Ba, Zn shows also striking chemical affiliations with other elements, such as Nb (R=0.97), Ta (R=0.82), Rb (R=0.90), and Li (R=0.76) in the aplites, aploids and pegmatites, manifesting that Zn is part of the pegmatitic mobilization process. Conclusively, Zn is a common element in the HPPP felsic mobilizates, structurally controlled as to its variation but unrelated to granites.

Unlike zinc, no data on the regional distribution can be presented for arsenic and bismuth in the NE Bavarian basement and the immediate surroundings of the HPPP. Solely for arsenic reliable data can be presented on the interrelationship with other elements anomalously enriched in the pegmatites in NE Bavaria. Arsenic is positively correlated with phosphate (R=0.77) and chlorine (R=0.73).

A comprehensive overview of source rocks of elements found in pegmatites and aplites in the NE Bavarian Basement is given in Table 3.5.

3.3 Geophysical Surveys in the Pegmatite-Aplite Target Areas of the NE Bavarian Basement

The geochemical surveys covering the NE Bavarian Basement and the chemical investigations focusing on those elements being present in the pegmatites and aplites in anomalously high amounts offer an insight into the pathways of element migration (Sect. 3.2). They skim the surface only. The real insight into the Earth's interior can only be achieved by means of geophysical surveys being of assistance in modeling the 3-D representation of the various lithologies at depth and the driving sources for the element migration (Sect. 3.3).

In the run-up to the site selection for the Continental Deep Drilling Program of the Federal Republic of Germany seismic, gravimetric, magnetic and geoelectric

Table 3.5 Summarizing potential sources of elements found in pegmatites and aplites in the NE Bavarian Basement

Element	Lithology-structure	Level	Area
Lithium	Metapelite	Intracrustal	Source: Moldanubian Region
	Granites	Intracrustal	Intermediate repository: Saxothurinngian-Compression
	Granite plugs	Intracrustal	Structurbound offshoots: penetrate the Moldanubian Region
Fluorine	Metapelite	Intracrustal	Source: Moldanubian Region
	Granites	Intracrustal	Intermediate repository: Saxothurinngian-Compression
	Granite plugs	Intracrustal	Structurbound offshoots: penetrate the Moldanubian Region
Tin	Granites	Intracrustal	Intermediate repository: Saxothurinngian-Compression
Uranium	Granites	Intracrustal	Intermediate repository: Saxothurinngian-Compression
	Granite plugs	Intracrustal	Structurbound offshoots: penetrate the Moldanubian Region
Rubidium-barium-zirconium	Granites		Magma-induced differentiation inside the granite complexes with no impact on the pegmatites
Niobium-tantalum	(meta)lamprophyres	Subcrustal	Mantle derived basic dikes interacting with country rocks of the Moldanubian on their ascent
Beryllium	(meta)lamprophyres (?)	Subcrustal	See above
	Cordierite-sillimanite gneiss	Intermediate repository	Source: Moldanubian Region
	Granites	Intermediate repository	Moldanubian Region
Boron	Gneiss	Intracrustal	Source: (1) Moldanubian H-T metamorphics (2) Tepla-Barrandian unit
Phosphorus	Metabiolites	Intracrustal	Source: Moldanubian Region
	Granite	Intracrustal	Intermediate repository: Saxothurinngian-Compression
	Granite plugs	Intracrustal	Structurbound offshoots: penetrate the Moldanubian Region
Phosphorus + manganese	Pegmatites	Intracrustal	Moldanubian Region
Rare-earth elements (REE), yttrium and thorium	Metapelite	Subcrustal	Source: Moldanubian Region
		Intermediate repository	
	Shearing induced	Subcrustal	Moldanubian Region
		Intermediate repository	

(continued)

Table 3.5 (continued)

Element	Lithology-structure	Level	Area
Arsenic	See phosphorus	Intracrustal	See phosphorus
Bismuth	Unknown	Subcrustal (?)	Unknown
Zinc	Metapelite	Subcrustal	Unknown
		Intermediate repository	
	Structurebound	Subcrustal	Moldanubian
		Intermediate repository	

surveys were carried out extensively in the potential target area in the Black Forest and in the NE Bavarian Basement (Fig. 2.1d). The NE Bavarian Basement took the win in front of the Black Forest, sparking another round of large-scale local geophysical measurements which also covered the HPPP and offered a good tool to identify the deep-crustal structures relevant for the emplacement of the HPPP pegmatites and aplites.

3.3.1 Gravimetric Survey

The Federal Republic of Germany has been studied in great detail as to the gravity field and the results were published as a map at scale of 1: 500,000 (Plaumann 1995). A large-scale gravimetry map was presented at the KTB meeting at Seeheim for the drill site selection in 1986. It showed the Bouguer anomalies with 2 mgal-spacing of the contour lines for the NE Bavarian Basement (Fig. 3.9a). The gravimetry map corresponds to the topographic map 1: 250,000 which is also available as a geological map. The two gravity highs between Erbendorf (ERB) and Weiden (WEI) coincide with the Zone of Erbendorf-Vohenstrauß. Numerous amphibolites and metaultrabasic rocks occur in this nappe complex. The preponderance of these basic and ultrabasic metamorphic rocks is so strong that in this NW-SE to N-S striking zone the impact of felsic metapegmatites on the gravity field is suppressed. The area where several pegmatites are exposed in the Falkenberg/Tirschenreuth Granite is also shown by a red framed area (Fig. 3.9a-1). Only its northeastern-most part coincides with a gravity low delineated by the – 46 isoline near Tirschenreuth (TIR). One of the most striking features well expressed in the map of the Bouguer anomalies are the NW-SE trending anomalies. At the southeastern end of such an elongated anomaly which runs between the Flossenbürg and Bärnau Granite, the HPPP sits east of Vohenstrauß (VOH) on gravity low marked by the −38 isoline (Fig. 3.9a-2). The geophysical base maps used for the interpretation have been created by Plaumann (1976, 1986).

The various metamorphic rocks of the Zone of Erbendorf-Vohenstrauß stand out by specific gravities of 2.82 g/cm^3, some basic rocks in this zone even achieve

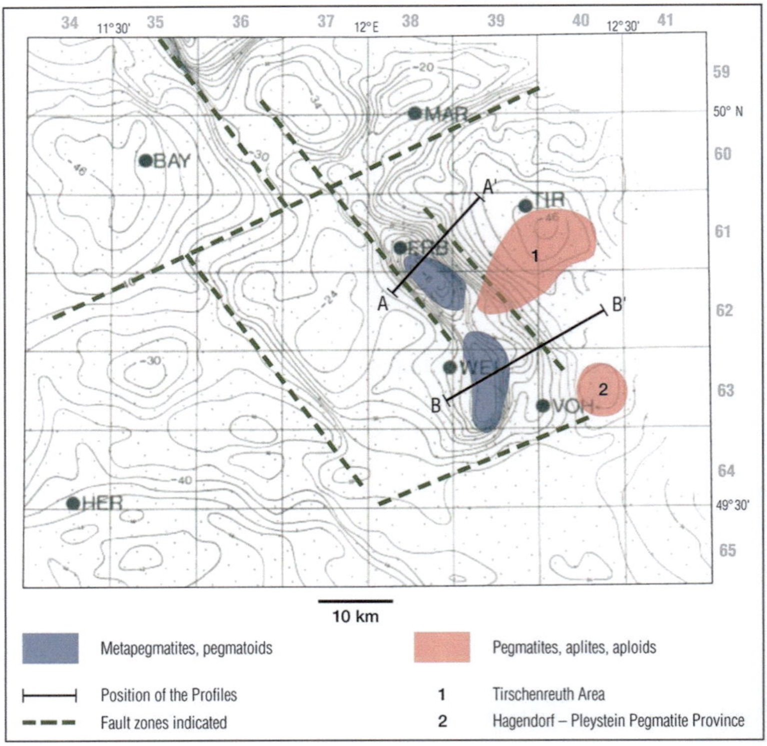

Fig. 3.9a Map to show the Bouguer gravity anomalies at the boundary between the Saxo-Thuringian and Moldanubian Zones (Plaumann 1986; Pucher et al. 1990). The HPPP is marked by the framed area and the names of towns abbreviated to a triplet in capital letters: *TIR* Tirschenreuth, *VOH* Vohenstrauß, *WEI* Weiden, *HER* Hersbruck, *ERB* Erbendorf, *BAY* Bayreuth, *MAR* Marktredwitz. Two transects are shown and marked with *A-A′* and *B-B′* – for more details see the following figures

values of 2.96 g/cm³. The Oberpfalz granites, represented in the transect A-A′ by the Falkenberg-Tirschenreuth, gently dip towards the NE and ENE and can be identified by its low specific gravities of 2.60 g/cm³, bifurcating in the dip direction in an upper and lower layer, which dips down to 7000 m below ground (Fig. 3.9a). The overall thickness of the granite sheet measures approximately 5000 m. The pegmatites one located near Schönficht within the granite and another E of the Tirschenreuth Granites near Marchaney can clearly be distinguished from their host rocks and well delineated in shape by their specific gravity of 2.85 g/cm³ which lies well above the density of the host granites and the surrounding gneisses (Fig. 3.9a). The term aplite was used by the geophysists who had no additional mineralogical and geological information at hands when they carried out their field work.

The Marchaney pegmatite is an Al-enriched P-B pegmatite with dumortierite, schorl, garnet, staurolite, apatite, rockbridgeite, lazulite, strunzite, beraunite, vivianite, metavivianite, and santabarbaraite. The Schönficht pegmatite placed upon the Tirschenreuth Granite is an Be-Nb-P pegmatite with beryl, tourmaline,

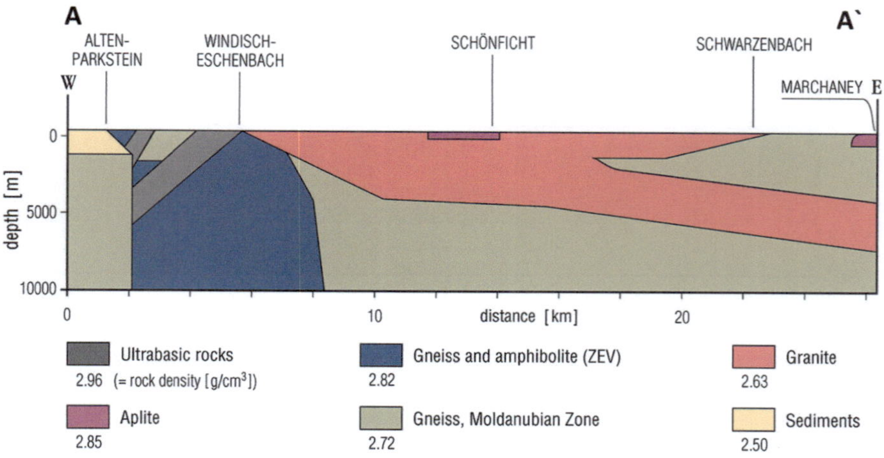

Fig. 3.9b Gravimetric profile *A-A′* with specific rock densities

columbite, apatite and zircon. In the adjacent gravimetric transect further south only three lithologies can be identified based upon the different specific gravities (Fig. 3.9b).

The Flossenbürg Granite dips away from the ZEV metamorphics towards the E, similar to its northern counterpart, the Tirschenreuth Granite. It widens to at least 5000 m in thickness. Aplites or pegmatites are neither identified within the granite nor within its metamorphic wall rocks. Only some basic metamorphic rocks have been identified near the surface. Small-scale gravimetric survey are of assistance in detecting deep-seated fault zones while gravimetric measurements along transects can even be used to localize pegmatites and aplites, because of their stark contrast with granites and gneisses of the country rocks.

3.3.2 Magnetic Survey

The geomagnetic field in the Oberpfälzer Wald was extensively investigated by Pucher (1986), Pucher and Wonik (1990) and Pucher et al. (1990) und used as a pre-well site study of the Continental Deep Drilling Program of the Federal Republic Germany (Fig. 3.10). To avail of these data also during the current studies on granites and the HPPP felsic mobilizates in the most suitable way, an overlay shaded in bright red was created to outline the exposure of the Oberpfalz granites. A set of arrowheads pinpoint the position of the main pegmatites and aplites of the HPPP on map of the magnetic field in the northern Oberpfalz (Fig. 3.10).

The granites are almost amagnetic in terms of the total intensity of the Earth's magnetic field, positive magnetic anomalies are seldom, and if present at all show a random orientation. Only at the eastern margin of the Leuchtenberg Granite NW-SE

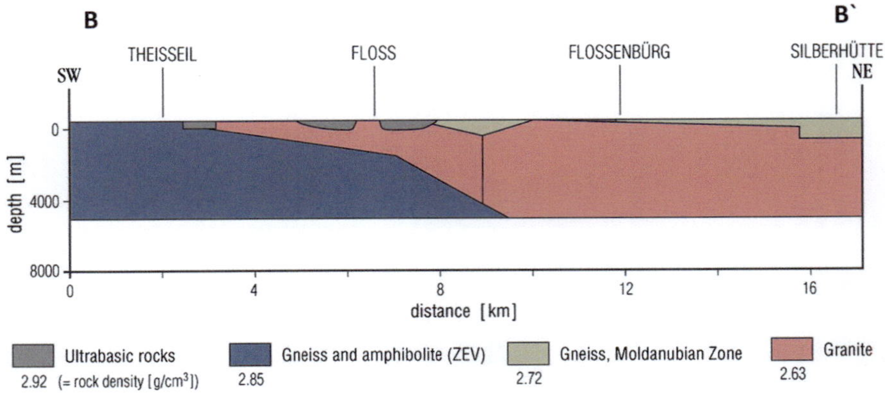

Fig. 3.9c Gravimetric profile *B-B′* with specific rock densities

trending positive anomalies can be spotted in the map. Heading further south, this random pattern of contour lines gradually turns into a more organized one with the magnetic anomalies striking preferably NE-SW in the western part of the HPPP. The magnetic anomalies clustering SW of the HPPP become more pronounced in terms of intensity (see +745 anomaly) and more accentuated by the linear arrangement of the contour lines. The eastern hemisphere of the HPPP is less well-structured than its western part but a vague tendency of anomalies is still to be seen there, marking a NNW-SSE trend among the anomalies. The strong magnetic anomalies are caused by deep-seated metaultrabasic and metabasic igneous rocks which may be traced in northeasterly direction up to Czech-German border. The elongated positive anomalies are induced by a swarm of basic or ultrabasic deep-seated magmatic rocks delimitating the felsic bodies of the HPPP towards the NW and less well expressed also towards the NE. South of the Oberpfalz granites, accentuated by the reddish overlay a ENE to NE trending deep structure, called Luhe Line exists. It is marked by ultrabasic to basic rocks of mantle affiliation. At the point of intersection, where it is crossed by another deep structure, this strong linear magnetic anomaly is perforated by a couple of negative rounded to oval-shaped magnetic anomalies like a "Swiss Cheese". The zone of intersection coincides with the cluster pegmatites and aplites of the HPPP- see the red and blue pattern (Fig. 3.10). For comparison see the green linear patterns in the map illustrating the gravimetry anomalies in Figs. 3.9a, 3.9b, and 3.9c.

3.3.3 Geoelectric Survey

Geophysical studies were also extended to geoelectric surveys encompassing among others the measurements of the electrical resistivity of rocks a great depth. As they were conducted in the region west of the HPPP they are a good tool to supplement

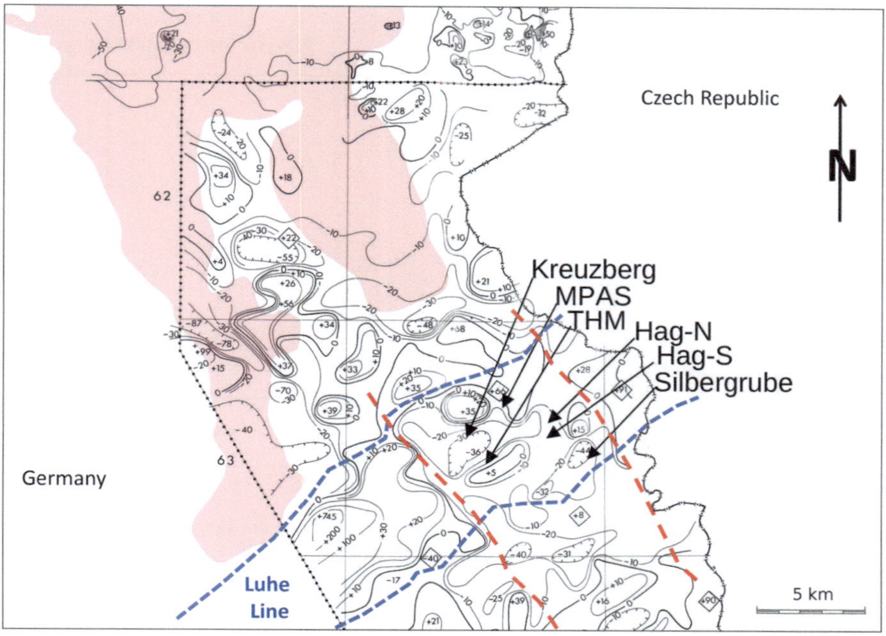

Fig. 3.10 The geomagnetic field in the Oberpfälzer Wald determined and processed by Pucher (1986) and Pucher et al. (1990). The overlay colored in *pink* is representative of the Oberpfalz granites. In addition to those granite complexes, the main pegmatites and aplites from the HPPP are also shown on the geophysical map to facilitate correlation with the magnetic anomalies. *Kreuzberg* Kreuzberg Pegmatite at Pleystein, *MPAS* Mießbrunn Pegmatite-Aplite Swarm, *THM* Trutzhofmühle Aploid, *Hag-N* Hagendorf-North Pegmatite, *Hag-S* Hagendorf-South Pegmatite, *Silbergrube* Silbergrube Aplite near Waidhaus. The Luhe Line is marked in *blue*, the deep-seated-lineamentary structure zones are marked in *red*

the gravimetric and magnetic surveys (LOTEM-Group 1986a, b; Haak 1989). Six different electromagnetic methods were applied during the pre-well site studies by the geophysists to compensate the weak points of one method by the strong points of another one during deep-sounding in the crystalline basement: (1) Magnetotelluric measurements (MT), (2) medium-magnetotelluric (MMT), (3) audio-magnetotelluric measurements (AMT), (4) controlled source audio magnetotelluric measurements (CSAMT), (5) time domain electromagnetic method (LOTEM), (6) Schlumberger direct current method (DCR). Even if many of the data collected in the area are difficult to explain and still pose some enigmas to the interpreter, the nearness of station points to some sites mineralized with pegmatites such as at Püllersreuth, Beidl, and Schönficht raise hopes that these data in context with the afore-mentioned geophysical data sets might help elucidate the deeper parts underneath the pegmatites of the HPPP and some other sites mineralized with pegmatites (Fig. 3.11a). The 1-D measurement shows the resistivity to be lower underneath the Nb-Be (meta) pegmatite at Püllersreuth where it was detected at a shallower depth than underneath

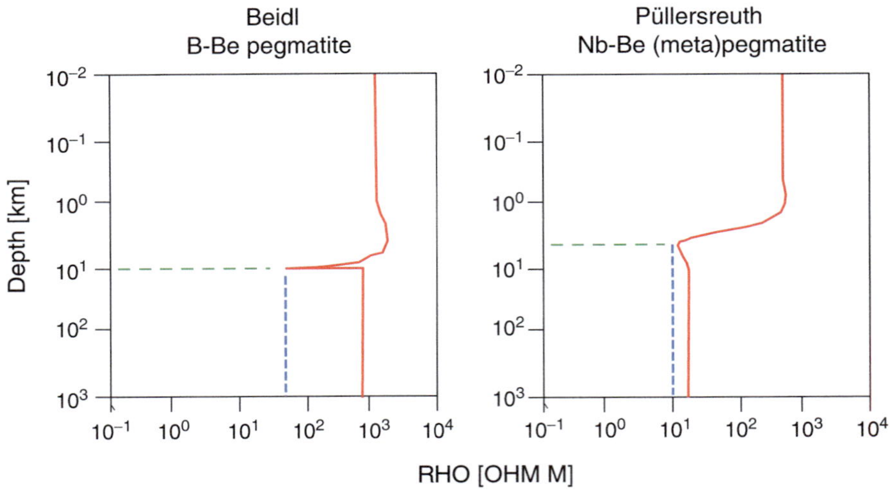

Fig. 3.11a Resistivity-depth function measured by MMT (=medium-magnetotelluric) in areas with exposures of different pegmatitic rocks (Modified from Haak 1989)

the Beidl Be-B pegmatite. The 2-D results obtained by means of the various methods were presented in a colored line drawing showing the resistivity data as a function of depth along a W-E transect across the basement boundary fault (Fig. 3.11b). The cross section of the rho-z model overlaps with the western half of the W-E transect A-A′ displaying the gravimetric results in Figs. 3.9a and 3.9b.

The low-resistivity zone known from the foreland dips down under the basement with a pronounced low-resistivity "hump" close to the basement boundary fault and it undercuts the ZEV where it is endowed with metapegmatites at Wendersreuth and the transitional Püllersreuth pegmatite-see also Fig. 3.11a. According to the geophysical working group, at a depth of about 10 km a layer of high conductivity with 0.1 Ωm has been identified by both LOTEM and MMT measurements. Another shallow conductor exists between 300 and 1000 m, according to AAMT well below 10 Ωm and according to AMT of 50–100 Ωm (Haak 1989). The low resistivity layer at shallow depth was interpreted as an accumulation of graphitic matter or highly saline waters. The deep layer is still enigmatic as to its composition. Graphitic material could also account for this low-resistivity zone.

Organic matter appears during the primary mineralization in the pegmatitic rocks, whereas saline fluids me be held accountable for at least part of the secondary mineralization converting the primary phosphates in a varied spectrum of secondary phosphates in the HPPP. Finding similar phosphate assemblages in the HPPP and in the foreland as well can consequently be explained by saline basinal brines infiltrating along NW-SE structures the NE Bavarian Basement (see sections 4, 5, 6).

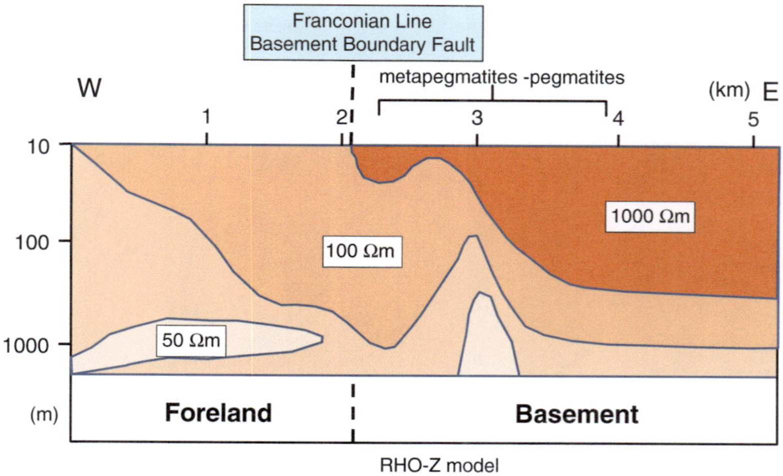

Fig. 3.11b Cross section of the rho-z model along a profile which overlaps with the western half of the W-E transect A-A′ plotted in Figs. 3.9a and 3.9b for positioning and correlation with the gravimetric data (Redrawn from Haak 1989). A low-resistivity zone identical with the foreland sediments (100 Ωm) to the west of the deep-seated basement boundary fault dips towards the east underneath the basement gneisses, undercutting the ZEV which is endowed with metapegmatites and the transitional Püllersreuth pegmatite (see also Fig. 3.11a)

3.3.4 Seismic Surveys and a Synopsis of the Geophysical Results in Terms of Pegmatitization

Seismic methods were extensively used during DEKORP and also during the pre-well-site studies to get a better understanding of zones of high reflectivity and the architecture of the upper crust in the northern Oberpfälzer Wald. The block diagram of Fig. 3.12 is based predominantly of these data and provides a synoptical overview of the subcrop in the Moldanubian Region and the adjacent Saxo-Thuringian Zone, published in Franke and Bram (1988). It was supplemented with some details pertinent to the evolution of the pegmatites in that region.

The Oberpfalz granites gently dip away from the ZEV towards the E (Figs. 3.9b and 3.9c). The pegmatites of the HPPP are not contingent upon these shallow granites. e.g., the Falkenberg and Flossenbürg Granites but related to high- velocity- layers developing a dome-like structure underneath the HPPP and at Püllersreuth, Coltan mineralization is only encountered within the basement block delimited by two NE-SW-trending deep-seated structures. Off-shoots of the lithium zone typical of the Saxo-Thuringian only exist along the NNW-SSE trending structure and its impact on the pegmatite mineralization tends to wane towards the South. Pegmatites tend to receive their rare elements from different source one extremely basic to

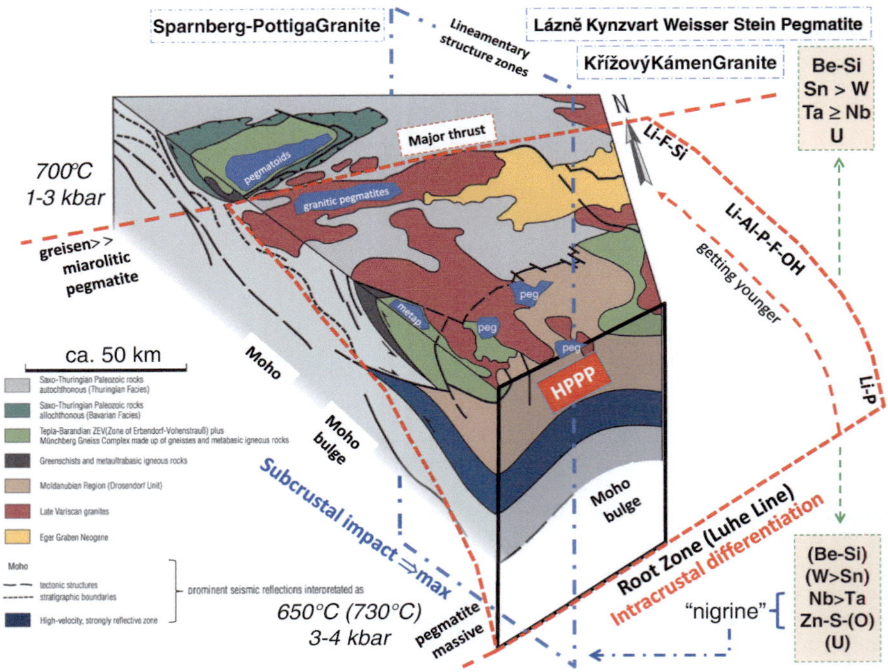

Fig. 3.12 Block diagram to show the geological setting based upon geophysical surveys (Modified and supplemented from Franke and Bram 1988). The 3-D representation was supplemented with those planar architectural elements from structural geology relevant to the explanation of the genesis of the pegmatites along the boundary between the Saxo-Thuringian and Moldanubian Zones and the contact between the allochthonous (root zone + major trust) and autochthonous units. The lineamentary structure zones striking around the N-S direction coincide with a bulge of the Moho. Element variation as a function of depth and geodynamic setting is given on the *right-hand side* and discussed in section 6. Granites and pegmatites which were vented and developed along these lineamentary structure zones are referred to at the top. For geology see also Figs. 1.1a and 2.1d. The various pegmatitic rocks are shown in *blue*, the Hagendorf-Pleystein Pegmatite Province (HPPP) marked by the *red box*. *metap.* metapegmatite, *peg* pegmatite. *Li-F-Si* lithium mica (e.g. zinnwaldite), *Li-Al-P-F-OH* lithium-bearing Al-phosphates (e.g. montebrasite-amblygonite), *Li-P* lithium-bearing phosphate (e.g. triphylite). The subcrustal impact on the pegmatite system increases towards the S along with a steepening of the thrust plane

ultrabasic the other strongly alkaline. The geophysical methods gravimetry, magmatic and electromagnetic measures respond in a different way to the various source rocks and their venting systems. A more comprehensive debate, considering also information gathered during mineralogical studies (section 4) and geological field work (section 5) is sparked in Sect. 6.4.1.

Chapter 4
Mineralogical Composition of Pegmatites and Aplites in the NE Bavarian Basement

Abstract The mineralogy of the pegmatites and aplites along the western edge of the Bohemian Massif cannot be sidelined. More than 250 minerals known from the most prominent members of these pegmatites and 272 species have been dealt with in this chapter. The minerals are treated irrespective of their siting within the Dana- or Strunz crystallographic systems and were grouped and discussed in accordance with the genetic evolution of the HPPP: Feldspar group, silica group, garnet s.s.s., aluminum silicates and corundum, zircon, phyllosilicates, niobium-, tantalum, tungsten and tin oxides, titanium minerals, molybdenite, carbon, calcium phosphates and calcium carbonates, aluminum phosphates with magnesium, iron, calcium and manganese, iron phosphates with magnesium, potassium and sodium, iron-manganese phosphates with magnesium, calcium, strontium, barium, potassium and sodium, manganese phosphates with calcium, manganese and iron oxides, sulfides and carbonates, arsenic minerals, bismuth minerals, copper minerals, halides, lithium minerals, rare earth element minerals, scandium minerals, beryllium minerals, boron silicates, uranium minerals, barium, lead, silver and antimony minerals, zinc minerals. The minerals were discussed as to their role as marker minerals or mineral associations to constrain the physical-chemical regime. They are used to unravel the pathway of their major elements from the source to the depocenter. Last but not least they play a vital role for mineralogical correlation and for the establishment of a minerostratigraphy within the HPPP and also across its boundaries. Some mineral groups play an outstanding role as to discrimination of the metamorphic and magmatic influence on the pegmatite evolution (columbite-tantalite s.s.s, zircon morphology). They can be applied for the depth zonation during the initial phases of the pegmatite's emplacement (phosphate vs. garnet) and during the alteration (Fe phosphate). K-Mn oxides and uranyl phosphates are key minerals for the chronological dating of the supergene alteration of the pegmatites. One phyllosilicate has been paid too little attention to during the recent past. Kaolinite group minerals bridge the gap between hydrothermal and supergene alteration. A diagnostic group is the "nigrine" mineral association which has derived from different parts of the roof rocks of the pegmatites and which can be found today in the stream sediments of the drainage systems intersecting the HPPP pegmatites.

H.G. Dill, *The Hagendorf-Pleystein Province: the Center of Pegmatites in an Ensialic Orogen*, Modern Approaches in Solid Earth Sciences 15,
DOI 10.1007/978-3-319-18806-5_4

From whatever angle you may look at the pegmatites and aplites along the western edge of the Bohemian Massif, you cannot sideline the mineralogy of these felsic rocks. It is too impressive to take a blind eye on that. More than 250 minerals known from the most prominent members of these pegmatites and from those emplaced along its edge have attracted mineralogists, crystallographers and rock hounds and mineral collectors alike (Table 4.1). 272 species have been listed in Table 4.1 and I am sure it will have to be added up with some more in the future. In the succeeding paragraphs the various minerals were dealt with in more detail with the focus placed upon the minerals from the HPPP and their adjacent mineral deposits in the Moldanubian region, the ZEV and the Saxo-Thuringian Zone so as to allow the readers themselves meditate on facies changes along a NW-SE section perpendicular to the boundaries of the geodynamic zones- for comparison and correlation see, e.g., Figs. 2.1c, d, 2.5b. For chemical analyses of the various minerals the reader is referred to the pertinent literature quoting the chemical composition of many of these minerals. This is also recommended to those interested in phase diagrams illustrating the physical-chemical regime. Space did not allow for presenting these more general issues. Whenever it has been possible, the morphology, color and texture of each mineral or mineral group is illustrated by a photograph on a macroscopic scale. From the numerous images taken under the SEM or EMP for mineralogical and textural analyses, only images have been selected for reproduction in this book if no appropriate photograph on a macroscopic scale has been available. There are countless publications to showcase minerals from pegmatites and supply the audience with lists of minerals. To place these minerals from the NE Bavarian Basement in a (field)geological context, Table 4.2, a synopsis, has been designed. It summarizes the data presented for each mineral or mineral group and outlines the facies, environment of formation and mineral association of the (Li)-Nb-P pegmatite system of the HPPP together with its incipient metamorpho-tectonic quartz-feldspar-muscovite (barren) pegmatoids, and Zn- and Be pegmatite systems. The metapegmatites were not considered in this list. The minerals are treated irrespective of their siting within the Dana- or Strunz crystallographic systems. These classification scheme at any rate merit to be used and without any doubt are helpful for any crystallographic study or systematics of mineral series but they are inappropriate for any study placing the formation of pegmatites into a geodynamic framework. The minerals were grouped and discussed in accordance with the genetic evolution of the HPPP.

4.1 Feldspar Group

Among the major mineralogical components making up the aplites and pegmatites in the HPPP, the feldspar-group minerals albite, orthoclase and microcline prevail not only among the silicates but predominate the entire mineralogical composition, excluding the Kreuzberg Pegmatite which suffered from strong alteration leaving behind a quartz pegmatite "ruin" that stands out from the present-day landscape in

Table 4.1 Minerals in pegmatitic and aplitic rocks in the NE Bavarian Basement

Mineral	Chemical composition	Section	Fichelgebirge granite+ granitic pegmatites	Püllersreuth	Hopfau	Marchaney	Wildenau-Plössberg	Peugenhammer	Brünst	Burkhardsrieth	Hagendorf-South	Hagendorf-North	Kreuzberg-Pleystein	Footwall Aplite Pleystein	Trutzhofmühle-Pleystein	Reinhardsrieth	Miesbrunn-Swarm	Silbergrube-Waidhaus	Hühnerkobel-Zwiesel	Birkhöhe-Zwiesel	Blötz-Bodenmais	Tittling granite pegmatite
Acanthite	Ag_2S	4.26									HS											
Aeschynite	$(Ca_{4.0})_{4.0}(Nb_{7.1}Ta_{0.2}Ti_{0.6}Fe_{0.25}W_{0.03})_{8.18}O_{24}$	4.8											(KB)									
Al strunzite	$(Mn_{0.65}Fe_{0.26}Zn_{0.08}Mg_{0.01})(Fe_{1.50}Al_{0.50})(PO_4)_2(OH)_2 \cdot 6H_2O$	4.12									HS							SG				
Albite	$NaAlSi_3O_8$	4.1					WP	PG			HS	HN	KB	NA	TM	RR	MB	SG	HK			TI
Allanite	$(Ce,Ca,Y)_2(Al,Fe)_3(SiO_4)_3(OH)$	4.21							BR													TI
Alluaudite "Hühnerkobelite"	$NaCaFe(Mn,Fe,Fe,Mg)_2(PO_4)_3$	4.13										HN							HK			
Anatas	TiO_2	4.9				MY								NA					HK			TI
Andalusite	Al_2SiO_5	4.4							BR												BB	
Apatite-(MnF)	$(Ca,Mn2+,Fe2+)_5(PO_4)_3(F,Cl,OH)$	4.10		PR	HO		WP		BR		HS	HN	KB		TM	RR	MB	SG	HK		BB	TI
Arrojadite-(BaFe)	$(Ba,K,Pb)Na_3(Ca,Sr)(Fe,Mg,Mn)_{14}Al(PO_4)_{11}(PO_3OH)(OH,F)_2$	4.13									HS											
Arrojadite-(KFe)	$KNa_2CaNa_2(Fe,Mn,Mg)_{13}Al(PO_4)_{11}(PO_3OH)(OH,F)_2$	4.13									HS											TI
Arrojadite-(Na)	$KNa_4Ca(Fe,Mn,Mg)_{13}Al(PO_4)_{11}(PO_3OH)(OH,F)_2$	4.13									HS								HK			
Arrojadite-(NaFe)	$NaNa_2CaNa_2(Fe,Mn,Mg)_{13}Al(PO_4)_{11}(PO_3OH)(OH,F)_2$	4.13									HS											
Arrojadite-(SrFe)	$SrFeNa_2Ca(Fe,Mn,Mg)_{13}Al(PO_4)_{11}(PO_3OH)(OH,F)_2$	4.13									HS											
Arsenopyrite	$FeAsS$	4.16	FG		HO		WP				HS		KB				MB	SG	HK			
Autunite	$Ca(UO_2)_2(PO_4)_2 \cdot 11H_2O$	4.25					WP				HS	HN	KB		TM			SG	HK		BB	
Barbosalite	$FeFe_2(PO_4)_2(OH)_2$	4.12									HS		KB									
Baryte	$BaSO_4$	4.26					WP						KB		TM	RR	MB					
Bassetite	$Fe(UO_2)_2(PO_4)_2 \cdot 8(H_2O)$	4.25									HS											
Bavenite	$Ca_4Be_2Al_2Si_9O_{26}(OH)_2$	4.23																				TI

(continued)

Table 4.1 (continued)

Bazzite	$Be_3(Sc,Al)_2Si_6O_{18}$	4.23																				TI
Benyacarite	$(H_2O,K)(Mn,Fe)_2(Fe,Ti)_2\ Ti[(O,F)_2\vert(PO_4)_4]\cdot 14H_2O$	4.9									HS		KB									
Beraunite	$FeFe_5(PO_4)_4(OH)_5\cdot 4H_2O$	4.12			HO		WP				HS	HN	KB		TM		MB	SG	HK			
Bermanite	$MnMn_2(PO_4)_2(OH)_2\bullet 4(H_2O)$	4.14									HS											
Berthierine	$(Fe,Fe,Al,Mg)_{2-3}(Si,Al)_2O_5(OH)_4$	4.15									HS											
Bertrandite	$Be_4(Si_2O_7)(OH)_2$	4.23	FG																HK			TI
Beryl	$Be_3Al_2Si_6O_{18}$	4.23	FG	PR			WP				HS								HK			TI
Betafite	$(Ca_{4.0}Na_{0.6}U_{1.6})_{6.2}(Nb_{3.6}Ta_{0.3}Ti_{3.2})_{7.1}O_{24}$	4.8											(KB)									
Beudandite	$PbFe_3(OH)_6SO_4AsO_4$	4.16	FG														MB					
Biotite	$K(Mg,Fe)_3(Si_3Al)O_{10}(OH,F)_2$	4.6		PR			WP			BH	HS	HN	KB				MB		HK			
Bismite	Bi_2O_3	4.17									HS		KB				MB					TI
Bismuth	Bi	4.17									HS		KB									
Bismuthinite	Bi_2S_3	4.17									HS	HN					MB					
Bityite	$CaLiAl_2(AlBeSi_2)O_{10}(OH)_2$	4.23																				TI
Brazilianite	$NaAl_3(PO_4)_2(OH)_4$	4.11									HS											
Brochantite	$Cu_4(SO_4)(OH)_6$	4.18									HS											
Brookite	TiO_2	4.9													TM							TI
Cacoxenite	$(Fe,Al)_{25}[O_6\vert(OH)_{12}\vert(PO_4)_{17}]\cdot 75\ H_2O$	4.11			HO		WP				HS	HN	KB	NA?	TM		MB	SG	HK			
Calcioferrite	$Ca_4Fe(Fe,Al)_4(PO_4)_6(OH)_4\bullet 12(H_2O)$	4.10									HS											
Calcite	$CaCO_3$	4.10									HS											TI
Carlhintzeite	$Ca_2AlF_7\cdot H_2O$	4.19									HS TYP	HN	KB									
Cassiterite	SnO_2	4.8	FG								HS		KB									
Cerite	$Ce_9Fe(SiO_4)_6[(SiO_3)(OH)](OH)_3$	4.21															MB					
Cerussite	$PbCO_3$	4.26																				TI
Chabazite	$(Ca_{0.5},Na,K)_4[Al_4Si_8O_{24}]\bullet 12H_2O$	4.7																				TI
Chalcanthite	$CuSO_4\bullet 5(H_2O)$	4.18									HS											
Chalcocite	Cu_2S	4.18									HS											
Chalcophanite	$(Zn,Fe,Mn)Mn_3O_7\bullet 3(H_2O)$	4.15									HS											
Chalcopyrite	$CuFeS_2$	4.18			HO						HS	HN						SG	HK			TI
Chalcosiderite	$CuFe_6 3+(PO_4)_4(OH)_8\cdot 4H_2O$	4.18									HS		KB									
Chamosite (Var. Thuringite)	$(Fe,Mg,Fe)_5Al(Si_3Al)O_{10}(OH,O)_8$	4.15									HS											
Cheralite	$(Ce,Ca,Th)(P,Si)O_4$	4.21									HS					RR						
Chernikovite	$(H_3O)_2(UO_2)_2(PO_4)_2\cdot 6H_2O$	4.25									HS	HN										
Childrenite	$FeAl(PO_4)(OH)_2\cdot (H_2O)$	4.11									HS		KB		TM	RR	MB	SG	HK			

Chlorite Group	$(Fe,Mg,Al,Zn)_6(Si,Al)_4O_{10}(OH)_8$	4.6					WP		BR		HS			NA				SG				TI
Chrysocolla	$(Cu,Al)_2H_2Si_2O_5(OH)_4•n(H_2O)$	4.18			HO						HS											TI
Churchite-(Y)	$YPO_4•2(H_2O)$	4.21	FG												TM							
Coffinite	$U(SiO_4)_{1-x}(OH)_{4x}$	4.25															MB					
Columbite-(Fe)	$FeNb_2O_6$	4.8	FG				WP				HS	HN	KB	NA	TM	RR	MB	SG	HK	BZ		TI
Connellite	$Cu_{19}Cl_4(SO_4)(OH)_{32}•3(H_2O)$	4.18									HS											
Copper	Cu	4.18									HS											
Corundum (sapphire)	Al_2O_3	4.4																			BB	
Covellite	CuS	4.18									HS		KB									
Crandallite	$CaAl_3(PO_4)_2(OH)_5•(H_2O)$	4.11										HN	KB	NA								
Cryptomelane	$K(Mn,Mn)_8O_{16}$	4.15					WP				HS		KB				MB	SG				
Cubanite	$CuFe_2S_3$	4.18									HS											
Cuprite	Cu_2O	4.18									HS											
Cuprobismutite	$Cu_{10}Bi_{12}S_{23}$	4.17									HS											
Cyrilovite	$NaFe_3(PO_4)_2(OH)_4•2(H_2O)$	4.12	FG								HS											
Devilline	$CaCu_4(SO_4)_2(OH)_6•3(H_2O)$	4.18									HS											
Diadochite	$Fe_2(PO_4)(SO_4)(OH)•6(H_2O)$	4.15									HS				TM				HK			
Dickinsonite-(KMn)	$KNa_4Ca(Mn,Fe)_{14}Al(PO_4)_{12}(OH)_3$	4.13									HS											
Dickinsonite-(NaMn)	$KNa_4Ca(Mn,Fe)_{14}Al(PO_4)_{12}(OH)_2$	4.13									HS											
Digenite	Cu_9S_5	4.18									HS											
Diopside	$CaMgSi_2O_6$	4.7								BH												
Djurleite	$Cu_{31}S_{16}$	4.18									HS											
Dufrénite	$Ca_{0.5}FeFe_5[(OH)_3\|(PO_4)_2]_2·2H_2O$	4.10									HS		KB						HK			
Dumortierite	$Al_{6.5-7}(BO_3)(SiO_4)_3(O,OH)_3$	4.24				MY																
Earlshannonite	$(Mn,Fe)Fe_2(PO_4)_2(OH)_2•4(H_2O)$	4.13									HS											
Emplectite	$CuBiS_2$	4.17									HS											
Eosphorite	$MnAl(PO_4)(OH)_2·H_2O$	4.11			HO						HS	HN	KB			RR	MB	SG	HK			
Epidote group	$Ca_2Fe3+_{2.25}\ Al_{0.75}(SiO_4)_3(OH)$	4.7								BH											BB	TI
Ernstite	$(Mn_{1-x}Fe_x)Al(PO_4)(OH)_{2-x}O_x$	4.11														RR						
Euclase	$BeAl(SiO_4)(OH)$	4.23	FG																			
Fairfieldite	$Ca_2(Mn,Fe)(PO_4)_2·2H_2O$	4.13									HS	HN	KB						HK			
Ferriallanite	$CaCe(Fe,Fe,Al)_3[SiO_4][Si_2O_7]O(OH)$	4.21															MB					
Ferrisicklerite	$Li(Fe,Mn)PO_4$	4.20			HO						HS	HN						SG	HK			
Ferristrunzite	$FeFe_2(PO_4)_2(OH)_3•5(H_2O)$	4.12													TM			SG				
Ferrolaueite	$FeFe_2(PO_4)_2(OH)_2•8(H_2O)$	4.12													TM							
Ferrostrunzite	$FeFe_2(PO_4)_2(OH)_2•6(H_2O)$	4.12																SG				

(continued)

Table 4.1 (continued)

Florencite-(Ce)	$CeAl_3(PO_4)_2(OH)_6$	4.21					WP															
Fluellite	$Al_2(PO_4)F_2(OH)\cdot 7H_2O$	4.11									HS		KB									
Fluorite	CaF_2	4.19	FG								HS											TI
Fluoroapatite	$Ca_5(PO_4)_3(F)$	4.10	FG											NA	TM			SG	HK			
Frondelite	$MnFe_4(PO_4)_3(OH)_5$	4.13									HS		KB		TM	RR			HK			
Gahnite	$ZnAl_2O_4$	4.27															MB					
Galena	PbS	4.26									HS											TI
Garnet s.s.s.	"Pyralspit" (Mg-Fe-Mn dominated)	4.3		PR			WP		BR	BH					TM		MB	SG	HK	BZ		
Gibbsite	$Al(OH)_3$	4.4												NA								
Goethite	alpha-FeO(OH)	4.15			HO		WP				HS	HN	KB		TM		MB	SG				
Gorceixite	$BaAl_3(PO_4)(PO_3OH)(OH)_6$	4.11	FG											NA								
Gordonite	$MgAl_2(PO_4)_2(OH)_2\bullet 8(H_2O)$	4.11													TM							
Goyazite	$SrAl_3(PO_4)(PO_3OH)(OH)_6$	4.11	FG								HS	HN	KB	NA								
Graftonite	$(Fe,Mn,Ca)_3(PO4)_2$	4.13									HS								HK			
Graphite	C	4.10									HS											
Greenockite	CdS	4.27									HS	HN										
Gypsum	$CaSO_4\cdot 2H_2O$	4.15									HS		KB									
Hagendorfite	$NaCaMn(Fe,Fe,Mg)_2(PO_4)_3$	4.13									HS TYP	HN										
Helvine	$Mn_4Be_3(SiO_4)_3S$	4.23																				TI
Hematite	Fe_2O_3	4.15			HO					BR	HS	HN	KB									TI
Hemimorphite	$Zn_4Si_2O_7(OH)_2\bullet(H_2O)$	4.27			HO						HS											
Herderite series	$CaBe(PO_4)F$	4.23	FG																			
Heterosite	$(Fe,Mn)(PO_4)$	4.13			HO						HS	HN	KB					SG	HK			TI
Hopeite	$Zn_3(PO_4)_2\cdot 4H_2O$	4.27									HS	HN	KB									
Hureaulite (+oxi-hureaulite)	$Mn_5(PO_3OH)_2(PO_4)_2\bullet 4(H_2O)$	4.14									HS	HN						SG	HK			TI
Ilmenite	$FeTiO_4$	4.9									HS		KB			RR	MB					
Jahnsite-(CaMnFe)	$CaMnFe_2Fe_2(PO_4)_4(OH)_2\bullet 8(H_2O)$	4.13									HS							SG				
Jarosite	$KFe_3(SO_4)_2(OH)_6$	4.15									HS											TI
Jungite	$Ca_2Zn_4Fe_8(PO_4)_9(OH)_9\bullet 16(H_2O)$	4.27									HS TYP											
Kaolinite	$Al_2Si_2O_5(OH)_4$	4.6					WP				HS	HN	KB	NA		RR	MB		HK			TI
Kastningite	$(Mn,Fe,Mg)Al_2(PO_4)_2(OH)_2\bullet 8(H_2O)$	4.11									HS							SG TYP				
Keckite	$CaMn(Fe,Mn)_2Fe_2(OH)_3(H_2O)_7(PO_4)_4$	4.13									HS TYP				TM			SG	HK			

Kidwellite	$NaFe_93+(PO_4)_6(OH)_{10}\cdot5H_2O$	4.12									HS											
Kingsmountite	$(Ca,Mn)_4(Fe,Mn)Al_4(PO_4)_6(OH)_4\bullet12(H_2O)$	4.11									HS							SG				
Kolbeckite	$ScPO_4\bullet2(H_2O)$	4.22													TM							
Kryzhanovskite	$MnFe_2(PO_4)_2(OH)_2\bullet(H_2O)$	4.13									HS	HN										
Kyanite	Al_2SiO_5	4.4																			BB	
Landesite	$(Mn,Mg)_9Fe_3(PO_4)_8(OH)_3\cdot9(H_2O)$	4.14									HS	HN						SG				
Langite	$Cu_4(SO_4)(OH)_6\bullet2(H_2O)$	4.18									HS											
Laueite	$MnFe_2(PO_4)_2(OH)_2\bullet8(H_2O)$	4.13			HO						HS TYP							SG	HK			
Laumontite	$CaAl_2Si_4O_{12}\bullet4(H_2O)$	4.7																				TI
Lazulite	$MgAl_2(PO_4)_2(OH)_2$	4.11													TM	RR	MB					
Lehnerite	$Mn[UO_2/PO_4]_2\bullet8(H_2O)$	4.25									HS TYP											
Lepidocrocite	$FeO(OH)$	4.15									HS											
Lermontovite	$U(PO_4)(OH)\bullet(H_2O)$(?)	4.25													TM							
Leucophosphite	$KFe_2(PO_4)_2(OH)\cdot2(H_2O)$	4.12			HO						HS		KB					SG				
Libethenite	$Cu_2(PO_4)(OH)$	4.18									HS											
'Limonite'	$FeOOH$	4.15									HS	HN	KB	NA	TM		MB	SG				
Lipscombite	$(Fe,Mn)Fe_2(PO_4)_2(OH)_2$	4.13									HS											
Ludlamite	$(Fe,Mg,Mn)_3(PO_4)_2\cdot4H_2O$	4.13									HS	HN						SG	HK			
Mackinawite	$(Fe,Ni)S_{0.9}$	4.15									HS											
Magnetite	$FeFe_2O_4$	4.15							BR		HS											TI
Malachite	$Cu_2(CO_3)(OH)_2$	4.18									HS											TI
Mangangordonite	$(Mn,Fe,Mg)Al_2(PO_4)_2(OH)_2\bullet8(H_2O)$	4.11									HS							SG				
Mantienneite	$KMg_2Al_2Ti(PO_4)_4(OH)_3\bullet15(H_2O)$	4.9																SG				
Marcasite	FeS_2	4.15									HS						MB	SG				
Matildite/Schapbachite	$AgBiS_2$	4.26									HS											
Matulaite	$CaAl_{18}(PO_4)_{12}(OH)_{20}\bullet28(H_2O)$	4.11									HS											
Mesolite	$Na_2Ca_2[Al_6Si_9O_{30}]\cdot8H_2O$	4.7											KB									
Messelite	$Ca_2(Fe,Mn)(PO_4)_2\bullet2(H_2O)$	4.10									HS								HK			
Metaautunite	$Ca(UO_2)_2(PO_4)_2\bullet2\text{-}6(H_2O)$	4.25									HS								HK			
Metaswitzerite	$Mn_3(PO_4)_2\bullet4(H_2O)$	4.14									HS					RR						
Metatorbernite	$Cu(UO_2)_2(PO_4)_2\bullet8(H_2O)$	4.25									HS											
Metavivianite	$(Fe_{3-x},Fe_x)(PO_4)_2(OH)_x\bullet8\text{-}x(H_2O),x=0.5$	4.12													TM		MB					
Meurigite-K	$KFe_7(PO_4)_5(OH)_7\cdot8(H_2O)$	4.12									HS		KB					SG				
Microcline	$KAlSi_3O_8$	4.1			HO		WP	PG	BR	BH	HS	HN							HK	BZ	BB	TI

(continued)

Table 4.1 (continued)

Milarite	$K_2Ca_4Al_2Be_4(Si_{24}O_{60})•(H_2O)$	4.23																				TI
Mitridatite	$Ca_2Fe_3(PO_4)_3O_2•3(H_2O)$	4.10			HO						HS				TM		MB	SG	HK			
Molybdenite	MoS_2	4.10			HO						HS	HN										
Monazite	(Ce,La,Nd,Th)PO_4	4.21					WP		BR		HS		KB		TM	RR	MB					
Montgomeryite	$Ca_4MgAl_4(PO_4)_6(OH)_4•12(H_2O)$	4.11																	HK			
Morinite	$NaCa_2Al_2(PO_4)_2(F,OH)_5·2(H_2O)$	4.11									HS		KB									
Muscovite	$KAl_2(Si_3Al)O_{10}(OH,F)_2$	4.6		PR	HO		WP		BR	BH	HS	HN	KB	NA	TM	RR	MB	SG	HK	BZ	BB	
Natrodufrénite	Na(Fe,Fe)(Fe,Al)$_5(PO_4)_4(OH)_6•2(H_2O)$	4.12	FG														MB					
Nontronite	$Na_{0.3}Fe_2(Si,Al)_4O_{10}(OH)_2•n(H_2O)$	4.6									HS		KB									
Nordgauite	$MnAl_2(PO_4)_2(F,OH)_2 . 5H_2O$	4.11									HS TYP							SG				
Opaline silica	$SiO_2.nH_2O$	4.2												NA								
Orthoclase	$KAlSi_3O_8$	4.1									HS	HN	KB	NA	MB	RR	MB	SG				
Oxiberaunite	(Fe,Mn)$Fe_5(PO_4)_4O(OH)_4•6(H_2O)$	4.13									HS	HN	KB		TM		MB	SG				
Pachnolite	$NaCaAlF_6·H_2O$	4.19									HS		KB									
Parahopeite	$Zn_3(PO_4)_2·4H_2O$	4.27									HS	HN	KB									
Parascholzite	$CaZn_2(PO_4)_2•2(H_2O)$	4.27									HS TYP	HN										
Paravauxite	$FeAl_2(PO_4)_2(OH)_2•8(H_2O)$	4.11									HS							SG				
Paulkerrite	K(Mg,Mn)$_2$(Fe,Al)$_2Ti(PO_4)_4(OH)_3•15(H_2O)$	4.9					WP															
Pavonite	(Ag,Cu)(Bi,Pb)$_3S_5$	4.17									HS											
Perloffite	Ba(Mn,Fe)$_2Fe_2(PO_4)_3(OH)_3$	4.13									HS											
Petitjeanite	$Bi_3O(OH)(PO_4)_2$	4.17											KB									
Petscheckite	UFe(Nb,Ta)$_2O_8$	4.8									HS											
Pharmakosiderite	$KFe_4(AsO_4)_3(OH)_4·6-7(H_2O)$	4.16			HO								KB									
Phenakite	$Be_2(SiO_4)$	4.23	FG																			
Phosphoferrite	(Fe,Mn)$_3$(PO4)$_2·3(H_2O)$	4.13									HS	HN TYP			TM				HK			
Phosphophyllite	Zn_2(Fe,Mn)$(PO_4)_2·4(H_2O)$	4.27									HS	HN TYP	KB									
Phosphoscorodite	Fe(P,As)$O_4•2(H_2O)$	4.16															MB					
Phosphosiderite	$FePO_4·2(H_2O)$	4.12			HO		WP				HS	HN	KB		TM			SG	HK			
Posnjakite	$Cu_4(SO_4)(OH)_6•(H_2O)$	4.18									HS											
Pseudolaueite	$MnFe_2(PO_4)_2(OH)_2•7-8(H_2O)$	4.13			HO						HS TYP											
Pseudomalachite	$Cu_5(PO_4)_2(OH)_4$	4.18	FG								HS											

Purpurite	$MnPO_4$	4.14									HS								HK			
Pyrite	FeS_2	4.15			HO	MY	WP				HS	HN	KB		TM		MB	SG	HK		BB	
Pyrochlore s.s.s. (unknown)	$(Si, Fe,Al,P,U, Ca)_3(Nb,Ti,Ta)_8O_{24}$	4.8											(KB)									
Pyrolusite	MnO_2	4.15									HS	HN	KB						HK			
Pyrosmalite	$(Fe,Mn)_8Si_6O_{15}(OH,Cl)_{10}$	4.15									HS											
Pyrrhotite	$Fe_{1-x}S$	4.15		PR		MY					HS	HN							HK		BB	
Quartz	SiO_2	4.2	FG	PR	HO	MY	WP	PG	BR	BR	HS	HN	KB	NA	TM	RR	MB	SG	HK	BZ	BB	TI
Reddingite	$Mn_3(PO_4)_2•3(H_2O)$	4.14			HO						HS	HN										
Rhodochrosite	$MnCO_3$	4.15									HS											
Riomarinaite(?)	$Bi(OH)SO_4•H_2O$	4.17															MB					
Rittmannite	$MnMnFeAl_2(OH)_2(PO_4)_4•8(H_2O)$	4.11									HS							SG				
Robertsite	$Ca_6Mn_9(PO_4)_9O_6(H_2O)_6•3(H_2O)$	4.14									HS								HK			
Rockbridgeite	$(Fe,Mn)Fe_4(PO_4)_3(OH)_5$	4.13			HO	MY	WP				HS	HN	KB		TM	RR	MB	SG	HK			
Rosasite	$(Cu,Zn)_2(CO_3)(OH)_2$	4.18									HS											
Rozenite	$FeSO_4•4(H_2O)$	4.15									HS											
Rutile	TiO_2	4.9				MY					HS											TI
Samarskite-(Y) Var. Uransamarskite	$(Y,Fe,U)(Nb,Ta)_5O_4$	4.8									HS										BB	TI
Santabarbaraite	$Fe_3(PO_4)_2(OH)_3•5(H_2O)$	4.12															MB					
Sarcopside	$(Fe,Mn,Mg)_3(PO_4)_2$	4.13									HS											
Sc phosphate (minx)	$(K,Ba)(Sc,Zr)_{2.5}(PO_4)_3$	4.22													TM							
Sc silicate-phosphate (minx)	$ZrSc(SiO_4)(PO_4) (H_2O)$	4.22													TM							
Scheelite	$Ca(WO_4)$	4.8	FG				WP															
Scholzite	$CaZn_2(PO_4)_2·2(H_2O)$	4.27									HS	HN TYP										
Schoonerite	$Fe_2ZnMnFe(PO_4)_3(OH)_2·9(H_2O)$	4.27			HO						HS		KB									
Scorodite	$FeAsO_4•2(H_2O)$	4.16	FG		HO		WP										MB					
Scorzalite	$(Fe,Mg)Al_2(PO_4)_2(OH)_2$	4.11									HS						MB					
Sicklerite	$Li(Mn,Fe)PO_4$	4.20																	HK			
Siderite	$FeCO_3$	4.15									HS		KB									
Sillimanite	Al_2SiO_5	4.4				MY			BR									SG		BZ	BB	
Silver	Ag	4.26									HS											
Sphalerite	$(Zn,Fe)_S$	4.27		PR	HO	MY				BH	HS	HN	KB				MB	SG	HK		BB	
Stannite	Cu_2FeSnS_4	4.8	FG								HS											
Stellerite	$CaAl_2Si_7O_{18}•7(H_2O)$	4.7																				TI

(continued)

Table 4.1 (continued)

Sternbergite	$AgFe_2S_3$	4.26									HS											
Stewartite	$MnFe_2(PO_4)_2(OH)_2{\bullet}8(H_2O)$	4.13			HO						HS							SG	HK			
Stibnite	Sb_2S_3	4.26															MB					
Stilpnomelane	$K(Fe,Al)_{10}Si_{12}O_{30}(OH)_{13}$	4.15									HS											
Strengite	$FePO_4{\cdot}2(H_2O)$	4.12			HO		WP				HS	HN	KB		TM		MB	SG	HK			
Stromeyerite	AgCuS	4.26									HS						MB					
Strunzite	$MnFe_2(PO_4)_2(OH)_2{\cdot}6(H_2O)$	4.13			HO		WP				HS TYP	HN	KB		TM		MB	SG	HK			
Sulphur	S	4.15									HS						MB					
Switzerite	$(Mn,Fe2)_3(PO_4)_2{\bullet}7(H_2O)$	4.13									HS								HK			
Tainiolite	$KLiMg_2Si_4O_{10}F_2$	4.20									HS											
Tavorite	$LiFe(PO_4)(OH)$	4.20									HS											
Titanite	$CaTiSiO_5$	4.9							BR													TI
Todorokite Co-bearing	$(Na,Ca,K)_2(Mn,Mn)_6O_{12}{\bullet}3\text{-}4.5(H_2O)$	4.15												NA								
Topaz	$Al_2(F,OH)_2\vert SiO_4$	4.4	FG		HO														HK			
Torbernite	$Cu(UO_2)_2(PO_4)_2{\cdot}12(H_2O)$	4.25									HS	HN	KB		TM			SG				TI
Tourmaline	Schorl-Dravite	4.24				MY	WP	PG	BR					NA	TM		MB	SG	HK	BZ	BB	TI
Triphylite	$LiFePO_4$	4.20			HO						HS	HN	KB					SG	HK			TI
Triplite	$(Mn,Fe,Mg,Ca)_2(PO_4)(F,OH)$	4.13					WP				HS	HN	KB		TM				HK			
Triploidite	$(Mn,Fe)_2(PO_4)(OH)$	4.13													TM			SG				
Turquoise	$Cu(Al,Fe)_6(PO_4)_4(OH)_8{\cdot}4(H_2O)$	4.18									HS		KB									
UNK9	$(Fe_{1.5\text{-}1.2},Mn_{0.3\text{-}0.5})(Fe_{2.7\text{-}2.9},Al_{0.0\text{-}0.3})(PO_4)_3(OH)_4(H_2O)$	4.13														RR						
Uralolite	$Ca_2Be_4(PO_4)_3(OH)_3{\bullet}5(H_2O)$	4.23																				TI
Uraninite	UO_2	4.25		PR							HS	HN			TM				HK			
Uranocircite	$Ba(UO_2)_2(PO_4)_2{\cdot}12(H_2O)$	4.25					WP					HN										
Uranophane	$Ca(UO_2)_2SiO_3(OH)_2{\cdot}5(H_2O)$	4.25									HS											TI
Uranosphaerite	$Bi(UO_2)O_2(OH)$	4.25									HS											
Uranpyrochlore	$(U,Ca,Ce)_2(Nb,Ta)_2O_6(OH,F)$	4.8									HS											
Valleriite	$4(Fe,Cu)S{\bullet}3(Mg,Al)(OH)_2$	4.18									HS											
Vandendriesscheite	$Pb(UO_2)_{10}O_6(OH)_{11}{\bullet}11(H_2O)$	4.25									HS											
Variscite	$AlPO_4.2(H_2O)$	4.11									HS						MB	SG				
Vauxite	$FeAl_2(PO_4)_2(OH)_2{\bullet}6(H_2O)$	4.11														RR						
Vivianite	$Fe_3(PO_4)_2{\cdot}8H_2O$	4.12				MY					HS	HN	KB		TM	RR		SG	HK			
Vochtenite	$(Fe,Mg)Fe[(UO_2)(PO_4)]_4(OH).12\text{-}13(H_2O)$	4.25													TM							
Vyacheslavite	$U(PO_4)(OH){\bullet}2.5(H_2O)$	4.25													TM							
"Wad" /"manganomelane"	Hydrated Mn oxides	4.15									HS		KB									

Wagnerite	$(Mg,Fe)_2(PO4)F$	4.12															MB					
Wavellite	$Al_3(PO_4)_2(OH,F)_3{\cdot}5(H_2O)$	4.11			HO		WP				HS	HN	KB		TM			SG				
Waylandite	$BiAl_3(PO_4)_2(OH)_6$	4.17					WP						KB	NA								
Whiteite-(CaMnMn)	$Ca(Fe,Mn)Mn_2Al_2(PO_4)_4(OH)_2.8(H_2O)$	4.11									HS TYP											
Whitlockite	$Ca_9(Mg,Fe)(PO_4)_6(PO_3OH)$	4.10									HS											
Whitmoreite	$FeFe_2(PO_4)_2(OH)_2{\cdot}4(H_2O)$	4.12									HS		KB		TM	RR	MB	SG	HK			
Wilhelmvierlingite	$CaMnFe(PO_4)_2(OH).2(H_2O)$	4.13									HS TYP											
Wittichenite	Cu_3BiS_3	4.17									HS											
Wolfeite	$(Fe,Mn)_2(PO_4)(OH)$	4.13									HS	HN	KB		TM							
Wolframite	$(Fe,Mn)WO_4$	4.8	FG										(KB)									
Xanthoxenite	$Ca_4Fe_2(PO_4)_4(OH)_2{\cdot}3(H_2O)$	4.10									HS	HN	KB						HK			
Xenotime-(Y)	YPO_4	4.21									HS					RR	MB					
Zairite	$Bi(Fe,Al)_3(PO_4)_2(OH)_6$	4.17											KB									
Ximengite (?)	$Bi(PO_4)$	4.17											KB									
Zircon	$ZrSiO_4$	4.5					WP		BR		HS	HN	KB	NA		RR	MB					
Zwieselite ("pseudo-zwieslite")	$(Fe,Mn)_2(PO_4)F$	4.13				MY	WP				HS	HN	KB			(RR)			HK	BZ	BB	

(continued)

Table 4.1 (continued)

Typ = type locality
? = uncertain
() = in stream sediments of the clastic aureole around the pegmatite
Compilation based upon data from Schlüter et al. (1999), Birch et al. (2011) Dill et al. (2006a, b, c, 2008a, b, c, 2009a, b, 2010a, b, c, 2012a, b, 2013a, b, c), Frondel (1958), Grey et al. (2010) Habel (2003), Habel and Habel (1991), Habel and Reinhardt (2011), Heinrich (1951), Hochleitner and Fehr (2010), Kastning and Schlüter (1994), Keck (2001), Laubmann and Steinmetz (1920), Moore and Ito (1980) Mücke (1977, 1978, 1980, 1983, 1987, 1988), Mücke and Keck (2008, 2011), Müllbauer (1925), Nägele (1982), Pfaffl (1966), Pöllmann et al. (2005), Redhammer et al. (2005), Robbins et al. (2008), Schnorrer et al. (2003), Strunz (1948, 1954a, b, 1956, 1961), Strunz and Tennyson (1956), Strunz and Fischer (1957), Strunz et al. (1975, 1976), Sturman et al. (1981), Tennyson (1954, 1958), Vochten et al. (1995), Walenta and Binder (1980), Weber (1977), Wilk (1959), Yakovenchuk et al. (2012)
The section in the book where the mineral is referred and discussed is given in the column 3 to the right of the chemical composition (e.g. 4.26). The various provinces are shaded in different colors.
Blue: Granitic pegmatites of the Fichtelgebirge Anticline
Grey: (Meta)pegmatite of the ZEV (Zone of Erbendorf-Vohenstrauß)
Brown: Pegmatites in the Moldanubian Region to the NW and N of the HPPP
Yellow: Pegmatites and aplites within the HPPP
Green: Pegmatites in the Moldanubian Region to the SE of the HPPP

Table 4.2 Facies, environment and mineral association of the (Zn-Be)-Li-Nb-P pegmatite system in the HPPP with its incipient metamorpho-tectonic quartz-feldspar-muscovite (barren) pegmatoids, and Zn- and Be pegmatite systems

Stage	Process	Eh/pH	T (°C)	Minerals	Site	Remarks
1	Metamorpho-tectonic syn- to late kinematic mobilization		≈600°C	Quartz, alkaline feldspar, muscovite	e.g. Zessmanssrieth, Peugenhammer, Schafbruck , Zengerhof (see also Fig. 5)	Aploid and pegmatoid mobilizates initial and marginal facies relative to the HPPP
2a	Late kinematic (to postkinematic) subcrustal magmatic mobilization		>500	Quartz, muscovite , (biotite), albite, K feldspar , apatite- (F), xenotime, monazite, cheralite (?), zircon I, tourmaline, magnetite, allanite , sillimanite, andalusite	Brünst, Hagendorf-South, Hagendorf-North, Pleystein-Kreuzberg, Burkhardsrieth, Miesbrunn (in parts)	Transitional between metamorphogenic and magmatogenic
2b	Late kinematic (to postkinematic) subcrustal magmatic mobilization		≈600	Beryl , ilmenite, rutile	Hagendorf-South, Plössberg	Transitional between metamorphogenic and magmatogenic
2c	Late kinematic (to postkinematic) subcrustal magmatic mobilization		≈400	Gahnite	Miesbrunn	Transitional between metamorphogenic and magmatogenic
2d	Late kinematic (to postkinematic) subcrustal magmatic mobilization		>210	Fe sphalerite (Cd-bearing) ,valleriite , cubanite , pyrrhotite	Hagendorf-South, Hagendorf-North, Pleystein-Kreuzberg, Burkhardsrieth,	Transitional between metamorphogenic and magmatogenic
3a	Postkinematic mobilization prevailing over late kinematic mobilization	<0	>400	Columbite-(Fe), petscheckite , samarskite, uranpyrochlor , uraninite, graphite,ferrialanite cerite , Ce-La-Nd silicate (unident.)	Hagendorf-South, Hagendorf-North, Pleystein-Kreuzberg, Trutzhofmühle, Pleystein-New Aplite, Reinhardsrieth, Waidhaus-Silbergrube, Miesbrunn, Wildenau-Plössberg	Transitional between metamorphogenic and magmatogenic
3b	Postkinematic mobilization > late kinematic		≈730 650–620 > 470	Quartz, "nigrine"A, B (plus Nb,U mineral inclusions)	Weißenstein dyke swarm (locus typicus)	Transitional between metamorphogenic and magmatogenic
4a	Postkinematic mobilization		550–500	Quartz	Hagendorf-South, Hagendorf-North, Pleystein-Kreuzberg,	Strong fractionation

(continued)

Table 4.2 (continued)

					Miesbrunn (in parts)	
4b	Postkinematic mobilization	< 0	600–500	Quartz , alkaline feldspar, triplite-zwieselite, „pseudo-zwieselite“, wolfeite , triphylite , triploidite, Mn apatite-(F), zircon II	Hagendorf-South, Hagendorf-North, Pleystein-Kreuzberg, Trutzhofmühle, Waidhaus-Silbergrube,	Granite may be present at depth. No correlation with any granite at outcrop
4c	Postkinematic mobilization		≈475	Scorzalite , lazulite-scorzalite , scheelite tainiolite (?)	Hagendorf-South, Miesbrunn, Wildenau-Plössberg, Trutzhofmühle, Reinhardsrieth	Contact metasomatic process along the margin with the metasedimentary rocks
4d	Postkinematic mobilization		500–410 Spess.	Zr-Sc -phosphate-silicate (unnamed) , K- Sc-Zr-Ba phosphate (unnamed), Zr-Sc -phosphate-silicate (unnamed)	Trutzhofmühle	Contact metasomatic process along the margin with the metasedimentary rocks (or mantle impact ?)
4e	Postkinematic mobilization	< 0	500–300	Graftonite, sarcopside , arrojadite s.s.s, dickinsonite s.s.s	Hagendorf-South,	Autometasomatism retrograde reaction
4f	Postkinematic mobilization	≤0	500–400	Hagendorfite , alluaudite	Hagendorf-South	Autometasomatism retrograde reaction
4g	Postkinematic mobilization	>0	500–300	Ferrosicklerite, heterosite–purpurite	Hagendorf-South	Autometasomatism retrograde reaction
4h	Postkinematic mobilization			Manganogordonite, gordonite,	Hagendorf-South, Trutzhofmühle,	Retrograde process along the margin with the metasedimentary rocks
5a	Epithermal mineralization I	<0	≈380	Arsenopyrite, pyrite, pyrrhotite , Fe chlorites chalcopyrite, cassiterite , stannite , bismuthinite bismuth+silver ,mackinawite, magnetite+, hematite, stilpnomelane , pyrosmalite , Mn-Fe carbonates, fluorite, sulfur (?)	Pleystein-Kreuzberg, Trutzhofmühle, Pleystein-New Aplite, Wildenau-Plössberg	Latest stock-like two-mica granites, shallow intrusions
5b	Epithermal mineralization I	<0	210–120 300–200	Coffinite , coffinite-zircon, molybdenite, calcite, graphite (bitumen ?)	Miesbrunn, Hagendorf-South	Latest stock-like two-mica granites, shallow

			calcite/C			intrusions
5c	Epithermal mineralization I	<0 >0	110–90	Ag-Bi-Cu-sulfides, galena , sphalerite (Fe poor), stibnite, barite	Hagendorf-South, Hagendorf-North, Trutzhofmühle Wildenau-Plössberg	Alteration and transition into post-Variscan unconformity-related vein-type mineralization
5d	Epithermal mineralization I	>0/ <6	<300 (?)	Goyazite I, variscite , nontronite I (X), brazilianite	Hagendorf-South,	Related to the late Variscan peneplain
5e	Epithermal mineralization I	>0/ <5	<390/ 1000 bar	Kaolinite	Pleystein-Kreuzberg, New Aplite (in parts supergene), Pleystein Weiss Mine, Hagendorf-South (only minor amouns)	Onset of the decomposition of the feldspar rim
6a	Hydrothermal mineralization (early stage)	<0	<300 >200	Mn vivianite , reddingite ludlamite (+Zn-bearing), phosphoferrite (+Ca-bearing), eosphorite-childrenite ,fairfieldite , metaswitzerite (?), lermontovite, vyacheslavite (?), vauxite, pyrite, berthierine-chamosite, siderite	Hagendorf-South, Hagendorf-North, Miesbrunn, Reinhardsrieth, Trutzhofmühle, Waidhaus-Silbergrube, Pleystein-Kreuzberg	Autohydrothermal alteration
6b	Hydrothermal mineralization (early stage)	≤0	>200	Kryzhanovskite	Hagendorf-South, Hagendorf-North	Autohydrothermal alteration
6c	Hydrothermal mineralization (early stage)		?	Parahopeite , hopeite	Hagendorf-South, Hagendorf-North, Pleystein-Kreuzberg	Autohydrothermal alteration Zn sequence
7a	Hydrothermal mineralization (late stage)	≤0	<200	Vivianite , meta-vivianite, rockbridgeite (+Al-bearing, +Zn-bearing (plimerite ??)), frondelite, lipscombite , perloffite , earlshannonite , apatite-(OH), fluorite, tavorite , barbosalite, paulkerrite, "UNK 9" , whitmoreite, (ferro)laueite,	Hagendorf-South, Hagendorf-North, Pleystein-Kreuzberg, Miesbrunn, Wildenau-Plössberg, Reinhardsrieth, Trutzhofmühle,	Autohydrothermal to externally driven hydrothermal processes
7b	Hydrothermal	≤0	<200	Kolbeckite	Trutzhofmühle	Autohydrothermal to

(continued)

Table 4.2 (continued)

	mineralization (late stage)					externally driven hydrothermal processes affecting Sc minerals
7c	Hydrothermal mineralization (late stage)	≤0	<200	Parascholzite, scholzite , phosphophyllite Rockbridgeite, (Zn-bearing) , schoonerite	Hagendorf-South, Hagendorf-North, Pleystein-Kreuzberg	Autohydrothermal alteration Zn sequence
7d	Hydrothermal mineralization (late stage)	≤0	<200	Whitlockite, dufrénite(+ natrod.), calcioferrite, messelite, mitridatite , “meta-mitridatite”	Trutzhofmühle	Autohydrothermal - topomineralic alteration decomposition of Ca minerals grading into the epithermal stage II
8a	Epithermal mineralization II		<200	Jungite	Hagendorf-South	Epithermal- topomineralic alteration Zn sequence
8b	Epithermal mineralization II	<0	<200 < 100	Hureaulite, wilhelmvierlingite , landesite , switzerite , rittmannite , robertsite , whiteite , meurigite-K, cyrilovite , bermanite (?),kidwellite (?), benyacarite , carlhintzeite , pachnolite,	Hagendorf-South, Hagendorf-North, Pleystein-Kreuzberg, Waidhaus-Silbergrube	Related to the late alpine peneplain, characterized by relapses in the redox system, in parts affecting also the late hydrothermal stage
8c	Epithermal mineralization II		<200	Nordgauite, kingsmountite, kastningite, paravauxite	Hagendorf-South, Hagendorf-North, Pleystein-Kreuzberg, Waidhaus-Silbergrube	Epithermal- topomineralic alteration decomposition of al minerals
8d	Epithermal mineralization II		<200	Jahnsite-whiteite series, keckite	Waidhaus-Silbergrube, Trutzhofmühle, Hagendorf-South	Epithermal- topomineralic alteration decomposition of Al - and Ca minerals
8e	Epithermal mineralization II	≤0 >0	<100 (F.I. strengite)	Greenockite, acanthite (?),covellite , Cu sulfide (sec), copper , marcasite, cuprite , pyrolusite, strengite, morinite , whitmoreite, leucophosphite , strunzite(+Al-bearing), laueite, pseudolaueite, stewartite , phosphosiderite , fluellite , xanthoxenite, ernstite, matulaite (?)	Hagendorf-South, Hagendorf-North, Miesbrunn, Wildenau, Plössberg, Reinhardsrieth, Trutzhofmühle, Pleystein-Kreuzberg	Related to the late alpine peneplain, characterized by relapses in the redox system
9a	Supergene alteration	>0/	Near	Silica (opaline material/ opal CT)	Pleystein New Aplite	Chemical weathering

		<7	ambient			(early stage)
9b	Supergene alteration	>0/ <5	Near ambient	Bismite, kaolinite, cryptomelaneI+II , todorokite, "wad ", goethite "limonite",crandallite ,goyazite II , gorceixite , waylandite , florencite-(Ce-La), zairite , lithiophorite , ximengite (?), cacoxenite , anatas, beraunite , oxi-beraunite , chalcophanite , vandendriesscheite, (meta)-autunite , uranosphaerite , (meta)-torbernite, bassetite , chernikovite , lehnerite uranophane, uranocircite, wavellite, hemimorphite , libethenite , pseudomalachite , nontronite (earthy), chrysocolla , chalcosiderite , turquoise , variscite , malachite, rosasite, scorodite , phosphoscorodite riomarinaite (?), beudantite, kolbeckite II, P-bearing Fe-Mn " limonite ", Sc-bearing vochtenite (?), Sc-bearing churchite-(Y)	Pleystein New Aplite, Hagendorf-South, Pleystein-Kreuzberg, Hagendorf-North, Miesbrunn, Wildenau-Plössberg, Reinardsrieth, Trutzhofmühle, Waidhaus-Silbergrube,	Chemical weathering (main stage)
9c	Supergene alteration	>0/ 5.5–11.5 >0/ ≥8 <0	Near ambient	Silica (opaline material/ opal CT), muscovite-illite chlorite, smectite, "wad" vivianite (earthy)	Pleystein new Aplite, Hagendorf-South	Chemical weathering (late stage)
10	Supergene alteration	>0/ <<7	Near ambient	Posnjakite, brochantite, langite, devilline, chalcanthite, connellite, diadochite, jarosite, rozenite, gypsum	Hagendorf-South,	Post-mining mineralization anthropogenic

Between stages 7a and 8a redox conditions fluctuate on a small scale and cause oscillating intergrowth of minerals among the phosphates
Question marks signify that the position within the succession is uncertain. The interpretation of physical-chemical regime (temperature, Eh, pH) is based upon the discussion in Sect. 4.4

Fig. 4.1a Orthoclase twins at the boundary between the quartz core and the K feldspar rim. Hagendorf-South

the city-center of Pleystein (Table 4.1, Fig. 1.3). Feldspar was in the past and is still today the staple product of exploitation in the HPPP (Sect. 1.3.2.1). The plagioclase composition is rather monotonous and made up of ordered albite to albite-oligoclase ($<An_{25}$). Near the gneissic wall rocks of the aplites and pegmatites chessboard albite developed at Miesbrunn. Potassium feldspar is by far the most widespread feldspar and developed, in places, well-shaped crystals in open-space-fillings near the boundary between the quartz core and the feldspar rim of the Hagendorf-South pegmatite stock (Fig. 4.1a).

The feldspar-group minerals are randomly intergrown with quartz in the tabular aplites and pegmatites, e.g., Reinhardsrieth, because they lack any pronounced zonation (Dill and Skoda 2015). By contrast, in the stock-like and zoned pegmatites, such as Hagendorf and Pleystein, members of the feldspar group are mainly concentrated in the hanging wall rocks forming a series of massive onion-shell-like zones around the quartz core towards the edge of the pegmatite: K-feldspar rim, quartz-feldspar eutectic zone, aplitic wall zone (Fig. 4.1b). Underneath the quartz core an aplitic zone has been proved by drilling at Hagendorf-South and by underground mapping the stopes in a gallery drifted below the quartz core of the Kreuzberg Pegmatite (Figs. 1.3, 5.3b). A closer look at the textural and morphological variation of feldspar in the study area reveals a varied spectrum of graphic intergrowth, perthite, anti-perthite, K- feldspar with myrmekitic rims and plates of cleavelandite a peculiar habit typical of albite. The latter is a reflection of fractionation trends between K- and Na feldspar, including quartz, that were interpreted in some classical papers as thermally-driven diffusion gradients by Jahns and Burnham (1969) and Orville (1963).

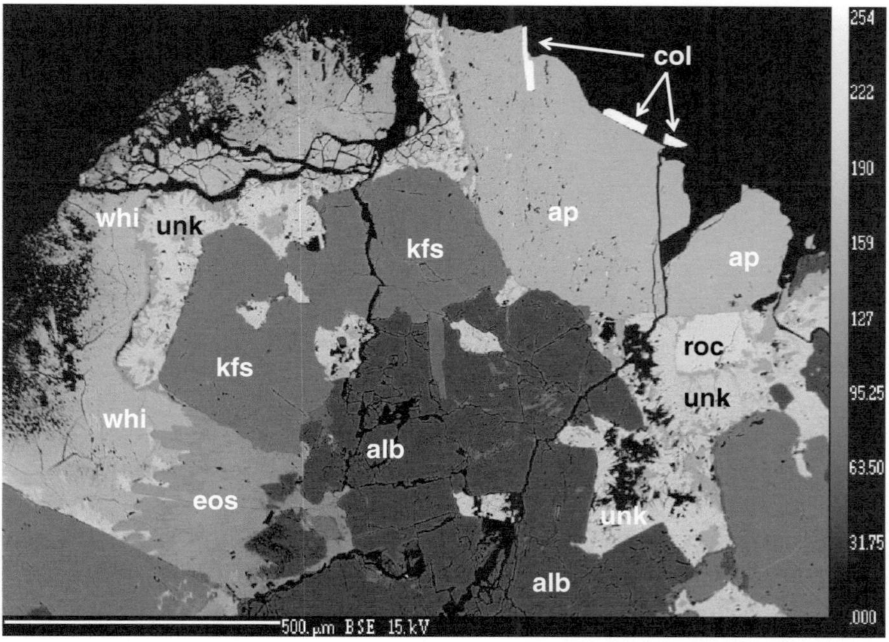

Fig. 4.1b Feldspar intergrown with phosphates and columbite in the Reinhardsrieth tabular pegmatite. Lath-shaped columbite-(Fe) (*col*) in the phosphate-bearing aplite, replaced by apatite-(Mn) (*ap*). The central part of the BSE image is made up of K feldspar (*kfs*) and albite (*alb*) which are both surrounded by phosphates: eosphorite (*eos*), whitmoreite (*whi*), rockbridgeite (*roc*), unknown phosphate (*unk*). BSE image/back-scattered electron image, electron microprobe (EMP)

Where does all the phosphorus to build up the great variety of phosphates in the aplites and pegmatites come from? The question has already been addressed in Sect. 3.2.9. Feldspar-group minerals were identified as a significant source of phosphorus in pegmatites s.st. while their next of kin, the pegmatoids and metapegmatites are conspicuously low in phosphorus and in K feldspars. Potassium feldspar abundant in pegmatites experienced a tendency to average higher P_2O_5 contents than Na-enriched plagioclase and can be made accountable for the presence or absence of phosphate minerals in the wake of feldspar formation (Tables 1.2 and 4.3).

4.2 Silica Group

Members of the silica group are ubiquitous in the HPPP and its adjacent pegmatites. While quartz is present in all pegmatites and aplites listed in Table 4.1, opaline silica has only been determined in samples from the New Aplite at

Table 4.3 Feldspar from pegmatites and aplites of the HPPP and adjacent feldspar pegmatites

Deposit										
Element	Plößberg	Brünst near Waidhaus	Silbergrube	Silbergrube	Silbergrube	Pleystein	Pleystein	Hagendorf 49 m-level	Hagendorf 109 m-level	Hagendorf 109 m-level
SiO_2	73.09	71.82	72.69	76.31	72.99	77.81	74.24	64.69	63.89	69.11
Al_2O_3	15.05	16.30	16.06	12.57	14.92	13.01	13.85	19.62	19.01	18.52
Fe_2O_3	0.85	0.95	0.30	0.55	0.50	0.76	1.04	0.31	0.04	0.04
CaO	0.69	0.50	0.20	0.41	0.09	0.36	0.97	0.16	0.07	0.43
MgO	0.22	0.25	0.08	0.12	0.01	0.17	0.11	<0.01	<0.01	<0.01
MnO	0.02	0.92	0.05	0.00	0.07	0.03	0.25	0.02	0.00	0.01
K_2O	3.72	4.62	4.80	3.71	3.71	2.90	4.40	8.83	13.49	0.33
Na_2O	4.83	4.38	5.40	4.14	5.02	3.38	1.60	4.56	1.39	9.95
P_2O_5	0.38	0.34	0.60	0.20	0.43	0.60	1.52	0.68	0.90	0.90
Na_2O/K_2O	1.30	0.95	1.13	1.12	1.35	1.17	0.36	0.52	0.10	30.15

Table from Dill et al. (2011c)

Fig. 4.2a Pinkish rose quartz from the Kreuzberg Pegmatite at Pleystein, the "City on Pegmatite"

Pleystein, which is kaolinized almost to completeness. Apart from the well-crystallized "rock crystal" ("Bergkristall" in German) which can be observed in vugs scattered across the felsic intrusives, there are another three types of quartz which warrant mentioning from the HPPP pegmatites. The quartz within the core of the pegmatite stocks and tabular pegmatites is a milky opaque quartz, locally assuming the outward appearance of a smoky quartz. The pinkish rose quartz of the Kreuzberg Pegmatite has become a symbol for the city of Pleystein ("Rosenquarz-Stadt" in German) (Fig. 4.2a). A quartz with an outstanding morphology and extraordinary size has been observed in many pegmatites of the HPPP (Hagendorf-South, Hagendorf-North, Pleystein, Zessmannsrieth, eastern Miesbrunn Pegmatite-Aplite Swarm) and found indicative of the border zone of the pegmatite proper towards its country rocks. A similar type of quartz has also been recorded by Weber (2013) from outside the HPPP, from the Muglhof pegmatoid, which is intercalated into amphibolites. Due to its large stubby crystals, that measure several centimeter in length and grow freely into nest and vugs of the aplite, this type of quartz was given a name of its own by the local geologists and miners and called "Sprossquarz" (English growing = German: sprossen) (Fig. 4.2b, c). The terms "sprouting quartz" or "artichoke quartz" reflecting peculiar sorts of split growth do not exactly describe the growth phenomena of the "Sprossquarz" in the HPPP (Rykart 1995). This is also the case with the so-called "Kappenquarz" that reflect hiatuses during its precipitation when non-siliceous compounds came to rest upon the newly-developed crystal faces during

Fig. 4.2b "Sprossquarz" from the Zessmannsrieth pegmatite with its pyramided faces strewn with flakes of muscovite

Fig. 4.2c "Sprossquarz" from the Kreuzberg pegmatite coated with hematite and argillaceous material of unknown composition

the waning stages of crystallization. Although far from being fully understood as to its origin this "Sprossquarz" morphology marks perfectly well the boundary of pegmatitization within the country rocks. It reflects abrupt changes or a very steep gradient in the physical-chemical regime during precipitation of the pegmatite quartz.

To get an idea of the temperature of formation and provide additional evidence for the coloration of the pinkish rose quartz at Pleystein, quartz samples from different pegmatites, aplites, and dykes whose sampling sites straddle the Czech-German border have been analyzed for their trace elements (Table 4.4) (Dill et al. 2012a, 2013a, b, c). The Ti-in-quartz geothermometer of Wark and Watson (2006) has been used as a tool to determine the quartz crystallization temperature based on the temperature dependence of the Ti^{4+}-Si^{4+} substitution in quartz and given the presence of rutile or, in its absence, an estimate of the TiO_2 activity of the system. This geothermometer can be applied to temperatures up to as high as 1000 °C. There is no extraordinary value among trace elements that might lend support to the idea that special chromophores are responsible for the pinkish color of the Pleystein rose quartz, so that mineralogical (e.g. lattice defects ?) or gas inclusions might have a stronger impact on the mineral color of rose quartz.

While the trace element chemistry of quartz does elucidate mineralogical features like its mineral color, it broadens our knowledge as to its environment of formation prevalently the temperature of formation of the quartz pegmatites under study. According to Table 4.4, quartz dykes formed at the higher temperatures than the pegmatites and aplites in the region which feature a wide temperature range down into the subcritical part of hydrothermal solutions. Quartz from the Miesbrunn pegmatite-aplite swarm crystallized around 660 °C (T_{max}: 677 °C, T_{mean}: 666 °C, T_{min}: 656 °C) (Table 4.4). This temperature is almost identical with the metamorphic temperature and the NW-SE striking quartz dikes west of the Hagendorf – Pleystein Pegmatite Province which intersect the metamorphic country rocks at Weißenstein (T_{max}: 645 °C, T_{mean}: 635 °C, T_{min}: 624 °C). Quartz from the neighboring quartz cores of Hagendorf-South and Pleystein pegmatites, taken for reference, shows temperatures of formation intermediate to the afore-mentioned data, falling in the range 525–475 °C. The temperature of formation during emplacement of pegmatite bodies are quite similar in the range 500–550 °C.

In conclusion, the Miesbrunn pegmatite-aplite swarm is a high-temperature predecessor of the Hagendorf-Pleystein pegmatites which are located in the center of the pegmatite province. Towards the rim and depth of the province, the Miesbrunn pegmatite-aplite swarm fades out into pure quartz dikes which were intruded at the same temperature of formation as the aplitic and pegmatitic mobilizates of the Miesbrunn pegmatite-aplite swarm. The T values form among other geothermometers the basis as it comes to create a model of the emplacement of the pegmatites during the waning stages of the Variscan orogeny- see final Sects. 4.5 and 4.6.

Table 4.4 Trace element analyses of quartz samples (ppm) from different pegmatites, aplites and quartz dykes on both sides of the Czech-German border

	Li	Be	B	Mn	Ge	Rb	Sr	Sb	Al	P	K	Ti	Fe	Zn	Ga	Max	Mean	Min
Hagendorf-South	9.86	0.73	0.00	0.00	0.20	5.87	0.13	0.00	699.39	12.29	547.43	3.99	3.05	1.63	0.14	474.4	466.8	459.2
Pegmatite	18.84	0.43	6.51	2.04	1.03	5.05	0.02	0.00	739.67	7.70	417.01	5.84	27.40	6.18	0.25	499.7	491.7	483.7
	7.40	0.32	3.85	3.54	1.23	24.02	0.03	0.01	1190.67	0.00	592.70	5.70	38.37	18.08	0.54	498.0	490.1	482.1
Germany	44.64	0.79	6.69	0.39	2.39	0.10	0.08	0.00	390.27	6.22	2.55	15.93	2.68	1.73	0.00	575.1	565.9	556.9
	39.43	0.80	7.76	0.23	2.14	0.12	0.00	0.05	326.79	0.00	2.86	12.01	1.19	0.78	0.00	552.4	543.6	534.9
	18.18	0.27	0.83	0.22	2.22	0.00	0.04	0.04	143.30	0.00	6.49	2.58	0.92	3.19	0.00	447.6	440.3	433.1
Hagendorf-North	13.23	0.60	8.28	1.16	1.52	5.31	0.13	0.00	1655.60	14.16	1116.87	4.21	68.52	8.97	0.61	477.9	470.2	462.5
Pegmatite	7.36	0.47	5.29	4.20	1.89	17.22	0.04	0.03	3457.51	15.45	1871.73	3.28	98.67	9.23	0.81	462.2	454.7	447.3
Germany	9.10	0.84	7.82	5.74	1.81	28.18	0.45	0.04	2109.15	7.26	1800.56	7.38	145.54	10.79	0.61	516.1	507.8	499.6
Miesbrunn	1.71	0.12	1.40	11.19	0.30	1.45	12.68	0.03	156.49	0.00	245.80	1.16	37.69	3.98	0.03	402.9	396.2	389.6
Pegmatite-Aplite	3.64	0.07	1.48	12.95	0.32	1.88	12.76	0.00	193.25	0.00	301.60	1.81	58.91	12.38	0.00	427.1	420.1	413.2
Germany	3.57	0.07	1.32	11.68	0.33	2.21	13.34	0.00	135.27	0.00	257.15	0.39	43.83	7.81	0.00	349.8	343.8	337.9
	10.39	0.00	29.85	0.93	1.93	0.06	0.02	0.07	475.09	0.00	1.06	69.62	6.27	1.52	0.00	717.0	705.6	694.4
	7.53	0.19	11.80	0.70	2.19	0.05	0.03	0.02	355.95	0.57	0.00	44.04	4.69	0.63	0.00	668.1	657.5	647.0
	7.64	0.21	17.36	0.48	2.84	0.00	0.01	0.03	301.04	0.00	1.52	35.27	6.06	0.94	0.00	646.1	635.9	625.7
Pleystein	21.33	1.06	5.98	0.68	2.83	0.22	0.01	0.00	545.14	2.82	11.34	5.46	3.11	1.35	0.00	495.1	487.2	479.3
Pegmatite	26.89	0.13	8.41	0.06	2.53	0.00	0.10	0.01	460.13	5.92	3.82	6.39	2.61	1.09	0.00	505.9	497.8	489.8
Germany	21.93	0.55	4.78	0.27	2.67	0.10	0.03	0.00	483.00	4.15	9.98	6.63	1.99	1.57	0.15	508.5	500.3	492.2
	18.24	0.34	22.00	0.84	2.34	1.46	0.32	1.06	2584.11	15.11	172.40	9.31	13.47	7.35	0.00	533.0	524.5	516.1
	9.08	0.13	6.34	1.07	2.51	1.08	1.16	1.15	2124.73	10.51	183.33	10.61	11.80	8.78	0.24	542.8	534.2	525.6
	16.47	0.16	10.31	0.75	1.53	1.70	0.18	0.26	2372.49	14.80	148.93	4.96	10.22	6.76	0.36	488.6	480.8	473.0
Waidhaus	19.93	0.20	2.82	0.22	2.80	0.24	0.07	0.56	438.87	3.48	19.77	11.83	2.88	2.12	0.00	551.2	542.4	533.7

Aplite(granite)	15.27	0.38	5.59	0.84	3.01	1.52	1.37	0.28	440.25	0.00	78.60	12.76	5.23	3.55	0.14	557.1	548.3	539.5
Germany	16.71	0.22	3.60	0.20	2.28	0.05	0.03	0.19	493.13	0.00	10.64	9.13	3.19	2.09	0.00	531.5	523.1	514.7
Ploessberg	4.44	0.02	0.99	7.39	1.12	0.50	4.66	0.09	569.07	0.00	260.89	1.17	69.76	4.04	0.03	403.2	396.5	389.9
Pegmatite	2.14	0.12	1.21	2.46	1.40	2.04	0.47	0.07	469.12	0.00	165.59	2.05	40.97	2.12	0.10	434.1	427.0	420.0
Germany	6.51	0.04	1.15	3.21	1.40	0.40	1.01	0.17	468.68	6.70	135.25	4.88	66.58	2.91	0.13	487.7	479.8	472.0
Beidl	16.02	1.07	6.41	0.00	2.20	0.07	0.28	0.01	241.82	0.00	4.59	17.71	3.79	0.82	0.10	583.9	574.6	565.4
Pegmatite	13.22	0.76	3.15	0.39	1.44	0.27	0.17	0.02	321.80	3.29	18.54	15.71	2.57	1.78	0.00	573.9	564.8	555.8
Germany	10.05	1.67	8.59	0.29	2.11	0.03	0.05	0.00	459.02	8.28	18.22	20.57	6.17	2.58	0.15	596.7	587.2	577.9
Weissenstein	4.56	0.08	14.34	0.58	0.94	0.17	0.14	0.05	839.37	0.00	4.42	78.62	10.24	3.92	0.27	730.9	719.2	707.7
Quartz dyke	4.93	0.00	5.67	0.00	0.30	0.00	0.08	0.00	99.60	0.00	5.35	4.75	1.02	1.27	0.08	485.8	478.0	470.2
Germany	5.52	0.00	28.99	0.25	0.75	0.01	0.24	0.08	569.21	0.21	4.53	71.06	9.09	3.78	0.22	719.3	707.9	696.6
Jedlina	38.24	0.13	1.34	21.28	0.57	3.09	20.91	0.09	1271.38	0.00	129.66	2.25	58.87	7.52	0.00	439.6	432.5	425.4
Quartz dyke	7.37	0.65	1.34	23.63	0.40	4.11	14.93	0.09	1381.00	2.14	732.15	2.37	41.29	9.65	0.00	442.6	435.4	428.3
Czech	6.24	0.00	1.03	9.36	0.30	1.69	9.30	0.06	700.98	1.48	383.72	1.42	61.92	8.14	0.13	413.6	406.8	400.0
Republic	11.83	0.00	0.52	3.45	0.34	0.64	3.56	0.16	932.95	9.97	331.99	2.17	357.62	7.39	0.04	437.4	430.2	423.1
Ostruvek	24.15	0.17	1.65	0.37	0.76	0.00	0.24	0.18	551.56	0.00	10.43	32.45	7.18	5.24	0.26	638.1	628.0	618.0
Quartz dyke	21.46	0.22	1.71	0.09	0.87	0.00	0.10	0.07	819.51	16.50	34.32	296.98	4.90	4.58	0.00	912.1	897.2	882.6
Czech Republic	26.12	0.35	0.97	4.03	0.87	0.03	0.43	0.07	1133.98	0.00	25.51	142.80	84.85	6.18	0.52	804.9	792.0	779.2
Rozvadov	5.94	0.15	70.95	0.70	1.14	0.04	0.06	0.57	1223.39	1.73	11.69	150.36	31.24	1.89	0.19	811.8	798.8	785.9
Quartz dyke	7.25	0.14	20.10	0.02	0.95	0.18	0.69	0.37	870.88	8.64	72.79	45.79	16.53	2.24	0.00	672.1	661.4	650.9
Czech Republic	8.59	0.22	33.00	7.30	1.21	0.58	0.34	0.67	1303.14	0.00	459.24	168.70	59.13	4.70	0.26	827.5	814.2	801.1

The temperatures of formation (maximum, mean, minimum values) given in °C were calculated based upon the titanium thermometer (see text) (Dill et al. 2012a, 2013a, b, c)

4.3 Garnet s.s.s.

Garnet is of widespread occurrence in the metamorphic country rocks and also in some of the felsic mobilizates of the HPPP, with each of its solid solution series (s.s.s.) characterizing a particular type of rock: Garnet s.s.s. enriched in spessartite and almandine = pegmatites and aplites, garnet s.s.s. enriched in grossularite = skarn and calcsilicate rocks, garnet s.s.s. enriched in almandine = metasedimentary rocks, enriched in pyrope = metabasic igneous rocks (Fig. 4.3a, b, c, d, Tables 4.5 and 4.6). Porphyroblasts of garnet in the country rock are anhedral to subhedral, attaining an average size of 1 cm, whereas the maximum grain-size of the euhedral crystals of garnet within the marginal zones of aplites can exceed, in places, 5 cm in size. These well-shaped crystals in the majority of cases display the rhombic dodecahedron {110}, rarely the icositetrahedron {211}, using the Miller's indices for labeling of the faces. A combination of the two morphological types is on display in Fig. 4.3e.

Chemical mapping of garnet in the aplite using a geoscanner and line scans performed by EMPA across the garnet crystals reveal a marked zonation, with spessartite and grossularite components being enriched in the core zone, whereas pyrope and, to a lesser degree, almandine tend to increase toward the rim of the garnet. Dodecahedral surfaces {110} dominate during the development of the spessartite-grossularite core zone, but complex face combination can also be expected, as demonstrated by the samples from the Wilma metapegmatite deposit (Fig. 4.3d).

As garnet is one of the most common heavy mineral in the alluvial-fluvial placer deposits of the region, exploration geologists can harness their mineral color and shape as an ore guide to localize potential areas underlain by pegmatites or skarn

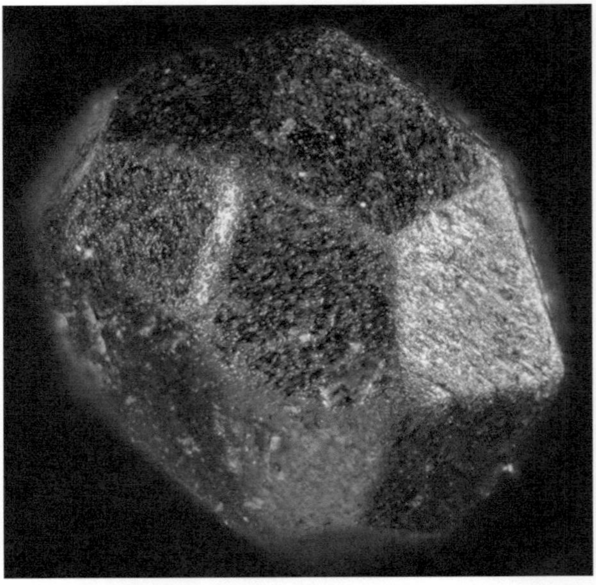

Fig. 4.3a Garnet in situ and reworked (63–400 μm if not stated otherwise). Almandine-pyrope garnet s.s.s. derived from metabasic rocks and panned from stream sediments of fluvial placer deposits of the HPPP

Fig. 4.3b Spessartite-enriched almandine garnet s.s.s. derived from pegmatitic and aplitic rocks at the edge of the HPPP reworked in fluvial placer deposits of the HPPP

Fig. 4.3c Almandine garnet derived from metasedimentary rocks in fluvial placer deposits of the HPPP

Fig. 4.3d Almandine garnet from the metapegmatite of the Wilma deposit near Püllersreuth (Weber 1978a) size: 2 cm

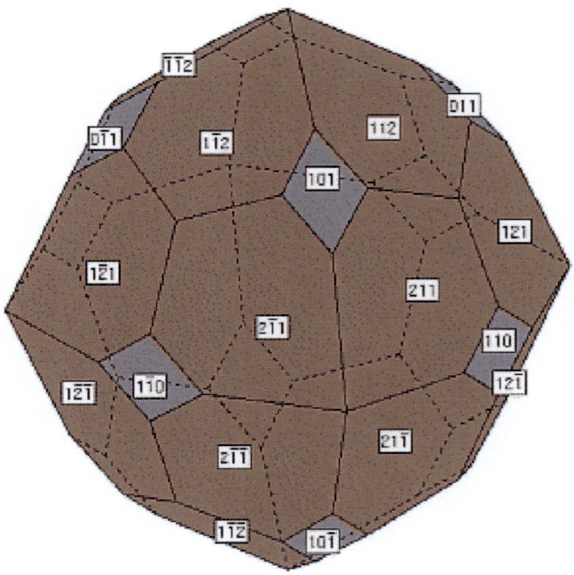

Fig. 4.3e Cartoon to display the complex face combination of the crystal morphology of the almandine garnet of Fig. 4.3d. The faces are denoted using the common Miller's indices for notation (Weber 1978a). The *gray faces* refer to the dodecahedron {110}, while the *brown faces* are characteristic of the icositetrahedron {211}.

Table 4.5 The chemical composition of garnet s.s.s. from the HPPP

	wt. %	SiO_2	TiO_2	Al_2O_3	FeO	MnO	MgO	CaO
Almandine	Min	37.579	0.032	20.655	18.361	0.366	2.519	4.854
	Mean	38.724	0.074	21.526	24.862	0.777	6.910	6.953
	Max	40.055	0.137	22.561	30.990	1.877	11.926	10.823
Spessartite	Min	36.453	0.000	20.416	19.567	0.788	1.980	0.420
	Mean	36.948	0.072	20.777	26.546	11.859	2.637	1.012
	Max	38.592	0.284	21.693	35.781	19.307	8.112	4.361
Grossularite	Min	37.187	0.094	5.176	9.379	0.205	0.943	5.956
	Mean	41.138	0.169	17.603	23.406	2.855	4.207	8.670
	Max	51.577	0.399	20.974	28.447	3.422	13.282	16.257

Table 4.6 The distribution of pyrope, almandine, spessartite and grossularite as function of source rocks (pegmatites plus aplites, gneiss, calcsilicate rocks)

Mineral	Pegmatite + aplite			Gneiss			Calcsilicate rock		
wt.%	Min	Mean	Max	Min	Mean	Max	Min	Mean	Max
Pyrope	8.00	10.70	31.30	9.96	26.42	44.53	12.10	12.24	12.54
Almandine	42.20	59.40	82.30	37.86	52.98	67.24	61.37	62.15	63.47
Spessartite	1.73	27.40	45.10	0.77	1.74	4.22	7.56	7.67	7.77
Grossularite	0.00	2.56	13.60	10.70	18.87	29.61	16.74	17.94	19.24

The major components in each s.s.s. are framed with a bold-faced line

deposits (Fig. 4.3). As early as 1925, Scholz made an attempt to classify in his "Untersuchungen über die Mineralführung und Mineralgenese der bayerischen Pegmatite", the pegmatites into Mn garnet pegmatites, which according to the present-day mapping are identical with the pegmatites emplaced at the edge of the pegmatite province and Mn phosphate pegmatites that are representative of the pegmatites hosting the varied spectrum of manganiferous phosphates (Table 4.1). Translated into the depth of formation, the mostly tabular pegmatites and aplites bearing spessartite developed at greater depth than the pegmatites enriched in Mn-bearing phosphates only- see also Fig. 5.5d-B.

Aplites and pegmatites irrespective of their tabular or stock-like shape around Pleystein and Hagendorf, in the central zone, are barren as to spessartite enriched garnet. The pegmatites in the Moldanubian Zone south of the HPPP, e.g., Hühnerkobel, were emplaced at a deeper crustal level than the pegmatites and aplites of the HPPP. On increasing differentiation and towards shallower depth of the intrusion, manganese was preferentially accommodated in the structure of apatite (see Sect. 4.10). On the contrary, toward a deeper level of intrusion, as was the case in the Bayerischer Wald, almandine became gradually enriched in Mn, leading at the very end to a garnet s.s.s dominated by spessartite (Schaaf et al. 2008).

Almandine acts as a sort of matrix with no preferred concentration within the garnet crystal. Phosphate, a critical element for the denomination of the aplites and pegmatites was also investigated as to its spatial distribution within the garnet s.s.s., yet did not show any preferred association with one of the afore-mentioned end members of the garnet s.s.s. It is bound to cracks and isolated nests within the garnet suggesting infiltration of phosphorus along fissures rather than forcing garnets into the role of a source mineral for phosphate in the HPPP.

A detailed investigation of the fractionation in rare-element granitic pegmatites has been conducted by Černý et al. (1985) addressing the Mn/Fe ratio which is the most important one in the pegmatite-related almandine- and spessartite-enriched garnet s.s.s. Moretz et al. (2013) who went through a vast collection of garnets derived from across the globe and investigated the composition of garnet as an indicator of rare metal (Li) mineralization in granitic pegmatites gave a wide range for the Mn/Fe ratio. The authors interpreted this wide range of compositions as reflecting a low-to-high degree of melt evolution. Their statement that Li-poor, NYF pegmatites typically have garnet with the lowest Mn and highest Fe contents, whereas garnet in Li-rich, LCT pegmatites has the highest Mn and lowest Fe contents seems very smart but cannot be proven by detailed analyses in the pegmatite provinces at the western edge of the Bohemian Massif. Heimann (1997) came to a similar conclusion. Li-poor, NYF pegmatites typically have garnet with the lowest Mn and highest Fe contents, whereas garnet in Li-rich, LCT pegmatites has the highest Mn and lowest Fe contents. This suggests that the major element composition of garnet can likely be used to distinguish pegmatites with Li mineralization from those with low Li contents. In this case, Hagendorf-South and Pleystein should have the most elevated contents of spessartite-enriched garnet. They do not have any, while Miesbrunn and Trutzhofmühle, devoid of Li have the highest spessartite-contents in the garnet.

It is true that pegmatite-related garnet is specialized in spessartite, but *per se* is no marker for lithium as numerous examples show. Garnet with from the Bayerischer Wald with Spess $_{57}$ through Spess $_{67}$ are unrelated to any lithium concentration (Schaaf et al. 2008). By contrast the most elevated Li concentration at Hagendorf is not correlated with Mn-dominated garnet. Manganiferous garnet may point to higher temperatures in the study area, but has to be treated in the pegmatite province, proper, together with apatite and in the marginal zone looked at *vis-à-vis* minerals like biotite. Spessartite-bearing almandine s.s.s. coupled with the above phosphates and silicates are marker for the depth of formation or the present-day level of erosion and they might be used to predict the presence of Mn-bearing phosphates in a pegmatite province. As the Mn is captured throughout the crystallization of silicates, e.g., garnet, in the course of the evolution of the melt, the systems gets depleted in Mn and rules out the precipitation of a wide range and high quantity of primary or secondary Mn phosphates during subsequent stages (Table 4.1).

Garnet grains in the HPPP belong to the "pyralspite" s.s.s. (pyrope-almandine-spessartite) (Matthes 1961). For the pure spessartite composition, the lower reaction limit at pressures between about 200 and 1500 atm. is at 410 °C. For spessartite-almandine mixed crystals the limit rose with increasing almandine content from 410 °C (spessartite $_{90}$ almandine $_{10}$) to 500 °C (spessartite $_{50}$ almandine $_{10}$) according to the author mentioned above.

4.4 Aluminum Silicates and Corundum

Although not a first-order chemical compound in pegmatites, Al_2SiO_5 is present with all of its three polymorphs sillimanite, kyanite, and andalusite in the pegmatites of the NE Bavarian Basement (Table 4.1). It is safe to say only for andalusite that this polymorph of Al_2SiO_5 is related genetically to the emplacement of the pegmatites and aplites. Sillimanite and kyanite are common constituents in the metamorphic country rocks and their geological and mineralogical setting does not always indicate whether they developed together with a pegmatitic rock nearby or were simply incorporated into the melt. The polymorphism of Al_2SiO_5 is a textbook example of mineral transformation controlled by temperature and pressure variation during metamorphism (Winkler 1967, 1976; Richardson et al. 1969; Bohlen et al. 1991). Andalusite is known to be the low-grade polymorph of Al_2SiO_5 which is stable under low-P- and low-T conditions as they occur during the andalusite facies of contact metamorphism and shallow regional metamorphism named Abukuma Facies Series. On increasing temperature above approx. 500 °C sillimanite is more stable relative to andalusite. These minerals are widespread in medium to high-grade regionally and contact metamorphic rocks. Kyanite is the high-pressure polymorph of the two Al_2SiO_5 polymorphs and absent from the suite of pegmatites and aplites in the HPPP.

Andalusite is another example which on one hand supports the idea of zonation within a pegmatite province and on the other marks the gradual passage into the surrounding metasedimentary country rocks (Table 4.1, Fig. 4.4a). Schaaf et al. (2008) attempted to chronologically constrain the age of this alumosilicate, calculating a two-point isochrone for the couple andalusite-muscovite from the Schwarzeneck and Blötz pegmatites. Their investigations yielded two Variscan ages of 346 ± 10 Ma and 330 ± 7 Ma, respectively, and place these pegmatites at the beginning of the Variscan mineralizing event prior to the intrusion of the granites and more or less concomitant with the thrusting in the Central Bohemian Massif (Fig. 3.1a). Another Rb-Sr isotope ratio was interpreted as a geological age of 127 ± 3 Ma. The authors were uncertain as to whether his age was caused by a later hydrothermal event, affecting the muscovite or reflecting a disturbance of the Rb-Sr system in the course of supergene kaolinization. The latter opinion seems the most plausible interpretation. During the early Cretaceous the NE Bavarian Foreland Basin fell dry, the sediments and crystalline basement rocks were cut by an unconformity/peneplain and suffered from chemical weathering and erosion likewise.

Novák et al. (1998b) investigate such peraluminous pegmatites in the central Bohemian Massif and concluded that the pegmatites were intruded into the brittle granulite host under a low-pressure regime characterized by PT conditions of ~2 kbar and 500 °C.

Apart from a few peraluminous andalusite-bearing pegmatites in the northwestern Saxo-Thuringian Zone (Wintersberg-Katharinenberg near Wunsiedel, Dillenberg (Tillenberg) near Waldsassen), the majority of sites known for their andalusite-bearing pegmatites are situated in the Moldanubian Zone of the Bayerischer and Böhmischer Wald (Döfering near Cham, Lengau and Herzogau near Waldmünchen,

Fig. 4.4a Andalusite from the exocontact zone of the NE Bavarian pegmatites. Cordierite-andalusite hornfels at the edge of the Leuchtenberg Granite to the west of the HPPP

Fig. 4.4b Muscovite-quartz-garnet-andalusite pegmatite from the Dillenberg (also Tillenberg) near Waldsassen, to the north of the HPPP

Hörlberg near Lam, Blötz near Bodenmais, Frath Mine, Taferlhöhe near Oberfrauenau, Wimhof near Vilshofen). The pegmatitic rocks take on a more and more aluminous character the deeper and older the they are. In the south-easternmost parts of the study area, this process was taken to the extreme, resulting in the formation of blue corundum, which comes close to what might be called sapphire (Schwarzeck near Lam-, Blötz near Bodenmais in gneiss together with Mn garnet, Arnbruck, Drachselrieth "In der Frathau" with Mn garnet). Corundum is in all sites mentioned above associated with andalusite, locally supplement with Mn garnet and in one places, hosted by marble and gneiss (Fig. 4.4b).

Corundum, or in chemical terms α-Al_2O_3, rarely goes together with quartz in felsic intrusive rocks, such as pegmatites. Its blue (sub)gem variety is less frequently than the opaque material, as it needs particular chromophores such Cr plus Fe to produce this outstanding color. Aluminum trioxide requires parent rocks of sufficiently high aluminum contents or its host rocks need to undergo some desilification in the course of metasomatic processes to achieve molar proportions of Al_2O_3 higher than that of SiO_2. At equal molar ratios both chemical compounds would go together to form andalusite or any other polymorph. This is at least in parts the case in some of the pegmatites under study in the SE Moldanubian Zone. A typical corundum-bearing pegmatite occurs at Dac Lac in southern Vietnam. Its origin lies in the field of migmatization and metamorphic remobilization and falls far apart from what has been described from true (granitic) pegmatites. In Madagascar, the circulation of fluids along discontinuities allowed in-situ alkaline metasomatism, forming corundum host rocks related to desilicified granites, biotitites, and some peculiar rocks such as sakenites and corundumites (Rakotondrazafy et al. 2008). At Mahenge, Tanzania, corundum was found in pegmatites bearing tourmaline as well as in marbles (Hauzenberger et al. 2005). Although making up only a rather subordinate fraction of the pegmatitic material under study at the western edge of the Bohemian Massif, the presence of blue opaque corundum reflects a deeper level relative to that of the northern Moldanubian and the Saxo-Thuringian Zones, where the HPPP pegmatites occur. Based upon the experimental data published by Chatterjee (1976) a lower temperature of formation for this mineralization may be assumed at 550 °C.

Unlike andalusite, topaz contains highly volatile compounds such as hydroxyl groups and fluorine and as such does not follow strictly this string of processes described for andalusite and corundum, albeit it does not fully escape from this geodynamically driven scheme of enrichment to be observed among these alumosilicates. Topaz predominates in the true granitic pegmatites of the Saxo-Thuringian Zone, near the zone of subfluence (Fig. 2.1), where it is encountered along the NE-SW striking Fichtelgebirge-Erzgebirge Anticline, in vugs and miaroles of the granites of the Epprechtstein and Kleiner Kornberg, Gregnitzgrund near Nagel, in the highly-fractionated Tin Granite G4 at Fuchsbau and Rudolphstein, near Schönlindt and at Hopfau (Table 4.1). Excluding sporadic finds at Silbergrube, the HPPP is a barren ground for this F-bearing mineral. Heading further south, topaz only sporadically shows up in the Hühnerkobel and Schwarzeck pegmatites. In this respect the variation of topaz is in line with that of the overall alumosilicates, being representative of the youngest and most apical parts of the pegmatites systems along the western edge of the Bohemian Massif and proximal to zone of subfluence, where the granite pegmatites *sensu stricto* developed by granitic fractionation. Off this deep-seated tectonic line the impact of fractionation on the formation of pegmatites diminished and topaz fade from the mineral assemblage of pegmatites. The miarolitic granitic pegmatites in this region were dealt with by Thomas et al. (2009). According to these authors melt inclusions to contain topaz crystals were accidentally co-trapped with the melt. Topaz was stable at the crystal growth temperature of smoky quartz (≈650 °C). At lower temperatures topaz is replaced by another mineral accommodating (OH)- and F- in its lattice. The fine-grained flaky muscovites powdering adjacent rock-forming minerals in pegmatites was given a name of its own, called "gilbertite". More frequently, topaz is released from the pegmatites and tin granites as the source rocks underwent erosion and chemical weathering. It was washed into the alluvial-fluvial placer system, where it can amount to almost 90 vol. % of the heavy mineral suite within a 5-km radius around the granites and the most widespread heavy mineral in the stream sediments of the clastic apron, guiding exploration geologists under the most favorable circumstances to these strongly fractionated rare metal granites and their granitic pegmatites.

A newcomer among the Al-bearing minerals from the HPPP, gibbsite was only recently discovered but not unexpectedly in the New Aplite at Pleystein. The aplite is pervasively altered and exposed as a earthy mass of kaolinite which, in places, was even deprived of its silica as the $\log_{H4SiO4}$ drops down to below –4.5, resulting in the formation of pockets of gibbsite. The same results may be achieved, as the ubiquitous muscovite in the aplite got rid of its K^+. The process, whose relics have only been preserved around Pleystein, marks a strong pervasive supergene alteration under tropical climatic conditions with an annual rainfall exceeding 200 cm. The pre-existing hypogene kaolinization, well-preserved in different more sheltered sites at Pleystein does not contain this mineral and thereby becomes a diagnostic mineral to distinguish both types of kaolinization.

4.5 Zircon

Picking up again the thread of an ore guide to rare metal granites and pegmatites, zircon plays a more prominent role than topaz, mentioned in the previous section as a potential marker mineral. Besides, rutile and tourmaline s.s.s., zircon is the most resistant mineral to chemical weathering, enabling its heavy mineral grains to travel far off the source area and survive even harsh conditions of weathering. Zircon has another edge over its associates rutile and tourmaline among the ultrastable heavy minerals, while providing the student of this Zr-bearing heavy mineral with a wide range of crystal morphologies that are controlled by the physical-chemical regime controlling the formation of its source rocks, be it granitic, pegmatitic or metamorphic. Pupin (1980), Pupin and Turco (1981), Bossart et al. (1986) Benisek and Finger (1993) and Bingen et al. (2001) investigated the extraordinary variability of the crystal morphology of zircon and correlated its crystal habit with the environment of formation. By this approach, they paved the way for its crystal morphology to be used in continental placer deposits as an ore guide and as a provenance marker

Fig. 4.5a Bipyramidal opaque brown zircon grains in stream sediments of fluvial placers concentrated by panning. Locality: Pleystein-Lerau drainage systems

Fig. 4.5b Stubby crystals of brown opaque zircon from Hagendorf-South pegmatite

to rare-element pegmatites (Dill et al. 2012b) (Fig. 4.5). Focusing on the HPPP, three morphological types can be singled out exclusive to the stock-like pegmatites Hagendorf-South, Hagendorf-North, the Kreuzberg pegmatite and the sheet-like pegmatite-aplite swarm at Miesbrunn (Fig. 4.5a, b, c, d, e). These morphological types significantly differ from another more elongated zircon the habit of which is

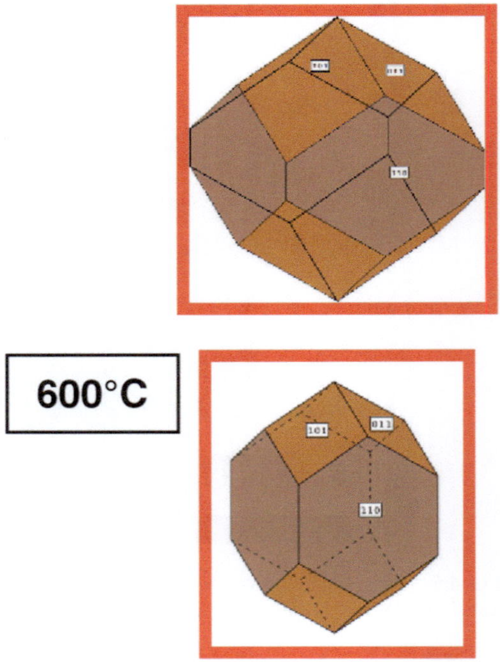

Fig. 4.5c Cartoon to show the various crystal habits of zircon type 4.5b dominated by the faces {110}. The temperature of formation is given in the rectangular box

Fig. 4.5d Bipyramidal brown opaque zircon from Hagendorf-North pegmatite. The *inset* shows the same type of bipyramidal zircon from the Miesbrunn tabular pegmatite-aplite swarm

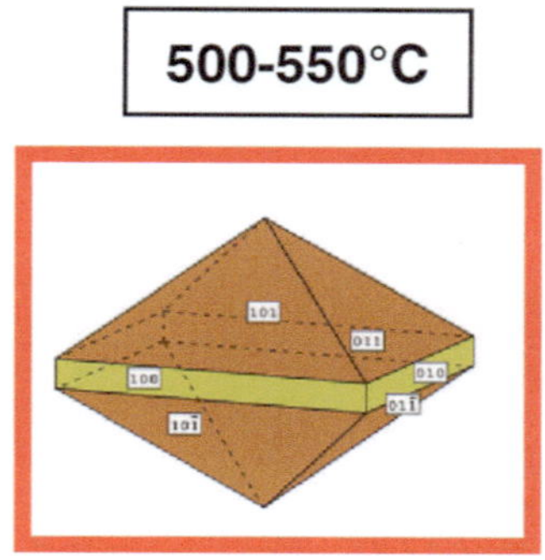

Fig. 4.5e Cartoon to show the crystal habit of zircon type 4.5d dominated by the faces {101}. The temperature of formation is given in the *rectangular box*

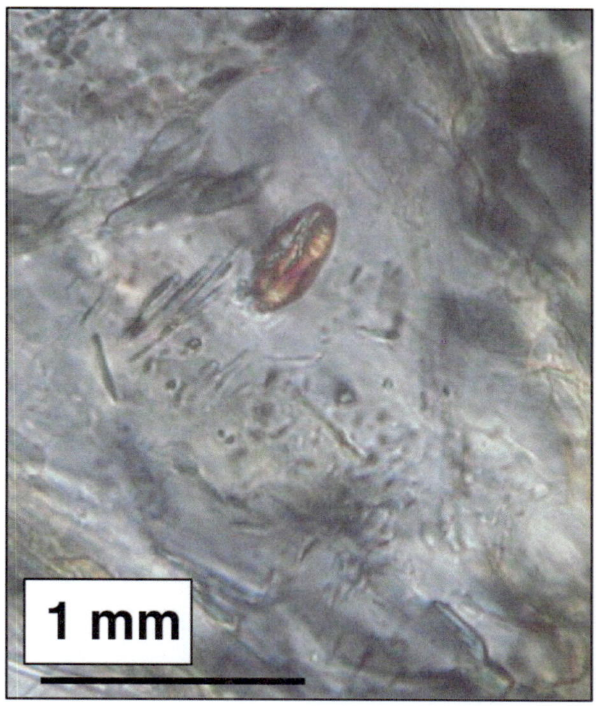

Fig. 4.5f Zircon grain under the petrographic microscope (Nic+) from a cordierite-sillimanite-biotite gneiss of the country rock included by quartz and fibrous sillimanite. N of Pleystein

diagnostic of the metasedimentary country rocks (Fig. 4.5f) but was also found in the tabular Trutzhofmühle aplite (Fig. 4.5g, h). Looking over the edge and casting aside for a while are focus on the HPPP will help us draw a more general picture of this mineral and its role it plays as a marker mineral. Examples from the pegmatites in northern Norway, from Malawi and the San Luis pegmatite province in NW Argentina, corroborate this idea that these types in Fig. 4.5c, e are applicable as an

Fig. 4.5g Prismatic zircon in quartz of the Trutzhofmühle tabular aplite

Fig. 4.5h Cartoon to show the crystal habit of zircon type in Fig. 4.5g with an elongated prism and a steep pyramid terminating the crystal on *top* and at the *bottom*

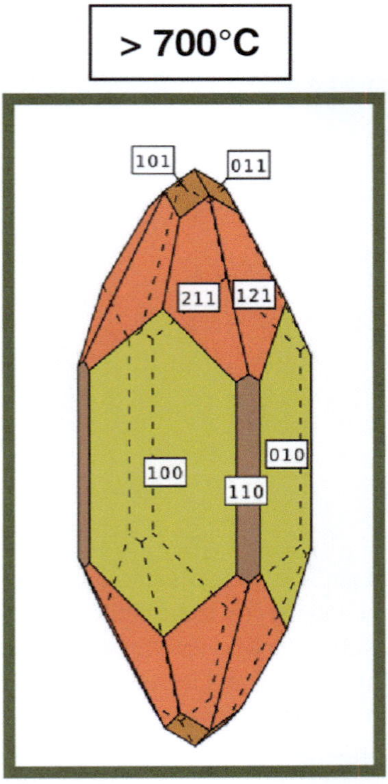

ore guide to pegmatite provinces (Pedersen et al. 1989; Roberts et al. 2006; Dill 2007; Oyarzábal et al. 2009; Dill et al. 2011a). The cordierite-sillimanite-biotite gneisses and the Trutzhofmühle aplite, both contain a morphological type, which developed at more elevated temperatures of >700 °C than the pegmatite-hosted zircon, whose morphology is characteristic for a temperature range of 500–600 °C. The question remains whether the Trutzhofmühle aplite formed at a much higher temperature or has incorporated the high-T zircon from the metasedimentary country rocks. The question is answered in favor of an "undigested" zircon within the aplite taken by heredity from the enclosing country rocks. Zircon is an accessory mineral and not of economic significance in the HPPP which appeared very early in the succession of minerals irrespective of its derivation as an undigested relic from the country rocks or as a newly-formed silicate in the pegmatite.

4.6 Phyllosilicates

Accumulation of phyllosilicates is nothing unusual along the western edge of the Bohemian Massif, whether in the basement, itself, or in its foreland immediately located to the west of it. In a great many places almost monomineralic deposits of mainly kaolin and bentonite developed from the late Paleozoic through the Neogene in such a way that these commodities could be mined or still be operated at a profit: Kaolin (Hirschau-Schnaittenbach, Tirschenreuth), bentonite and special clays (Molasse Basin, Naab Valley), talc/steatite (Göpfersgrün) (Schmid and Weinelt 1978; Dill et al. 2008a- further literature is cited in these comprehensive overviews).

In the NE Bavarian pegmatite deposits, the phyllosilicates make up a group of minerals that is very heterogeneous by quantity as well as by quality, one reason more why these minerals never were in the limelight (Table 4.1). On the other hand there is no other group of minerals which has had a stronger impact on the preservation of the pegmatites in the area. No greater example of this statement could have been given than quartz pegmatite at Pleystein where the feldspar rim was eaten away by kaolinization of hypogene and supergene type (Table 4.2, Fig. 1.3). Today we are still attracted by the quartz reef and nobody asked about who was the culprit for all that. As we no longer see this phyllosilicate as it was present in the geological past- see also Fig. 4.6a- our attention is drawn to the ubiquitous mica-type phyllosilicates which did not share the fate of the kaolinite-group minerals and can easily be spotted all across the HPPP.

Micaceous phyllosilicates are abundant in the pegmatites and represented by two principal types, <u>muscovite</u> with its alteration product <u>illite</u> and the <u>biotite</u> s.s.s (Fig. 4.6a, b).

<u>Kaolinite</u> or kandite-group sheet silicates are second in abundance in the HPPP, with the exception of the Pleystein area, where a pervasive kaolinization has affected almost all aplites and pegmatites being exposed at an altitude of higher than 500 m above mean sea level (a.m.s.l) as mentioned above (Fig. 4.6c, d). The process leading

Fig. 4.6a Muscovite rosettes from the Hagendorf-North pegmatite

Fig. 4.6b Aggregates of biotite at the contact between aplite (in the foreground) and aplite granite (in the background). Schafbruck aploid near Pleystein

Fig. 4.6c Aplite kaolinized almost to completeness from the Pleystein pegmatite mine. Only schorl has survived this strong alteration unharmed (black elongated crystals)

Fig. 4.6d Kaolinite concertina textures (kao) besides some muscovite-illite (mus) under the SEM form the New Aplite at Pleystein (Dill et al. 2015)

Fig. 4.6e Massive green pure nontronite from the Pleystein pegmatite mine

Fig. 4.6f Hedgehog of brown nontronite from Hagendorf-South

to this argillitisation is discussed in Sects. 4.5 and 4.6, attention will be drawn especially to the mineralogical features and peculiarities but not on the regional processes in this section. The extent of this argillitisation is unusual for pegmatites and aplites, not only for the NE Bavarian Basement but for the entire Bohemian Massif, although a moderate kaolinization is also known from some pegmatites of the Bayerischer and Böhmer Wald further south.

Smectite-group phyllosilicates, represented by its dioctahedral Fe-bearing member nontronite are confined to the major pegmatite stocks at Pleystein and Hagendorf-South. It is mainly an earthy dioctahedral Fe smectite and a well-crystallized modification that can be reported from the pegmatite stocks (Fig. 4.6e, g). Dioctahedral smectites enriched in Fe (nontronites) of supergene origin formed

Fig. 4.6g Aggregates of curled nontronite flakes from the same deposit under the SEM

during the late Neogene in Central Europe under (sub)tropical climatic conditions (Dill et al. 2010c). They are represented in the present study by the earthy subtyp.

Nontronite also formed during low-temperature alteration of sulfide-bearing felsic igneous and in epithermal copper-gold-silver deposits of high- and low sulfidation type (Ioannou and Spooner 2000; Dill 2001; Fernandez-Caliani et al. 2004). In these mineral deposits oxidative dissolution of pyrite and hydrolysis of feldspar is supposed to be the common mode of nontronite formation. APS minerals such as goyazite or alunite may appear in these Cu-Au-Ag deposits at shallow depths under increasing oxygen fugacities. Hypogene nontronite is the type clay mineral among the smectite-group which comes into existence under elevated oxygen fugacity, near surface, as primary sulfides turn into APS minerals and lead to what may be categorized as an "epithermal high-sulfidation" ore deposit. A similar textural scenario may be reported at Hagendorf-South for goyazite (Sect. 4.11), Conclusively, acicular nontronite similar to the large crystals goyazite are representative of an epithermal system, one at the end of the late Paleozoic and one during the Cenozoic (Fig. 4.6f).

Chlorite-group minerals bearing Fe and Mg are frequently in risk to be overlooked among the wealth of minerals and often mistaken for dark mica which they have derived from in the course of alteration.

During the initial phases of the pegmatite emplacement micaceous minerals and Mg-Fe chlorites show up in a characteristic way, as documented for the Hagendorf-North pegmatite. Approaching from the aplitic sillimanite-cordierite and biotite gneisses in the immediate vicinity of Hagendorf-North deposit the pegmatite, proper, the preferred orientation of phyllosilicates is gradually obliterated by a random arrangement of biotite (92.9 vol. %), chlorite (3.5 vol. %) and minor muscovite (2.0 vol. %), so that the name aplitic gneiss is the most appropriate term to describe the fabric changes in context with the mineralogical ones.

In the fine-grained aplite granite the biotite content significantly decreased relative to the gneiss, whereas the quantity of muscovite and chlorite increase. The Fe content of the chlorite attests to a Fe-enriched chlorite relative to a Fe-Mg chlorite in the gneisses. Despite a significant increase of the contents of chlorite (22.6 vol. %) and of muscovite (15.1 vol. %), biotite (47.2 vol. %) is still dominant. Its chemical composition is characterized by 7397 ppm Rb, the highest Rb contents ever determined from a dark mica of the HPPP.

In the aplite, next to the feldspar pegmatite, the biotite (1 M type) is still present while the chlorite content decreased to almost nil and muscovite (80.0 vol. %) becomes the prevailing phyllosilicate present in large rosettes at the inner zone which is transitional into the pegmatite (Fig. 4.6b). A reduction of the Fe content in the biotite-muscovite zone is compensated by a slight increase in the zone of muscovite which is now the dominating phyllosilicate. The muscovite belongs to the 2 M1 and 3 T polytypes according to the results obtained from XRD analyses. White mica in the zone of argillitisation may be determined also in infra-red spectra (IR). Different OH-stretching spectra of samples result from larger illite/muscovite contents and assist in diagnosing these phyllosilicates among prevailing kaolinite in these samples.

The core of the Hagendorf-North deposit is formed by a pegmatitic body composed of large tabular crystals of K feldspar and massive milky quartz with little muscovite. The ensuing core of the Hagendorf stock-like deposits is formed by a pegmatitic body composed of large tabular crystals of K feldspar and massive milky quartz but with little muscovite only.

The textural and mineralogical changes among the suite of mica and chlorite reflects some retrograde metamorphic overprinting of the country rocks which is expressed by an increase of muscovite and Fe-(Mg) chlorite, the latter replacing the primary biotite in these country rocks. Strong shearing along the contact strongly increased the quantity of chlorite at the expense of biotite, as it is the case at the Plössberg pegmatite. Muscovite is the only micaceous phyllosilicate along with kaolinite spotted in the white kaolinized aplites barren as to the primary feldspar, biotite and chlorites. Thermodynamic calculations constrain the physical-chemical regime of this 10-Å phyllosilicate to a pH range 5.5–12.5 and a log a $_{SiO2(aq)}$ in the range –6.0 to –3.5. The physical-chemical regime of the accompanying chlorite group minerals is difficult to be established due to the presence of bivalent and trivalent Fe accommodated into the lattice which has not been determined during routine XRD. Chlorite group phyllosilicates used to form at $7 < pH < 11$, dependent upon the valence state of Fe ($Eh > 0$ or < 0) accommodated into the lattice of chlorite and a log a $_{Al+++/(H+).3}$ of 7.5.

It goes without saying that biotite and muscovite are the prime target minerals when it comes to chronologically constrain the age of the emplacement of pegmatites as shown earlier in this book, in Sect. 3.1, see, e.g., Glodny et al. (1998). This fact has also strong implication as to the temperature decrease and cooling behavior of a pegmatitic melt. We selected in this paragraph a muscovite from the NE-Bavarian pegmatite province to get an idea when muscovite went through the 425 °C barrier up on cooling (Harrison et al. 2009; Duvall et al. 2011). The muscovite cooling age lies at 305.3 ± 3.4. The chemical composition of micaceous pegmatite-related phyllosilicates was among others investigated by Alfonso et al. (2003). Microprobe analyses of rock-forming minerals from the Trutzhofmühle Aploid yielded the following mean values: muscovite (0.07 wt. % P_2O_5), albite (0.25 wt. % P_2O_5), K-feldspar (0.70 wt. % P_2O_5). It does not place the micaceous phyllosilicates very high as the typical pegmatite minerals of the HPPP are ranked as a potential source of phosphate. Though not considered as a potential source of P, the proximal position of secondary phosphates, such as rockbridgeite, which are often intimately intergrown with this white mica, attest to a transfer of hydrous components from a primary rock-forming phyllosilicate to a secondary phosphate. Two trace elements warrant a more detailed treatment for their involvement in the evolution of pegmatite in NE Bavaria. Monovalent rubidium can substitute for monovalent potassium and niobium plus tantalum are the most diagnostic elements for the HPPP (see Sect. 4.8). To document increasing fractionation of a melt from the pegmatite wall zone to the core various elements have been selected, among others Rb and K, and their ratios plotted to show if rare element granites or pegmatites might be expected, e.g., Kuznetsova (2013). In the pegmatitic setting under study, two groups significantly different as to the Rb/K ratio occur. Highly differentiated pegmatites and aplites have Rb/K*1000 ratios of biotite and muscovite in the range 104–126, exceeding the same ratios of aploids and pegmatoids the ratios of which fall in the range <10. Biotite deems to be more susceptible to this differentiation than muscovite in the same deposit. Shifting to the Nb contents yields the same pattern. Highly differentiated pegmatites and aplites contain Nb in the range 635–693 ppm and barren equivalent felsic rocks have Nb in the range 32–33 ppm. Tantalum which does not play a similar prominent role as Nb in the HPPP, does not allow for such a differentiation (see Sect. 4.8) The Rb/K and Nb spectra for the unmineralized pegmatites and aplites are equal or only slightly above what is known from the granites in NE Bavaria. It is worth noting, that even the harshest supergene kaolinization cannot fully eradicate these Nb contents and enable us to decide on whether the primary pegmatite or aplite was a rare element pegmatite or a barren one.

Kaolinite and nontronite are dealt with together despite of their different mineralogical kindreds, the first one belonging to the 7-Å phyllosilicates and the second one to the expandable 14-Å phyllosilicates (Fig. 4.6c, d, e, f, g). For a precise diagnostic study of the kaolinized pegmatites apart from the common XRD, IR and DSC analyses were conducted, supplemented by CEC analyses (cation exchange capacity) to assess the amount of swelling phyllosilicates in the kaolinized rocks (Dill et al. 2015). For some typical examples this analytical approach to investigate the mineralogical composition and the degree of crystallinity of kaolinite is given in the diagrams of Fig. 4.6h, i, j. In the graph of Fig. 4.6i, displaying the IR spectra, the four characteristic OH-stretching vibrations originating from the different AlAlOH bondings are assigned v1–v4. The v1 through v3 vibrations result from inner surface hydroxyls whereas v4 corresponds to inner hydroxyls. Kaolinite with so called

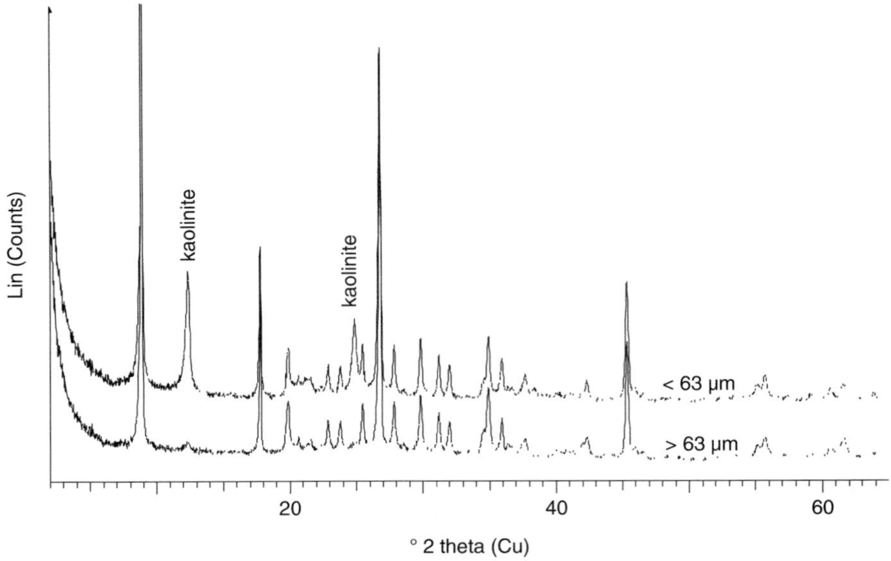

Fig. 4.6h XRD runs of the kaolin from the Kreuzberg Pegmatite to show muscovite and kaolinite to be the only phyllosilicates and kaolinite being enriched in the <63 µm fraction

b-axis disorder (Russell and Fraser 1994) lacks the v2 stretching vibration and in the case of halloysite, both, v2 and v3, are absent. Results obtained by IR spectroscopy confirmed, only traces of kaolinite in the >63 fraction. Comparing the IR data of kaolin (<63 µm fractions) of the fault gouge from site 2 to the kaolin from the New Aplite at site 1 reveals different degrees of crystallinity. The different OH-stretching spectra of both samples result from the larger illite/muscovite contents of the sample taken at site 2 and from a lower degree of crystallinity of the kaolinite in this sample. The appearance or absence or even the intensity ratio of the 3670 and 3650 cm^{-1} band are often considered reflecting the crystallinity (degree of structural order) of the kaolinite (Russell and Fraser 1994). Accordingly, the low intensity of the band at 3670 cm^{-1} indicates a lesser degree of crystallinity compared to the samples at site 2. Another applicable method to derive some information about structural order of the kaolinites from, IR spectra was provided by Brindley et al. (1986). These authors established an empirical relation of the position of v1 and the famous Hinkley index which is determined by XRD to characterize the degree of structural order. The v1 position of all samples ranges from 3695 to 3697 which corresponds to a Hinkley index of 0.5–1.0. The DSC curves of the three samples resemble each other indicating a similar phyllosilicate composition (Fig. 4.6j). The curves are dominated by the thermal reactions of kaolinite. Only a very weak endothermic peak of sample 11304735 indicates the presence of some swelling clay minerals in the core zone of the aplite, a fact which has not been unraveled by the previous methods applied to the kaolin. This is particularly evident from the mass spectrometer curve of evolved water showing two- step-dehydration. The pronounced

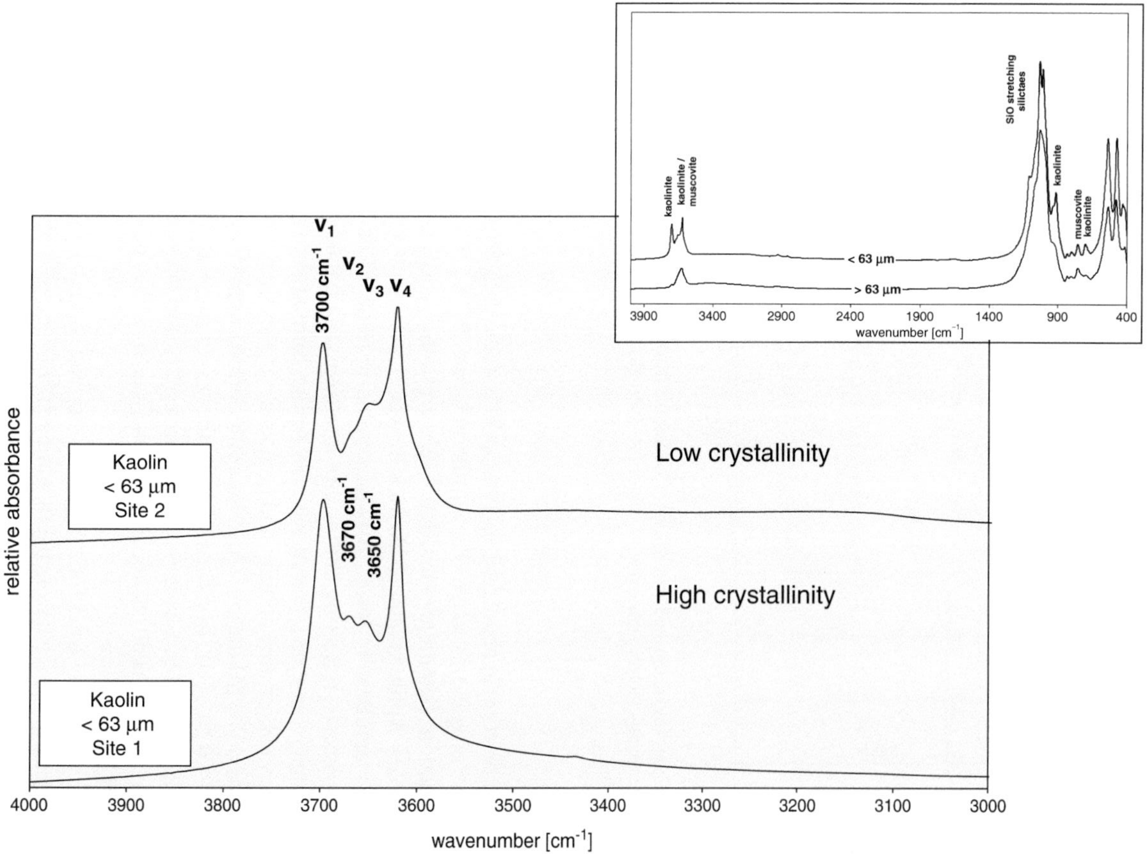

Fig. 4.6i IR spectra of the Pleystein kaolin. The inset on the *top right* illustrates the relative abundance of kaolinite in the various grain .size fraction the main plot reveals different degrees of crystallinity for kaolinite at site 2 (Kreuzberg Pegmatite) and site 1 (New Aplite). The four characteristic OH-stretching vibrations originating from the different AlAlOH bondings and are assigned v1–v4

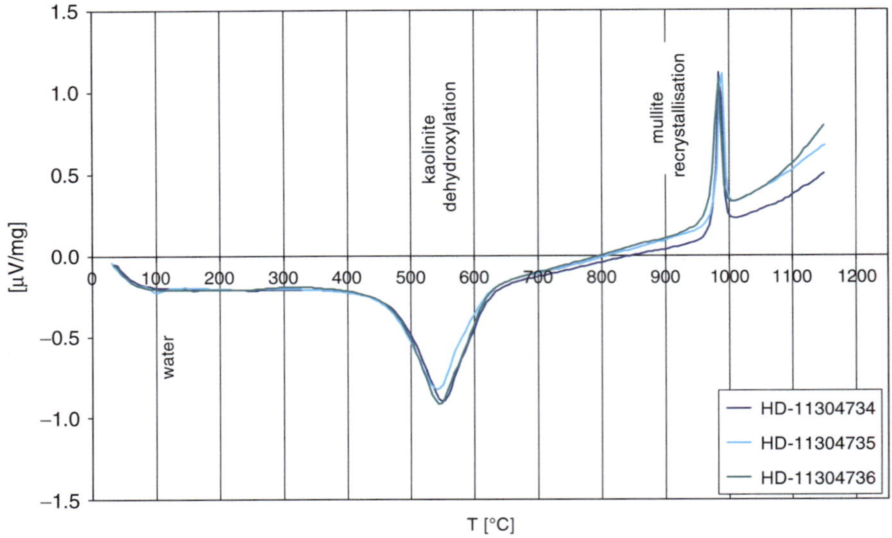

Fig. 4.6j DSC curves of the <63 µm fractions of kaolin from the New Aplite at Pleystein. The inset shows the mass spectrometer curve of water (evolved gas analysis)

endothermic reaction at about 550 °C corresponds to the dehydroxylation of kaolinite, typically being in this range. The exothermic peak at 983 °C corresponds to the ceramic reaction (kaolinite → mullite). The crystallinity of the kaolinite from Pleystein's New Aplite is almost identically, whereas the fault bound kaolin of the Kreuzberg is significantly lower. It is interpreted in terms of an ongoing aging or kaolinite diagenesis, reflecting in the current setting a more moderate diagenesis of obviously younger age than that of the New Aplite. Similar thermo-dynamic calculations performed for the kaolinite and nontronite at Pleystein yielded the succeeding parameters. The pH during which kaolinite formed runs from 4.5 to 6.5 at a log a $_{SiO2(aq)}$ of −4.0 to −3.0. Iron-bearing smectite-group minerals of the nontronite-saponite series formed in a wide range of sedimentary and magmatic rocks (Grauby et al. 1994). The physical-chemical conditions under which nontronite came into being are $pH \geq 8$ and $Eh > 0$.

A rather exotic depocenter for a pegmatite-derived kaolinite and some associated phyllosilicates was reported from the Pflaumbach drainage-system around Pleystein. Kaolinitic clay minerals occur in two different modifications as internal sediments in cassiterite aggregates which have definitely been originated from the Kreuzberg Pegmatite (Dill et al. 2006a). One forms well-defined flakes of sheet silicates, the other is present in lumps. Illite is not very much different from kaolinite with respect to its grain and aggregate morphology (Fig. 4.8c). Almost all phyllosilicates mentioned above were identified on microscopic scale in a succession of minerals by EMPA. Potassium, magnesium and iron present in these sheet-like aggregates and lumps are indicative of illite. Calcium being present in the aggregates points to some mixed-layers with smectite. Smectite occurs also as a phyllosilicate on its own. Small amounts of Fe-Mg chlorite and vermiculite are also present in the hexagonal voids in cassiterite. An unknown Si-Al phase was spotted in the EDS spectrum. Its Al/Si ratio of 1:3 strongly contrasts with the Al/Si ratio of kaolinite (1:1) and illite (3:4). The ratio would well fit with pyrophyllite (?) but could not be proved by an independent method.

4.7 Miscellaneous Silicates

Investigations in the previous sections dedicated to the tecto- and sheet silicates which make up the lion share of the rock-forming minerals of the pegmatites and aplites may bear the risk that other silicates are eclipsed. Some of them are only "mineralogical curiosities", such as mesolite whose genetic position is not well established and as such does not significantly contribute to the determination of the physical-chemical regime of these felsic rocks (Table 4.1).

Two minerals, epidote s.s.s. and diopside, albeit present only in minor amounts in those aplites and pegmatites in NE Bavaria, deserve a more detailed treatment here as they behave as missing links between the felsic mobilizates on one side and the skarn deposits on the other (Table 4.1, Fig. 4.7a, b). The pegmatites and aplites of both Ca silicates are barren as to rare metals, such as Burkhardsrieth and a few other aplites in the environs of Pleystein. Diopside may be traced back to Ca-enriched wall rocks and was inherited into the endocontact of the respective pegmatite.

Fig. 4.7a Two generations of epidote in the marginal aplite of the Burkhardsrieth pegmatite. The first generation of epidote is disseminated in the aplite the second one lines the walls of joints intersecting the contact zone aplite

Fig. 4.7b An aplite west of Pleystein show a stockwork-like mineralization made up of green Fe-enriched epidote. The inset provides an outline of a monomineralic concentrate composed of grains of green clinozoisite-epidote s.s.s. from the Plaumbach drainage system W of Pleystein (<1 mm)

Epidote goes a different way from diopside in its emplacement. Two generations of epidote were determined in the marginal aplite of the Burkhardsrieth pegmatite. The first generation of epidote is disseminated in the aplite the second one lines the walls of joints intersecting the wall aplite, another epidote known from an aplite west of Pleystein shows a stockwork-like mineralization made up of green Fe-enriched epidote, which can easily be spotted with the unarmed eye not only in the primary host but also in the stream sediments of the drainage system cutting into the foot slope of this rock exposure (Fig. 4.7a, b).

Epidote-group minerals need to be discussed also as the basic host of REE elements, leading at the very end to allanite s.s.s. a mineral present also in the HPPP the Bayerischer Wald. Since the REE issue is the one that makes these epidote-group minerals so attractive it is treated together with the REE minerals in Sect. 4.21. Allanite-(Ce) is a member of the epidote group, allanite subgroup originating from clinozoisite by homovalent substitutions and one coupled heterovalent substitution (Armbruster et al. 2006). Epidote is characteristic of the low-grade stage of regional and contact metamorphism which lies dependent upon the pressure regime between approximately 400 °C and 500 °C. Following the studies of Cota et al. (2010), magmatic epidote may crystallize in tonalite and granodiorite at sufficiently high pressures (≥5 kb), where it is less common due to lower CaO. Epidote is inferred to have crystallized at relatively low pressure (<2 kb) in felsic intrusive rocks and encountered in miarolitic cavities and fractures, similar to what has been described from the HPPP in Fig. 4.7a and interpreted to be magmatic in origin. Strmić Palinkaš et al. (2012) provided an in-depth study of an epidote-bearing pegmatite, at Čanište, Republic of Macedonia and determined the physical-chemical regime of the intermediate zone enriched in epidote. Cooling of the granodioritic initial melt below 550 °C at 5.2 kbar provoked the precipitation of epidote besides hematite, garnet, muscovite, and quartz. Our epidote-clinozoisite s.s.s. are almost barren as to REE (<1000 ppm REE_{tot}). Their Fe_2O_3/CaO correlative with the position in the epidote-clinozoisite s.s.s. has a spread from 0.67

through 0.73, bearing evidence of a clinozoisite-dominated s.s.s. For comparison the ideal composition of epidote in nature stands at 1.60. Cota et al. (2010) reported observations that are compatible to the aplites in the HPPP. Aplites hosting epidote are poor in phosphate, even the common apatite-(F) is scarce and the bulk composition ranges from 0.024 to 0.112 wt. % P_2O_5. Calcium has used to form epidote rather than apatite. Epidote is a negative marker, indicative of a P-poor environment. Its Ca analogue clinozoisite is placed outside the epidote-bearing wall rock aplites and found either in calcsilicate bands intercalated into the amphibolites or mineralized skarn deposits (Sect. 5.1.1).

Heading further south, we will still see the epidote to be one of the accessory silicates in the pegmatites. In these sites it is accompanied in its host pegmatites by phosphates such as zwieselite at Blötz, or at Oberfrauenau, where zwieselite, vivianite, scorzalite and lazulite formed in the same pegmatite. It is true that these Ca-bearing silicates are absent from the stock-like representatives of P-bearing pegmatites, e.g., Hühnerkobel, and their analogues of the HPPP, Hagendorf and Pleystein.

Moreover in the south of the working area, zeolites, enriched in Ca, chabazite, laumontite, and stellerite occur together with some prehnite, an association not out of the ordinary in a series of basic or ultrabasic rock such as the ophiolite suite in the Troodos Mts., Cyprus, but at odds with the ideas of the emplacement of a felsic magma (Dill et al. 2007a, b, c). Stellerite and laumontite are present at the contact between the Basal Group and the Intrusive Complex underneath. Laumontite and stilbite typical of subfacies I originated from Ca released during breakdown of plagioclase in the sheeted dyke dolerites. At low p_{H2O} laumontite comes into existence, at higher p_{H2O} stilbite may form instead according to the following reactions: anorthite + quartz + $4H_2O \Rightarrow$ laumontite and anorthite + quartz + $7H_2O \Rightarrow$ stilbite. In a P–T diagram showing the experimentally determined relations between Ca-zeolites, stilbite is stable besides laumontite below 140 °C and below 700 bar (Liou et al. 1991; Frey et al. 1991). The presence of prehnite in this subfacies indicates maximum temperatures between 200 and 280 °C. Laumontite, prehnite and clinozoisite-epidote s.s. are stable at temperatures around 260 °C at pressure greater than 500 bar. From whatever angle one may look at the P-T regime, there is one prerequisite which must be fulfilled prior to the coming-out of Ca silicates, the contamination of the felsic magma with a basic country rock by metasomatic processes. Junks of basic country rocks within the endo- and the exocontact of pegmatites in the metapsammopelitic of the NE Bavarian Basement, ranging in their lithology from calcsilicates, through amphibolite to true marble provoked locally an increase of Ca and gave rise to the formation of Ca silicates as a function of the P-T regime, either of the very-low or low-grade stage. In both cases the outlook as to the presence of rare element phosphate-bearing pegmatite will not be very much promising if these minerals increase and get into conflict with apatite for calcium.

This negative result may not always be expected when zeolitization takes place during the formation of pegmatites and, locally, can point to the presence of gemstones, e.g., chiavennite or spodumene- see Sect. 1.2.1 (Dill 2010). There is no pendant in the NE Bavarian Basement to that for geodynamic reasons.

4.8 Niobium-, Tantalum, Tungsten and Tin Oxides

There is only one ore mineral, genetically related to the Ca silicates of the clinozoisite-epidote s.s.s. and diopside discussed in the previous section. Scheelite is a rare ore mineral known only from the Wildenau-Plösberg pegmatite at subeconomic grade (Table 4.1). Quite a different story is the scheelite mineralization to the west of Pleystein, at Gsteinach. This skarn-type mineralization is genetically related to the emplacement of the pegmatites of the HPPP, but does not form part of the formation of the pegmatite themselves (see Sect. 5.1.1). Therefore its mineral association has not been considered in Table 4.1.

Being always closely related to the late Variscan granites, wolframite cannot be considered as a rare ore mineral along the western edge of the Bohemian Massif. However within the exo- or endocontact of the pegmatites in the region it has not yet been recorded. Its occurrence in the HPPP is rather unusual, being present as a rare constituent of the "nigrine" mineral aggregates in the stream sediments of the drainage system that incised into the pegmatites and its wall rocks of the Pleystein area. Its intimate spatial relationship with columbite-(Fe) points to a source rock area fertile for pegmatites. Outside the HPPP it mineralizes quartz veins, but has so far not been discovered in the external contact zone of the pegmatites of the HPPP. To understand the origin of wolframite and the role it plays as an ore guide to rare-metal pegmatites this W minerals has to be dealt with together with "nigrine" (Sect. 4.9).

It is written in many textbooks that Sn oxides are chemically related to W minerals in highly fractionated granitic rocks such as those from the Fichtelgebirge-Erzgebirge Anticline in the Saxo-Thuringian Zone where Sn mining persisted for more than a century. Tin minerals were also found in the much stronger fractionated zoned pegmatites of the HPPP in the Moldanubian Zone. Mücke and Keck (2008) recorded cassiterite and stannite from the Hagendorf-South pegmatite although at a quantity far beyond economic significance. After cassiterite was found to be one of the most widespread heavy minerals in the "nigrine" alluvial-fluvial placer deposits in the drainage-system around Pleystein by Dill et al. (2006b), it was not really a surprise, as we found the primary Sn source of the placers in the so-called sulfide mineral association – see Table 4.2- of the Kreuzberg in the center of Pleystein (Fig. 4.8a, b, c) (Dill and Weber 2014a). The cassiterite grains show the morphological features well-known from high-temperature pegmatitic Sn deposits where the faces of the prism are reduced to almost nil, approximating in their outward appearance the morphological zircon type illustrated in the cartoon of Fig. 4.5e. Apart from zircon- Sect. 4.5- cassiterite is another example of how the crystal morphology varies as a function of the temperature of formation. As the temperature drops the faces of the prisms of cassiterite become more and more pronounced relative to the pyramid leading to what might be called the "Cornish Type". A further drop in the temperature of formation gives rise to more elongated and slender crystals leading at the end to what is called "needle" tin. A botryoidal modification called "wood tin" marks the end of the sequence the entire sequence of which was

Fig. 4.8a Subrounded cassiterite aggregate made up of twinned cassiterite crystals from the alluvial-fluvial placer deposits of the drainage-system in the environs of the Pleystein pegmatite which have been derived from the HPPP pegmatite stocks

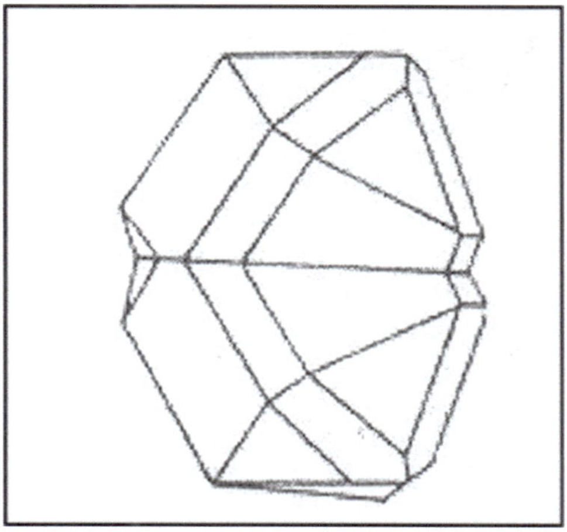

Fig. 4.8b Cartoon to show an idealized intergrowth of twinned cassiterite crystals called "Visiergraupen" from the pegmatitic deposits

encountered in the roof rocks of the Late Paleozoic Sparnberg-Pottiga Granite, at the outer perimeter of the Fichtelgebirge-Erzgebirge Granite Line and shown by a series of polished section under the ore microscope in plates 33 and 34 by Dill (1985a). The sequence is indicative of low-temperature mineral assemblages decreasing from 200 °C to below 100 °C. Neither morphological type is represented in the HPPP pegmatites and the presence of a low-T hydrothermal Sn remobilization during subsequent stage of pegmatite evolution similar to what has been described from Cornwall and the apical parts of the Sparnberg-Pottiga Granite can

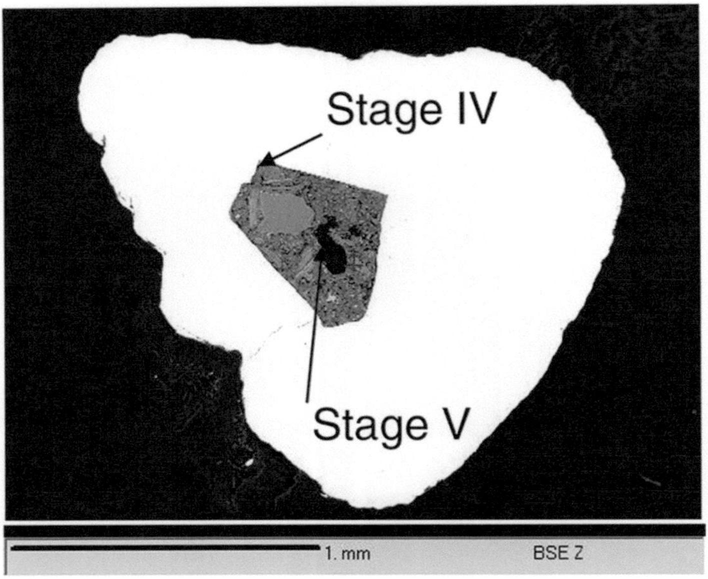

Fig. 4.8c Subrounded cassiterite aggregate whose voids are filled with minerals of supergene origin, concentrated during various stages of weathering. They are denoted by the *arrowheads* I and II. Mineral association I is characterized by anatase, kaolinite and APS minerals, all of which are lining the cavity. The remaining part is filled with kaolinite, kaolinite-illite mixed layers, quartz and Ca smectite. It was labeled association II. These internal sediments reflect the environmental changes in the aftermaths of the weathering, erosion and transport of the cassiterite from the primary pegmatitic source to the depocenter (BSE image from EMPA Dill et al. 2006b)

definitely be ruled out. Cassiterite from Pleystein Kreuzberg pegmatite is depleted in minor elements. Our investigation revealed that the cassiterite clusters are almost chemically pure SnO_2 with low concentration of FeO (<0.14 wt. %) and Nb_2O_5 (<0.5 wt. %), and Ta, V, Mn and W below the detection. TiO_2 is slightly enriched and its contents range from 0.4 to 0.8 wt.%. TiO_2.

In the course of studying the chemical composition of cassiterite another solid solution series of oxides has been touched whose significance for the genetic investigation of the HPPP pegmatites cannot be overestimated. The Nb-Ta-Mn-Fe minerals pertaining to the columbite group or columbite-tapiolite series are present in almost all pegmatites and aplites, excluding Brünst, and Peugenhammer (Table 4.1, Fig. 4.8d, e, f, g, h). The minerals of the orthorhombic oxidic Nb-Ta s.s.s. mostly occurs in crystals flattened to the prism faces $\{01\bar{1}0\}$ and they are intergrown with quartz, feldspar and sulfides of high-T origin, close to the core-rim contact in the zoned pegmatite stocks as well as in the tabular aplites and pegmatites. The latter hosts saw these Nb-Ta minerals to be randomly distributed among minerals of the initial stages of pegmatitization, such as apatite or monazite (Fig. 4.8e). The chemical composition of the s.s.s. from the HPPP is typical of a ferrocolumbite or columbite-(Fe) *sensu* Černý (1992) and Černý (1989) as the Fe/Mn and Nb/Ta ratios are

Fig. 4.8d Plates of columbite-(Fe) epitactically intergrown with octahedra of uraninite

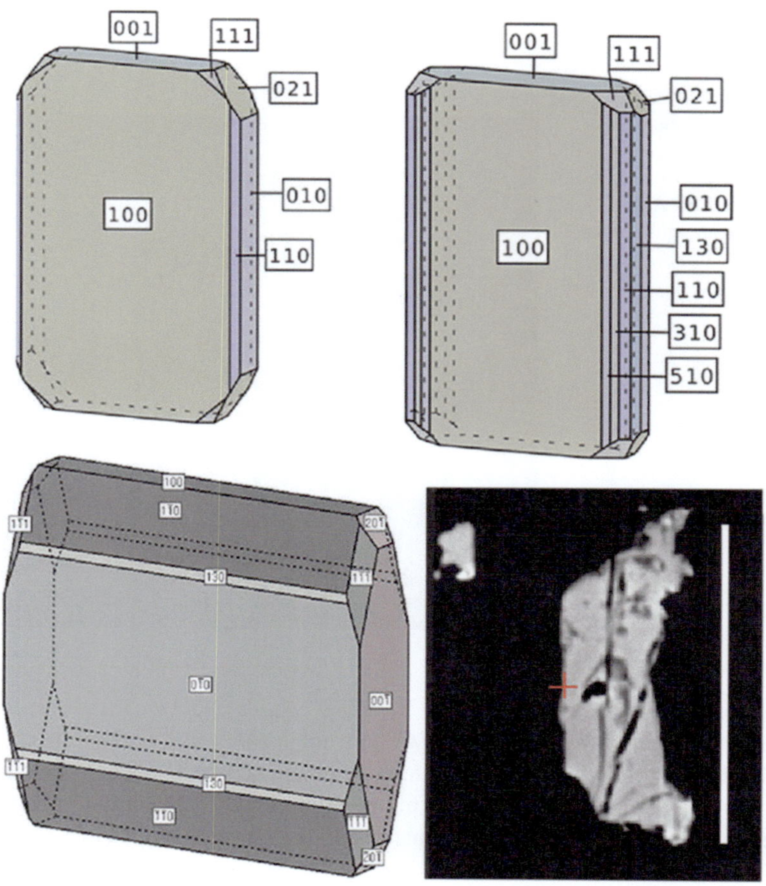

Fig. 4.8e Cartoon to illustrate the typical morphologies of columbite-(Fe)

Fig. 4.8f Plate of columbite-(Fe) replaced along a hydro-fracture by benyacarite at Pleystein (SEM)

Fig. 4.8g Columbite-(Fe) with a Nb-enriched core surrounded by a Ta-enriched rim from the New Aplite, Pleystein, in an argillaceous matrix (SEM)

well above 1 in all samples under study. There are unzoned and zoned mineral grains of columbite s.s.s (Fig. 4.8g, h). Their minor element contents may be very variable.

Columbite-(Fe) from the Trutzhofmühle aplite and the Miesbrunn pegmatites-aplite swarm contains moderate amounts of Ti (1.2–1.9 wt. % Ti), Sc (<0.5 wt. % Sc)

Fig. 4.8h Zoned columbite-(Fe). The variation of the Nb/Ta ratio is given in the framed boxes. By visual examination of several columbite grains, the conclusion may be drawn, that Ta-enriched domains in the columbite s.s.s. are more vulnerable to supergene alteration than those enriched in Nb (EMPA/BSE image). New Aplite, Pleystein

and it is markedly enriched in W (2.7–2.9 wt. % W). Its plates are lined up along flakes of chloritized biotite and its growth can still proved to be post-kinematic relative to the stress-strain pattern of the host rocks. At Otov, Czech Republic, which is the nearest occurrence of columbite further to the East, or in other word, towards the center of the Bohemian Massif, columbite forms exsolution lamellae in cassiterite (pers. com. M. Novak). There are elevated contents of WO_3 which are positively correlate with Nb_2O_5 (R = 0.72) and TiO_2 (R = 0.70) contents in the columbite-(Fe) from Reinhardsrieth. Zirconium, uranium and tin reveal a close genetic interrelationship documented by the highly positive correlation coefficients: $R_{ZrO2\ vs.\ TiO2} = 0.91$, $R_{ZrO2\ vs.\ UO2} = 0.99$, $R_{TiO2\ vs.\ UO2} = 0.94$. Not only elevated Sn and W contents have been reported in context with columbite, but the abnormally high values of Sn and W proved by the presence of cassiterite and wolframite inclusions, too (Strunz and Tennyson 1975). The Hühnerkobel pegmatite did not only give host to the two W and Sn oxides but was also the *locus* where Rose (1845) discovered niobium in 1844 in a Ta-enriched member of the columbite s.s.s.

According to Aleksandrov (1963) and Linnen and Keppler (1997) columbite s.s.s. reveal characteristic changes upon evolution of pegmatites with their Nb/Ta ratio decreasing along with fractionation. Zoned columbite s.s.s. investigated under the EMP used to show their outer or younger parts to be enriched in Ta (Fig. 4.8g, h). The cross plot of Ta_2O_5*100/(Ta_2O_5+Nb_2O_5) vs. MnO*100 (MnO+FeO) portrayed in Fig. 4.8i proved the most efficient way to show the genetic trends of consanguineous pegmatites and aplites in an ensialic orogen like at the western edge of the Bohemian Massif.

The chemical data of the columbite s.s.s. from the tabular and stock-like pegmatites and aplites of the HPPP form the centerpiece of the cross plot in Fig. 4.8i. Columbite included in "nigrine" – Sect. 4.9- is an armored relic. Its data cluster at very low Ta and Mn contents, they are the most primitive and least fractionated pegmatite-related but not pegmatite-hosted columbites in the study area. These data were determined in columbite contained in "nigrine" hosted by the gneissic wall rocks of pegmatites where the mineral aggregates were released from into the alluvial-fluvial placer deposits after unroofing of the pegmatites by denudation. Püllersreuth columbite is hosted by a metapegmatite and documented by the subhorizontal trend line does not furnish evidence of any fractionation of Ta and Nb either (see Sect. 3.1.2). A set of data found in columbite of the Kreuzberg pegmatite at Pleystein shows an almost identical trend and consequently reflects a strong metamorphic impact. The anticlockwise trend from metamorphic to magmatic felsic mobilizates (metapegmatites, intermediate (meta) pegmatites and magmatic felsic mobilizates) reveals a diminishing impact of metamorphic processes onto the pegmatites from Püllersreuth, through Kreuzberg, Reinhardsrieth I, and Trutzhofmühle. They are arranged in order of increasing magmatic impact that is represented by the slope of the positive trend lines. The columbites from the Tepla-Barrandian in the Czech Territory have metamorphic and magmatic signals. The magmatic trend is attested to by the vertical to subvertical trend lines of the Pleystein and Hagendorf trends with almost no Fe-Mn fractionation. According to Černý et al. (1986) such a trend is characteristic of less fractionated beryl- and spodumene-bearing pegmatites. These HPPP pegmatites have only minor contents of beryl and no spodumene and belong to most the strongly fractionated pegmatites. The data array of the columbites from the Bayerischer and Böhmer Wald further to the south of the HPPP and located at a deeper level of erosion than the HPPP overlap with the data of the older magmatic Pleystein trend mixed up with some offshoots reflecting a metamorphic component on the formation of pegmatites equal or slightly greater than that of the HPPP. These data fully agree with the evolution of the ensialic orogen and the geodynamic siting of the various pegmatites (Sects. 2.3, 3.1, and 3.3).

Thinking a bit more outside the box and taking a glimpse at granites and granitic pegmatites of the northern Fichtelgebirge, reveals there a presence of columbite besides tapiolite in the Waldstein (G3) and Rudolfstein (G4) granite massifs. Tapiolite is chemically very close to columbite and tantalite but structurally different. Solid-solution series in the columbite-tapiolite series were discussed by Schroecke (1966).

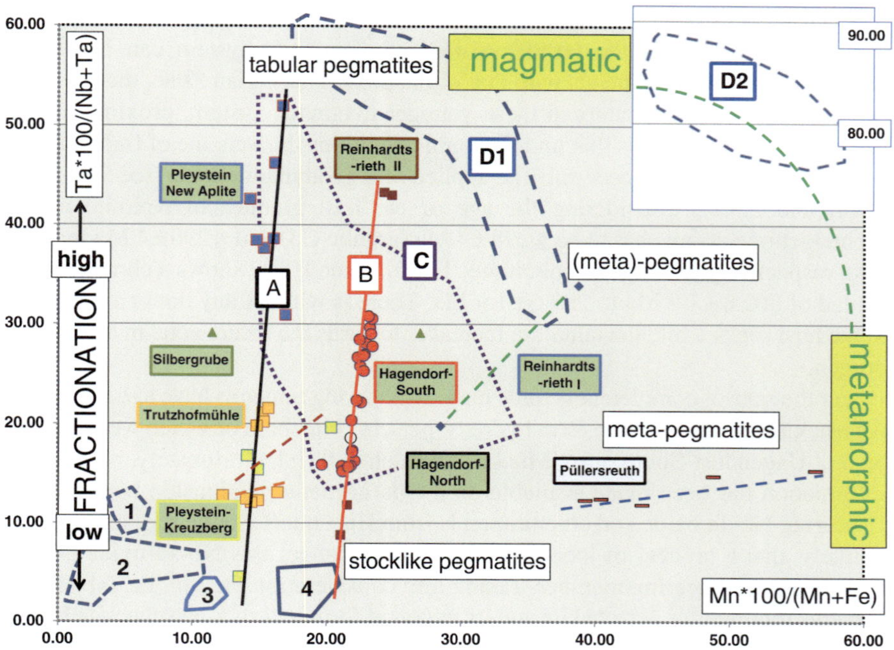

Fig. 4.8i Cross plot to show the Ta_2O_5*100/(Ta_2O_5+Nb_2O_5) vs. MnO*100 (MnO+FeO) of columbite s.s.s. from metamorphic to magmatic columbite mineralization. The data arrays of the tabular and stock-like pegmatites of the HPPP are given in the rectangular boxes. The *framed boxes* and the *data symbols* are plotted in the same color (e.g. Hagendorf-South in *red*). Area framed with dash-point lines terminated the data arrays of columbite s.s.s. included in "nigrine" (see Sect. 4.9) *1*: Deggendorf "nigrine" in gneiss, *2*: Deggendorf "nigrine" in alluvial-fluvial placer deposits, *3*: Iglersreuth "nigrine" in alluvial-fluvial placer deposits, *4*: Pingermühle: "nigrine" in alluvial-fluvial placer deposits *A* gives the Pleystein trend, *B* the Hagendorf trend, *C* shows the data array of the pegmatites in the Bayerische-Böhmer Wald including data from the Hühnerkobel, Pochermühle, Blötz, Schwarzeck, Schwarzenbach, Birkhöhe, Pauliberg, Kautzenbach pegmatites (source: Schaaf et al. 2008), *D1* and *D2* are two discrete data arrays of columbite s.s.s. from Otov, Czech Republic, immediately E of the Czech-German border (Source: M. Novak, Brno). D2 is located above Ta_2O_5*100/(Ta_2O_5+Nb_2O_5)=60.00 and fit into the base diagram as to the MnO*100 (MnO+FeO) ratio 50.00 to 60.00. The trend lines of the various data *arrays* in the diagram follow two different trends as far as the Nb-Ta mineralization is concerned (**a**) an anticlockwise trend from **metamorphic to magmatic felsic mobilizates** (metapegmatites, intermediate (meta) pegmatites and magmatic felsic mobilizates). Such trends are recognized at Püllersreuth, Kreuzberg, Reinhardsrieth I, Trutzhofmühle (arranged in order of increasing magmatic impact, represented by the slope of the positive trend lines) (**b**) a magmatic trend showing the **degree of fractionation** from low to high. The most primitive columbites evolved in the "nigrine" aggregates, with *1* and *2* apparently derived from metamorphic processes and *3* and *4* signaling the initial level of magmatic fractionation

Even if these Nb-Ta minerals are a bit out of the ordinary among the mineral associations there and the HPPP's (Li)-Nb-P pegmatite system can hardly be claimed to be consanguineous with that of the Saxo-Thuringian Zone, the presence of Ta-enriched end members in these younger magmatic system, proximal to the suture zone bears witness that under certain conditions this scheme of fractionation of Nb and Ta can also successfully be applied on a geodynamic scale (see Sects. 2.1, 2.3.1, and 2.3.2). Considering the age of the host granites of tapiolite in the Fichtelgebirge, its age has to be $\leq 298 \pm 4$ Ma (granite G3) and $\leq 290 \pm 4$ Ma (granite G4), respectively. The age of columbite-(Fe) from the HPPP shows a chronological spread of 302.8 ± 1.9 Ma to 299.6 ± 1.9 Ma. There is without any doubt a younging in the host rocks while the tantalum increases towards the suture zone in an ensialic orogen.

For the sake of completeness three more Nb-bearing minerals have to be reported, petschekite and its hydrated form hydroxy-petscheckite, both of which were discovered at Hagendorf-South only (Mücke and Keck 2008). Unfortunately, no detailed information has been made available on the paragenetic relationship between this U-bearing Nb-Ta oxide and other minerals from Hagendorf-South, e.g., uraninite. It is likely, that a process of local scale has brought about this Nb-Ta oxide and its hydroxyl-group-bearing member. Taking into consideration the Mn, Fe, Nb and Ta contents reported for petschekite and its hydrated form hydroxy-petscheckite by the above authors place the data points of petschekite in the cross plot of Fig. 4.8i between the data array of columbite from the Trutzhofmühle Aplite and the Kreuzberg Pegmatite. The data of (O,OH)-petschekite, however, plot in the "Hagendorf Trend" of Fig. 4.8i. While the Nb-Ta ratio is almost constant excluding any fractionation, the Mn-Fe ratio significantly varies. To approximate the environment of formation of these rare mineral petschekite relative to columbite and for lack of any chronological data the conditions of formation are likely to have been as follows. The petschekite mineralization points to a pegmatite system of a moderate degree of fractionation ($Ta^*100/(Nb+Ta) \approx 15.5$–$16.4$). Its $Mn^*100/(Mn+Fe)$ ratio falls in the range ≈ 12.2–31.1 which would mean a relic metamorphic impact (?).

Samarskite-(Y), also bearing U, is likely to have formed under the same restricted physical-chemical regime at Hagendorf-South. Others than petschekite, samarskite-group minerals are more widespread in the southern Moldanubian Zone.

Although the author did not examine any sample from the HPPP containing uraniferous pyrochlore [$(U,Ca,Ce)_2(Nb,Ta)_2O_6(OH,F)$], it has to be mentioned here for the sake of completeness. Hagendorf-South is said to host this rare mineral (Kastning and Schlüter 1994). In this paragraph pyrochlore form the "nigrine"-type B – see succeeding section- is used to describe the interrelationship with its neighbors and to draw a more detailed picture of the chemical composition of this Nb-Ta oxide (Dill et al. 2007b). Its mineral grains are very small measuring as much as 30×5 µm across and used to replace columbite-(Fe). Pyrochlore s.s.s have the general structure: $A_{2-}mB2X6[(O,OH,F)=Y]_{1-n}*pH_2O$ (Hogarth 1977). The average and maximum contents – in brackets- are as follows (wt. %): 76 (78)% Nb_2O_5; 2.7 (3.1)% Ta_2O_5; 4.1 (5.1)% TiO_2; 0.5 (0.9)% WO_3; 1.4 (1.7)% FeO; 18.0 (18.2)% CaO. Trace amounts of V_2O_3 (to 0.06 %), MnO (0.15 %), SnO (0.27 %), Sc_2O_3 (0.09 %),

Al_2O_3 (0.03 %), Na_2O (0.02 %), UO_2 (0.14 %) and PbO (0.3 %) were determined. The chemical requirements to qualify for a member of the pyrochlore solid solution series (s.s.s.) Nb + Ta > 2Ti and Nb > Ta. are met by these analyses. Pyrochlore group minerals with an established average formula of $(Ca_{4.0})_{\Sigma 4.0}(Nb_{7.1}Ta_{0.2}Ti_{0.6}Fe^{3+}{}_{0.25}W_{0.03})_{\Sigma 8.18}O_{24}$ are close to the aeschynite-type (Piantone et al. 1995) and occur in ilmenite in close association with columbite. In the same sample uraniferous betafite was found $(Ca_{4.0}Na_{0.6}U_{1.6})_{\Sigma 6.2}(Nb_{3.6}Ta_{0.3}Ti_{3.2})_{\Sigma 7.1}O_{24}$. Betafite is replaced at the margin by a hitherto unknown phase. The results obtained from EMP analysis are 42.7 % Nb_2O_5, 20.9 % TiO_2, 13.6 % Ta_2O_5, 4.7 % SiO_2, 4.1 % FeO, 3.5 % UO_2, 3.0 % P_2O_5, 2.2 % Al_2O_3 and 0.3 % CaO with traces of PbO (1 %), SnO (0.3 %), V_2O_3 (9.2 %) and MnO (0.1 %). Due to the totals of 96.85 wt. %, H_2O or $(OH)^-$ may be expected in the formula. A simplified calculation, ignoring hydrogen, yields a formula of $(Si,Fe,Al,P,U,Ca)_3(Nb,Ti,Ta)_8O_{24}$. Relative to its parent mineral uraniferous betafite, Nb and Ta are enriched at the expense of Ca and U in the altered product. It might be some sort of a hydrated betafite (?) for which no special name has yet been coined.

4.9 Titanium Minerals

Little attention is paid to titanium minerals by "pegmatologists", since most representatives of this element are located in the wall rock of this felsic intrusive rather than the pegmatites themselves and appear very late either as supergene or hypogene alteration minerals during the evolution of pegmatites. Those who look more closely at this group of minerals will come to the conclusion that Ti minerals can significantly contribute to the genetic part of economic geology as well as applied economic geology of pegmatitic rocks if other geoscientific disciplines such as geomorphology, and sedimentology used in addition to mineralogy for discussion. Most Ti minerals are stable to ultrastable in terms of their vulnerability during chemical weathering. They are transported far off the pegmatitic source and those being able to "read the minerals" will understand the nature of pegmatites and their geological setting much better before getting in touch with them physically – see Figs. 4.8i and 4.6a.

Brookite and anatase are both low-T and low-P polymorphs of rutile (Dachille et al. 1968). They appear in the pegmatite when Ti-bearing primary minerals, such as mica, ilmenite or sphene undergo supergene weathering and/or hypogene alteration. Based upon its morphology two types of anatase may be recognized (Fig. 4.9a, b). Type-I anatase typically forms plates with {001} faces prevailing over {101} faces, some anatase aggregates displaying a combination of {001} and {105} faces, intersecting each other at an obtuse angle (Fig. 4.9b). Type-II anatase displays the reverse order of face dominance associated with anatase I, with prevalently {101} and {110} faces. In cavities within corroded ilmenite or biotite subjected to chloritization, bipyramidal crystal aggregates and sprays of acicular anatase were seen (Fig. 4.9a). The first-mentioned type has been derived from

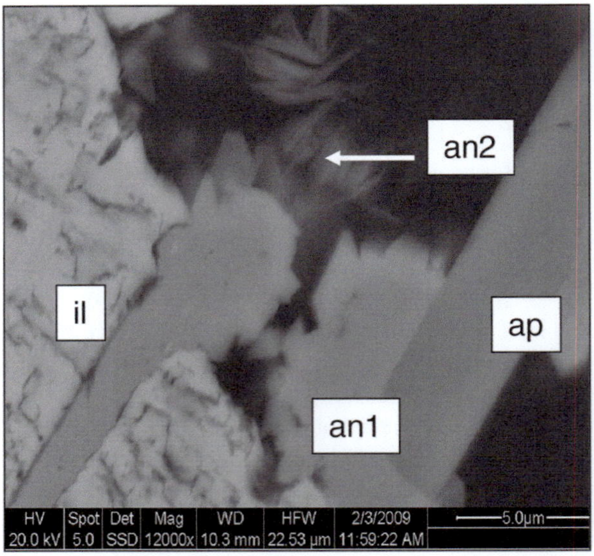

Fig. 4.9a Bipyramidal to long prismatic/acicular anatase growing into caverns of ilmenite postdating apatite: *an1* bipyramidal anatase, *an2* acicular anatase. Both crystal morphologies are frequently found in autohydrothermally altered rocks and hydrothermal alteration zones of felsic igneous rocks. Biotite gneiss

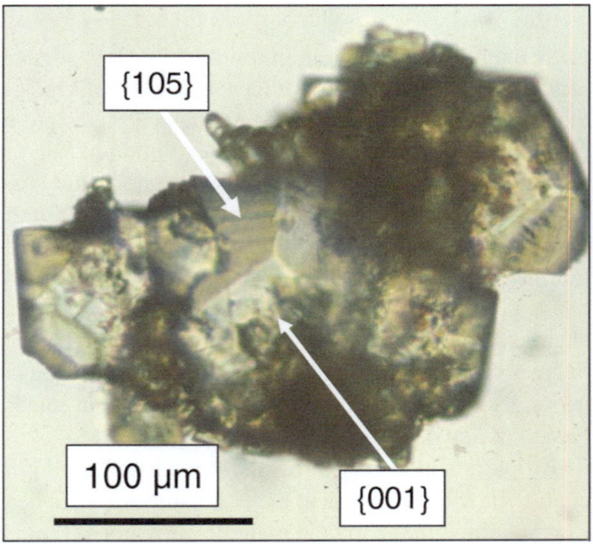

Fig. 4.9b Tabular aggregates of anatase forming a composite of crystal faces {001} and {105} which is typical of anatase formed in the course of supergene alteration in saprolites and paleosols. (*thin section* plane-polarized light)

chemical weathering, the second one from hydrothermal alteration of biotite, amphibole and ilmenite. Anatase is the only TiO_2 polymorph in the strongly kaolinized New Aplite at Pleystein.

Rutile and ilmenite are common rock-forming minerals in many sedimentary, magmatic and metamorphic lithologies but in the HPPP they were encountered only as accessory minerals (Table 4.1). Nb- and/or Ta-rich rutile is not a common mineral in rare-element pegmatites themselves, because Ti concentrations in pegmatite melts are usually low, and the available Ti is incorporated either into biotite, or into early columbite, tapiolite or minerals of the liandradite-petscheckite series as shown in the HPPP. Rutile and ilmenite are more widespread in the roof and wall rocks of the pegmatites either as individual minerals or intimately intergrown with each other, forming aggregates of ilmenite and rutile (mainly niobian rutile) called "nigrine". In the colluvial, alluvial and fluvial placer deposits around Pleystein, these "nigrine" pebbles become a suitable marker-mineral aggregate due to its conspicuous black metallic luster which eases its discovery among accompanying heavy minerals and its variegated spectrum of mineral inclusions which allow for singling out positive and negative markers (Dill et al. 2007a, b, c, 2014). "Nigrine" (amount of rutile > ilmenite) and "antinigrine" (amount of ilmenite > rutile) contain inclusions of, e.g., columbite-(Fe), pyrochlore s.s.s., wolframite s.s.s., monazite, zircon, xenotime, Fe oxides, sphalerite, pyrrhotite as well as alteration minerals such as pseudorutile and Fe-Ti-Nb-Ta-Al-P compounds (Tables 4.1 and 4.7). Based upon their, textural, chemical and mineralogical variation they may be subjected to a tripartite subdivision into types A, B and C (Fig. 4.9c).

A-type "nigrine" has been encountered mainly in gneisses, in silicification zones, and in quartz dykes (Fig. 4.9d). The quartz-bearing fault system hosting "nigrine" in the NE Bavarian Basement forms part of the strain-stress system of the "Pfahl" zone (Great Bavarian Quartz Lode). The northwest-striking "Pfahl" zone is a mylonitic shear zone that is associated with brittle-ductile deformation fabrics and a conspicuous hydrothermal quartz mineralization. It has also some *en echelon* faults running N-S to the main NW-SE quartz dyke cutting through granites aged between 321 and 329 Ma and two granodiorites aged 325 ± 3 Ma and 326 ± 3 Ma, respectively (Siebel et al. 2006). The maximum quartz temperature is close to 730 °C. Type-A "nigrine" is pre- to synkinematic relative to the shearing in the quartz-bearing fault system. Its samples are also ubiquitous in the stream sediments of the NE-Bavarian Basement. The temperature of formation coincides with the data array obtained for the temperature of formation of the quartz dykes at Rozvadov and Ostruvek, being located immediately east of the Czech-German border proximal to the HPPP. Type-A "nigrine" formed at very low Eh << 0 with pyrrhotite (and little sphalerite). There is little trivalent Fe present in this Fe-Ti system, although rutile may be doped with ferric iron, involving substitution of Fe^{3+} onto octahedral Ti^{4+} sites and charge-balanced by oxygen vacancies (Bromiley and Hilaret 2005).

Type-B "nigrine" is widespread in stream sediments of pegmatite-aplite provinces. Its grains do not show any deformation and, hence, have to be described as post-kinematic relative to the last tectonic activities of the Variscan orogeny. It is the most diagnostic type of "nigrine" in terms of the fertility of a region as to pegmatites which contains pyrochlore and betafite, both replacing columbite-(Fe). Type B-"nigrine" is indicative of the waning stages of the Variscan heat event. In type-B "nigrine" bivalent and trivalent Fe in magnetite, passing eventually into hematite reflects an intermediate stage to be attained as to the Eh regime.

Table 4.7 Chemical and mineralogical composition of "nigrine" and "antinigrine" types A, B and C ("nigrine" in red and "antinigrine" in blue) in the NE-Bavarian Basement

Sampling location	Type	Range	Size	Rutile core		Rutile lamellae/grains		Rutile in fissures		Ilmenite		Inclusions					I+A	Alteration		
			mm	Nb_2O_5	FeO	Nb_2O_5	FeO	Nb_2O_5	FeO	MnO	Nb_2O_5	Columbite	Pyrochlore	Monazite-xenotime	Zircon	Minor constituents	Hematite-magnetite	Pseudorutile	Ti-Al-P-Fe ox	Nb-Ti-Ta-P
Weißenstein Pflaumbach Steininger Loh Pingermühle Deggendorf Zwiesel	A	Min	0.4	0.3	0.2	1.1	0.5	0.0	0.0	0.8	0.0			x		sph, po		x	x	
		Max	11.0	6.6	6.6	10.0	4.8	12.6	1.0	4.0	0.5									
Pingermühle Iglersreuth Tirschenreuther Zwiesel Deggendorf	B	Min	3.0	0.3	0.4	0.0	0.1	0.3	0.0	1.14	0.0	x	x		x	U-pyr.	x	x	x	x
		Max	6.0	0.4	0.8	26.6	15.0	30.9	0.9	9.0	1.2									
Pflaumbach Deggendorf	C	Min	0.5	0.2	0.5	0.2	0.0			2.1	0.0									
		Max	8.0			0.3	2.0			2.2	2.0									

The particle size is given in columns 3 and 4. The main chemical constituents of the rutile-ilmenite intergrowth in columns 5 through 12. The column "Inclusions" reports on minerals of the primary environment of formation and may also be used as a tool for exploration, while the column "Alteration" provides the secondary minerals filling vugs and cavities of the heavy mineral aggregates. These secondary minerals are indicative of the chemical weathering and the regime on transport

Sph sphalerite, *po* pyrrhotite, *U-pyr* uraniferous pyrochlore, *Ti-Al-P-Fe ox* "leucoxene"+"limonite"+ aluminum phosphates, *I+A* inclusion plus alteration

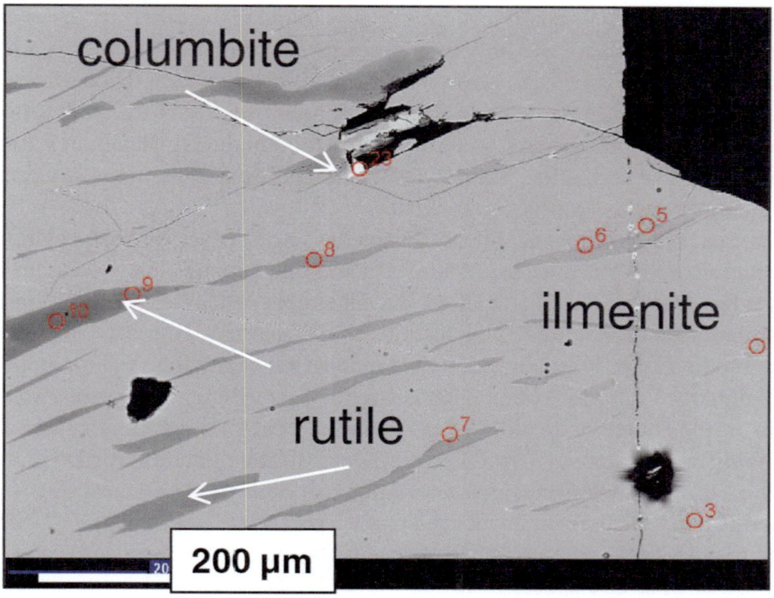

Fig. 4.9c "Nigrine" type B bearing slender lamellae of niobian rutile intergrown with ilmenite. Small irregularly –shaped grains of columbite-(Fe) are visible. Locality Zwiesel (EMPA-BSE image)

Fig. 4.9d "Nigrine" type A intergrown with milky quartz in a pegmatitic dyke from the Kühbühl near Pleystein

Type-C "nigrine" (W-bearing "nigrine") was found in the quartz dykes as well as in the streams sediments, yet only to a minor extent compared to types A and B. It is undeformed like type B and postkinematic as to the structural disturbances of its host rocks. It postdates the youngest structural processes along the quartz mineralization during the late Variscan tectonic deformation in the NE Bavarian Basement. Its temperature of formation is as low as 470 °C. Type C is the most strongly oxidized member of the "nigrine" -aggregates.

Minerals listed in Table 4.7 under "Inclusions" depict the (primary) environment of formation of the "nigrine" aggregates, whereas the minerals in the columns under the header "Alteration" characterize the supergene alteration on transport and deposition of them. For a detailed description of "internal sediments" the reader is referred to Sect. 4.6. Pseudorutile, uncommon to the pegmatite system, proper, but abundant in the "nigrine" compounds is a relic of the alteration of primary Ti compounds in the source area, still preserved in the heavy mineral aggregates during weathering and transport. Al- and Fe-bearing phosphate minerals are also present in "nigrine". Similar to variscite and cacoxenite, met during the youngest stages of the pegmatitic mineralization both phosphates are indicative of pH values ≤ 4 (Table 4.2).

The relation of the three different types of "nigrine" to the pegmatites need to be explained in time and space, from the initial stages of emplacement through the unroofing and destruction of the pegmatite system and thus has been depicted in a cartoon in Fig. 4.9k. "Nigrine" is a textbook example for a marker mineral aggregate disseminated in the primary dispersion aureole which on account of the low susceptibility to chemical weathering and attrition of its major oxidic Ti minerals became a prominent member in the secondary dispersion aureole too. The primary source of these titaniferous mineral aggregates lies within the gneissic rocks of the crystalline basement proximal to or within shear zones mineralized with silica.

Moreover these sites are lined up along the boundary between the autochthonous basement rocks of the Moldanubian and the allochthonous nappe of the Bohemicum (Teplá–Barrandian) (Fig. 2.1). Nigrinisation as it is referred to in this book is an alteration process exclusive to the gneissic country rocks between 329 and 321 Ma. The shear-zone-related mobilization started with the primitive Type A around 730 °C. Emplacement of pegmatite was accompanied by the more complex type B, abundant in Nb, between 500 and 600 °C. Type B reflects an advanced level of fractionation at a more distal position relative to the shear zone but at a more proximal position relative to the present-day *loci* of pegmatites and aplites enriched in rare metals, prevalently niobium. Its columbite degree of fractionation comes close to the least fractionated columbites of the "Pleystein Trend" (Fig. 4.8i). At shallow depth, within the shear zones, W-bearing type C developed around 400 °C as the youngest type of Ti mobilization, and still reflecting the granitic affiliation by its abnormally high Nb and W contents. "Nigrine" and "antinigrine" form part of the primary dispersion aureole around pegmatites within the crystalline rocks and can be used as a lithochemical tool for hidden pegmatite deposits, with type B "nigrine" developing most proximal to Nb-bearing pegmatites. The secondary aureole of dispersion developed when the channels of the modern drainage system cut into the primary dispersion aureole surrounding the pegmatites (Fig. 4.9k). Ti-bearing min-

Fig. 4.9e Benyacarite (be) under the stereomicroscope surrounded by meurigite-(K). Hagendorf-South

erals and mineral aggregates got released from their source and became a tool to track down the pegmatites in the catchment area of drainage systems. "Nigrine" and "antinigrine" are proximity indicators being located most proximal to the pegmatitic target. Columbite-(Fe) is a proximity indicator efficacious at a more distal position, which comes into effect only when the "armored relics" and Ti-bearing HM aggregates were destroyed on transport. The HM aggregates act as a shelter and, thereby, enlarge the mechanical dispersion aureole around pegmatites by some orders of magnitude. If columbite-(Fe) had to travel freely in the alluvial-fluvial drainage systems its transport range would be considerable confined to less than 2 km around the pegmatite leaving little chance for exploration geologists to localize the pegmatite.

While the afore-mentioned Ti oxides have a rather simple chemical composition and are very popular with geologists, mantienneite, paulkerrite and benyacarite, members of the paulkerrite-group, are rare constituents of the mineral assemblage of phosphate pegmatites and the pursuit of a handful of mineral collectors focusing on pegmatites (Table 4.1). Figure 4.9e, f display benyacarite from two different locations of the HPPP, one from Hagendorf-South and the other from Pleystein, under the stereomicroscope and scanning electron microscope. A local concentration of Ti, F and K are decisive for the built-up of this titaniferous phosphate. A simple break-down of Ti-bearing mica accompanied by the decomposition of apatite would have provoked a more widespread occurrence. Hydrofracturing of columbite s.s.s. in a matrix of white mica is a prerequisite for the formation of this group of Ti-K-F-bearing phosphates. Columbite is responsible for the provision of Mn, Fe and Ti and muscovite adds up K and F to the alteration process to produce eventually benyacarite, as the redox regime is close at the brink of shifting from Eh$<$0 to Eh$>$0 (Figs. 4.8f and 4.9g). Hydraulic fracturing, also known as fracking by drilling

Fig. 4.9f Benyacarite associated with meurigite-(K) under the SEM. Kreuzberg Pegmatite, Pleystein

Fig. 4.9g Hydraulic fracturing of columbite-(Fe) in a matrix of muscovite. It kick starts the formation of benyacarite at Pleystein

Fig. 4.9h Postkinematic paulkerrite in an argillaceous shear zone from the Plößberg pegmatite

engineers, played a significant role during this initial opening and filling structures in the pegmatite attesting to a closed system with the roof and wall rocks still sealing the pegmatites (Fig. 4.9g). Fragments of fracked minerals are unrotated, due to the fact that this hydraulic fracturing is unrelated to any kinematic process. It is younger than the structural disturbances along with the emplacement of the pegmatite and younger than the brittle deformation under subcritical hydrothermal conditions (Baumgärtner and Zoback 1989). Widening of the intergranular space led to a more advanced structural type and to keep these rock fragments floating implies a more highly viscous fluid. Such cut-and-fill structure were caused by tectonic fracturing and one can artificially create these structures in boreholes even at shallow depth down to 1000 m (Zoback et al. 1986). The chemical system is not very much tolerant as to changes of the fluid system, and quickly responds to a drop in the fluorine content and an increase of the magnesium content by generating paulkerrite instead of benyacarite, as it is the case with the Ploessberg pegmatite (Fig. 4.9h). All the ingredients known from the Kreuzberg Pegmatite at Pleystein to form titaniferous phosphate can also be observed in the Ploessberg pegmatite. Its contact zone is, however, enriched in Mg and impoverished in F, due to the fact that clinochlore derived from biotite is more widespread in the sheared contact zone than muscovite (Table 4.8). The pegmatite system is still in the state of self-supply as to the element budget. The structure of benyacarite reported by Robbins et al. (2008) from Hagendorf is devoid of Mg. Our predecessors, once they were studying the NE

Table 4.8 Chemical composition of the Plößberg pegmatite (mean of 7 samples) and the sheared alteration zone at the contact between the pegmatite, proper, and the metasedimentary wall rocks

Element		Pegmatite			Sheared contact zone between the pegmatite and the country rocks		
		Max	Mean	Min	Max	Mean	Min
SiO_2	%	88.37	81.30	73.09	69.37	59.82	47.94
TiO_2	%	0.04	0.02	0.00	0.85	0.63	0.20
Al_2O_3	%	15.05	9.92	4.17	17.36	14.76	11.55
Fe_2O_3	%	0.85	0.52	0.24	9.92	5.94	2.53
MnO	%	0.02	0.02	0.01	0.21	0.11	0.05
MgO	%	0.48	0.24	0.03	13.78	5.80	1.03
CaO	%	2.16	1.10	0.46	7.51	3.97	1.83
Na_2O	%	4.83	3.55	1.49	3.13	1.95	0.31
K_2O	%	3.72	1.67	0.43	4.20	2.20	0.99
P_2O_5	%	1.62	0.79	0.35	0.44	0.34	0.23

Data are given in wt. %

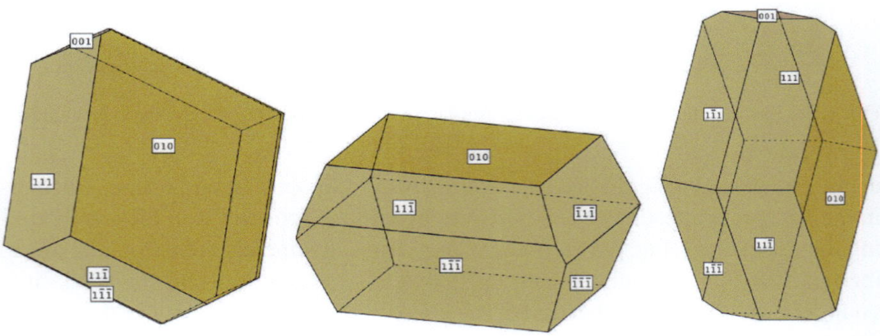

Fig. 4.9i Cartoon to illustrate the morphological types of paulkerrite in the shear zone at Plößberg pegmatite

Bavarian pegmatites alleged to have discovered a new mineral called by them "angelardite" which was never approved by IMA and later turned out to be a mixture of the Fe phosphates such as strengite and phosphosiderite accompanied by paulkerrite (Fig. 4.9j). The crystal habit of paulkerrite is not at variance with that described by Demartin et al. (1997) for benyacarite, so that the morphological types reported for paulkerrite can also be used to characterize those of benyacarite (Fig. 4.9i). Strongly fluctuating oxygen pressures gave rise to a suite of phosphates accommodating elements in their lattice very much different in their valence state. Relative to Pleystein the system has been shifted to a more oxidizing fluid system. Nevertheless paulkerrite neither formed during supergene nor during early pegmatitic stages. This Ti phosphate takes an intermediate position among the phosphates

Fig. 4.9j "Angelardite" a mineral aggregate of strengite, phosphosiderite and paulkerrite from the sheared contact of the Plössberg pegmatite

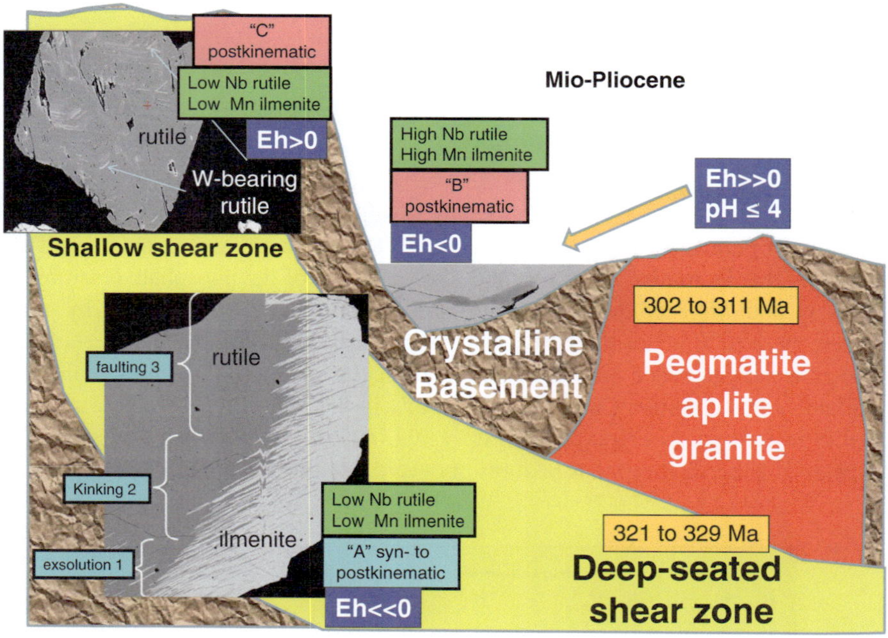

Fig. 4.9k Synoptical overview of the built -up of the primary dispersion halo, its destruction and conversion into a secondary dispersion aureole. Type A developed along deep-seated lineamentary shear zones such as the Great Bavarian Quartz Lode. The micrograph (BSE image) shows the syn- to postkinematic nature of this Ti mineral aggregate. The onset of shearing and mobilization took place between 321 and 329 Ma. Type B developed in the wall- and roof rocks of granites, pegmatites and aplites which were intruded between 302 and 311 Ma. Type C developed near the granites and shear zones as part of a shallow shear-zone-hosted mineralization. From the Mio-Pliocene onward chemical weathering and erosion shaped the landscape of the NE Bavarian Basement and caused the titaniferous mineral aggregates released from their source rocks to be accumulated in creeks and rivulets around the exposed pegmatites and aplites

in the pegmatite as to the succession of minerals. Titanium phosphates precipitated during the late hydrothermal stage at shallow depth and were termed "epithermal" by analogy with similar mineralizing processes and geological settings known from the Cu-Au deposits. Paulkerrite seems to be a proximity indicator, typical of the pegmatitic contact zone to Mg- and Ti-bearing wall rocks. It marks a more advanced state of fracturing than at Pleystein (benyacarite) with its formation still post-kinematic. The titaniferous phosphates can be used as sensitive markers for physical-chemical and kinematic changes in an intermediate stage of alteration of phosphate-bearing pegmatites.

While titanite is a very rare mineral in the NE Bavarian pegmatite, in the rivulets and creeks its abundance may feign higher amounts than actually present in the source area to the student (Fig. 4.9k). This selection has been done by chemical weathering and transport both of which foster a relative concentration of this Ti silicate in the streams sediments. At Brünst titanite presumably has derived from the inclosing country rocks where it is common to the calcsilicate-amphibolite series within the metamorphic country rocks. Its value as an ore guide or to constrain the physical-chemical regime is limited in the HPPP.

4.10 Molybdenite, Carbon, Calcium Phosphates and Calcium Carbonates

Normally it is not an easy task to identify apatite among the major rock-forming minerals of granitic or pegmatitic rocks in NE Bavaria with the unarmed eye. This is mainly due its minute grain size and often pale tint of the phosphate resembling the gray and white colors of the rock-forming minerals of the matrix. In some places, the host aplite of the HPPP takes a greenish tint caused by the presence of pale green apatite added up with some rockbridgeite and thereby enables geologists to identify this common Ca phosphate already in the field. Primary acicular green apatite in feldspar is displayed in Fig. 4.10a, while secondary apatite is present forming either collomorphous textures or developing transparent hexagonal prisms,

Fig. 4.10a Primary green Mn-bearing apatite embedded in feldspar. Hagendorf-South

Fig. 4.10b Secondary transparent, colorless prisms of apatite. Hagendorf-South

Fig. 4.10c Secondary "scepter apatite", slender corroded prism of colorless apatite are terminated by a flat prism of apatite, which is uncorroded. Kreuzberg Pleystein

locally, also as scepter apatite (Fig. 4.10b, c, d). The latter modification of apatite has a corroded prism at the base and a flawless flat prism on top indicative of a hiatus during the growth of secondary apatite. Apatite occurs as F-bearing apatite in most of the felsic intrusive rocks adjacent to the HPPP, whereas apatite in the HPPP, proper, is well-known for its elevated Mn contents. Apatite rarely is zoned and

Fig. 4.10d Vugs filled with secondary apatite of globular texture, "peppered" with *green* prisms of rockbridgeite. Trutzhofmühle

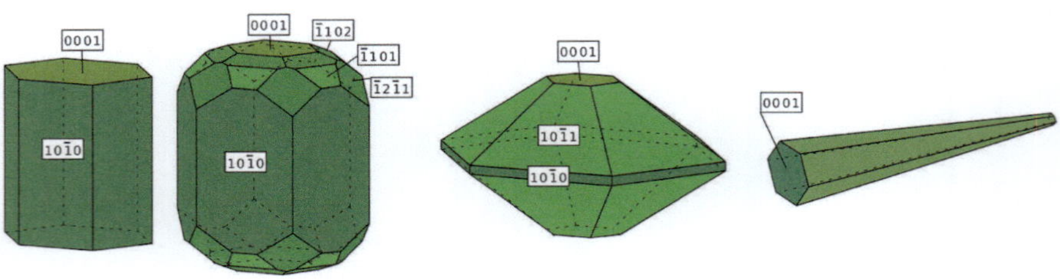

Fig. 4.10e Cartoon to show the idealized crystal morphologies of apatite in the HPPP

contains hydroxyl groups, only in the apatite common to the later mineralizing stages (Table 4.2).

In the Trutzhofmühle aplite, F- and Cl-free hyroxylapatite occurs as anhedral grains developing into radiating aggregates of slender prisms, which amalgamate to globular crusts lining the walls of druses and vugs. Another type of apatite has Mn contents in the range 2.4–3.4 wt. % MnO. Manganoapatite is not uncommon in granitic pegmatites (Cruft 1966; Keller and von Knorring 1989; Pieczka 2007). Manganiferous apatite has been described from zoned lithium-rich pegmatites and the highest MnO concentration ever was reported to stand at 10.3 wt. % MnO (Falster et al. 1988). At Miesbrunn manganiferous apatite-(F) is older than Mn-bearing garnet suggesting a deepening of the level of emplacement. With this peculiar element fractionation in mind, the tabular aplites at Trutzhofmühle and Miesbrunn are considered as predecessors of the stock-like bodies of the HPPP (e.g. Hagendorf South)- see also Sect. 4.3 and Fig. 5.5b.

The depth-related variation of Mn in almandine-spessartite s.s.s and Ca phosphate, where Mn and Fe decrease with depth, encouraged to design a simplified diagram plotting the sum of FeO + MnO (wt.%) vs. CaO (wt. %) in apatite to get an idea of the relative depth zonation in a consanguineous pegmatite province like the HPPP (Fig. 4.10n). The Trutzhofmühle is used as the base line in this diagram for geological, mineralogical and geomorphological grounds which will be discussed later during this study -Sect. 5.1.4. The depth of the other pegmatites and aplites is not given in absolute figures but referred to this tabular aploid in the diagram to provide a relative depth-zonation in the HPPP (Dill and Skoda 2015).

All those sites being placed above the reference line of the Trutzhofmühle trend formed at a shallower depth relative to the Trutzhofmühle tabular aplite. The depth relation is confined to the HPPP and to the stage of pegmatization when the primary manganoan apatite was precipitated together with the bulk of silicates. Reinhardsrieth and Waidhaus tabular aplites are the shallowest of all pegmatites and aplites under study, being located at the periphery of the HPPP, while the large Hagendorf-South stocklike pegmatite formed a bit deeper than these tabular aplites but at a shallower level than its "little brother" Hagendorf-North. The samples of Hagendorf-South have been taken from Mücke and Keck (2011) which separated an upper pegmatite from a lower pegmatite. The different data arrays overlap each other and their statistical parameters (trend line) do not significantly differ from each other. The Miesbrunn pegmatite-aplite swarm comes closest to the Hagendorf- North level as to the depth of emplacement and lies next to the Trutzhofmühle reference aplite which is the

Fig. 4.10f Whitlockite from the Hagendorf-South pegmatite

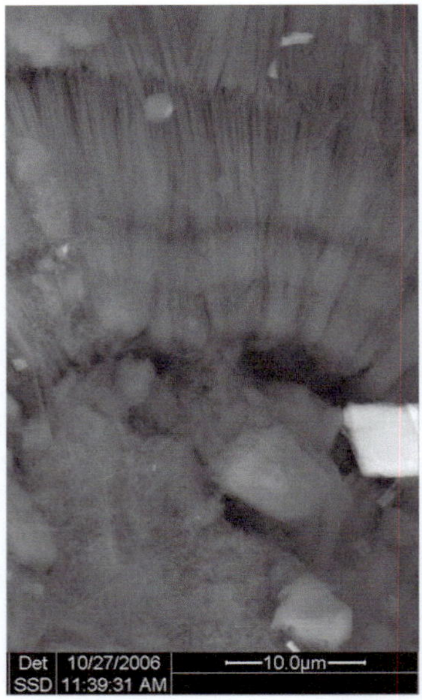

Fig. 4.10g Dufrénite from Hagendorf-South pegmatite under the SEM

Fig. 4.10h Mitridatite together with beraunite from Hagendorf-South

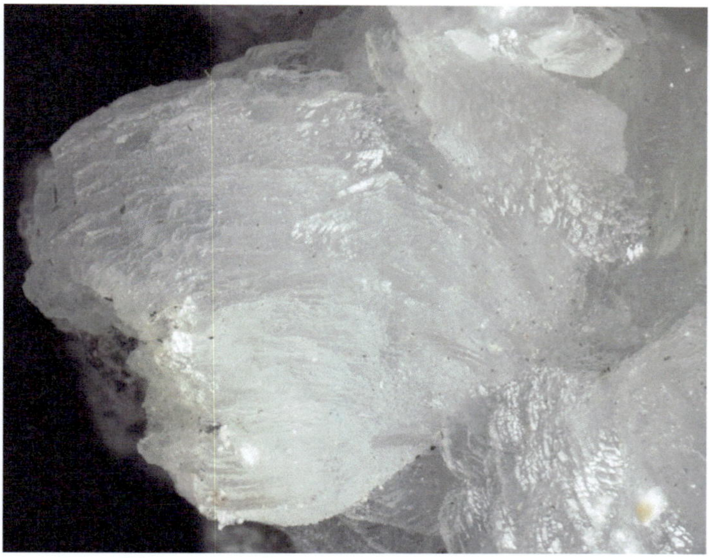

Fig. 4.10i Messelite from the Kreuzberg pegmatite at Pleystein

deepest tabular felsic intrusion (Fig. 4.10n). Judging by the position of the pegmatites and aplites the deepest level of intrusion lies in the south-western part of the HPPP, near Trutzhofmühle in the "Pleystein Trend" and the shallowest level of intrusion in the north-eastern and eastern parts of the HPPP, respectively (Fig. 1.1b).

As far as the derivation of calcium is concerned the conflicting processes of Ca silicates, e.g., epidote and Ca phosphate, capturing the limited quantity of Ca in felsic melt, needs to be repeated herein (see also Sect. 4.7 for a more detailed discussion).

The Ca-Fe phosphates whitlockite, dufrénite, xanthoxenite and mitridatite are not so widespread as the pure Ca phosphates in the HPPP (Fig. 4.10f, g, h). Mitridatite is a Ca-Fe phosphate rather common to pegmatites and non-pegmatitic environments following decomposition of primary phosphates such as apatite (Rogers and Brown 1979; Medrano and Piper 1997; Galliski et al. 1998; Nizamoff et al. 2004). Mitridatite is associated at Hagendorf with triphylite, zwieselite and arrojadite s.s.s. The Ca-Fe phosphates take an intermediate position in the succession of minerals known from the HPPP and appear as the Ca^{2+} and water contents increase during ongoing mineralization. The Ca-Fe phosphates are held to have formed by hydrothermal replacement of primary triphylite and apatite. The most common Ca-Fe phosphate of this group is mitridatite (Table 4.1). To get straight to the point, dufrénite can be considered as some sort of a missing link to an hitherto unknown Fe-Mn phosphate which for the first

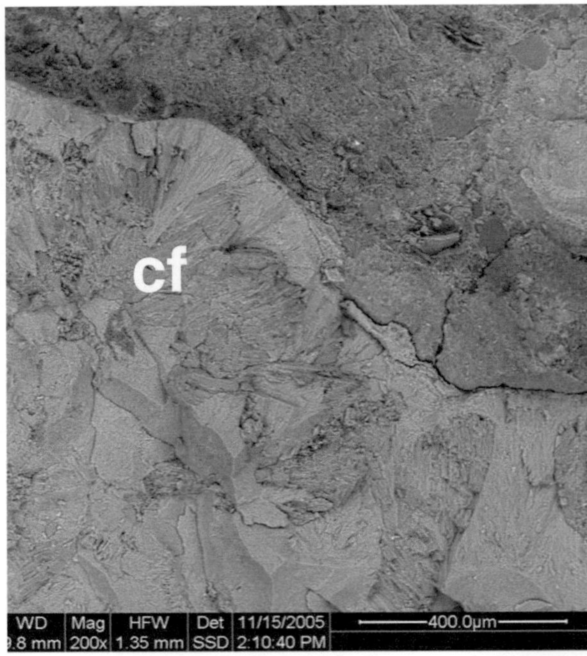

Fig. 4.10j Acicular aggregates of calcioferrite (cf) from the Trutzhofmühle

Fig. 4.10k Xanthoxenite crystals from the Kreuzberg Pegmatite at Pleystein (SEM)

time has been identified in the Reinhardsrieth aplite in the HPPP and is discussed in Sect. 4.13 (UNK 9). Xanthoxenite is restricted to the three triphylite-bearing pegmatite stocks of the HPPP (Fig. 4.10k). It shows up rather late among the epithermal minerals.

Messelite and calcioferrite are rare Ca-bearing phosphates and texturally quite similar in their outward appearance as shown by the radiating to acicular crystal aggregates in the images of Fig. 4.10i, j. Messelite has been among others described from the Kerch Fe ore Basin in Eastern Crimea by Chukanov (2005).

To get an idea, why these secondary Ca-bearing phosphates were emplaced an isolated discussion of these phosphates will not lead to the desired result. Therefore another couple of minerals, albeit present of subordinate amounts, has to be studied in this context. It is the couple calcite and carbon, the latter was described as graphite by Mücke and Keck (2008) that is crucial to the understanding of why Ca-bearing phosphate or carbonate minerals come into existence during the evolution in an otherwise carbonate-poor environment. Carbon/carbonate and phosphate compete for the favor of calcium. The competing reaction to constrain the physical regime under which these minerals developed is controlled by the Eh, pH and the soluble compounds HCO_3^-, HPO_4^{2-} and Ca^{2+}. In the Eh-pH calculations the dissolved species were used as $\log_{activity}$, the temperature given in °C and the Eh in volts. At unrealistically low temperatures of 25 °C, equivalent to a supergene alteration, only Ca phosphates would come into existence, using the dissolved species HCO_3^-, HPO_4^{2-} and Ca^{2+} as −2. Raising the temperature to e.g. 100 °C does impact on the stability fields of the mineral components involved. Only a drastic lowering of the amount of HPO_4^{2-} along with an increase of the amount of HCO_3^- to −1 fosters the presence of calcite at pH > 11. Further increase to 200 °C only slightly shifts the pH to 11.5 at higher temperature calcite get the prevailing mineral which is stable at pH > 4.5. Excluding the hydrocarbons present in the system at Eh < 0, the above competitive reactions are not Eh sensitive. By and large the striving of HCO_3^- and HPO_4^{2-} to be accommodated together with Ca^{2+} in a crystal structure will result in the appearance of Ca carbonate only at very low concentrations of phosphate and high quantities of carbonate in the system, a setting only locally represented during the emplacement of the Hagendorf-South phosphate pegmatite at temperature around 200–300 °C. But why opts calcium to go along with carbonate? A quick look at the quartzose vein-type deposits on both sides of the Czech-German border gives the answer to this question. At Wäldel and Höhensteinweg U deposits in Germany and at Dylen, Czech Republic metamorphosed bitumen are intimately intergrown with U oxide, U silicates, U titanates and molybdenite, denominated as "monotonous U association" to emphasis the striking differences to the "variegated U association" which contains among others arsenopyrite, scheelite, native gold, bismuth, pyrite, chalcopyrite and Bi-Pb selenides besides of pitchblende as the sole U host (Ruzicka 1971; Dill 1982, 1983a, b, c, d, 1986; Dill and Weiser 1981). In the uraniferous quartz veins epi-, meso- and kataimpsonite infiltrated the interstices of the quartz mosaic

Fig. 4.10l Plates of molybdenite at Hagendorf-South

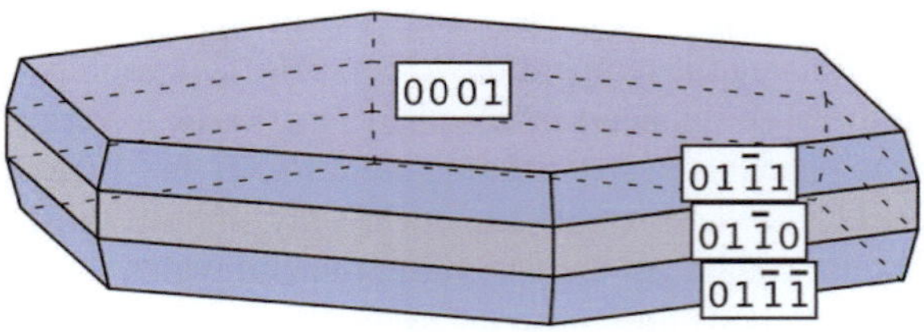

Fig. 4.10m The ideal crystal morphology of the molybdenite from Hagendorf-South.

(Jacob 1967). The afore-mentioned succession of pyrobitumen reflects an increasing temperature of maturity or diagenesis. Semigraphite may be found in the uraniferous quartz veins but no well-ordered graphite as in the pegmatite at Hagendorf. With this in mind Hagendorf-South may be in some parts of its mineralization described as a "graphite pegmatite" (Chessboard classification scheme code 52a D), provoked by the infiltration of bitumens along deep-seated fractures similar to what has been known from the alkaline pegmatites (Jaszczak et al. 2007). The "monotonous U association" which occurs side-by-side with the "variegated U association" next to the HPPP is held to be of subcrustal origin, whereas the "variegated U association" is supposed to be of granitic derivation. By analogy, the scarce presence of carbonate derived primarily from the immigrated hydrocarbons are a signal of a local and minor subcrustal impact on the formation of the Hagendorf-

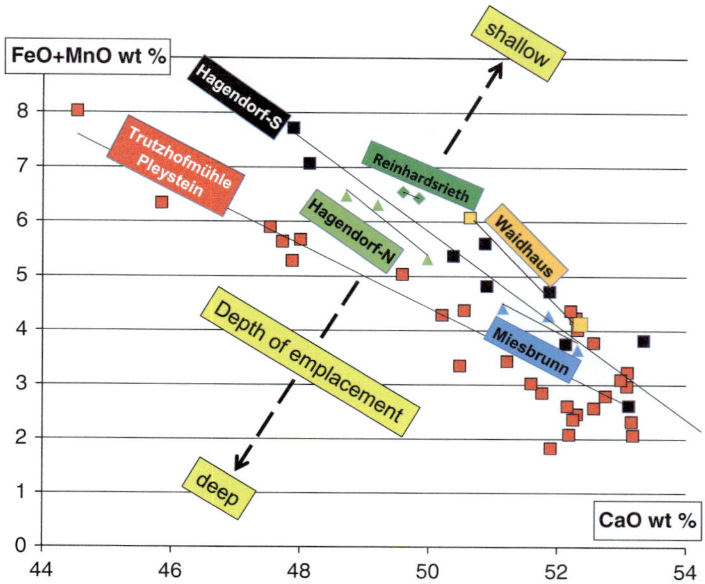

Fig. 4.10n Cross plot of FeO + MnO vs. CaO in apatite (Dill and Skoda 2015). Apatite from the Trutzhofmühle is used as base line which other pegmatites and aplites are referred to in the diagram. All those pegmatites and aplites being placed above the reference line of the Trutzhofmühle trend formed at a shallower depth relative to the Trutzhofmühle tabular aplite. The depth relation is confined to the HPPP and to the stage of pegmatization when the primary manganoan apatite precipitated. Data of Hagendorf-South taken from the literature (Mücke and Keck 2011). The trend lines reflect the depth relative to the deepest felsic mobilizate the Trutzhofmühle Aploid

South pegmatite that gave also rise to molybdenite which prefers to precipitate next to the phosphates at Hagendorf-South and in the uraniferous quartz veins, exemplified by the Wäldel uranium deposit near Mähring, Germany (Dill 1983d) (Fig. 4.10l, m). Its neighbors although devoid of graphite and calcite may in places contain well-crystallized molybdenite which is common to shear zones cutting through amphibolites with carbonaceous fault gouge in the Zone of Erbendorf-Vohenstrauß (Dill and Weber 2010a). Aluminum and magnesium accommodated in minor quantities in the structure of calcioferrite and whitlockite, respectively, have derived from the feldspar in and around the pegmatites and the biotite which finds its way from the biotite-sillimanite gneisses well into the pegmatites proper, yet with decreasing amounts towards the center.

In this peculiar case the reader is again referred to the Sect. 3.3.3 where a deep-seated low-resistivity layer has been detected by geoelectric methods and interpreted as a horizon abundant in graphite.

Fig. 4.11a Lazulite-scorzalite s.s.s. from the border zone of the Miesbrunn pegmatite aplite swarm where this Al-enriched phosphate formed as result of phosphate-bearing solutions reacting with Mg-bearing phyllosilicates

4.11 Aluminum Phosphates with Magnesium, Iron, Calcium and Manganese

Aluminum phosphates locally have very complex chemical compositions in the NE Bavarian basement and take up a wide range of bivalent elements, such as Mg, Na, Ba, Ca, Sr, Mn and Fe as charge balance among the cations and supplemented, in places by hydroxyl groups and fluorine (Table 4.1).

Lazulite and scorzalite appear together very early in the mineral succession and are restricted to the tabular aplites where the minerals of the lazulite-scorcalite.s.s.s occur near the contact with the enclosing biotite-sillimanite gneisses. Minerals of the lazulite-scorzalite s.s.s *sensu* Schmid-Beurmann et al. (1999) are concentrated in lens-shaped bodies which are aligned subparallel to the foliation of the gneissic wall rocks of the aplites at Miesbrunn. The contents of bivalent Fe lie in the range 5.97–8.40 wt. % FeO, those of Mg in the range 8.12–9.63 wt. % MgO. Lazulite-scorzalite s.s.s. formed isolated from any other phosphate minerals. It is intergrown with degraded biotite and muscovite and genetically related to the decomposition of Mg-bearing phyllosilicates (biotite, chlorite). Its formation resulted from a reaction between phosphate-bearing fluids, which were created during melting, and phyllosilicates, which were generated within the metamorphic country rocks. The physico-chemical conditions leading to the precipitation of lazulite have been investigated experimentally by Brunet et al. (2004). In an environment abundant in tourmaline as

Fig. 4.11b Brazilianite (*br*) intergrown with eosphorite (*eo*) from Hagendorf-South

it is the case with the Miesbrunn Pegmatite Aplite Swarm, the temperature for scorzalite–lazulite s.s.s stands at 475 °C at a pressure of 3.8 kbar (Fig. 4.11a). The Fe-enriched end member scorzalite also occurs in the Hagendorf-South pegmatite stock, where Fe prevails over Mg. The lazulite-scorcalite.s.s.s is not a solid solution series restricted to the outer part of thin tabular aplites or pegmatites but was also encountered in phosphatic iron ores, where both minerals are typical of the early epigenetic stage (Robertson 1982).

Another Al phosphate present in the HPPP has also been recorded from the epigenetic mineral assemblage of an iron deposit by Robertson (1982). Brazilianite has only been found at Hagendorf-South in the HPPP (Mücke 1987) (Fig. 4.11b). It needs sodium to build up this phosphate and may be considered as the more central Al phase relative to scorzalite-lazulite. Bilal et al. (2012) have described this phosphate in the wake of an albitization process associated with apatite. In some places, brazilianite [$NaAl_3(PO_4)_2(OH)_4$] is also associated with lacroixite [$NaAl(PO_4)F$] (Němec 1998). The chemical compositions of this mineral Na-Al phosphate couple direct our thoughts to a mineral, rather common to the Saxo-Thuringian granites but absent from the Moldanubian region. It is topaz [$Al_2SiO_4(F,OH)_2$], that is also enriched in Al and accommodates fluorine and hydroxyl groups in its lattice. The F-dominated modifications occur in vugs and veinlets within the granite and granite pegmatites in the Erzgebirge-Fichtelgebirge Anticline. Hydroxyl group-dominant members have been described by Zhang et al. (2002) associated with woodhouseite as a product of retrograde metamorphism of kyanite-enriched quartzites. A similar relation can also be

Fig. 4.11c White coating of nordgauite from Hagendorf-South

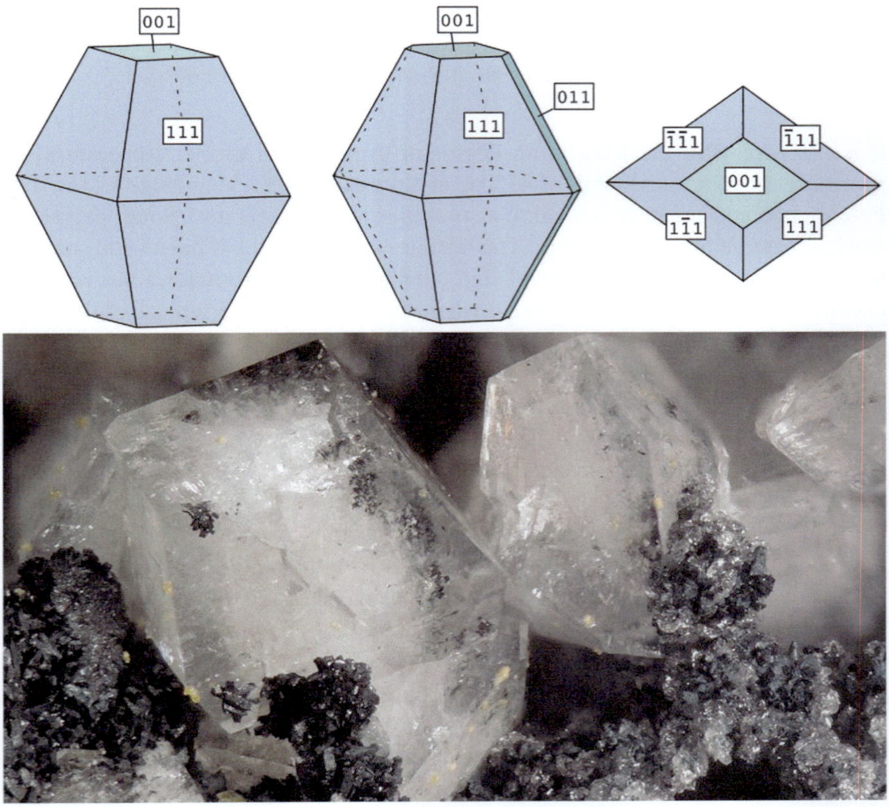

Fig. 4.11d Fluellite (obsolete term "kreuzbergite", "pleysteinite") from the Kreuzberg Pegmatite Pleystein, with a *bar* on *top* of the image illustrating the various crystal morphologies

established for the amblygonite [$LiAlPO_4F$] – montebrasite [$LiAlPO_4OH$] relationship-see also Fig. 2.4b. The enigmatic position of brazilianite in the HPPP is interpreted in the same way as a low-temperature alteration of the primary mineralization immediately in the aftermaths of the formation of the primary phosphates. The Na-Al phosphate is unrelated to the APS minerals of the crandallite s.s.s. Via the chemical relationship between brazilianite and lacroixite discussed above, this Na-Al phosphate demonstrates that there was a moderate impact on the pegmatites of the HPPP from the Saxo-Thuringian realm, via "relay pegmatite stocks", e.g., the Weisser Stein Pegmatite near Lázně Kynžvart, Czech Republic, which is situated approximately 45 km NNE of the HPPP- see also Sect. 5.2. As many other phosphate minerals present only at a subordinate level these external impacts could by neglected relative to the overall concentration of pegmatite minerals in the HPPP.

Two Al phosphates, fluellite, which was described from the Kreuzberg Pegmatite in the center of the city of Pleystein, and erroneously named as "kreuzbergite" and "pleysteinite" in the past and nordgauite, a phosphate discovered and approved by IMA only recently, have one thing in common, fluorine (Birch et al. 2011) (Fig. 4.11c, d). Both phosphates did not receive their F content during the early stages of pegmatite formation but rather late as the apatite-(F) decomposed and secondary fluorine-poor phase formed instead. The phosphate contents of primary

Fig. 4.11e Deep blue gordonite (see *arrowhead*) from the Trutzhofmühle Aploid

Fig. 4.11f Collomorphous matulaite at Hagendorf-South

Fig. 4.11g Kastningite from the type locality Silbergrube near Waidhaus with a cartoon to illustrate its ideal crystal morphology

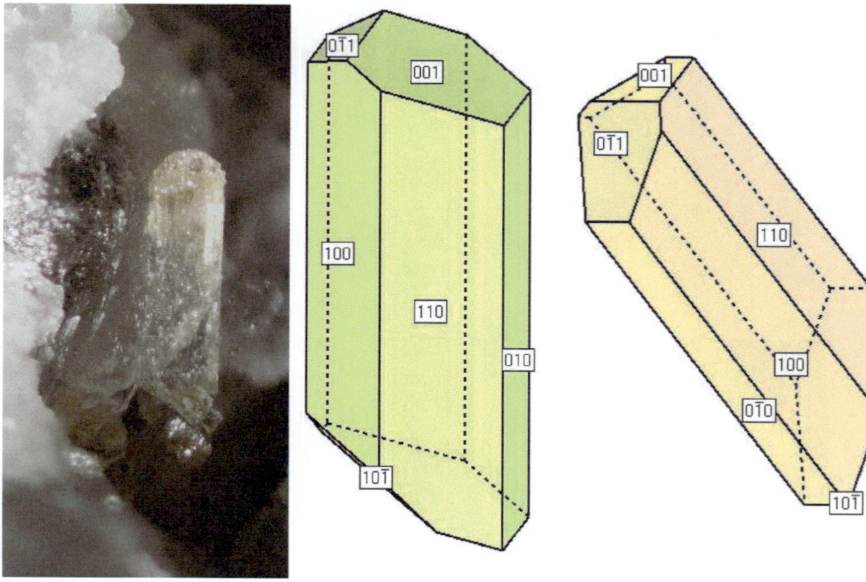

Fig. 4.11h Paravauxite from the Silbergrube near Waidhaus with two cartoons to show its ideal crystal morphology

phosphates, such as apatite-(F) and zwieselite-triplite were the primary source of F to produce nordgauite which according to Birch et al. (2011) also contains trace amount of Zn. Fluellite is also stable in non-pegmatitic environments such as at Krásno Sn-W district, Czech Republic, where in addition to apatite-(F) topaz and fluorite are abundant minerals (Beran and Sekora 2006).

Gordonite is the Mg-bearing analogue to nordgauite in the contact zone to the Mg-bearing wall rocks; it resembles lazulite with respect to its bluish tint, yet it is free of Fe in the Trutzhofmühle aplite. Manganogordonite is known from Silbergrube and from Hagendorf-South. Morinite, albeit bearing some fluorine similar to fluellite and nordgauite, developed at a higher level of hydration. Morinite is not a typical member of the so-called APS that come into existence as primary phosphates undergo supergene alteration (Dill 2001). Pirard et al. (2007) depicted the process of its formation in the kaolin zone of the Montebras Pegmatite, France. According to the authors the phosphate formed during the waning stages of the alteration at rather low pH, a fact also supported by the overall presence of kaolinite (Fig. 4.11e, f).

Similar to manganogordonite, kastningite accommodates also bivalent Mg, Fe and Mn in its structure (Table 4.1, Fig. 4.11g). It was recorded from Silbergrube and Hagendorf-South and investigated in terms of its crystallography by Schlüter et al. (1999). Another Mg- and Fe-free aluminum phosphate, present in both localities is paravauxite (Table 4.1, Fig. 4.11h). Only recently in the newly discovered Reinhardsrieth aplite, vauxite was encountered and proved by EMP. Vauxite appears very early among the phosphate minerals at Reinhardsrieth, where it is closely

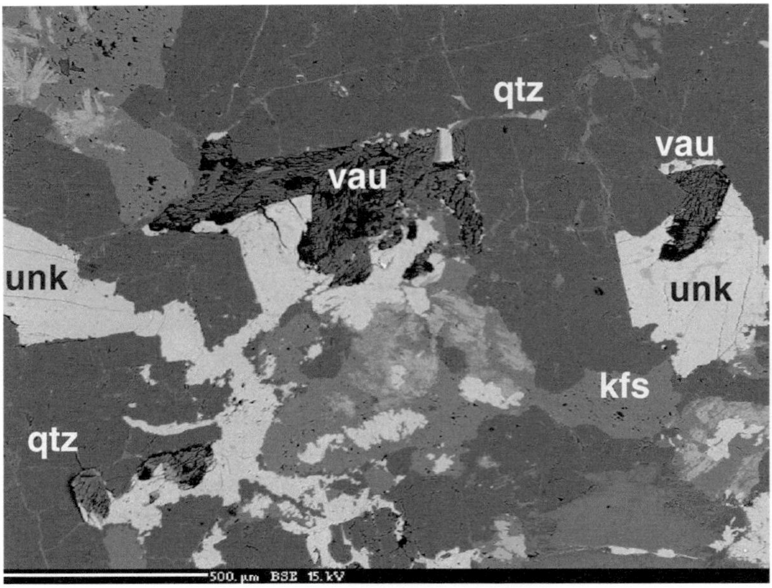

Fig. 4.11i Vauxite (*vau*) infills the interstices of a aplitic groundmass of K feldspar (*kfs*) and quartz (*qtz*). It is replaced by an unknown Fe-bearing phosphate (*unk*) from the Reinhardsrieth aplite. BSE image/back-scattered electron image electron microprobe (Dill and Skoda 2015)

Fig. 4.11j Tabular crystals of an eosphorite-enriched childrenite-eosphorite s.s.s. overgrown and replaced by ernstite at Reinhardsrieth. SEM image

Fig. 4.11k Childrenite-enriched end member of the childrenite-eosphorite s.s.s. from Waidhaus and a cartoon displaying the ideal crystal morphology of childrenite

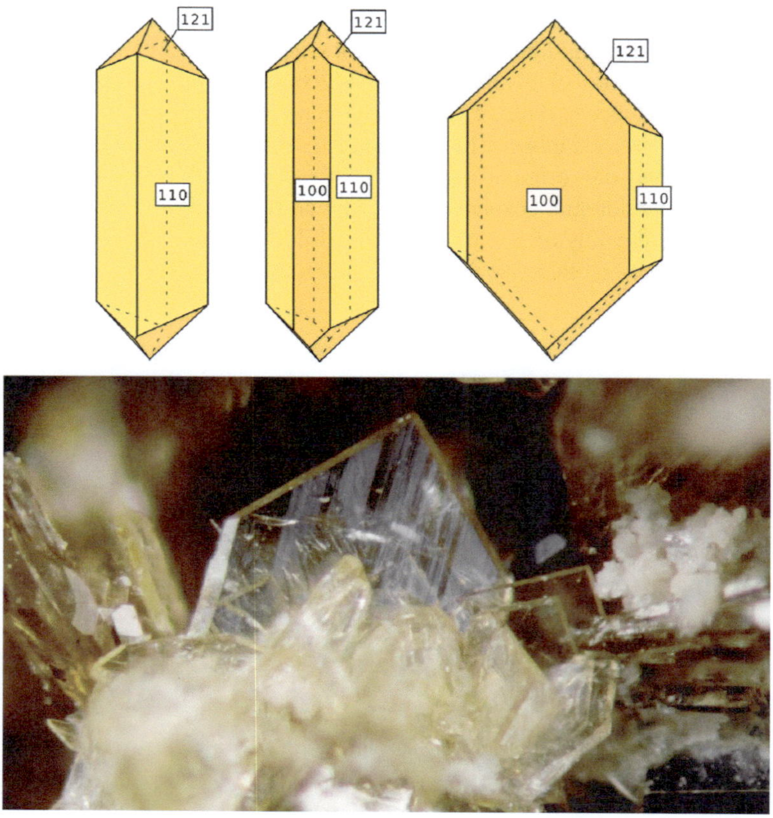

Fig. 4.11l Eosphorite-enriched end member of the childrenite-eosphorite s.s.s. from Hagendorf-South and a series of cartoon displaying the ideal crystal morphology of eosphorite

intergrown with K feldspar and quartz (Fig. 4.11i). A phosphate structurally quite similar to vauxite, the Mn analogue called nordgauite, has been described as a new mineral from Hagendorf-South by Birch et al. (2011). The chemical composition of vauxite has only little of its bivalent Fe converted into trivalent iron and is devoid of Zn unlike childrenite-eosphorite s.s.s., which is by far the most widespread Al phosphate in the HPPP (Table 4.1, Fig. 4.11j).

Childrenite-eosphorite series minerals resulted from phosphate-bearing solutions reacting with aluminum from the host aplite/pegmatite and to a lesser degree from the metasedimentary wall rocks (Fig. 4.11k, l). The ore texture definitely points to a hypogene rather than supergene mineralization, but to give more precise data than <300 °C on the temperature of formation is not possible with the data available from the sites under study (Dill et al. 2009a, b). Eosphorite-childrenite series associated with Li-Mn-Fe phosphate minerals are common to phosphate-bearing pegmatites, e.g., Buranga, Rwanda (Fransolet 1980). The manganiferous end member is a low-T metasomatic mineral in LCT granitic pegmatites where it infills open cavities in cleavelandite (Simmons et al. 2003). In the current situation part of the bivalent Fe diagnostic for the childrenite-enriched end member is converted into trivalent Fe at Reinhardsrieth, so that the name ernstite, introduced by Seeliger and Mücke (1970) for phosphates of the childrenite-eosphorite s.s.s. containing trivalent Fe is justified. The partial conversion of bivalent iron into trivalent iron was also recorded by Braithwaite and Cooper (1982). The youngest generation of the childrenite-eosphorite s.s.s being converted into ernstite is displayed in Fig. 4.11j, where subdomains of ernstite overgrow the original crystals of the childrenite-eosphorite s.s.s. At Reinhardsrieth, childrenite-eosphorite s.s.s take an outstanding position within this s.s.s. as they are the only ones showing elevated ZnO contents in a pegmatitic environment of deposition otherwise devoid of true Zn minerals.

Fig. 4.11m Rittmannite crystals (*brown*) surrounded by a spray of white kingsmountite at Waidhaus

Fig. 4.11n Cacoxenite from Silbergrube Mine at Waidhaus

Hagendorf-South is the type locality of whiteite-(CaMnMn) (Yakovenchuk et al. 2012). According to the authors, whiteite-(CaMnMn) crystallized on the walls of voids within altered zwieselite crystals or form coronas (up to 1 mm in diameter) around cubic crystals of uraninite. Neither zwieselite, nor uraninite may account for why this Al phosphate abundant in Mn precipitated in this site. The only thing for sure is that it formed in the central facies of pegmatites where Mn is abundant and rather early within the mineral succession when the Eh value of the redox regime was rather low. Devoid of Ca but similar with regard to its chemical composition the Mn-enriched Al phosphate rittmannite was recorded from Hagendorf-South and from Silbergrube, where it is engulfed by another Al-bearing phosphate, named kingsmountite (Fig. 4.11m). It was first described from the Mangualde Pegmatite, Portugal, by Marzoni Fecia Di Cossato et al. (1989) where the afore-mentioned phosphate, a member of the whiteite group, was found coating kryzhanovskite, frondelite, hureaulite and adularia in a vein-type pegmatite.

Cacoxenite is the only Fe-Al phosphate in the HPPP, accommodating only trivalent iron in its lattice (Fig. 4.11n). Cacoxenite is one of the latest phosphate in the mineral succession and resides on muscovite and replaces rockbridgeite at Trutzhofmühle. The physical-chemical conditions under which this Fe-Al phosphate came into being is similar to what has been determined for the APS minerals discussed in the succeeding section.

Montgomeryite, matulaite, goyazite, gorceixite and crandallite are typical APS minerals which rarely form end member phases of their own. The APS minerals mainly occur as s.s.s. in a wide range of environments, provided the pH is sufficiently low, a fact also marked by the presence of ubiquitous kaolinite in the regolith of the

Fig. 4.11o Acicular aggregates of crandallite-goyazite s.s.s. in K feldspar from the New Aplite, Pleystein (SEM)

New Aplite at Pleystein (Fig. 4.11f, o, p). In and around Pleystein, there are variable phosphate minerals pertaining to the APS minerals (aluminum-sulfate-phosphate) *sensu* Dill (2001) always associated with kaolinite-group minerals. It is the area most strongly affected by this type of alteration in the HPPP. These APS minerals constitute a complex s.s.s. accommodating various cations such as Ca^{2+}, Sr^{2+}, Ba^{2+}, and REE^{3+} in their lattice (Dill et al. 1997). At the Kreuzberg, the crandallite s.s.s. contains considerable amounts of strontium (up to 13.6 wt. % Sr) and hence, has to be denominated as goyazite-crandallite s.s.s. Moving downhill to the New Aplite, another s.s.s. among the APS mineral group has been discovered, containing as major component gorceixite (up to 6.5 wt. % Ba), goyazite (up to 1.9 wt. % Sr) and crandallite (up to 1.5 wt. Ca). The goyazite- crandallite-s.s.s. and gorceixite-goyazite- crandallite-s.s.s. formed from acidic meteoric fluids in the range pH 2–5, at a log activity of HPO_4^{2-} of – 5 and a log a $_{SiO2(aq)}$ < 1.5. Goyazite also appears as single crystals of up to 1000 µm (Mücke and Keck 2011). Although being emplaced under the same physico-chemical conditions, the unique goyazite and the goyazite s.s.s. are the result of two processes different in time and space. Goyazite I mega crystals resulted from a topomineralic mineralization – high-sulfidation-type *sensu lato*- during the late Paleozoic, whereas the goyazite s.s.s. were brought about by Cenozoic supergene alteration.

While all the afore-mentioned aluminum phosphates have Ca, Mn, Mg, and Fe in its trivalent and bivalent state for reasons of charge balance, wavellite and variscite do not need accommodate such cations in their lattice and can be described as aluminum phosphate in the strict sense (Fig. 4.11q, r). The environment of formation of variscite is identical to what has been depicted for the APS minerals in the previous paragraph. Wavellite is a member of the variegated group of Al phosphates which evolve in soils and duricrusts under near ambient conditions under low phosphate concentrations (Nriagu 1976; Dill 2001). Generalized stability relations show

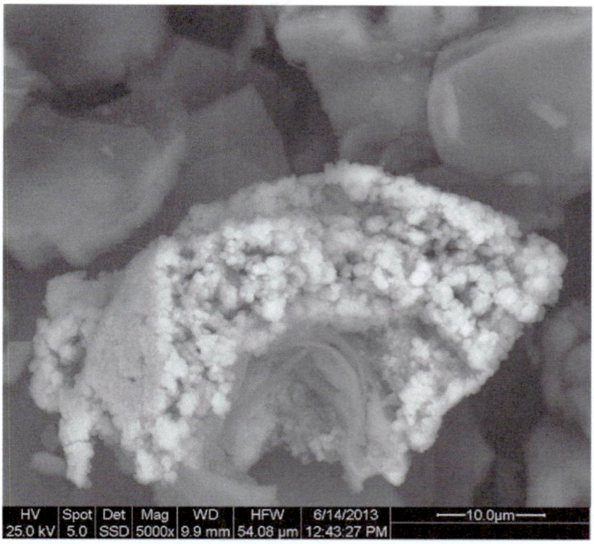

Fig. 4.11p Aggregates of APS minerals. The pseudocubic crystals clearly bear evidence of the existence of APS minerals, in this case of a s.s.s. of gorceixite (Ba) with crandallite (Ca) and goyazite (Sr). Pleystein regolith on the New Aplite

Fig. 4.11q Cavities filled with variscite at the Silbergrube Mine at Waidhaus

Fig. 4.11r Rosettes of wavellite from the Kreuzberg Pegmatite, Pleystein

wavellite to form at pH below 7. Acidic conditions are also indicated by the presence of the phyllosilicate kaolinite. Increasing acidity of the pore solution and a lowering of the pH value down to 4 causes variscite to come into existence dependant on the quantity of H_3PO_4 (log a $H_3PO_4 = -2.75$).

4.12 Iron Phosphates with Magnesium, Potassium and Sodium

The Fe-bearing phosphates in the HPPP rank among the most prominent group of minerals and constitute a very complex series of minerals, to be present as pure Fe phosphates, accommodating bivalent and trivalent at different quantities into their lattice or as mixed phosphates taking up bivalent and monovalent elements such as magnesium, potassium and sodium for charge balance (Table 4.1).

Wagnerite is a rare Mg- and Fe-bearing phosphate mineral in the Miesbrunn pegmatite-aplite swarm, a series of tabular felsic mobilizates intercalated into Mg- and Fe-bearing metasediments which had a strong impact on the element composition of the felsic mobilizates and the phosphate minerals (Fig. 4.12a). Unlike lazulite and scorzalite (Sect. 4.11), wagnerite occurs sporadically in the central parts of the

Fig. 4.12a Massive wagnerite aggregate forming the central parts of the aplite in mineralization D at Miesbrunn

aplite separated from the Fe-Mg-bearing phyllosilicates. Nevertheless, there is hardly any other source of Mg than the Mg-bearing phyllosilicates to account for the Mg contents of this Fe phosphate. Wagnerite was found in high-MgAl granulite near Anakapalle, ca. 40 km W of Vishakhapatnam, East-India. The rocks experienced Grenvillian ultra-high-temperature (UHT) metamorphism (8 kbar; T > 900 °C) (Simmat and Rickers 2000). It is the only P-bearing phase, while apatite is absent or wagnerite reacted with the intruding felsic melt and is now rimmed by apatite. Pitra et al. (2008) reported an equilibration temperature of 550 °C and a pressure of lower than 4 kbar for wagnerite in a cordierite-gedrite gneiss. Even if it is difficult to constrain the precise T- and P conditions of its formation in the HPPP mainly due to its small amount, wagnerite is a high-T primary phosphate that formed in the center of tabular pegmatites or aplites at more elevated temperatures than lazulite or scorzalite typical of the marginal facies of these tabular pegmatites and aplites.

In contrast to wagnerite a rare phosphate that appeared very early, vivianite and its meta-form (meta-vivianite) are rather common to the various felsic mobilizates of the HPPP (Fig. 4.12b, d). It has only bivalent iron in its structure, locally, with minor but remarkable amounts of Mn and it is a secondary phosphate indicative of reducing conditions. In the HPPP three different modifications can be distinguished from each other. Two well-crystallized morphologically different types, one enriched in Mn and another depleted in this element follow immediately after the primary Mn-bearing phosphates precipitated. The third type is poorly crystallized and despite the fact that it contains bivalent Fe indicative of a reducing regime held to be part of the supergene alteration.

Fig. 4.12b Well-crystallized blue vivianite (early generation) from Hagendorf-South (see also Fig. 4.12d)

Fig. 4.12c The author in the Silbergrube open pit at Waidhaus with a block of aplite granite coated with late-stage earthy blue vivianite along fractures

Fig. 4.12d Cartoon to show the ideal crystal morphology of the early generation of vivianite

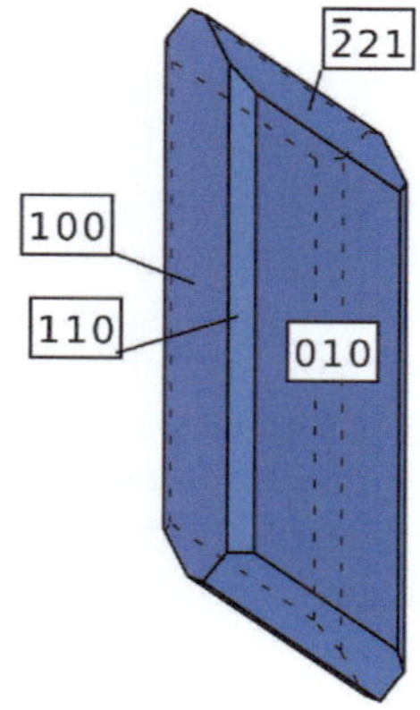

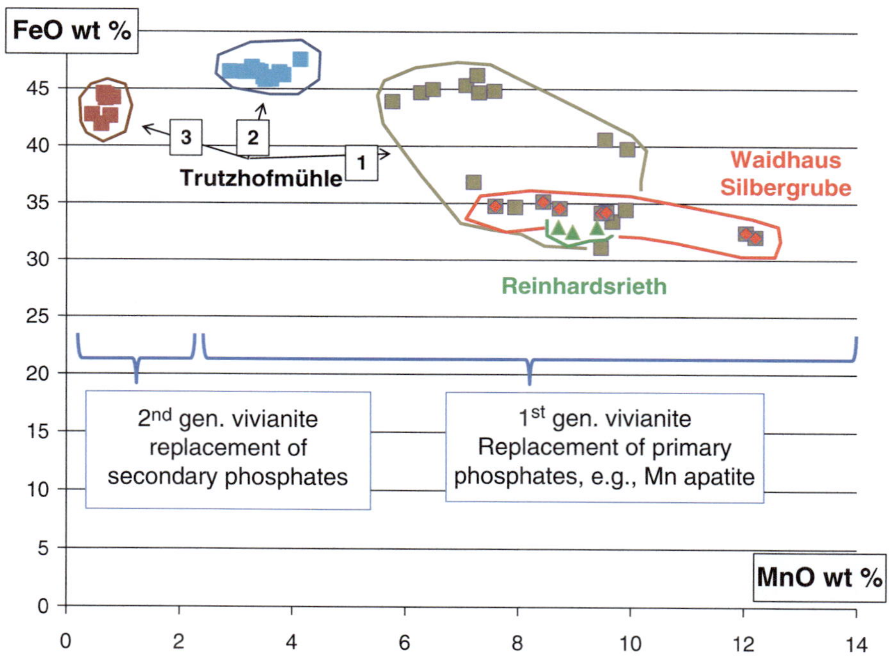

Fig. 4.12e Different generations of vivianite of the HPPP in an FeO vs. MnO cross plot. The Trutzhofmühle Aploid contains three distinct types of vivianite, each labeled in *boxes* with Arabic numerals (Dill and Skoda 2015)

Samples of early vivianite $(Fe_{2.4}Mn_{0.5}Mg_{0.1})(PO_4)_2 \cdot 8(H_2O)$ taken at the Silbergrube are enriched in Mn. Vivianite is stable over a wide range and also occurs late in the mineralization forming earthy encrustations (German: "Blaueisenerde" = blue colored clays) which is low in Mn (Fig. 4.12c, d). In those vivianite samples derived from replacement of primary Mn-bearing apatite, Mn may run up to 9.40 wt. % MnO, leading to the denomination of this Fe phosphate as manganiferous vivianite. In Fig. 4.12e the FeO/ MnO ratio is illustrated in an x-y plot for the tabular aplites at Reinhardsrieth, Waidhaus-Silbergrube and Trutzhofmühle, all of which are part of the HPPP. At Trutzhofmühle three distinct data arrays may be subdivided from each other based on the FeO/ MnO ratio. One vivianite from Reinhardsrieth analyzed by means of EMPA also yielded elevated Zn contents of 0.14 wt. % Zn in its lattice. In the majority of cases Mn vivianite replaced Mn apatite while being itself replaced by a wide range of Fe-Mn phosphates also pertaining to the group of secondary phosphates. Another approach similar to what has been done for the Mn apatite, has been taken to get an idea of the depth regime during the hydrothermal alteration of the pegmatites and aplites. Manganese and iron, although chemically very closely related with each other tend to split apart on decreasing temperatures and changing redox conditions. While the Eh increases, Fe is more swiftly oxidized into its trivalent state than Mn and hence no longer available for being accommodated in the lattice of vivianite. Higher Mn contents in the vivianite may signal a higher Eh and a shallower depth of hydrothermal alteration (Fig. 4.12e).

There is a wide range of Fe-only phosphates which differ from each other by their Fe $^{2+}$/Fe $^{3+}$ ratios and the degree of hydration, e.g., ferristrunzite, whitmoreite, beraunite, ferrolaueite, barbosalite, ferrostrunzite. The Fe^{2+}/Fe^{3+} ratios are listed below in decreasing order of. Strunzite *sensu lato* occurs in two different modification as to the valence state of Fe and was named in honor of the late Professor Dr. Karl Hugo Strunz, who was born and raised in the Oberpfalz not far away from Hagendorf, where he spent a great deal of his life studying the Hagendorf pegmatite (Fig. 4.12f). Laueite, hagendorfite, and petscheckite were found by him and

Fig. 4.12f The "strunzite story" a homage to the late Prof. Dr. Karl Hugo Strunz

Fig. 4.12g Radiating aggregates of ferristrunzite from the Trutzhofmühle under the SEM

Hagendorf-South became thereby type locality of these phosphates. Ferrostrunzite is dominated by trivalent Fe, whereas ferristrunzite is *per definitionem* a Fe-III phosphate (Fig. 4.12g).

Fe phosphate	FeO (wt.%)	Fe_2O_3 (wt.%)	Fe III/Fe II
Ferristrunzite	0	46.37	∞
Whitmoreite	4.90	40.00	8.16
Beraunite	8.24	45.78	5.56
Ferrolaueite	6.90	30.80	4.46
Barbosalite	18.35	40.79	2.22
Ferrostrunzite	14.38	31.96	2.22

Beraunite has the highest content of trivalent Fe and is the most common Fe phosphate of this group of secondary phosphates in the latest stages where it has been encountered together with strengite and cacoxenite and with phosphate-bearing Fe-Mn oxide-hydroxides whose precise composition is hard to determined and kaolinite (Fig. 4.12h). This mineral association reflects a lowering of the pH ($pH \leq 4$) and Eh values exceeding 0.4 mV. The final stage of alteration in sequence 2 at Trutzhofmühle is conducive to a stage characterized by P-bearing oxyhydroxides and kaolinite, which both suggest strongly oxidizing and acidic conditions and characterized the simultaneous adsorption of phosphate onto limonitic material.

Whitmoreite has yet a more elevated Fe III/Fe II ratio and also contains Zn of as much as 0.96 wt. % ZnO. Strengite formed filamentous and acicular aggregates around whitmoreite (Fig. 4.12i, j). It is one of the late minerals, showing up when the oxidizing regime starts to sweep its way through the primary and secondary phosphates of the pegmatite.

Fig. 4.12h Acicular beraunite from Pleystein. The elongated prismatic crystal shows the ideal crystal morphology of beraunite

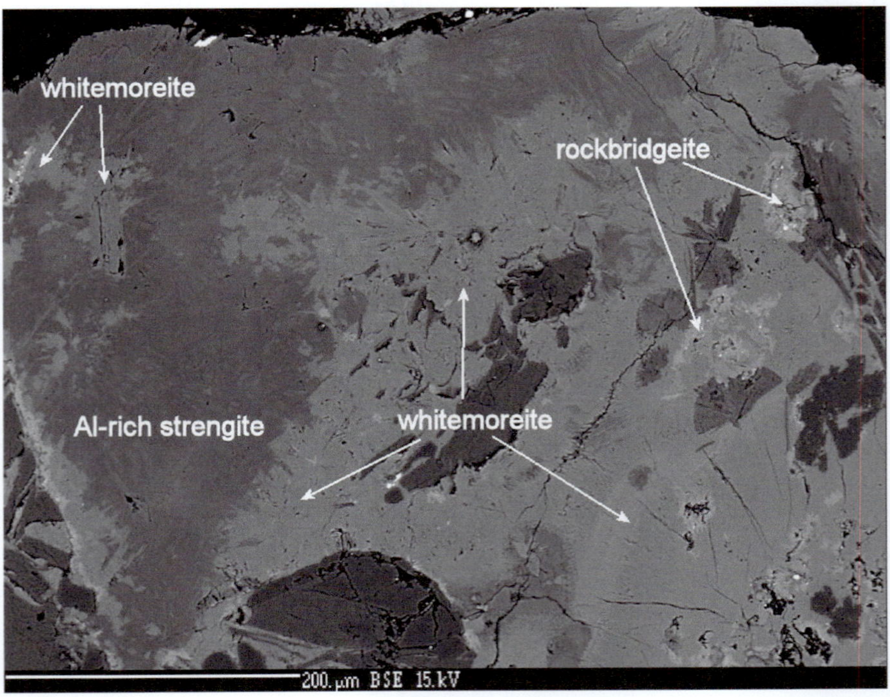

Fig. 4.12i Iron phosphates at intergrowth. Rockbridgeite is replaced by whitmoreite, and itself replaced by Al strengite at Vorderberg near Miesbrunn. (EMPA, BSE)

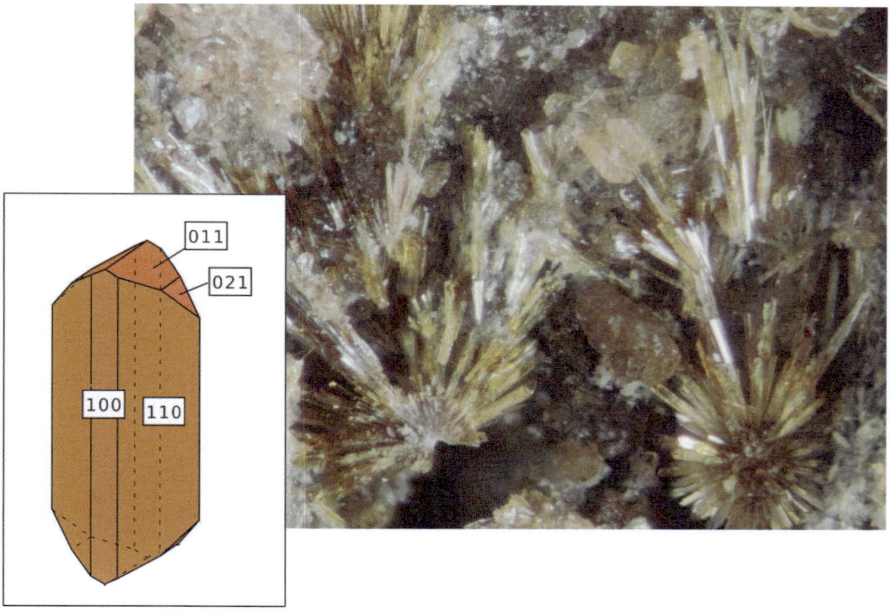

Fig. 4.12j Acicular aggregates of whitmoreite from Hagendorf-South with a cartoon showing its ideal crystal shape

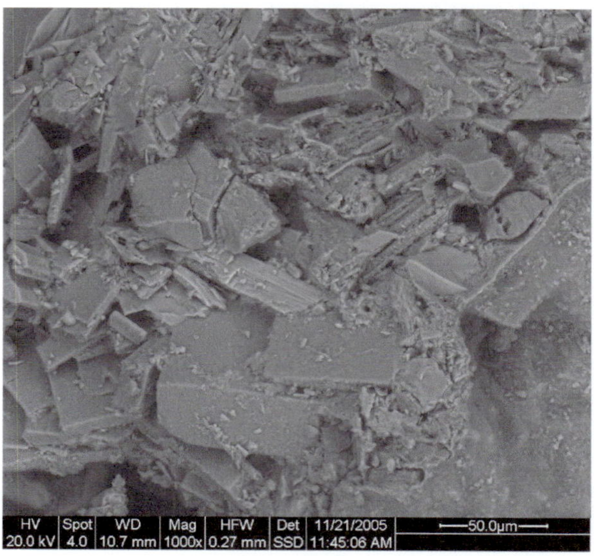

Fig. 4.12k Ferrolaueite plates of mineralizing stage IV between muscovite are corroded by supergene fluids at Trutzhofmühle (SEM, BSE image)

Ferrolaueite is much lower as to the Fe III/Fe II ratio than the afore-mentioned Fe-only phosphates (Fig. 4.12k). In a transect showing the variation of FeO, P_2O_5, Al_2O_3, MnO and Na_2O a gradual changes from Fe-enriched rockbridgeite, through Al-bearing Fe rockbridgeite into ferrolaueite and childrenite could be traced at

Fig. 4.12l Natrodufrenite from Miesbrunn (SEM, BSE image)

Fig. 4.12m Cyrilovite from Hagendorf-South with two cartoons showing its ideal crystal morphology

Trutzhofmühle. Barbosalite is another Fe-only phosphate, known from the most prominent pegmatite stocks at Hagendorf-South and Kreuzberg in Pleystein. Both phosphates formed during a late low-temperature hydrothermal stage as the pegmatites underwent strong alteration of their primary and early secondary minerals.

It should be noted here that the phosphate- and Fe-bearing solutions during the later stages of the pegmatite evolution also interacted with the rock-forming minerals K- and Na feldspar. The feldspar got locally decomposed and its alkaline elements were used to build up cyrilovite, natrodufrénite, leucophosphite, kidwellite and meurigite-K (Fig. 4.12l, m, n, o, p, q, r). Unlike the Fe-only phosphates, these phosphates accept trivalent Fe at a rather high amount, excluding

Fig. 4.12n Leucophosphite from Hagendorf-South with two cartoons showing its ideal crystal morphology

Fig. 4.12o Columbite-(Fe) (white mineral grain of high reflectivity in the center of the image) surrounded by well-shaped leucophosphite from Hagendorf-South (60-m level) (SEM, BSE image)

Fig. 4.12p Leucophosphite crystals overgrown by collomorphous frondelite. Hagendorf-South (SEM, BSE image)

Fig. 4.12q Meurigite-K forms the platform for autunite (greenish-yellow) to grow upon (Hagendorf-South)

Fig. 4.12r Rosettes of kidwellite from Hagendorf-South

natrodufrénite that has a moderate amount of FeO accommodated into its lattice (Fig. 4.12l). The most recent alteration stages of the pegmatite, also called epithermal, to show its shallow and transitional character among the mineral assemblages lead to the formation of phosphate species under the influence of meteoric and hydrothermal fluids at rather high Eh values. Therefore leucophosphite is found accompanied by phosphosiderite and even in parts by Fe-Mn bearing oxides (Fig. 4.12n, o, p). It grew upon collomorphous frondelite (Fig. 4.12p). A similar setting was also recorded from the rare-element pegmatites in the Olary Block, South Australia by Lottermoser and Lu (1997) for the mineral cyrilovite (Fig. 4.12m). The most elevated quantity of trivalent Fe among the Fe phosphates may be reported for kidwellite (Fig. 4.12r).

Fe phosphate with alkaline elements	FeO (wt.%)	Fe_2O_3 Wt. %
Cyrilovite		49.44
Natrodufrénite	0.42 %	46.21
Leucophosphite		40.55
Kidwellite		52.52
Meurigite-K		47.71

The ultimate stage of oxidation of iron can be recognized among the polymorphs strengite and phosphosiderite as well as in the mineral santabarbaraite (Fig. 4.12s, t, u, v). Those phosphates formed under the most elevated Eh values resultant from

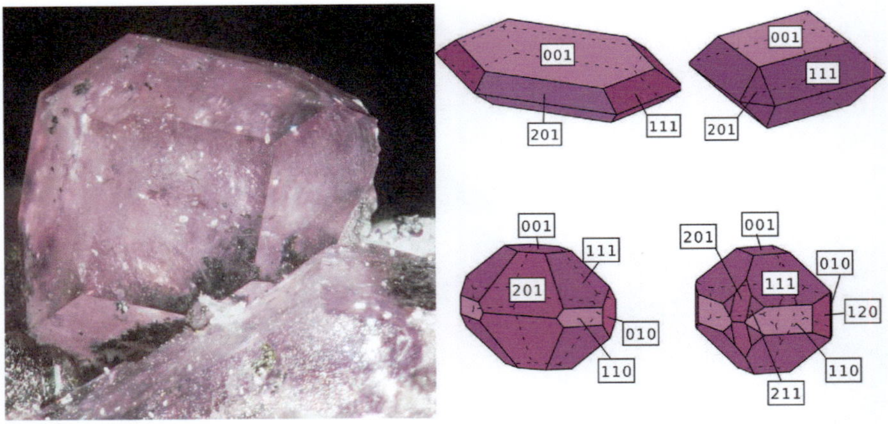

Fig. 4.12s Strengite form the Kreuzberg pegmatite at Pleystein with four cartoons showing its ideal crystal morphology

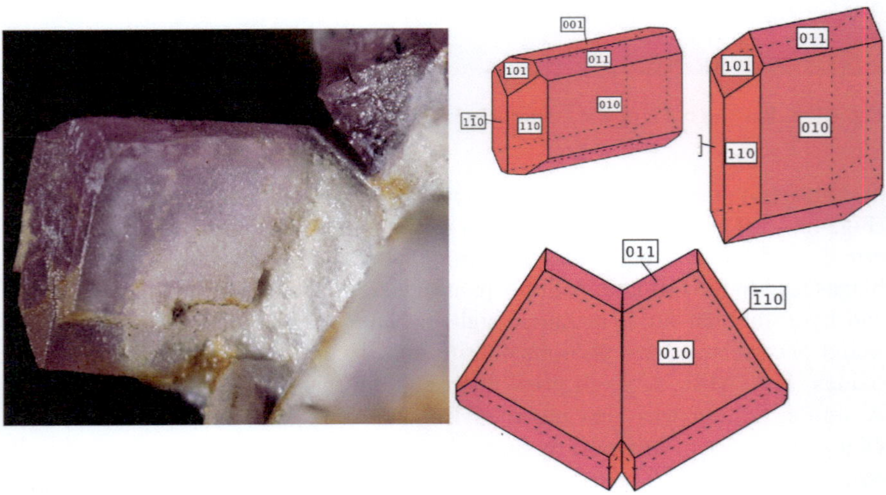

Fig. 4.12t Phosphosiderite form the Kreuzberg pegmatite at Pleystein with three cartoons showing its ideal crystal morphology as single crystal and twinned crystal aggregates

early chemical weathering and late hydrothermal alteration in the HPPP. Strengite and phosphosiderite formed very late relative to the afore-mentioned phosphates, being present in cellular structures and lining solution vugs, where they can develop excellent crystal shapes very much attractive to mineral collectors (Fig. 4.12s, t, u). Experimental results by Warry and Kramer (1976) showed the iron-oxyhydroxide-phosphate system produced strengite and phosphosiderite under oxidizing conditions and temperatures around 100 °C. This could also be proved by fluid inclusion studies at Miesbrunn which suggest temperatures of precipitation $<$ 100 °C (Dill

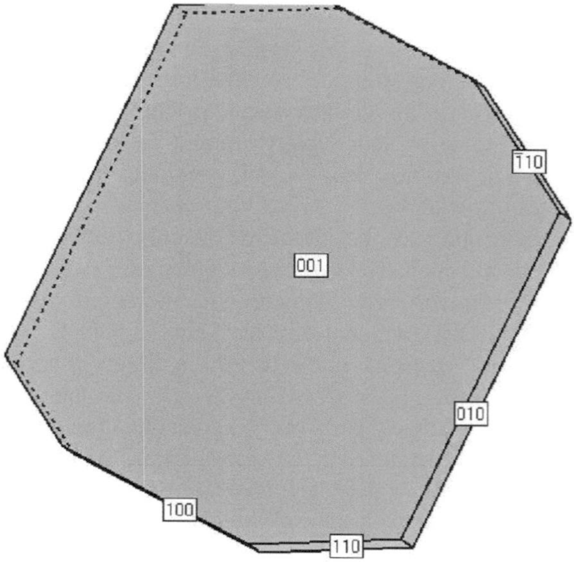

Fig. 4.12u A cartoon to show the tabular crystals of phosphosiderite from the Silbergrube at Waidhaus

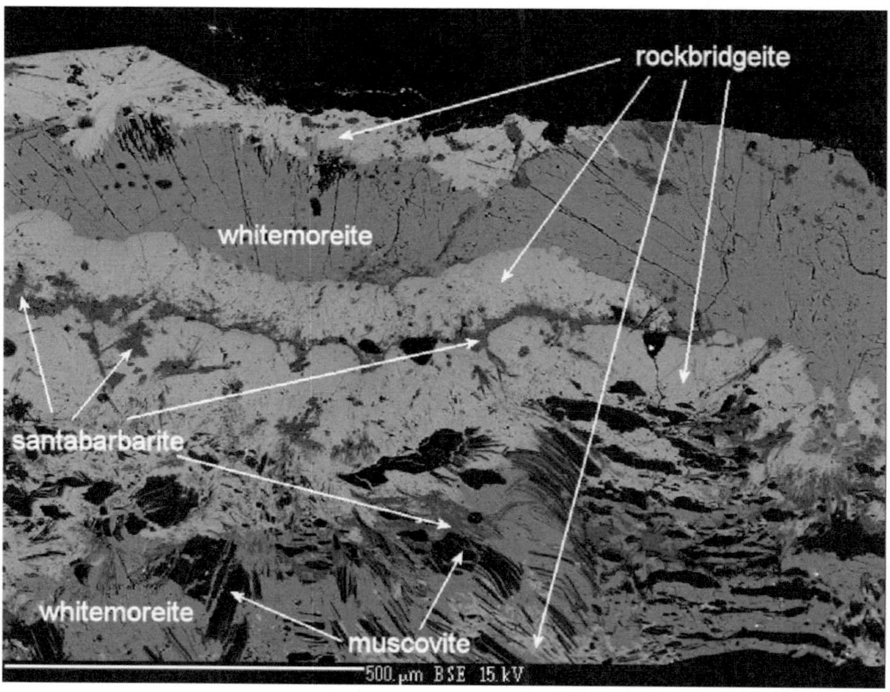

Fig. 4.12v Rhythmic banding of whitmoreite, santabarbaraite and rockbridgeite in the Miesbrunn pegmatite-aplite swarm on muscovite (EMPA)

et al. 2012a, b). Upon a steady increase in aluminum during the weathering and a concomitant decrease in iron, strengite gradually converts into variscite. There are also aggregates of aluminous strengite which convert into strengite *sensu stricto*. In the same near-surface setting, santabarbaraite is widespread, whereas cacoxenite and natrodufrénite are only sporadically present in the most recent mineral assemblages of the Miesbrunn aplites and pegmatites (Fig. 4.12v). Strengite replaces aggregates of rockbridgeite and whitmoreite and used to occur as an open space filling. Strengite and phosphosiderite are the only representatives of Fe phosphates known to contain exclusively trivalent Fe and also developed in significant quantities in a non-pegmatitic environment, e.g., the neighboring Fe deposits at Auerbach (Dill et al. 2009b), which is discussed later in Sect. 5.2 for chronological or "minero-stratigraphic" reasons. In the HPPP the dissolved species are assumed as follows: log $_{aFe}=-4$, log $_{aHPO4}=-2$ and log $_{aHCO3}=-4$ to approximate the redox conditions for the most youngest parts of the pegmatite alteration. At 25 °C which is the most reasonable temperature for the near-ambient conditions strengite develops above $Eh=0.2$ (volt) and a $pH<6$. Above this pH and below the assumed Eh value Fe and phosphate ions precipitate as vivianite. The late-stage earthy vivianite is a striking example of local reducing conditions in the vadose aquatic system. Considering the temperature during formation of the early/1st generation Mn-bearing vivianite from Mn-bearing apatite, the temperature was certainly far below 300 °C, somewhere around 100 °C, because of the considerable reduction of the stability field of this Fe-II phosphate upon the increase of temperature of formation.

4.13 Iron-Manganese Phosphates with Magnesium, Calcium, Strontium, Barium, Potassium, Fluorine and Sodium

Among the phosphate minerals of the HPPP, the group of Fe-Mn phosphates is unrivaled as to its chemical variety and quantity even if in the majority of cases the Fe-Mn-P minerals are exclusive to the large Hagendorf-South pegmatite (Table 4.1).

Wolfeite and zwieselite are the most common primary Fe-Mn phosphate minerals in the stock-like and tabular pegmatite and aplites alike (Fig. 4.13a, b). They may be discussed in a s.s.s. together with triplite and triploidite, which are also formed very early among the primary phosphates of the HPPP (Fig. 4.13c) (Keller et al. 1994). Fluorine is an essential element typical of these Fe-Mn phosphates which formed under reducing conditions during the initial phases of the pegmatite emplacement. Manganiferous apatite-(F) is another phosphate associated with these Fe-Mn phosphates and bearing witness of the high F contents in the melt. At Hagendorf-South, wolfeite is replaced by sarcopside (Mücke et al. 1990). While this reaction has almost no impact on the Fe/Mn ratio, it is accompanied by a strong dehydration and depletion of elements others than Fe and Mn. The process takes place still under reducing conditions prior to the hydrothermal stage which is

Fig. 4.13a Massive zwieselite from the Kreuzberg pegmatite at Pleystein

Fig. 4.13b Wolfeite crystal aggregates from the Kreuzberg pegmatite at Pleystein

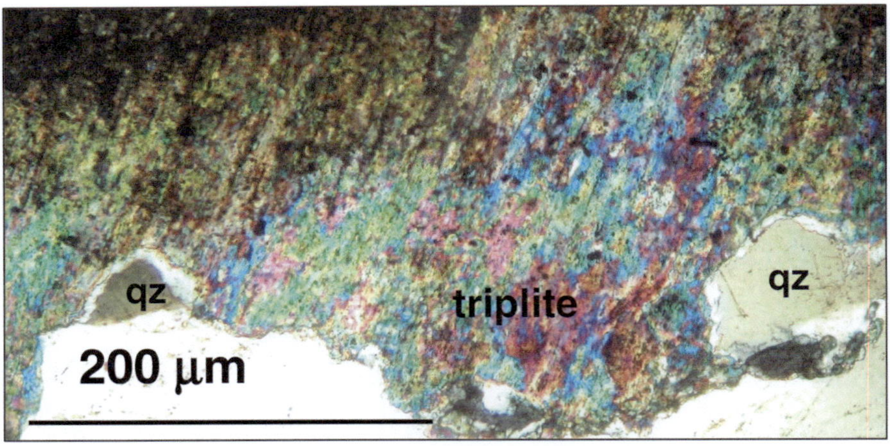

Fig. 4.13c Triplite enrichment in a triplite-zwieselite s.s.s. engulfing quartz (*qz*). Hagendorf-North. *Thin section*, crossed polars

Fig. 4.13d Graftonite under the SEM. Hagendorf

marked by a series of hydrated Fe-Mn phosphates. Experimental work conducted by Ericsson et al. (1986) and Charalampides et al. (1988) suggests that sarcopside may be stable at low subsolidus temperatures, in the range between 300 and 500 °C. Hatert et al. (2007) performed hydrothermal experiments between 400 and 700 °C (1 kbar) to determine the stability field of the triphylite plus sarcopside assemblage. Graftonite has been derived from the primary Li phosphate triphylite as being depleted in Fe and Li (Mücke et al. 1990) (Fig. 4.13d). Vignola et al. (2011) recorded the alteration of triphylite into sarcopside and graftonite. Exsolution lamellae of the latter mineral are not known from Hagendorf-South.

In conclusion, zwieselite and triplite are Fe-Mn phosphates which are representative of the central phosphate facies of the pegmatites during the initial phases of the emplacement of pegmatite stocks at Hagendorf-South, Hagendorf-North and Pleystein (Table 4.1). A similar mineral association may be envisaged for the pegmatites Hühnerkobel and Birkhöhe near Zwiesel in the Bayerischer-Böhmer Wald. They bear also a Fe-Mn-F-P system with manganiferous apatite as a predecessor. Slight variations among the pH, in the intensity of hydration and changes in the F availability do not alter the Fe-Mn ratio but may lead to another suite of Fe-Mn phosphates such as the more strongly-hydrated phosphates wolfeite and triploidite in the tabular aplites and aplite granites at Trutzhofmühle and Waidhaus. These satellite intrusive deposits of the HPPP show all hallmarks of a stronger interaction between the siliceous melt and the gneissic wall rocks than the stock-like pegmatite bodies (external fractionation). The short-distance or in-situ alteration is restricted to the stocklike pegmatites only where a local deficiency in the fluidal components (H_2O, F) was responsible for the precipitation of graftonite and sarcopside. The Fe-Mn phosphates may have formed at temperatures >500 °C. The presence of Fe-Mn at such an amount in a felsic melt can only be accounted for by considering the Fe-Mn system in context with apatite and garnet. The phosphate-silicate system is known to be governed by the depth of intrusion. At deeper level garnet prevalently accommodates Mn in its lattice. That is one plausible explanation why the deeper pegmatites in the Bayerischer-Böhmer Wald near Zwiesel and Bodenmais are significantly depleted in this group of Fe-Mn phosphates and contain manganiferous garnet instead (Table 4.1). An unknown Mn (II) – Fe (II) phosphate ("pseudo-zwieselite") replaced manganiferous apatite in the Reinhardsrieth aplite. Due to the presence of subordinate amounts of aluminum in the chemical composition of this hitherto unknown phosphate, whose denomination is provisional, minor amounts of

Fig. 4.13e Massive aggregate of arrojadite (bright *greenish* to *yellow*), hagendorfite (*reddish brown*) and *dark gray* triplite in a specimen from Hagendorf-South

the iron present in the structure of pseudo-zwieselite is expectedly present as trivalent iron. Hydroxyl-groups are supposed to be accommodated in the lattice whereas fluorine could not be detected in the samples from Reinhardsrieth. It is the most distal mineral of this group of primary Fe-Mn phosphates, located at the north-eastern edge within the marginal facies of the HPPP. Some other pegmatites still further north, such as Brünst and Peugenhammer lack these Fe-Mn phosphates on account of its different structural and geochronological position.

The members of the arrojadite- and dickinsonite series in the HPPP-see Table 4.1- have a very complex chemical composition (Fig. 4.13e). Depending upon the availability of bivalent Fe and Mn in the Hagendorf pegmatite, either arrojadite (23.1–33.0 wt. % FeO vs. 0.6–14.7 wt. % MnO) or dickinsonite came into existence (13.1 wt. % FeO vs. 31.9 wt. % MnO). Chemical analyses carried out by Haase and Keck and published in the paper of Mücke and Keck (2011) yielded Fe contents in the range 23.01–28.25 wt. FeO and Mn contents in the range (18.41–24.27 wt. % Mn). In addition to the hydroxyl-groups and the remarkable fluorine contents, both known from the zwieselite-triplite series, each of the above phosphates contains alkaline and/or earth alkaline elements in its structure, indicative of an intimate interaction of phosphate with the felsic melt during the incipient stages of the emplacement of the pegmatite stocks. None of the arrojadite-series members was observed in any of the tabular pegmatites or aplites, and they are exclusive to the type localities of both provinces in the northern Oberpfälzer Wald and the Bayerischer-Böhmer Wald, in the Hagendorf-South and Hühnerkobel pegmatite stocks. The structurally-controlled Fe-Mn phosphate mineralization, which has already been discussed previously, becomes more conspicuous when considering these complex Fe-Mn phosphates which due to the presence of bivalent Fe and Mn formed under reducing conditions similar to their predecessors of the triplite-zwieselite series. Browsing the pertinent literature, minerals of the arrojadite and dickinsonite seem to be not only rare phosphates in the HPPP. Chopin et al. (2006) and Demartin et al. (1996) reported minerals of the arrojadite-series from green schist- to amphibolite-facies metamorphic quartzites. In hydrothermal veins cross-cutting shales and ironstones of very low metamorphic grade Robinson et al. (1992) recorded their presence, while Yakubovich et al. (1986) synthesized ferrian arrojadite at low temperatures (450 °C). The complex arrojadite-series suffers the same fate as many other minerals which were held to be exclusive to the "treasure box" pegmatite. Careful and ongoing studies of certain minerals disclosed a lot of environments also suitable to develop these exotic minerals. The only statement which is for sure in this place is that the Fe (II) -Mn (II) phosphates are restricted to the initial phase of emplacement and the central part of pegmatites and phosphates, such as sarcopside, graftonite and phosphates of the arrojadite-dickinsonite series. They are exclusive to large stock-like bodies where they reflect some kind of retrograde adjustment within the pegmatite system, involving slight changes of fluid migration and reaction with the Na-K-Ca silica system. The outcome of this process resembles the wide range of barren aploids and pegmatoids in metamorphic terrains subjected to diapthoresis. All of these phosphate minerals tend to form massive

Fig. 4.13f Prismatic blue vivianite is overgrown with fairfieldite. Hagendorf

Fig. 4.13g Bands of black rockbridgeite alternating with bright brown wilhelmvierlingite. Hagendorf

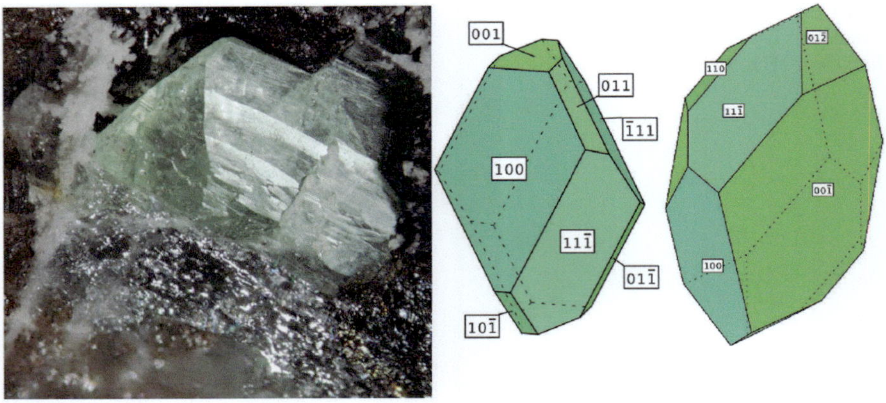

Fig. 4.13h Ludlamite. Hagendorf with two cartoons showing its ideal crystal morphology

Fig. 4.13i Phosphoferrite from Hagendorf with three cartoons showing its ideal crystal morphology

textures ("closed texture") with little vugs reflecting (auto)metasomatic processes rather than open space fillings in sealed pegmatite bodies.

Quite similar as to their chemical composition but different with regard to the degree of hydration, another group of Fe-Mn phosphates came into existence made up of the following phosphate minerals: Ludlamite, phosphoferrite, wilhelmvierlingite, fairfieldite, switzerite and landesite (Fig. 4.13f, g, h, i, j, k). Fairfieldite and wilhelmvierlingite both contain calcium but occur in a distinct structural settings. Fairfieldite grows freely in open vugs and on early-formed minerals such as vivianite (Fig. 4.13f). It typically formed an "open texture" compared to what has been told about the Fe-Mn phosphates in the previous paragraph. Bands of wilhelmvierlingite alternate with bands of black rockbridgeite at Hagendorf (Fig. 4.13g). While rockbridgeite is a common mineral in the HPPP and also known from sites outside

Fig. 4.13j Switzerite in a cavity of rockbridgeite from Hagendorf

Fig. 4.13k Landesite at Hagendorf

this mineral province, wilhelmvierlingite takes a similar exclusive position as the arrojadite-dickinsonite series. The well-known globular layers of rockbridgeite responded in a different way. The inner lamellae of rockbridgeite are replaced by wilhelmvierlingite, the outer parts are overgrown by it. Rockbridgeite has a FeO/MnO ratio of 0.329, wilhelmvierlingite of 0.889. It reflects a local relapse towards reducing conditions caused by Ca-bearing fluids in an environment otherwise heading to more oxidizing conditions. Calcium may have derived from Ca carbonate decomposition which took place during this period of evolution. The same holds true for switzerite as far as the relapse of the redox system is concerned. It occurs, however, at a later stage where open space filling rather than a lit-par-lit replacement was possible due to the open vugs provides by previous dissolution or a striking misbalance between accommodations space provided and mineralizing fluid available (Fig. 4.13j). Landesite a strongly hydrated Fe-Mn phosphate defines a similar physical-chemical regime as its predecessor in this section, where it replaces reddingite (Fig. 4.13k).

Ludlamite and phosphoferrite formed earlier than the afore-mentioned Fe-Mn phosphate and can be placed as to their redox and temperature regime close to vivianite (Fig. 4.13h, i). The thermo chemical behavior of ludlamite is similar to that of vivianite (Rodgers 1989; Rodgers and Henderson 1986). These Fe-II phosphates show a major differential thermal response spanning 65–315 °C attributable to an endothermic loss of structural water combined with the oxidation of Fe^{2+} and consequential breakdown of the original structure.

Hagendorfite, jahnsite-(CaMnFe), lipscombite, perloffite, rockbridgeite, earlshannonite and the unknown phosphate "UNK 9", have one thing in common. All of them accommodate into their lattice bivalent as well as trivalent iron (Table 4.1). Hagendorfite is one of the most remarkable phosphates of the HPPP (Fig. 4.13e). The Hagendorf-South pegmatite was namesake and type locality, all in one. Hagendorfite is the only phosphate which during the initial stage of pegmatite evolution has part of its Fe content oxidized with an FeO/Fe_2O_3 ratio of 1.237. Hagendorfite has a complex chemical composition but became not hydrated, as it replaced triphylite that lost its Li contents to completeness during this reaction. The phosphate which is considered as part of the in-situ retrograde adjustment within the stocklike pegmatites at Hagendorf-South and Hagendorf-North marks a break in the evolution of the pegmatites as the succeeding hydrothermal stages start off with

Fig. 4.13l Jahnsite on globular aggregates of rockbridgeite at Hagendorf

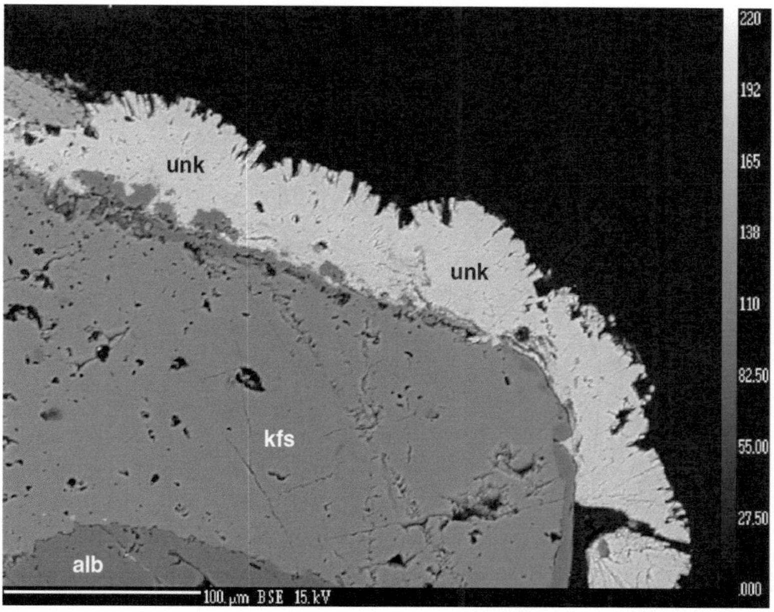

Fig. 4.13m K feldspar rimmed by acicular aggregates of an unknown Fe(II)-Fe(III)-Mn phosphate from Reinhardsrieth aplite "UNK9" (BSE image)

a series of Fe phosphates in another reducing regime. Fransolet et al. (2004) described in a sample from the Kibingo granitic pegmatite, Rwanda, hagendorfite transitional into alluaudite *sensu stricto*, and results from an oxidation, of $Fe^{2+} \rightarrow Fe^{3+}$. Cyrilovite and heterosite were recorded to accompany this phase transition. The authors estimated the temperature of formation of hagendorfite at about 600 °C and concluded its occurrence to be a primary one. Even if the temperature seems a bit too high for Hagendorf, Germany, it provides us with a rough idea of the temperature regime under which hagendorfite may have come into existence. Jahnsite-(CaMnFe) was found to crystallize on globular aggregates of rockbridgeite at Hagendorf (Fig. 4.13l). The jahnsite group consists of a very complex series of hydrated phosphates which will spark some more discussion. Although no precise temperature data can be provided the mineral is supposed to be a medium- to low temperature hydrothermal mineral (Weiß et al. 2004). At Reinhardsrieth another unknown Mn-Fe phosphate has been identified that unlike the so-called "pseudo-zwieselite", appears rather late in the mineral succession and has considerable amounts of trivalent Fe prevailing over bivalent Fe in its structure. It forms bundles of needle-shaped crystals and shows a globular outward appearance (Fig. 4.13m). There is neither an analogue in the HPPP nor does its chemical composition match any of the known Fe-Mn phosphate that have been accepted by IMA. The unknown phosphate is similar in its chemical composition and morphology to the "unknown mineral 9" (see UNK 9) described by Sekjora et al. (2006) from the 5th level of the Huber Shaft of the Krásno-Horní Slavkov area in western Bohemia, Czech Republic.

Fig. 4.13n Perloffite crystals from Hagendorf

Fig. 4.13o Rockbridgeite crystal from Hagendorf

Another occurrence well corresponding to the unnamed phosphate from Reinhardsrieth has been recorded from the Buranga pegmatite, Rwanda, by von Knorring and Sahama (1982), who called their find "Fe-Mn dufrenite-like phase". The authors pointed to triplite and rockbridgeite-frondelite series minerals to be intergrown with this unknown phosphate, suggesting that this phosphate developed at an intermediate stage of the mineral association. The two Fe-Mn phosphates lip-

Fig. 4.13p Radiating aggregates of yellow earlshannonite on black rockbridgeite from Hagendorf

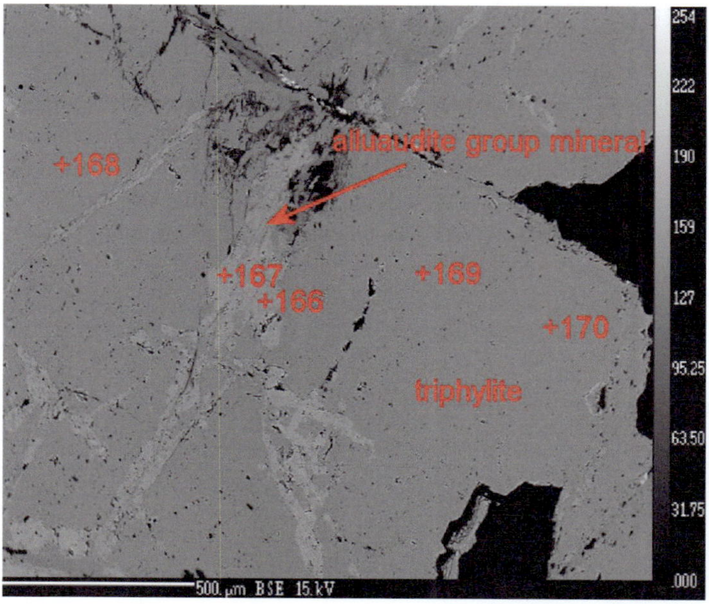

Fig. 4.13q Alluaudite replaces along cracks triphylite at Hagendorf-North. The *red spots* and numbers denote points of measurements, where chemical data were collected using the EMP (BSE image)

scombite and perloffite are only known from Hagendorf-South. Medrano and Piper (1997) gave an overview of the Fe-Ca-phosphate, Fe-silicate, and Mn-oxide minerals in concretions from the Monterey Formation, including lipscombite. Corresponding to their study, the Eh increased from the precipitation of vivianite to that of rockbridgeite/lipscombite. Perloffite is one of the few phosphates which contain barium the source of which may be looked for among the K-bearing rock-forming silicates, where Ba^{2+} can substitute for K^+ (Fig. 4.13n). Rockbridgeite is by far the most widespread mineral among the Fe phosphates in the HPPP and also known from vein-type mineralization, e.g., Auerbach Fe deposit, Germany (Figs. 4.1b, 4.10d, 4.12i, v, 4.13g, k, o). Rockbridgeite, which is the Fe-enriched member of the rockbridgeite-frondelite s.s.s formed under more oxidizing but fluctuating redox conditions ($Eh \leq 0$) at temperatures below 200 °C. Radiating aggregates of earlshannonite were found on rockbridgeite that reflects the consequent continuation of an increasing Eh during precipitation of the Fe(II)-Fe(III) phosphates (Fig. 4.13p).

By the end of this discussion alluaudite has to be mentioned (Fig. 4.13q). Frequently, the chemical composition is given with Fe in its trivalent and bivalent state $(Na,Ca)_2(Mn,Mg,Fe^{2+})(Fe^{3+},Mn^{2+})_2(PO_4)_3$, sometimes also missing the bivalent Fe (Moore 1971). There is crystallo-chemical and field evidence that alluaudite-group minerals are derivative of triphylite, lithiophyllite, ferrisicklerite, and heterosite. It is therefore common to see in polished section alluaudite replacing along fissures and cracks triphylite (Fig. 4.13q). The process may be described as metasomatic substitution of Na for Li and considered in context with what has been termed already as autometasomatism processes. This phosphate typical of the alteration is found only at Hagendorf-North (Dill et al. 2013a), one of the pegmatite stocks of the HPPP and near Zwiesel, where it was described as "huhnerkobelite", a term no longer in use, from the Hühnerkobel pegmatite stock. As alluaudite is associated with hagendorfite, it is held to be part of the supercritical retrograde autometasomatic process during which the relict fluids were accumulated in the large pegmatite stocks so as react with previous phosphates under moderately oxidizing conditions and decomposed in parts some of the primary phosphates prior to the subcritical hydrothermal stages responsible for the plethora of secondary phosphates.

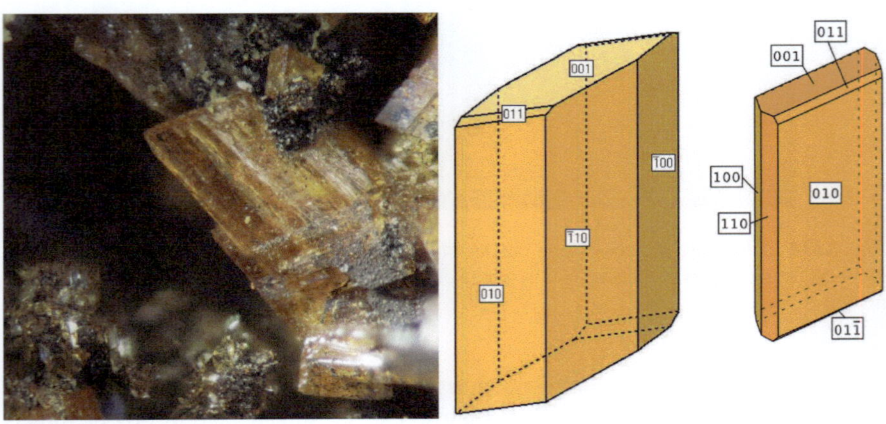

Fig. 4.13r Laueite from Hagendorf with two cartoons showing its ideal crystal morphology, on the *right-hand-side* for Waidhaus and on the *left-hand-side* for Hagendorf

Fig. 4.13s Stewartite with two cartoons showing its ideal crystal morphology from Hagendorf

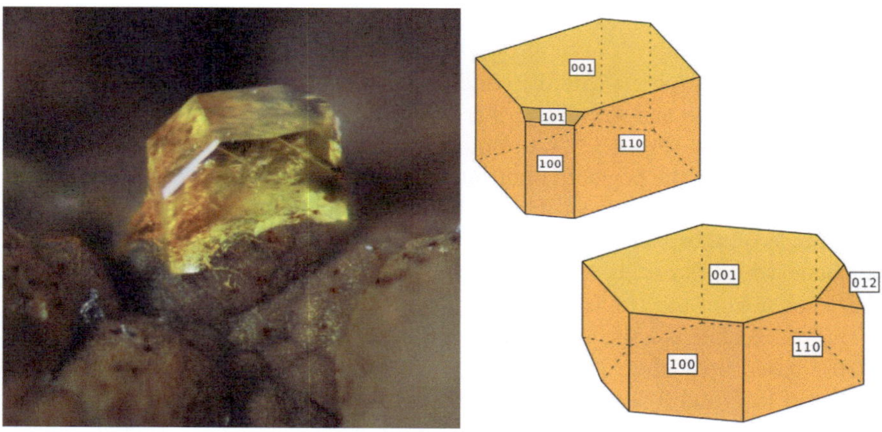

Fig. 4.13t Pseudolaueite with two cartoons showing its ideal crystal morphology from Hagendorf

Hatert et al. (2011b) revealed that at 1 kbar under an oxygen fugacity controlled by the Ni/NiO buffer, single-phase alluaudites crystallize at 400 and 500 °C.

Laueite, stewartite, pseudolaueite, keckite, strunzite, frondelite, kryzhanovskite, heterosite and oxiberaunite are the most strongly oxidized Fe-Mn phosphates reported from the HPPP. Some of them, e.g., heterosite and oxiberaunite, also contain trivalent Mn in their lattice. Many of these phosphates show up in nice micro mounts of showcase quality due to the fact that the texture is getting "more and more open" towards younger stages either by dissolution or mechanical processes thereby providing the space for the phosphates to grow freely and unimpeded by their neighbors.

Fig. 4.13u Keckite crystal coating rockbridgeite-frondelite s.s.s in the Trutzhofmühle Aploid

Laueite is dimorphous with the mineral stewartite and contributes together with strengite, strunzite and keckite to the group of phosphates infilling fissures and cracks at Waidhaus (Fig. 4.13r, s, t). Both minerals are found neither in the supergene regolith on top of the pegmatite nor intimately intergrown with the phosphates of the hydrothermal stage, proper. They belong to the epithermal phase which covers a wide range of phosphates, among others laueite and stewartite. The mineral pseudolaueite which is exclusive to the Hagendorf-South pegmatite in the HPPP and also known from the Hopfau pegmatite, pertains to the same phase of pegmatite alteration. Keckite is a special member of the jahnsite-group. Hochleitner and Fehr (2010) based upon their Mössbauer spectroscopic investigations of Fe-(III) and studies of its C and B site occupancies proposed a new formula: $CaMn(Fe^{3+},Mn)_2Fe^{3+}_2(OH)_3(H_2O)_7[PO_4]_4$ (Fig. 4.13u). Similar to the jahnsite-group, keckite spreads across the boundary late hydrothermal to epithermal which is not sharp one. Strunzite is one of the most common late-stage Fe-Mn phosphates and was already addressed together with the minerals ferri- and ferrostrunzite (Fig. 4.12f).

For strunzite an Al-bearing modification was recorded by Dill et al. (2008b) from Hagendorf-South and defined crystallographically by Grey et al. (2012). This modification of strunzite occurs as fibrous aggregates in a crystallographically oriented association with jahnsite on altered zwieselite. Although it is not a Fe-only phosphate but a Fe-Mn phosphate similar to its brethren strunzite it is listed in this place. It has been attributed to the transitional epithermal stage. Frondelite can simply be discussed together with rockbridgeite with which it makes up a s.s.s. (Fig. 4.13v). Kryzhanovskite is the oxidized equivalent of phosphoferrite (Lottermoser and Lu 1997) (Fig. 4.13w). The precise position of this rather strongly oxidized phosphate

Fig. 4.13v Frondelite-enriched s.s.s of the rockbridgeite-frondelite series at Hagendorf

Fig. 4.13w Kryzhanovskite crystals (*dark*) embedded into green ludlamite at Hagendorf

is difficult to constrain for the HPPP. Heterosite has the most elevated content of trivalent Mn among the Fe-Mn phosphates dealt with in Sect. 4.13 (Fig. 4.13x). Heterosite (–purpurite) s.s.s. is known to form the final product of a succession of minerals starting off with triphylite and going through ferrisicklerite-sicklerite, as

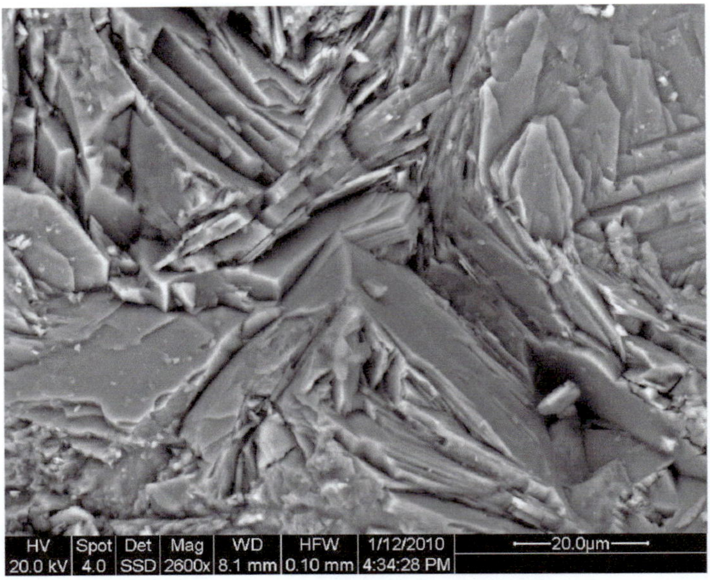

Fig. 4.13x Heterosite-enriched heterosite-purpurite s.s.s at Hagendorf-North (SEM)

Fig. 4.13y Needle-shaped aggregated of oxi-beraunite, the most strongly oxidized Fe-Mn phosphate from the HPPP

shown by the following reactions $Li(Fe^{2+},Mn^{2+})PO_4 \Rightarrow Li_{<1}$ $(Fe^{3+},Mn^{2+})PO_4 \Rightarrow (Fe^{3+},Mn^{3+})PO_4$ (Quensel-Mason Sequence) The alteration is coupled with a simultaneous Li-leaching and an oxidation affecting many rare-element pegmatites, according to the substitution mechanism $Li^+ + Fe^{2+} \rightarrow [\,] + Fe^{3+}$ at temperatures between 300 and 500 °C. One important question still remains to be answered: "When did this oxidation take place?". Heterosite is intimately intergrown with un-

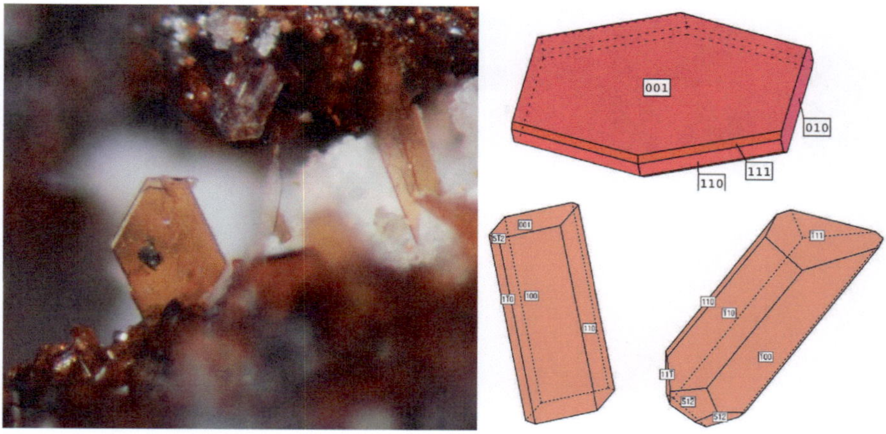

Fig. 4.14a Bermanite crystals from Hagendorf with a cartoon showing their ideal crystal morphology

hydrated Fe-Mn- and Li phosphates (no void filling), some of which were described in this section, and it is replaced by OH-enriched apatite. The ubiquitous rockbridgeite-frondelite s.s.s is also younger than the Fe(III)-Mn(III) phosphate under discussion. Oxiberaunite reflects a state of strong oxidation with all the iron converted into trivalent Fe. It marks a pre-stage of the supergene alteration but it is uncommon to the phosphates which developed during the Neogene in this region (Fig. 4.13y).

4.14 Manganese Phosphates with Calcium

The group of manganese-only phosphates in pegmatites is of lesser diversity than equivalent iron-only or mixed-type phosphates discussed in previous sections (Sects. 4.12 and 4.13). A representative of this group is the Mn phosphate bermanite which contains considerable amounts of trivalent Mn besides MnO and which is strongly hydrated (Fig. 4.14a). It is exclusive to the Li-enriched Hagendorf-South pegmatite, filling cracks as open space mineralization during the late-hydrothermal stage and it has been derived from primary triplite. Its precise position cannot be given in the minerals succession of Hagendorf-South. Contrary to the Mn(II)-Mn(III) phosphate, purpurite forms part of a s.s.s. with heterosite and appears rather early in the mineral successions in the HPPP, terminating the Quensel-Mason Sequence (Fig. 4.13x). The remaining Mn-only phosphates have only bivalent Mn in their lattice: Hureaulite, metaswitzerite, reddingite and robertsite (Fig. 4.14b). Hureaulite suffered the same fate as some other phosphates in the HPPP and was first described as “wenzelite” and “baldaufite”. A brief overview of naming and re-naming minerals in the HPPP is given below and sheds some light on the back and fro. Considering these actions

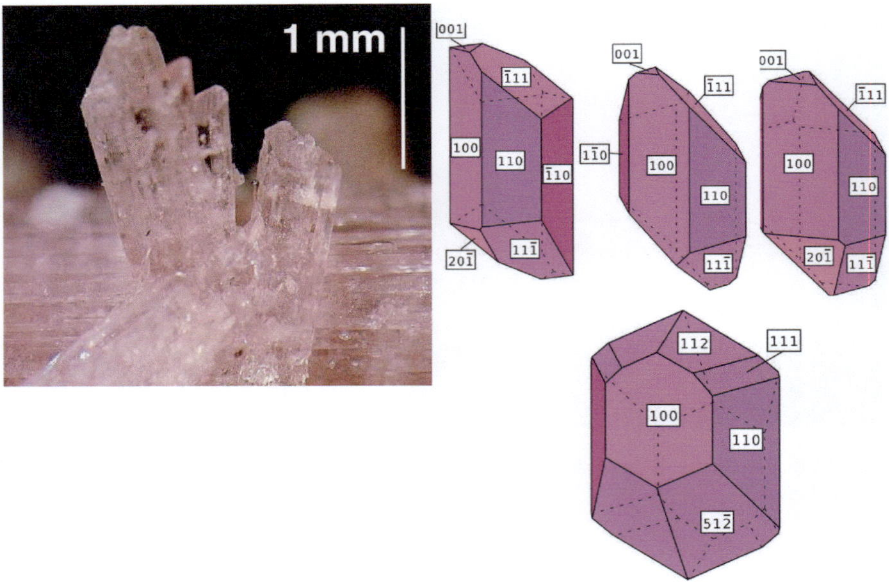

Fig. 4.14b Hureaulite crystal aggregate with six cartoons showing their ideal crystal morphology. The bright brown crystal morphologies are typical of the Waidhaus aplite, the pinkish ones of Hagendorf-South

may help us gain a different perspective of the world of minerals that is not only driven by pure natural sciences.

The publications of Laubmann and Steinmetz (1920), Laubmann (1924), Scholz (1925), and Müllbauer (1925) are of value especially when studying the history of mineralogy and for those interested in the ancient names and their sources, even if they were not approved by IMA: "Kraurit" (= rockbridgeite), "Pleysteinit" (= fluellite), "Kreuzbergit" (= fluellite), "Lehnerit" (= ludlamite), "Baldaufit" (= hureaulite), "Wenzelit" (= hureaulite), "Eleonorit" (= Oxi-beraunite), "Triplit" (= zwieselite), "Triploidit" (= wolfeite), "Purpurit" (= heterosite). "Cleveit" (= uraninite). The ancient studies should also be handled with great care and caution, because of dubious chemical compositions. Phosphophyllite was recorded from Hagendorf-North by Laubmann (1924) as a sulfate-phosphate. Some of these minerals have turned out later as wrong determination for Hagendorf-North but were later used for a different mineral, e.g. lehnerite or were approved by IMA for minerals of different chemical composition such as triplite, triploidite or purpurite.

Hureaulite grew in cavities left over after the precipitation of vein-type rockbridgeite, demonstrating as slight decrease in the oxygen partial pressure relative to its predecessor rockbridgeite which contains a significant amount of trivalent Fe. Oxi-hureaulite from Silbergrube which is a mirror image of the

Fig. 4.14c Globular aggregates of robertsite on rockbridgeite at Hagendorf

Fig. 4.14d Reddingite from Hagendorf with two cartoons showing its crystal morphology

Fig. 4.14e Metaswitzerite from the Reinhardsrieth aplite

variability of redox states expressed by its variation in color has a much wider stability field than the afore-described Fe-III phosphates and is stable under oxidizing to reducing conditions of Eh > −0.25 volts at temperature <100 °C and a pH > 6.

Robertsite is the only Mn-(II)-bearing phosphate which also took up Ca^{2+}. Upon replacement of apatite-(OH), the calcium released during this reaction was incorporated into robertsite, which pertains to the same physical-chemical regime as hureaulite (Fig. 4.14c). Reddingite and metaswitzerite are chemical resembling each other with respect to the major chemical components but are different as to the degree of hydration (Fig. 4.14d, e, Table 4.1). Reddingite precipitates as a cavity filling in an environment similar to what has been described for vivianite and has been derived from decomposition of triphylite. The position in time for metaswitzerite cannot precisely be given for the sites under investigation. Due to its close chemical resemblance with reddingite, one may assume that this Mn phosphate crystallized also at an early stage under reducing conditions.

4.15 Manganese and Iron Oxides, Sulfides, Sulfates and Carbonates

All oxidic Mn minerals, pyrolusite, chalcophanite, todorokite and cryptomelane in the HPPP were generated by supergene alteration (Fig. 4.15a, b, c). Their precise identification is, however, often fraught with difficulties. Manganese minerals show up in two different ways in the study area. Dull brown to black in color, Mn mineral

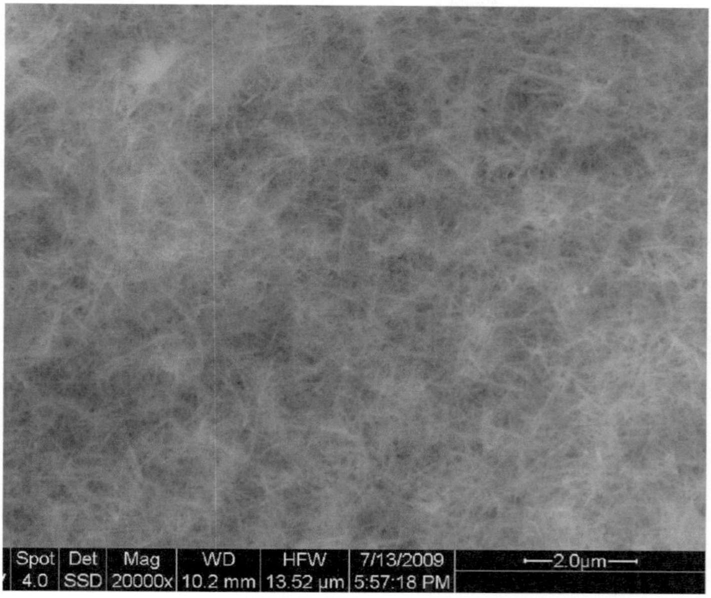

Fig. 4.15a A widely-spaced filamentary network of cryptomelane I composed of bundles of tiny fibers of Mn minerals. Specimen taken at 87 m below ground at Hagendorf-South. (SEM)

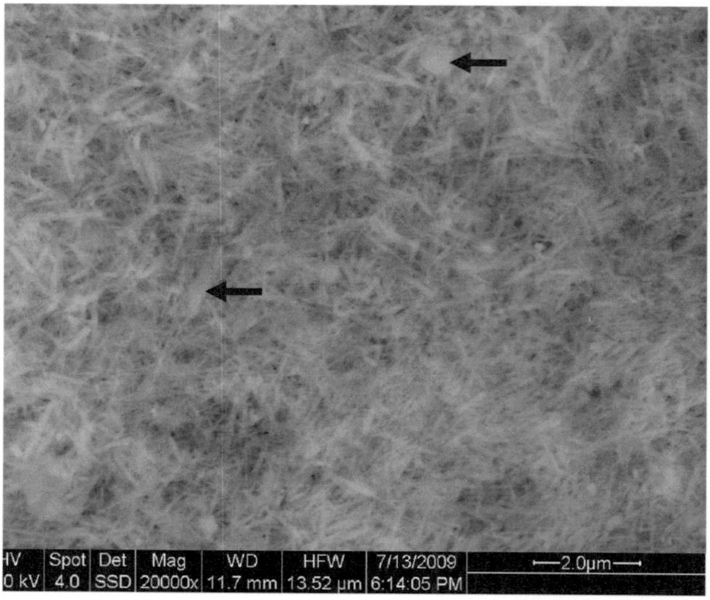

Fig. 4.15b A more densely-packed network of cryptomelane II composed of bundles of platy elongated crystals of cryptomelane I surrounding single prisms (*arrowheads*) of tetragonal morphology. Cryptomelane II reflects a more advanced stage of oxidation, which only can be recognized under the SEM at high magnification. Specimen taken at 76 m below ground at Hagendorf-South. (SEM)

Fig. 4.15c Manganese minerals on a macroscopic scale. Botryoidal black cryptomelane ("cy") encrustation coating brown beraunite ("be"). Pleystein Kreuzberg Pegmatite

aggregates may display an earthy outward appearance and, hence, were denominated in this paper as "wad". Some mineral assemblages occur as botryoidal aggregates lining the walls of fissures and druses in the feldspar-bearing rocks or coat Fe-Mn-bearing phosphates, e.g., beraunite and triphylite ("manganomelane") (Fig. 4.15c, Table 4.1).

Among the oxidic Mn minerals, cryptomelane is by far the most common and also the most attractive one, not by its outward appearance but chemical composition. It provides a clue to the age of supergene alteration affecting the HPPP pegmatites during the Cenozoic (Fig. 4.15a, b, c). Cryptomelane is a K-Mn-bearing oxide in the lowermost part of the weathering/ supergene alteration zone of the Hagendorf Pegmatite, SE Germany. This supergene Mn mineral has been used for K/Ar dating of weathering and paleohydraulic processes as young as Plio-Pleistocene (Dill et al. 2010b). Secondary dark brown manganese minerals are encountered from 60 m below the present-day surface down to a depth of 87 m below ground. Two type of cryptomelane have been observed in this depth range. Cryptomelane I from the 87-m-level displays a more open fabric or widely-spaced filamentary network of tiny fibers of Mn minerals. No enlarged crystals or clusters of crystals have been spotted under the SEM (Fig. 4.15a). The age obtained for this Mn oxide stands at 4.20 ± 0.33 Ma. Cryptomelane II taken at 76 m below ground shows a densely-packed network of cryptomelane with bundles of platy elongated crystals of cryptomelane II surrounding single prisms of manganese minerals forming faces with a tetragonal morphology like well-shaped crystals of pyrolusite. Cryptomelane II yielded an age of formation of 0.87 ± 0.04 Ma. Eh-pH diagrams of oxidizing Fe and Mn minerals in the weathering zone at Hagendorf, Germany were calculated for a temperature of

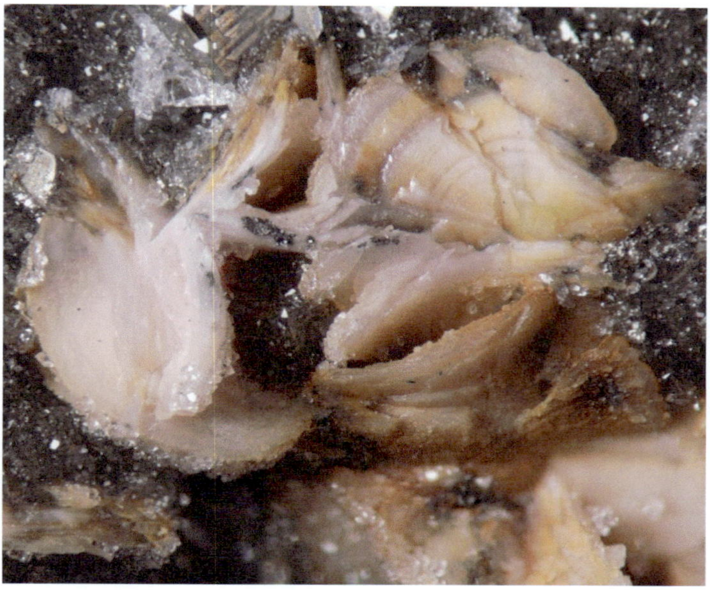

Fig. 4.15d Rhodochrosite-enriched carbonate (pinkish) piercing a matrix of Mn oxide-hydrates ("manganomelane"). Hagendorf-South

formation of 25 °C, using the dissolved species as log $_{aFe}=-6$ and as log $_{aMn}=-6$ (mol/lit). Iron and manganese oxide-hydrates are both stable under oxidizing conditions. To precipitate almost pure Mn minerals invokes a complete separation of Fe from Mn which may be achieved in the pH interval 6–10. While Fe is still fixed as oxide bearing Fe in its trivalent state (hematite, goethite), bivalent Mn is mobile and able to split apart and migrate away from the Fe-Mn system so as to enter Mn mineral species on increasing Eh- or pH values. The log f $_{O2}$ for pure cryptomelane was estimated to be at −17.8 to −18.2, pyrolusite intergrades in the cryptomelane network would signal a log f $_{O2}$ of roughly – 10.78 (Parc et al. 1982). At a depth of 87–76 m below ground there were moderately oxidizing and alkaline conditions. The separation of Fe and Mn within the mineralizing fluids took place within the percolation zone under slightly oxidizing conditions around pH=7 as phreatic conditions persisted, leaving behind Fe minerals containing trivalent iron (Dill et al. 2010a). Considering the splitting apart of Fe from Mn underneath in the infiltration zone at slightly reducing conditions and mildly acidic to neutral conditions has to invoke HPO_4^{++} as an additional constituent. In this case Mn-bearing phosphate is still stable whereas Fe is mobilized. Therefore I suggest Fe was mobilized in the infiltration zone as Fe-phosphate complex. On lowering the weathering or hydraulic front another cleansing of the system took place as Mn was mobilized and re-precipitated on increasing oxygen fugacity and more alkaline conditions. In conclusion, different cryptomelane age data found in the same weathering and paleohydraulic system resulted from continuous hydraulic processes in the weathering system rather than

Fig. 4.15e Intergrown tiny rhombs of yellow brown siderite resulting in these pinnacles. Hagendorf-South

from a stepwise uplift that shaped the landscape in the granitic area of the NE Bavarian Basement. Almost pure cryptomelane mineralization is indicative of the basal zone of supergene alteration, because a complete separation of Mn from Fe can favorably be achieved only in the infiltration zone under slightly reducing conditions or in the percolation zone under slightly oxidizing conditions.

In the pervasively kaolinized "New Aplite", at Pleystein, rhythmic layers made up of manganese, aluminum with traces of cobalt but with neither Si nor P present in this encrustation were interpreted as a member close to the todorokite series of oxidic Mn minerals.

A different story as to the siting of pegmatite minerals can be told for the rhodochrosite-enriched Fe-Mn carbonates, an example of which is illustrated in Fig. 4.15d. and the rare Mn silicate pyrosmalite (Table 4.1). They are representative of a topomineralic effect that is exclusive to the Hagendorf-South pegmatite, and not inherent to the emplacement of a rare-element pegmatite in general. The process may also be invoked to have taken place at Pleystein, yet its mineralogical imprint on the pegmatite has been eradicated almost to completeness by the pervasive chemical weathering and erosion, there. Only relics of siderite attest to this event.

For a genetic discussion, it does not make sense to discuss these carbonate s.s.s., characterized by its end members $FeCO_3$ (siderite) and $MnCO_3$ (rhodochrosite), under separate headings (Fig. 4.15e). The current debate of mineralogical results is based on the paper published by Mücke and Keck (2011) as far as the distribution of Fe-Mn carbonates at Hagendorf-South is concerned. The authors investigated 18 samples, 8 from the upper part of the pegmatite and 10 from the lower section of the Hagendorf-South pegmatite.

Fig. 4.15f Intergrowth of pyrite crystals giving rise to a face combination with hexahedron prevailing over the octahedron. Hagendorf-South

Fig. 4.15g Hematite and massive pyrite side-by-side. Hagendorf-South

The Fe-, Mn-, Ca-, Mg-, and Zn contents of the Fe-Mn carbonates s.s.s. of the upper pegmatite are as follows: Fe: 31.69–45.38 wt. % FeO, Mn: 14.87–29.53 wt. % MnO, Ca: 0.00–0.59 wt. % CaO, Mg: 0.05–0.86 wt. % MgO, Zn: 0.00–1.20 wt. % ZnO.

For the lower pegmatite equivalent data can be reported, yet they have to be split up into two different data arrays: I: Fe: 42.54–61.75 wt. % FeO, Mn: 0.76–18.04 wt. % MnO, Ca: 0.00–0.11 wt. % CaO, Mg: 0.00–0.23 wt. % MgO. Zinc could not be determined by means of EMP according to the authors.

Samples of the data array II have the following chemical composition: Fe: 1.17–7.14 wt. % FeO, Mn: 52.13–60.88 wt. % MnO, Ca: 0.17–1.73 wt. % CaO, Mg: 0.00–0.26 wt. % MgO, Zn: 0.00–0.06 wt. % ZnO.

The Fe-Mn carbonate s.s.s in the upper pegmatite are accompanied by Ca-Fe-Mn-Al phosphate minerals known to appear during the initial stages of pegmatite emplacement, such as apatite, triplite, zwieselite, hagendorfite, dickinsonite and arrojadite. The authors also mentioned eosphorite, goyazite and variscite.

The minerals associated with the Fe-Mn carbonates chemically represented by the array I of the lower pegmatite may be summarized as follows: zwieselite, pyrrhotite, viviante, ludlamite, pyrite, chalcopyrite, sphalerite, magnetite, stilpnomelane, pyrosmalite, hematite (Fig. 4.15f, g). Some of them may be well-known Fe-Mn phosphates, others like magnetite and pyrosmalite are "newcomers on the pegmatitic scene" and not typical of the pegmatites under study.

The minerals accompanying the Fe-Mn carbonates in the lower pegmatite and represented by the array II are rather simple as to the chemical composition but extraordinary as to the texture and can be characterized as following: "Muschketoffite" (magnetite pseudomorphosing hematite), hematite, quartz, apatite. These minerals have derived from the so-called "Muschketoffite Zone", a mineral association exclusive to the HPPP and quite exceptional for a rare-metal pegmatite like that.

As has been already mentioned above, it is a topomineralic process overprinting the formation of the pegmatite. A similar mineral association has been encountered in the Saxo-Thuringian Zone to the west of Hof (Dill 1985a, b). The various mineral assemblages there can be directly correlated with those discovered at Hagendorf-South.

In the mineral province west of Hof, vein-type deposit with fluorite, barite, siderite, and Cu-Bi-As-Zn-Pb sulfides have a precursor mineralization produced by a two-mica granite enriched in Sn and W the apical part of which was hit by drill holes at a depth of between 261 and 516 m below ground (von Horstig 1972). The oldest mineralization, named C1 by Dill (1985a, b) features skarn minerals, including scheelite, cassiterite, sphalerite, magnetite (locally present as "muschketoffite"), pyrrhotite, chalcopyrite, cubanite, mackinawite, hematite, and stilpnomelane. This C1 mineralization is correlative with the mineral association discovered in the lower pegmatite, what has been denominated as the "Muschketoffite Zone" at Hagendorf-South. The minerals were disseminated among these phosphates and penetrate in a stockwork-like way minerals at the boundary between the primary (e.g. zwieselite) and secondary phosphates (e.g. ludlamite, vivianite) which is also a kinematic boundary. The primary phosphates, in places, furnish evidence of structural disturbances, whereas the secondary phosphates do not show any evidence for such overprinting.

The upper pegmatite is significantly enriched in Mn relative to Fe, in Mg, Ca and Zn when compared with the data array I, a chemical scenario also known from the afore-mentioned reference mineral district near Hof, where siderite became gradually enriched in Mn towards younger mineralizing stages abundant in barite, fluorite and pyrolusite (reference stages: H, I, J, L according to Dill 1985a, b). In the upper pegmatite only primary Fe-Mn phosphates, e.g., triplite, zwieselite, hagendorfite, dickinsonite and arrojadite were altered by the Fe-, Mn- and HCO_3^- -bearing fluids. As a result of this mineralizing process an Al-rich mineralization with eosphorite, goyazite and variscite formed under a low pH from the preexisting phosphates, while separated by a sharp contact Mn-enriched siderite-rhodochrosite s.s.s. came into existence. This special mineralization is related to a late Variscan shallow two-mica granite stock bound to a deep-seated lineamentary fault zone. It furnish clear evidence, that the primary phosphate mineral assemblage predates this late Variscan heat event responsible for many of the above Fe-Mn minerals and that the late Paleozoic geomorphology (peneplain) and hydraulic system had already an impact on the initial stages of formation of the HPPP pegmatites. Age dating of U minerals provide another evidence that an early alteration of the late Variscan mineralization worked hand in hand with a progressive uplift of the basement and shaping of the landscape.

Chamosite, an Fe-enriched member of the chlorite group (variety thuringite) and berthierine, pertaining to the 7 Å serpentine or amesite subgroup, were only recorded from the pegmatite stock of Hagendorf-South, where they occur in densely packed aggregates of rosettes (Fig. 4.15k). This is also the case for stilpnomelane, which gives the name for a whole phyllosilicate group of its own. Chamosite, berthierine and stilpnomelane are present only at Hagendorf-South. The afore-mentioned Fe-bearing phyllosilicates are directly related to the topomineralic alteration referred to in the previous paragraph and correlated with the latest Variscan thermal event along the western edge of the Bohemian Massif. The question why these alteration is confined to the stocklike pegmatite at Hagendorf only and has no equivalent Fe-Mn mineralization in some of the tabular pegmatites and aplites may be given by

Fig. 4.15h Pyrrhotite in massive aggregates. Hagendorf-South

Fig. 4.15i Pyrite pseudomorphosed by goethite ("limonite") in a matrix of acicular rockbridgeite. Hagendorf-South

Fig. 4.15j Marcasite aggregates from Waidhaus Silbergrube Mine

a quick look at Fig. 5.5d that discloses the relative depth zonation of tabular and stock-like pegmatites in the host anticlines.

Goethite, lepidocrocite, hematite are a "jack of all trade". The occur as single monomineralic minerals or in aggregates with the precise chemical composition of the individual Fe oxide-hydrates left unidentified and, hence, given the common term "limonite". The Fe oxides occur predominantly in the topmost zone of the HPPP pegmatites and aplites as they were affected by chemical weathering or

Fig. 4.15k Ferroan variety of chamosite, called thuringite from Hagendorf-South

Fig. 4.15l Diadochite filling cavities of a phoscrete in the Trutzhofmühle Aploid

hydrothermal processes along fractures and joints (Fig. 4.15f, i). The presence of pyrite is often hard to pigeonhole to a certain stage as it spreads across the primary phosphate mineralization into the early stages where hydrated phosphates, such as ludlamite or vivianite precipitated under Eh < 0 (Fig. 4.15e, h). As the Eh starts to rise and rockbridgeite s.s.s began to appear pyrite hexahedra get pseudomorphosed by "limonite" (Fig. 4.15i). Pyrrhotite is restricted to the primary mineralization at Hagendorf-South (Fig. 4.15h), where mackinawite rounds off the series of Fe minerals. Marcasite, albeit pertaining to the series of Fe sulfide is encoun-

Fig. 4.15m Spray of gypsum needles in a solution cavity of the Kreuzberg Pegmatite at Pleystein.

tered as "Speerkies" (= "spearhead pyrite") rather late at the passage from the hypogene into the supergene mineralization (Fig. 4.15j). The existence of marcasite attests to reducing conditions with pH values fluctuating around 6 (Murowchick and Barnes 1986).

Iron sulfides are known to be very vulnerable to supergene alteration resultant in a wide range of sulfates by the oxidation of iron and sulfur in their parent sulfides. Although far from being an Fe-enriched deposits effervescences/sulcretes and cavity fillings with prevailing S-bearing minerals may be encountered in the run-off mine samples as well as on the dumps, where they came into being under the influence of the modern-day atmospheric impact. Diadochite acts as a link between the great variety of phosphates and the sulfates derived from the decomposition of predominantly Fe sulfides (Fig. 4.15l). The decomposition of Fe-bearing sulfides in an environment enriched in phosphate under oxidizing conditions can be described as follows:

$$FeS_2 + 3.75\,O_2 + 3.5\,H_2O \Rightarrow Fe(OH)_3 + 2\,SO_4^{2-} + 4\,H^+$$

Fe sulfide was oxidized, and the resultant sulfate reacted with apatite to form diadochite and gypsum, which was washed out from the soil (Fig. 4.15m).

Fig. 4.15n Fe-enriched black sphalerite from the Hagendorf-South Pegmatite in a matrix of muscovite. Cracks intersecting the Zn sulfide are filled with globules of green sulphur

$$FeS_2 + 3.75\,O_2 + 3.5\,H_2O \Rightarrow Fe(OH)_3 + 2\,SO_4^{\;2-} + 4\,H^+$$

$$Ca_3(PO_4)_3(OH) + 8\,SO_4^{\;2-} + 6\,Fe^{3+} + 30\,H_2O \Rightarrow 5\,CaSO_4.2\,H_2O + 3\,Fe_2(SO_4)(PO_4)(OH).6H_20 + 6\,H^+$$

Two sites can be cited to be favorable for the formation of diadochite. It is the soils on top of the pegmatites and mining induced processes responsible for the production of acid mine drainage (AMD) (Peacor et al. 1999). Predominantly under semiarid climatic conditions this sulfate-phosphate finds favorable conditions to precipitate, given a sufficiently high activity of HPO_4^{--} (García-Lorenzo et al. 2012). Another hydrated Fe-Al sulfate observed on tailings impoundment as a result of AMD is jarosite which was spotted at Hagendorf-South. Bigham et al. (1996) found jarosite at pH 2.6 in ochre samples from AMD. Its emplacement may be described by the reaction:

$$6\,FeOOH + 2\,K^+ + 2\,H^+ + 4\,HSO_4^{\;-} \Rightarrow 2\,KFe_3(SO_4)_2(OH)_6$$

Rozenite is an intermediate product in this anthropogenic cycle of mineralization which indicates a rather low pH around 4. Under the current climatic conditions it

Fig. 4.16a Isometric uncorroded crystal of arsenopyrite surrounded in a quartzose matrix. Hagendorf-South

would not form in nature. Sulfur is a rare constituent and its origin difficult to account for. As illustrated in Fig. 4.15n, native sulfur "droplets" fill fissures in primary sulfides, such as Fe-bearing sphalerite. All the reactants H_2S, CO_2 and CH_4 necessary to kickstart a thermochemical sulfate reduction resulting in the formation of native sulfur are present at Hagendorf-South. While the siting of the sulfur next to sulfides does not cause any problems, its temporal siting within the mineral succession is fraught with uncertainties and has still to be question marked. Considering the redox conditions, a thermochemical sulfate reduction could have taken place during the waning stages of the primary pegmatite evolution, as the Eh slightly increased and another reducing environment heralded the beginning of the secondary mineralization. At Miesbrunn native sulfur was found together with hydrated Bi sulfate. It is one of the youngest minerals there and evidence of a stagnant water level in what might be called a hydromorphic soil.

4.16 Arsenic Minerals

Arsenopyrite is the sole primary arsenide found in the pegmatites and aplites of the HPPP. Its position and outward appearance in the pegmatite stocks significantly differs from those of the tabular aplites and pegmatites, most notably in the Miesbrunn pegmatite-aplite swarm (Dill et al. 2012a) (Fig. 4.16a, b). At Hagendorf-South arsenopyrite developed well-shaped isometric crystals disseminated in a quartzose matrix. While it is uncorroded at Hagendorf, at Miesbrunn, arsenopyrite is corroded and oxidized almost completely to scorodite and displays some mimetic crystallization at the edge of the individual aplites and pegmatites, following the folded and kinkbanded muscovites. It is another example of structurally-related crystallization

Fig. 4.16b Anhedral aggregates of arsenopyrite (*as*) are aligned along the deformed flakes of muscovite (*ms*) in form of a mimetic crystallization. Arsenopyrite in this site suffered strongly from the supergene alteration resulting in the formation of scorodite (*sc*). Miesbrunn (SEM)

of a pegmatite mineral. There is no longer any doubt that the tabular felsic mobilizates intercalated in the metapelitic rocks are chemically and morphologically controlled by the enclosing country rocks. Similar to pyrite, arsenopyrite from Miesbrunn is impoverished in minor elements and, hence, may also be called barren as to "invisible gold", an element quite common along the western edge of the Bohemian Massif (Morávek and Pouba 1990; Morávek and Lehrberger 1997). Stratiform Au-As-Bi mineral associations were found in cordierite-sillimanite gneisses in the close vicinity of these pegmatites (Lehrberger et al. 1990). The regional metamorphism affecting the cordierite-sillimanite gneisses reached its peak temperature at about 600 °C accompanied by strong shearing and folding. Minerals such as loellingite and maldonite attest to a S-deficient environment of deposition and account for why arsenopyrites is of so widespread occurrence within the tabular pegmatites and aplites. They did not bring about any of these sulfur-deficient minerals and underscore that there were two discrete sulfur systems which did not correspond with each other at the time of the emplacement of the HPPP's pegmatites.

The trace element content of the pyrite-arsenopyrite association (due to the intimate intergrowth no precise data can be given for the individual sulfides) is as follows. The contents of In stand at 33 ppm, of Ag at 3 ppm, of Cd at 1 ppm and of Ga at 13 ppm.

A closer look at the granites and granite pegmatites of the Fichtelgebirge, in the Saxo-Thuringian Zone, immediately to the north of the study area reveals that arsenic

Fig. 4.16c Coatings of scorodite (SEM). The *inset* gives a scorodite crystal viewed under the stereomicroscope whose morphological type belongs to type 2 in Fig. 4.16e. Miesbrunn

Fig. 4.16d Aggregates of phosphoscorodite (SEM). Miesbrunn

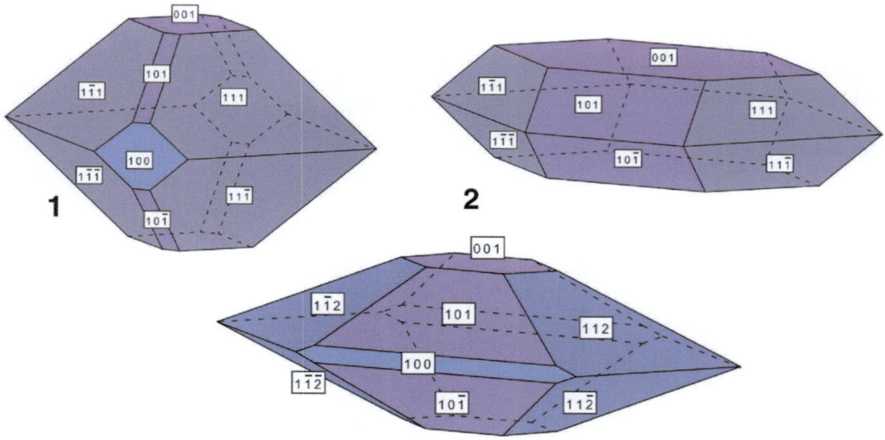

Fig. 4.16e Series of cartoons showing the morphological changes within the scorodite-strengite s.s.s. The three types referred to in the text are labeled with Arabic numerals. Crystal morphology of phosphoscorodite and scorodite (relative size of faces given by the Miller's indices) *1*: isometric phosphoscorodite {111} > {001}, *2*: tabular phosphoscorodite {001} > {111} ≈ {101}, *3*: rhombohedral scorodite {101} ≈ {112} > {001}. The phosphate content decreases from type 1 through type 3

is a common element there and accountable for a variegated mineral assemblage encompassing not only arsenopyrite, but also a varied spectrum of secondary minerals, some of which are also present in the HPPP. In the Fichtelgebirge granites of the Rudolfstein (G4) and the Waldstein (G3) massifs contain arseniosiderite, chalcophyllite, chenevixite, olivenite, pharmacosiderite, scorodite and zeunerite, the As-enriched analogue of torbernite. They are rather widespread an known to exceed locally the number of phosphates. Arsenic-bearing sulfides in NW- and NNW-trending mineralized structure zones are common to the Fichtelgebirge Anticline. Arsenopyrite, locally, abundant in pegmatites and aplites of the HPPP, e.g.,. the Miesbrunn pegmatites-aplite dyke swarm can thus be interpreted in terms of a offspring of the younger Saxo-Thuringian mineralizing event "invading" into the Moldanubian Region.

Arsenopyrite at Miesbrunn has been converted during the Neogene chemical weathering into scorodite and phosphoscorodite, containing up to 7 wt. % P_2O_5 in its structure and being present often in well-crystallized rhomb-shaped and tabular crystals (Fig. 4.16c). The lowest phosphate content determined in these scorodite-strengite s.s.s. lies around 1 wt. % P_2O_5 (Fig. 4.16d). Along with decreasing phosphate contents within the scorodite crystals, the morphology of the Fe arsenates changes when its {112} and {101} faces becoming more pronounced (Fig. 4.16e). The Fe arsenates form the cement of a fitting breccia at Miesbrunn. Neither the fragments of arsenopyrite nor the quartz veinlets intersecting the FeAsS are rotated against each other or were displaced along these stockwork-like mineralization. Phosphoscorodite is a solid solution series between scorodite and strengite, a structural change also reflected by a change in the crystal morphology. As the phosphate content increases, scorodite s.s.s. tends to take on a morphology typical of strengite. In the Eh-pH diagram a temperature of 40 °C was assumed as it well represents the

Fig. 4.17a Cobweb of acicular bismuthinite crystals enmeshing chalcopyrite (*cp*) and pyrite (*top margin* of image) from Hagendorf-South

maximum temperature to be reached during tropical climatic conditions of the late Cretaceous and the Cenozoic in this part of the Bohemian Massif. The temperature rate has been adjusted towards higher temperatures taking into consideration a slight increase by the oxidation of sulfides which cannot be neglected. Phosphoscorodite developed under strongly acidic conditions at a pH of less than 5 from strengite in the Eh range 0.25–0.75 (volt). More oxidizing conditions or slight changes of the pH towards more neutral conditions as it was the case during the more recent parts of the geological history rendered scorodite *sensu stricto* to become the more stable phase within this phosphate-arsenate system. This physico-chemical regime also allowed beudantite to precipitate. At Pleystein another arsenate, pharmacosiderite formed part of the supergene alteration.

4.17 Bismuth Minerals

In many of the tabular pegmatites around Miesbrunn, arsenopyrite can be encountered apart from pyrite – see Sect. 4.16- but genetically associated with bismuthinite, a Bi sulfide recorded from the Hagendorf-South and -North pegmatite stocks too (Fig. 4.17a). At Hagendorf Bi sulfide is intimately intergrown with chalcopyrite and pyrite (Fig. 4.17a). Further associates of bismuthinite are cassiterite, emplectite and native bismuth (Fig. 4.17b). Apart from the crystals of native bismuth the Cu-Bi-Sn mineral association is only present on a microscopic scale and the precise distinction of the Bi sulfides can only been achieved using the ore microscope or the scanning electron microscope. This is especially true for the Cu-Bi sulfides; not of all them have been spotted by the author, so that for some of these minerals their siting in time and space cannot be done properly, e.g., pavonite and wittichenite

Fig. 4.17b Native bismuth from Hagendorf-South

Fig. 4.17c Cuprobismutite crystal at Hagendorf-South (SEM)

reported by Rabe (1975) and listed in Kastning and Schlüter (1994). The minerals above infiltrate the quartz-feldspar-phosphate matrix along grain boundaries or form tiny veinlets. Rarely they form more massive patches in the early mineralization of Hagendorf-South. Cuprobismutite, locally, bearing silver, emplectite, and most likely pavonite and wittichenite are supposed to pertain to a younger mineralization relative to the bismuth-bismutinite mineralization, irrespective of being associated with arsenopyrite or cassiterite (Fig. 4.17c). The Bi-As mineral assemblage in the HPPP came into being at a rather high temperature. Quartz associated with the As-Bi sulfides at Miesbrunn developed at temperatures around 380 °C, slightly above the boundary between the sub- and supercritical state of water (Dill

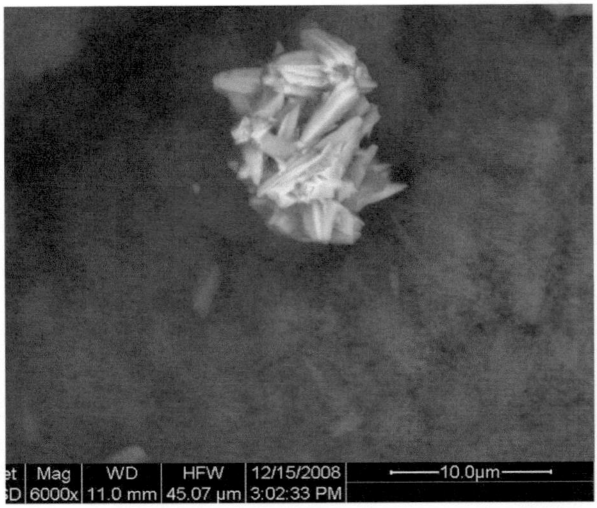

Fig. 4.17d Aggregate of bismite (Bi ochre) on zincian strengite at Pleystein (SEM)

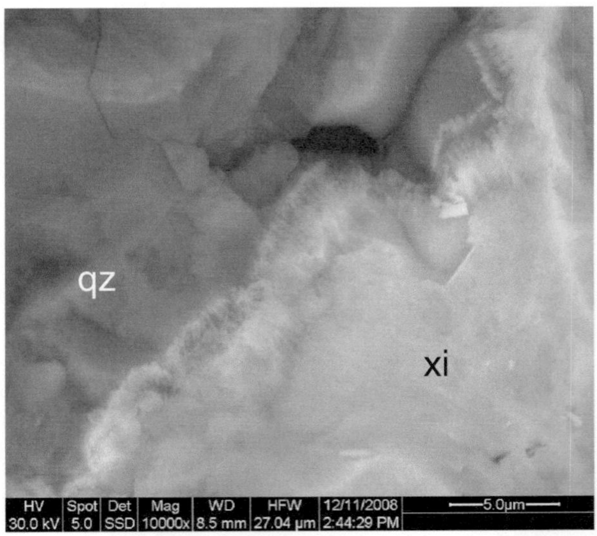

Fig. 4.17e Ximengite (?) (*xi*) on quartz (*qz*) at Pleystein (SEM)

et al. 2012a, b). Cu-Bi sulfides and bismuthinite plus native Bi are considered to have formed in response to hydrothermal activity associated with tectonic processes at temperatures between 90 °C and 110 °C (Dill et al. 2013a). This alteration only plays a minor part in the entire pegmatite evolution, judging by the amount of Bi-Cu sulfides relative to the overall pegmatite minerals (Dill et al. 2013b). Bismuth sulfides are converted into a varied spectrum of oxidic Bi minerals in the course of supergene alteration at Miesbrunn and to a larger extent at Pleystein.

Bismuthinite is the source of a series of Bi minerals, including Bi ochre or bismite, a bismuth oxide mineral close to bismuth trioxide, and also ximengite which cannot be proved yet with certainty (Fig. 4.17d, e). By means of SEM, the

Fig. 4.17f Waylandite from the Kreuzberg pegmatite at Pleystein

most common Bi mineral in the kaolinitic regolith on the Kreuzberg and to a larger extent on top of the "New Aplite" can be identified. It is the waylandite-zairite s.s.s. which was known from various pegmatites but has so far not discovered in the HPPP (von Knorring and Mrose 1963; Baldwin et al. 2000; Novák et al. 2001a, b) (Fig. 4.17f). There is waylandite with as much as 0.58 wt % Fe. Zairite [Bi(Fe,Al)$_3$ (PO$_4$)$_2$(OH)$_6$], the ferric analogue of waylandite has Fe contents ranging from 10.5 to 17.3 wt. % Fe. Petitjeanite does not develop well-shaped pseudohedral to -octahedral crystal like waylandite and appeared only as earthy coating under the SEM in specimens from Pleystein. REE-bearing APS minerals were not been identified in the kaolinitic regolith on top of the pegmatites at Pleystein. Not so at Plössberg, where one of the most recent phosphates is florencite (1.4 at. % Ce, 2.2 at. % La). In these earthy aggregates considerable amounts of bismuth (8.7 wt. % Bi) were determined. Due to the minute grain size and the intimate intergrowth of the mineral grains the true nature of the Bi host cannot be proven. Considering the wide range of elements to be incorporated into APS minerals it is therefore logical to assume in the present situation a florencite-waylandite s.s.s. Among the aluminum phosphates waylandite-gorceixite s.s.s. and waylandite-zairite s.s.s. play an exceptional role. Even if the parent sulfides have completely been eradicated by the alteration processes, at the very end under supergene conditions, their former presence may still be proved by the Bi-Al phosphates of their own or as a members in one of the APS s.s.s. Aside of this conclusion, Bi is considered as rather immobile during kaolinization and being retained in the topmost parts of the kaolinized aplites and pegmatites. Whatever APS s.s.s. you might look at, waylandite-gorceixite s.s.s or waylandite-zairite s.s.s. the physical-chemical regime may be characterized by the log activities as follows: $HPO_4^{2-} = -5$, pH 2–5, $SiO_2(aq) < 1.5$. Riomarinaite, a Bi bearing sulfate, was recorded from the Miesbrunn area. Owing to its fine-grained nature the presence of riomarinaite can only be claimed by SEM-EDX [Bi(OH) SO$_4$•H$_2$O]. It is the youngest mineral and found together with native sulfur.

Fig. 4.18a Relics of chalcopyrite (*cp*) *left over* after its replacement by rockbridgeite (*ro*) in a matrix of quartz (*qz*). The *arrowheads* denote the cracks along which chalcopyrite was replaced by the secondary phosphate

4.18 Copper Minerals

Glancing at the list of minerals in Table 4.1 might raise undue expectations as to the significance of Cu accumulation during the evolution of the pegmatites of the HPPP. The majority of Cu-bearing minerals is restricted to the pegmatite Hagendorf-South, where some of them may gain showcase quality but still are negligent by quantity. This valid for most of the pegmatites and aplites of the HPPP which are no real match to the Hagendorf-South Pegmatite in terms of the quality of Cu minerals. Only chalcopyrite is a mineral that is of rather widespread occurrence and so ranks first among the Cu hosts, as one may deduce from the discussion of minerals ahead (Figs. 4.17a and 4.18a, Sects. 4.15 and 4.17). The Cu-bearing minerals may be subdivided into four different groups: (1) primary Cu sulfides, (2) Cu minerals of the phreatic hydraulic zone, (3) Cu minerals of the vadose hydraulic zone, (4) Cu "minerals" produced by post-mining alteration. Inverted commas are justified to classify number 4 chemical compounds since not all geoscientists rank these man-made products among the minerals, proper.

The Cu sulfides cubanite and valleriite were listed by Strunz et al. (1975) without further notice as to their precise genetic position. With regard to the common chalcopyrite, they are interpreted as high-temperature relicts. Cubanite has two polymorphs Cubanite undergoes an irreversible phase transition to isocubanite at 210 °C (Berger et al. 2012). Below 210 °C isocubanite does not revert to cubanite but exsolves chalcopyrite and pyrrhotite, which both are present among the primary Fe-Cu sulfides at Hagendorf-South. Valleriite is a common phase of high-T ores. Iiishi et al. (1970) synthesized valleriite at T values as high as 700 °C, a temperature that opens the way for the sulfide up into the incipient pegmatitic stages. Valleriite occurs in different host rocks among others migmatites which used to spawn gra-

Fig. 4.18b Dark blue massive covellite from Hagendorf-South

Fig. 4.18c Native copper from Hagendorf-South

nitic melts (Talapatra 1968). Both Cu sulfides hallmark a small but significant Cu concentration for genetic interpretation during the early stages of the pegmatite emplacement. Chalcopyrite formed later, but temporarily not far from the secondary phosphates. The most common one, rockbridgeite, replaced chalcopyrite along hydraulic cracks and helps constrain the temporal position of this ubiquitous Cu sulfide among the mineral succession of Hagendorf-South (Fig. 4.18a).

Digenite, chalcocite, djurleite, covellite and native copper are minerals of the phreatic hydraulic zone underneath the oxidized part of the gossan (Fig. 4.18b, c). Age dating of cryptomelane, discussed in Sect. 4.15, has already unraveled that the supergene alteration in the pegmatite is rather complex and sharp hydraulic levels,

Fig. 4.18d Tiny globules of libethenite residing on chrysocolla at Hagendorf-South

Fig. 4.18e Malachite in cavities between quartz grains at Hagendorf-South

as they were known from vein-type or stratiform base metal and uranium deposits are not to be expected in these intrusive stocks.

Cuprite, malachite, rosasite, chrysocolla, libethenite, pseudomalachite, turquoise and chalcosiderite formed part of a mineral assemblage which is anything but the oxidized pendant of the primary Cu minerals mentioned in the previous paragraph (Fig. 4.18d, e, f, i). It is restricted to the vadose hydraulic zone, where cuprite used to bridge the gap to the underlying zone. According to the classification scheme established by Dill et al. (2013c), the orecretes-(Cu) consist of silcretes-(Cu) (chrysocolla), carbocretes-(Cu) (malachite, rosasite) and phoscretes-(Cu) (libethenite, pseudomalachite, turquoise and chalcosiderite) (Fig. 4.18m, n). Bearing in mind,

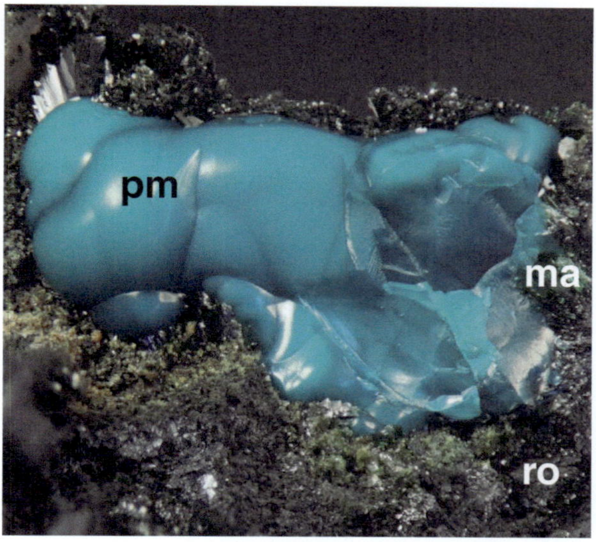

Fig. 4.18f Pseudomalachite (*ps*) and malachite (*ma*) on rockbridgeite (*ro*) at Hagendorf-South

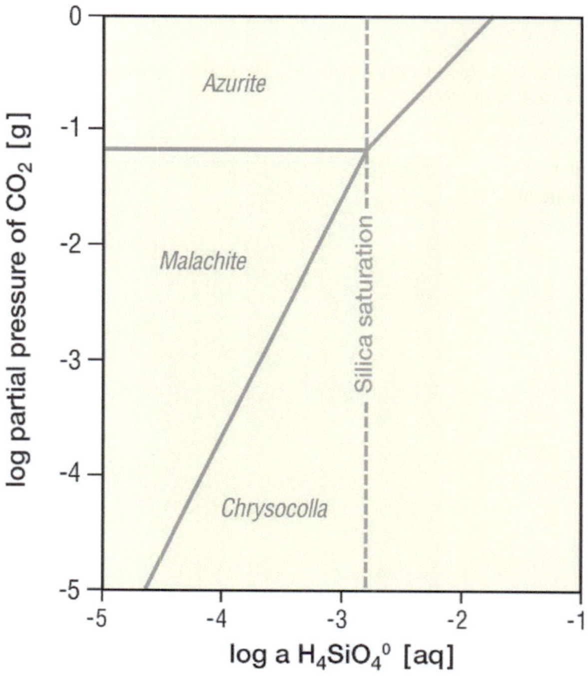

Fig. 4.18g Stability plots of chrysocolla relative to Cu carbonates at 25 °C after Crane et al. (2001)

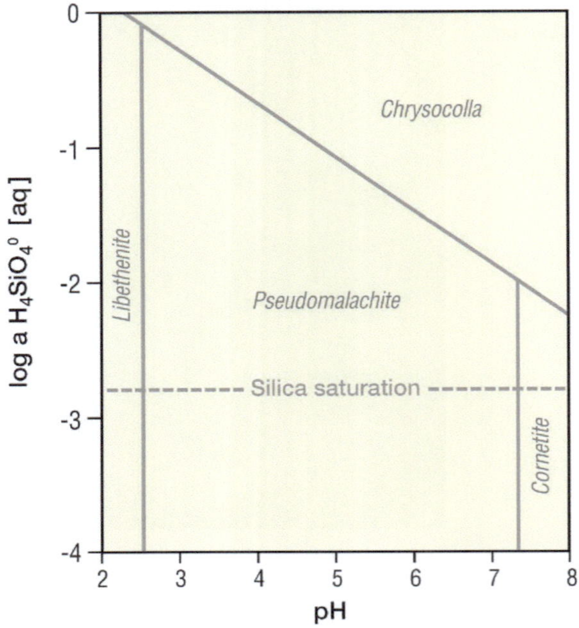

Fig. 4.18h The stability fields of chrysocolla vs. libethenite, pseudomalachite, and cornetite at 25 °C (Data from Crane et al. 2001)

Fig. 4.18i Globular aggregates of rosasite at Hagendorf-South

that the supergene alteration took place during the Plio-Pleistocene, we can assume a tropical wet-dry paleoclimate for this period of time at the paleolatitude and -longitude of the HPPP. The physico-chemical regime for the environment of deposition has been modeled for malachite, chrysocolla, libethenite, and pseudomalachite according to Crane et al. (2001) (Fig. 4.18d, e, f, g, h). The stability field of

Fig. 4.18j Sprays of lavender blue devilline needles Hagendorf-South

malachite stretches at 30 °C the pH range from 5.5 to 7.5. Cuprite which is also present in the mineral association is stable over a very wide pH range down to pH=4. Chrysocolla is not pH-sensitive as shown by Fig. 4.18h relative to Cu phosphates. At log $H_4SiO_4^o = -4$ the log partial pressure of CO_2 (g) has to exceed −4 to produce malachite, which responds rapidly to changes of the log P_{CO2} (Fig. 4.18g). The heterogeneity of the parent material as to the distribution of silicates, phosphates and Cu sulfides in the pegmatite rules out encrustations to develop at a certain depth or coherent orecretes as they may be met under near-ambient conditions in more homogeneous parent lithologies.

Brochantite, chalcantite, connellite, devilline, langite and posnjakite are post-mining mineralizations which sometimes may also appear on dumps (Fig. 4.18j, k, l, o). These sulfates -sulcretes-(Cu) – are younger than the afore-mentioned orecretes-(Cu) and rarely form under the current regime of chemical weathering. Brochantite whose stability field is not very much different from that of the Cu carbonates does not appear in Eh-pH diagrams until the sulfur concentration is as high as 10^{-1} M (Zhang 1994). A constant supply of sulfur can be accomplished in the natural aqueous solutions under hyperarid climatic conditions a physical situation which was neither realized during the Neogene, nor during the Quaternary. At 25 °C, brochantite is thermodynamically stable over posnjakite which formed first. Wroewolfeite is converted into posnjakite and brochantite (Dabinett et al. 2008). Therefore sulcretes-(Cu) used to evolve under a high- sulfidation regime only, which can, locally, be achieved in the pegmatite. Connellite a chlorine-bearing sulfate (halcrete plus sulcrete) is confined in nature to morphoclimatic conditions as they exist in the desert, e.g., Atacama Desert, Chile. They do not form under temperate climatic conditions where these Cu-bearing effervescence are dissolved on the spot by the strong annual rainfall. Dry conditions as in abandoned shafts and galleries, however, offer favorable conditions for this type of Cu mineral to grow and get preserved.

Fig. 4.18k Langite triplets in growing within solution cavities supplemented by a cartoon to show the ideal crystal morphology of the single crystals (*left*) and the triplets (*right*)

Fig. 4.18l Myriads of chalcosiderite crystals from Hagendorf-South

Fig. 4.18m Blue turquoise draped around columbite-(Fe) which is left uncorroded during this supergene alteration at Hagendorf-South

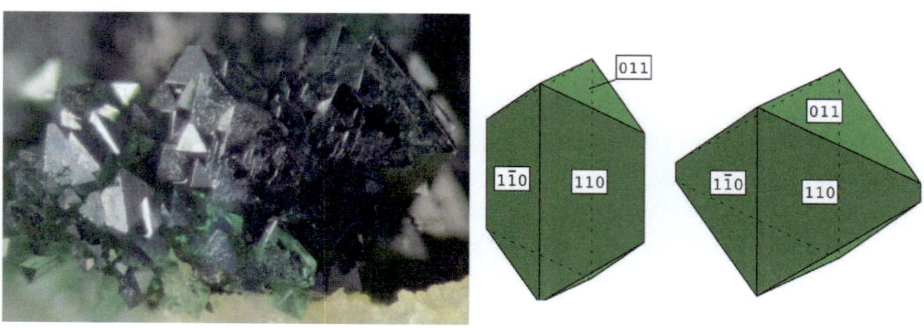

Fig. 4.18n Pseudooctahedral libethenite with two cartoons showing its ideal crystal morphology at Hagendorf-South

Fig. 4.18o Globular aggregates of connellite typical of the post-mining mineral association at Hagendorf-South

4.19 Halides

It is no exaggeration to state that fluorite takes a special position as some sort of a type mineral of the western edge of the Bohemian Massif, particularly of the Oberpfalz, where fluorite was mined near Regensburg, Lam, and Nittenau (Dill 1985a, b; Dill et al. 2011b). The Nabburg-Wölsendorf mining district in NE Bavaria, Germany, belongs to the most famous mining sites in terms of CaF_2 production, accounting for more than 10 % of the world fluorite production until 1987, when mining came to a halt (Bayerisches Staatsministerium für Wirtschaft und Verkehr 1978). This mining district is renowned for its colorful and morphologically diverse of fluorites (Dill and Weber 2010b, c). The most well-known of these fluorites is the black fetid fluorite (German: Stinkspat) with its pungent odor resulting from the "mineral" native fluorine which was only recently identified by Kraus (2014) but has obviously not yet gained mineral status. The lithochemical variation of fluoride shows a strong anomaly in the southeastern part of the NE Bavarian Basement (Fig. 3.3). All the more surprising, then, that there is no fluorite concentration in the HPPP, where fluorite was sporadically encountered proximal to altered zwieselite and in the primary sulfide-phosphate mineral associations at Hagendorf-South only (Keck 2001). The age of formation of the 1st generation fluorite at Nabburg-Wölsendorf which is concomitant with that of uraninite belongs to the late Variscan mineralization.

Another couple of halides or, more precisely described as aluminum fluoride hydrates, carlhintzeite and pachnolite are more widespread. Fluorine is substituting for the hydroxyl groups in various mica, it is common to the primary Fe-Mn phos-

Fig. 4.19a Carlhintzeite in a cavity of the Kreuzberg Pegmatite, Pleystein

Fig. 4.19b Pachnolite from Hagendorf-South

phates and found in manganiferous apatite too (Fig. 4.19a, b). As these minerals get decomposed and Al^{3+}, Na^{+} and Ca^{2+} made available by the concomitant destruction of alumosilicates the recorded halides come into being. It is mainly zwieselite that suffered from alteration into halides under a late hydrothermal regime. The redeposition of fluorine at Hagendorf is a self-contained process within the pegmatite stocks and obviously needs no external source for supply.

Lottermoser and Lu (1997) claimed for the existence of fluorite and pachnolite in a series of other secondary phosphates a low-temperature environment (<250 °C), high activity of HF, and a high Al mobility at a low pH in the pegmatites of the Olary Block, South Australia.

4.20 Lithium Minerals

All lithium minerals, excluding tainiolite a Li silicate, belong to the group of phosphates and they are confined to the three major pegmatite stocks, Hagendorf-South, Hagendorf-North and Pleystein only (Table 4.1). Tainiolite is a rare phyllosilicate of the biotite-group which was reported by Mücke (1977) who listed all those minerals known at that time. Others than in the northern Saxo-Thuringian zone, where lepidolite and polylithionite are widespread in the granitic pegmatites and Zinnwald became the namesake or *locus typicus* of zinnwaldite, lithium mica is a rarity in this pegmatite province dominated by phosphates. True Li silicates, such as spodumene and Li-bearing tourmaline are absent from the mineral associations of the HPPP.

The well-known transformation from triphylite via ferrisicklerite into heterosite has already been described in Sect. 4.13 and it is illustrated here only by a series of hand specimens in Fig. 4.20a. Tabular crystals of ferrisicklerite were also observed lining the walls of fractures and cavities at Hagendorf-South (Fig. 4.20b). Sicklerite is only known from the Hühnerkobel Pegmatite to the south of the HPPP. Exsolution textures, such as lamellar sarcopside in triphylite were not reported from the HPPP. Further replacement textures and an estimation of the temperature of formation has been given in Sect. 4.13. Triphylite one of the early primary phosphates in the HPPP also undergoes transformations different from those described by the Quensel-Mason Sequence resultant in the formation of, e.g., ludlamite, vivianite, barbosalite or the Li-bearing phosphate tavorite. The mentioned alteration processes went along with dissolution and/or fracturing creating the accommodation space necessary for the precipitation of these secondary minerals (Fig. 4.20c).

Taking into consideration the primary lithium minerals in the adjacent region to the NW of the HPPP, the following succession may be established from the root zone of thrustal plane to the frontal part of the collision zone in the Fichtelgebirge-Erzgebirge Anticline: Li-Fe-Mn phosphates ⟹ Li-Al phosphates ⟹ Li silicates (Li mica). Temperature and a striking preponderance of fluorine and water obviously impeded spodumene to develop at the end of this collisionally driven mineralizing process. These conditions were only realized at the opposite end of the Bohemian Massif, now reactivated within the Alpine Mountain Range.

Fig. 4.20 (a) Sequence of specimens to illustrate the replacement/oxidation of samples from Hagendorf-South: Triphylite plus K feldspar and apatite ⇒ ferrisicklerite ⇒ heterosite plus ferrisicklerite (b) Ferrisicklerite crystals lining a fracture (Hagendorf-South) (c) Bright grey tavorite on triphylite (Hagendorf-South)

4.21 Rare Earth Element Minerals

Rare earth elements (REE) constitute only a small percentage of the pie-chart diagrams of Fig. 2.5a and REE minerals are not expected to make up a great deal of the mineral assemblages in the HPPP pegmatites and aplites. Many of these phases are very small and the interrelationship with their neighboring minerals can only be disclosed by means of SEM or EMP. Xenotime, which is the only host of HREE in the HPPP is mainly concentrated in the tabular pegmatites and aplites, being located in the NNE part of the HPPP (Table 4.1, Fig. 4.21a, b). In the Hagendorf-South pegmatite stock it is associated with the primary phosphates, such as triplite, white mica, quartz and K feldspar as well as with secondary phosphates, such as

Fig. 4.21a Xenotime (white) in well-crystallized aggregates of leucophosphite from the 60 m-level of Hagendorf South. SEM

Fig. 4.21b Xenotime (white) surrounded by triplite at Hagendorf South. SEM

leucophosphite (Fig. 4.21a, b). The REE phosphate xenotime was never discovered within the pegmatites or aplites of the Miesbrunn pegmatites and aplites but it is characteristic for the contact zone to the garnet-bearing cordierite-sillimanite-biotite gneisses. Therefore the garnet-bearing cordierite-sillimanite-biotite gneisses of the wall rocks which also contains some xenotime show a slight increase in the HREE. In the Reinhardsrieth xenotime, formed during the initial stages of the mineral succession. Its amount tend to decrease from the gneiss towards the aplite, which is strikingly evidenced by the REE and Y variations.

Monazite and to a lesser amount xenotime grains were detected in the gneisses and aplitic gneisses but only sporadically observed within the aplite, proper, where a mineral of the cheralite group formed in their place (Fig. 4.21c). Its REE contents also are very low while the amounts of Th (3.79–5.00 at % Th) and U (1.09–1.45 at % U) are considerably high. Another couple of LREE-bearing minerals cerite and ferriallanite has been discovered in the contact zone of the Miesbrunn pegmatite swarm (Fig. 4.21d, e). Allanite-(Ce) is only known from the southern Bayerischer-Böhmer Wald, from the Tittling granite and from Brünst. There is no question that the REE content of the tabular pegmatites and aplites has derived from the surrounding country rocks. While in some cases the REE phosphates were incorporated into the felsic mobilizates, the LREE silicates have newly formed in the contact zone of the tabular pegmatites and aplites. Similar to zircon, where two basic types are diagnostic of the pegmatites, the crystal growth of monazite is also controlled by the environment of formation (Fig. 4.21f). Monazite which originated from granites, e.g., Tittling, developed well-shaped stubby crystals. The {100} and {101} faces are very well expressed. This is also the case at Miesbrunn at the gneiss-aplite contact with the trend of both faces to get reduced in favor of {11$^-$1}. How can we explain the presence of a granitic morphology of monazite in the contact zone of an aplite intercalated into biotite-cordierite-sillimanite gneisses ? Further information derived from the variation of the REE normalized to PAAS furnished evidence that these tabular aplites saw some granitization process prior to the emplacement of the aplite. In other words, the morphology of monazite in the gneisses of the HPPP and the REE_{tot} contents are neither a mirror image of the regional metamorphism nor an expression of the original heavy minerals of the psammopelitic parent rocks which the metasediments have derived from. Monazite crystals which formed *in-situ* during the initial stages of the emplacement of aplites, e.g., Trutzhofmühle and pegmatites such as Hagendorf-South show a distorted morphology and elongated crystals, thereby providing a tool to decide whether the minerals formed inside the felsic mobilizate or were incorporated into it from the country rocks (Fig. 4.21f). In the Fichtelgebirge granites, proximal to the collision zone, primary REE phosphate such as those described above from the Moldanubian Zone are missing. Instead of the REE-bearing phosphates a sporadic find of parisite-(Ce) $[Ca(Ce,La)_2(CO_3)_3F_2]$ in the Waldstein Massif furnish evidence of a more fluorine and hydrocarbonate-enriched environment as the REE were accumulated there. As the primary REE hosts were subjected to chemical weathering phosphate plays a more significant part and rhabdophane-(Ce) as well as churchite-(Y) appear in the saprock. Only the latter has an equivalent within the HPPP (Fig. 4.21g).

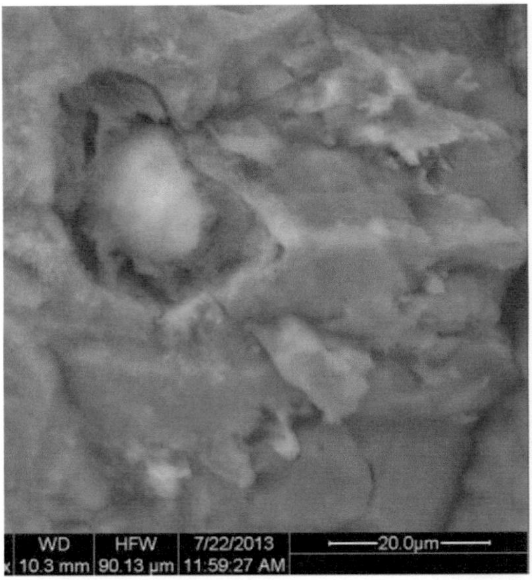

Fig. 4.21c Cheralite-group mineral in the Reinhardsrieth aplite. SEM

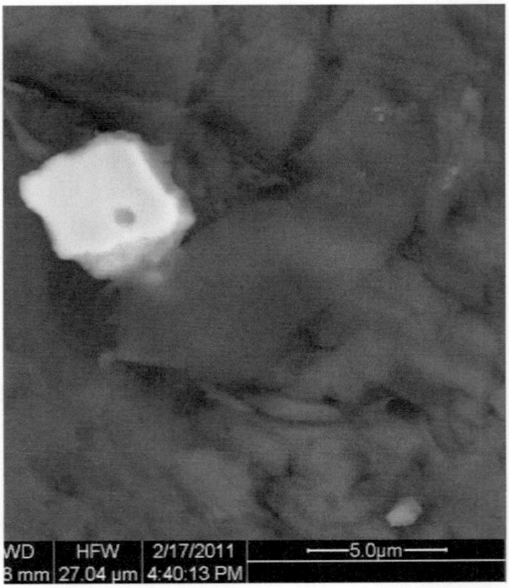

Fig. 4.21d Cerite in the contact zone of the pegmatite swarm at Miesbrunn.SEM

Churchite-(Y) from the Trutzhofmühle is in simple terms a hydrated xenotime and as such a host of HREE in the supergene regolith. On the opposite side florencite-(Ce) is representative of a LREE carrier in the supergene zone. Unlike churchite-(Y) it cannot be expressed as a "hydrated monazite". It is a mem-

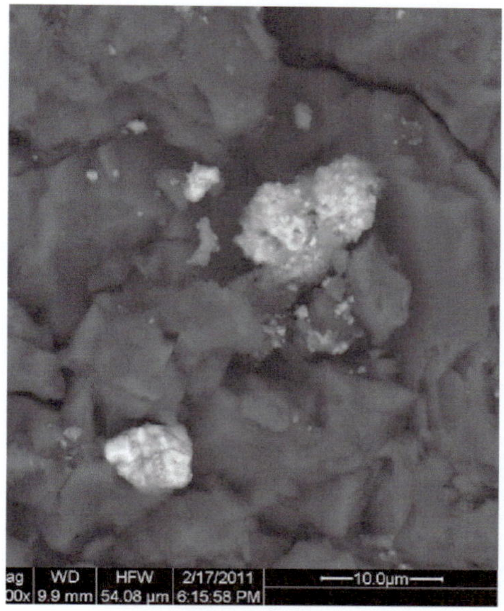

Fig. 4.21e Ferrialanite in the contact zone of the pegmatite swarm at Miesbrunn.SEM

Fig. 4.21f Monazite and columbite-(Fe) intimately intergrown with each other at Trutzhofmühle

ber of the APS minerals with which it forms a wide range of s.s.s., as at Plössberg Pegmatite, and a geoacidometer, indicative of a low pH. Moderately acidic to neutral meteoric fluids would give rise to rhabdophane in the weathering zone, which was not detected in the HPPP, but as mentioned above, is common to the granites and pegmatitic miaroles of the G 4 granite.

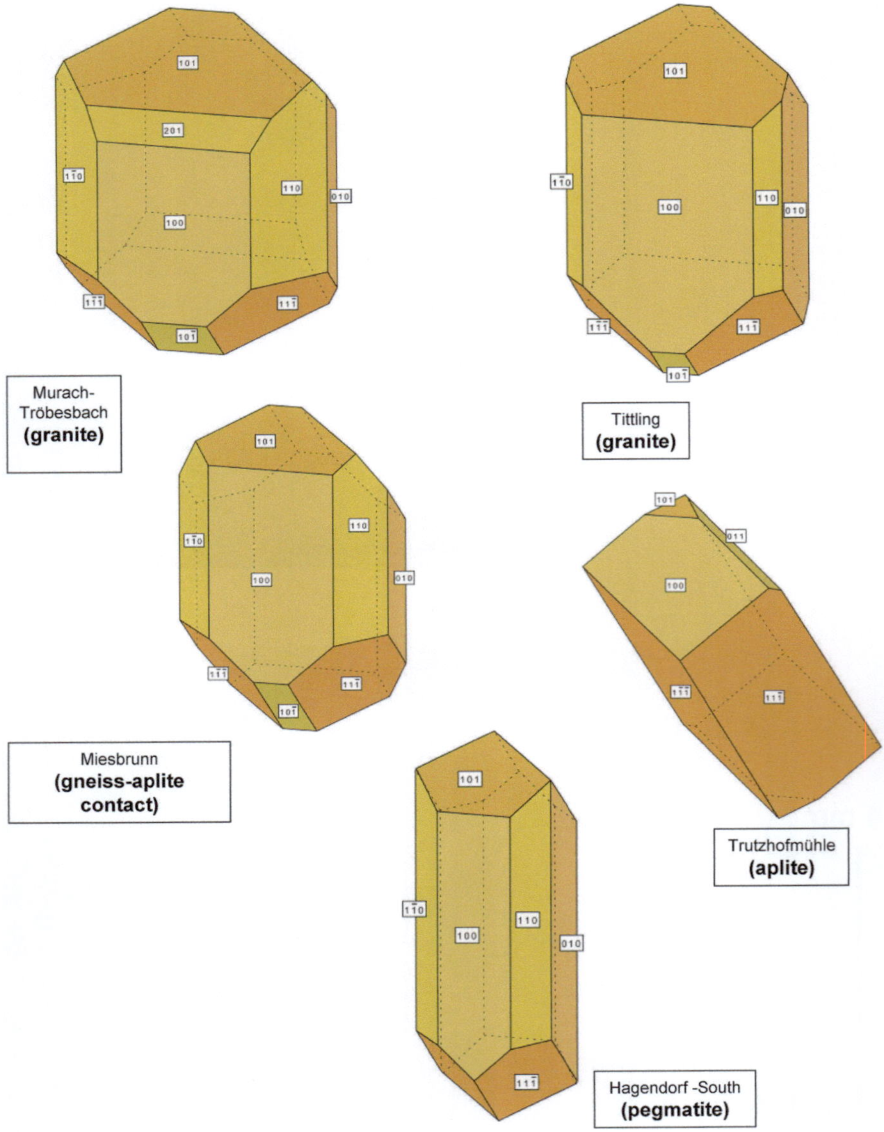

Fig. 4.21g The crystal morphology of monazite in relation to its host rocks. These different types may be found in alluvial-fluvial placer deposits in the environs of the pegmatite provinces

The older granite G 1 in the northern Fichtelgebirge brought about churchite-(Y) at Grossschloppen U deposit, indicative of a local HREE anomaly, whereas the younger granites G 3 and G 4 gave host to the rhabdophane-(Ce), attesting to a local LREE anomaly. In a transect perpendicular to the boundaries of the geodynamic zones, REE minerals are at variance as to the accompanying anions in the primary minerals. They act as geoacidometers attesting to different pH ranges, albeit undergoing the same sort of morphoclimatically induced chemical weathering during the Neogene and preserved their LREE/HREE ratio in the primary hosts even as being subjected to chemical weathering. REE show a strong persistence within the granitic pegmatites and the pegmatites *sensu stricto*, exemplified by the HPPP.

4.22 Scandium Minerals

From the chemical point of view, scandium cannot be attributed to the group of REE but it is often dealt with together with them as it is done with yttrium. Scandium minerals are seldom in nature, because their major component Sc does not hold a prominent position among the elements in the upper continental crust with an average concentration of 11 ppm Sc. Scandium minerals have been recognized for the first time in the HPPP from the HPPP in the Trutzhofmühle Aploid, that is conformably intercalated into the paragneisses near Pleystein. It is the hydrated Sc phosphate kolbeckite (Fig. 4.22a). No other felsic mobilizate in the HPPP has so far

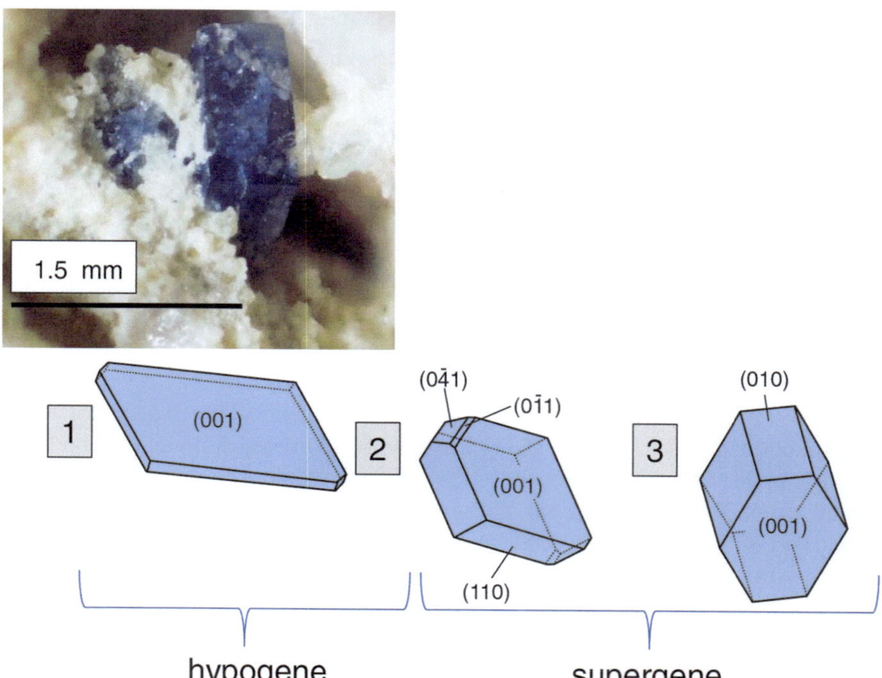

Fig. 4.22a Well-shaped kolbeckite crystal in a solution cavity in quartz at Trutzhofmühle. Below a sequence of cartoons shows the ideal crystal habit of the various morphological types of kolbeckite observed at Trutzhofmühle as a function of the environment of formation

stood out by another find of Sc minerals. Therefore one has to raise the question, where did the Sc come from for the Sc phosphate in an otherwise Sc poor environment. Kempe and Wolf (2006) reported up to 0.88 % Sc in wolframite, 1 % in cassiterite and zircon, and 0.39 % in scheelite from Sn-W deposits in the German Erzgebirge region, Kazakhstan and Mongolia. Compositional data suggest that columbite, pyrochlore and ferberite are the most probable source for Sc to form kolbeckite in the quartz veins of the study area. Coupled substitution may be assumed for these Ti- and Nb-bearing minerals according to the exchange: $Sc^{3+}+(Ti,Sn)^{4+} \Leftrightarrow (Fe,Mn)^{2+}+(Nb,Ta)^{5+}$. Elevated Sc contents were also reported from niobian rutile of the McGuire granitic pegmatite, Colorado-USA, by Černý et al. (1999). The exsolution products in the Ti-rich host are controlled by the (Fe+Mn+Sc)/ (Nb+Ta+W) ratios. Columbite in the Eptevann pegmatite (Norway) exsolved from a Fe^{3+} dominant niobian rutile and contain as much as 12.50 wt.% Sc_2O_3. The stability of these phosphates may be deduced from the experimental data published by Nriagu (1976). It depends on the activities of dissolved silica, $Al(OH)^{4-}$ and Ca^{2+}. When the activities of $Al(OH)_4^-$, Ca^{2+} and silica are set at 10^{-6}, 10^{-3} and $10^{-2.85}$, respectively, at phosphate activities of log a $_{HPO4\ 2-}$ of −1, variscite is stable at pH below 4.5.

$$Al(OH)_3 + HPO_4^{2-} + 2H^+ \Rightarrow \text{variscite} + H_2O$$

$$Sc(OH)_3 + HPO_4^{2-} + 2H^+ \Rightarrow \text{kolbeckite} + H_2O$$

Kolbeckite, whose formation may be described by the simple reaction above, is assumed to be stable under similar conditions.

Owing to the crystal habit three subtypes of kolbeckite can be established at Trutzhofmühle which contain as much as 2.66 wt. % Fe and 2.27 wt % Ca, respectively. Type I kolbeckite developed tabular crystals with the basal pinacoid {001} prevailing in size over {110} faces (Fig. 4.22a-1). No pronounced "limonitization" has been recognized around these crystals of type I. Stubby kolbeckite (type IIa) displays complex crystal aggregates with the faces {001}, {110}, {041}, {011} and {010} (Fig. 4.22a-2). It has to be noted, that faces {001} display a conspicuous lineation towards [100]. The crystal habit of type-IIb kolbeckite closely resembles type IIa but its faces {041} and {011} are downsized to almost nil and beyond recognition under the stereomicroscope (Fig. 4.22a-3). Kolbeckite present in tabular crystals (type I)- Fig. 4.22a-1- is intergrown with quartz, whereas stubby crystals (type II) -Fig. 4.22a-2 and 3- occur in quartz strongly corroded by dissolution and coated with "limonite". Thus, platy type I kolbeckite presumably formed under reducing conditions, e.g., late hydrothermal, whereas kolbeckite type II crystallized during strong "limonitization" in the course of supergene alteration.

Scandium accommodated in the unit cell of columbite together with the trace amounts found in the host rutile and ilmenite in the country rocks, were likely released into a weathering zone already strongly depleted in Fe and Al, and Sc could therefore form minerals of its own instead of being captured as a trace element in

Fig. 4.22b Rosettes of a Sc-U-Ca-bearing churchite-(Y) coating quartz. The phosphate formed during the Neogene chemical weathering. Trutzhofmühle (SEM)

Fe-rich phosphates. This Fe limitation was attained just after formation of mitridatite which is the only secondary hydroxide phosphate containing notable amounts of Sc of up to 0.13 wt.% Sc. There is little doubt that supergene kolbeckite II formed just after mitridatite, while kolbeckite I is a low-temperature hydrothermal mineral accompanying mitridatite. Kolbeckite is replaced along cracks by scandian churchite-(Y) containing variable contents of Ca (2.3–4.2 wt. %), Sc (2.5–3.3 wt. %) and Yb (3.2–4.7 wt. %) (Fig. 4.22b). In one mineral aggregate, 2.6 wt. % U were established in scandian churchite-(Y). Churchite and kolbeckite are isotypic and trivalent scandium and yttrium show a similar chemical behavior not very much different from that of other REE. Hence, scandium contents in churchite-(Y) are not surprising. A mineral was named Sc-bearing vochtenite while closely related in space with Sc-bearing churchite-(Y). It can only be quoted with question marks here due to technical limitation in the methods available to its determination.

Two new Sc phosphates have been encountered in a sample rich in manganoapatite. Mineral (1) comprises perfectly euhedral crystals, ca. 50 µm in size, of a complex K-Ba-Zr-Sc phosphate (Fig. 4.22c). Smaller (<20 µm) euhedral crystals are identified as Zr-Sc phosphates-silicates. Both minerals are further associated with chlorite, quartz, and a hydrous Fe-Al silicate. In back-scatter electron images, the K-Ba-Zr-Sc phosphate displays both, a sector and discontinuous growth zonation that is caused by variable Ba concentrations compensated by K, Zr, Mn and Fe. Compositional variations detected by electron microprobe are: 5.7–8.8 wt. % K_2O, 7.9–14.4 wt. % BaO, 5.9–9.1 wt. % ZrO_2, 24.2–26.7 wt. % Sc_2O_3 and 41.6–43.6 wt. % P_2O_5. FeO (2.6–3.7 wt. %), MnO (1.0–2.1 wt. %) and CaO (0.2–0.6 wt. %) are

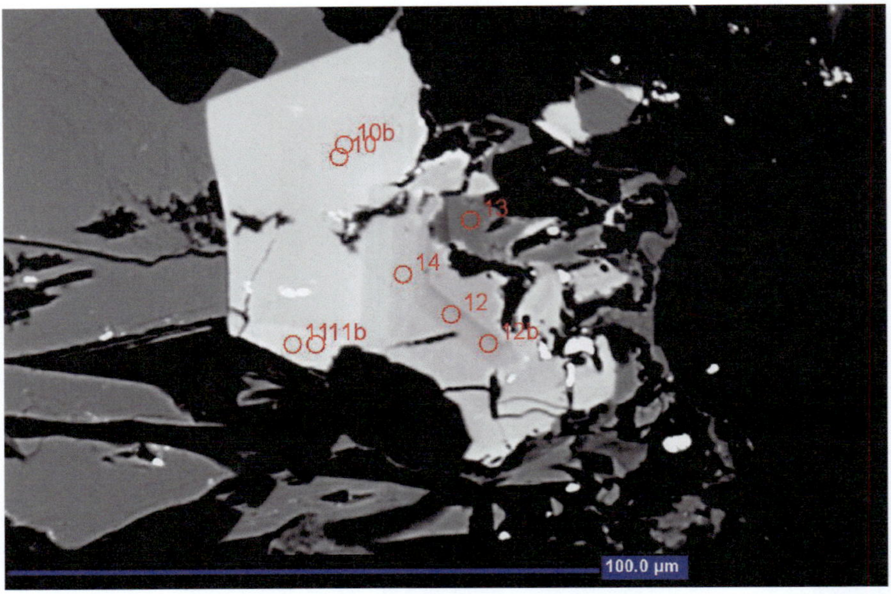

Fig. 4.22c A hitherto unknown scandium phosphate at Trutzhofmühle in a BSE image with station points for measurements

present in minor concentrations. The Zr-Sc phosphate-silicate has a more variable composition, ranging from 12.5 to 34.7 wt. % ZrO_2, 13.4–37.1 wt. % Sc_2O_3, 17.0–33.3 wt. % P_2O_5 and 6.5–15.4 wt. % SiO_2. Minor elements include CaO (1.6–1.9 wt. %), FeO (1.1–3.5 wt. %), Y_2O_3 (2.2–2.7 wt.%) and UO_2 (1.3–4.6 wt. %). The collection of all data so as to render the unknown Sc mineral acceptable by IMA is under way. The crystal structure has been determine but the determination of further characteristic features needed to gain acceptance are still in progress. Unlike kolbeckite, both unknown scandium minerals developed during the incipient stages of pegmatite evolution.

4.23 Beryllium Minerals

In Table 4.1 several Be minerals which were found in pegmatites along the western edge of the Bohemian Massif are listed but only one these minerals plays a significant role within the HPPP and at its outer rim. The majority of the Be minerals listed in Table 4.1 are hosted by pegmatites in the southern Moldanubian or northern Saxo-Thuringian zones. A glimpse at the list of Be minerals demonstrates that only beryl is of relevance in the northern Oberpfalz. Bertrandite, bityite, helvine, milarite, uralolite, bavenite and bazzite are exclusive to the Moldanubian Zone. In general, beryllium minerals hosted by pegmatites are abundant in the southern and eastern parts of the Bohemian Massif and in the Alpine-Carpathian Mountain Range (Fig. 2.5a). In the Saxo-Thuringian zone Be is mainly concentrated in cavities and miaroles of the late Variscan granites, some of which attain, in places, a pegmatitic outward appearance, yet do not reach the size of the pegmatites and aplites under consideration in the HPPP. The highly fractionated "Tin Granite" (No 4), the youngest intrusive complex of a fourfold series of granites, in addition to beryl, bears euclase and members of the herderite series, which were found by mineral collectors at the Rudolfstein, a granitic tor carved out by the chemical weathering during the Neogene and in the abandoned quarries at Fuchsbau. In the Epprechtstein Granite (No 3), its predecessor, bertrandite and phenakite formed in addition to the afore-mentioned Be minerals. Černý (2002) and Grew (2002) dealt with the mineralogy of beryllium in granitic pegmatites in general. An all-embracing discussion of Be mineralization in the pegmatites is beyond the scope of this book, but there may be questions about this conspicuous lack of Be minerals in the HPPP relative to its northern and southern neighbors that wait for an answer.

To get an idea of the phase relations in the Be-Si-Al system a compilation of reactions is given in Table 4.9 according to Barton (1986). In a water saturated system above 600 °C, beryl reacts with one of the Al_2SiO_5 polymorphs to produce chrysoberyl, a Be mineral that has not been spotted in any of the pegmatites or granites along the western edge of the Bohemian Massif so far. Based upon the relation chrysoberyl-out and beryl-in a clear upper boundary may be drawn as to the temperature of formation. Around 200 °C phenakite and bromellite hydrate to bertrandite and behoite, respectively, and pure beryl reacts to euclase + quartz + bertrandite or phenakite, which sporadically may be found in miaroles of the Fichtelgebirge granites but have not yet been encountered in the pegmatites of the HPPP.

With this in mind, beryl has to be attributed to the primary pegmatite mineralization in Hagendorf-South and Plössberg (Table 4.1). When the primary mineral

Table 4.9 Reaction in the system Be-Al-Si according to Barton (1986)

Be minerals: behoite ($Be(OH)_2$), bertrandite ($Be_4Si_20_7(OH)_2$), beryl ($Be_3Al_2Si_60_8$), bromellite (BeO), chrysoberyl ($BeAl_20_4$), euclase ($BeAlSiO_4OH$), and phenakite (Be_2Si0_4)
$Be_3Al_2Si_60_8 \Rightarrow BeAl_20_4 + Be_2Si0_4 + 5\ SiO_2$ (dry)
$20\ BeAlSi0_4OH \Rightarrow 3\ Be_3Al_2Si_60_{,8} + 7\ BeAl_20_4 + 2\ Be_2Si0_4 + 10\ H_20$
$4\ BeAlSiO_4OH + 2\ Si0_2 \Rightarrow Be_3Al_2Si_60_8 + BeAl_20_4 + 2\ H_20$ (T 300 °C)
$Be_3Al_2Si_60_8 + 2\ Al_2SiO_5 \Rightarrow 3\ BeAl_20_4 + 8\ Si0_2$ (water saturated – stable at T > 600 °C)
$2BeO + H_4Si0_4 \Rightarrow Be_2Si0_4 + 2\ H_20$
$4BeO + 2\ H_4Si0_4 \Rightarrow Be_4Si_20_7(OH)_2 + 3\ H_20$
$BeAl_20_4 + Be_2Si0_4 + 5\ H_4Si0_4 \Rightarrow Be_3Al_2Si_60_8 + 10\ H_20$
$3\ BeAl_20_4 + 8\ H_4Si0_4 \Rightarrow Be_3Al_2Si_60_8 + 2\ Al_2SiO_5 + 16\ H_20$
$3\ Be_2SiO_4 + 2Al_2SiO_5 + 7\ H_4Si0_4 \Rightarrow 2\ Be_3Al_2Si_60_8 + 14\ H_20$
$2\ Be_3Al_2Si_60_8 + 14\ H_20 \Rightarrow 4\ BeAlSiO_4OH + Be_2Si0_4 + 7\ H_4Si0_4$ (T = 200 °C)
$2\ BeAl_20_4 + Be_2Si0_4 + 3\ H_4Si0_4 \Rightarrow 4\ BeAlSiO_4OH + 4\ H_20$. (T = 400 °C).

Fig. 4.23a Green beryl in quartz from the Plössberg pegmatite with a cartoon to show its ideal crystal morphology

assemblage of the pegmatites hosting beryl was exposed to hydrothermal alteration the temperature did not drop below 200 °C, otherwise we might expect the above minerals. Granite pegmatites and granite greisen from Tanco to Strange Lake in Canada form in the temperature interval around 600 °C at pressures between 1.0 and 2.0 kb. Considering the salinity, the pegmatitic Be deposits overlap in the range 2–35 wt. % $NaCl_{eq}$ with many of the known Be deposits of non-pegmatitic origin (Barton 1986).

Even with the unarmed eye, the semitransparent crystals of green beryl may be detected in the Plössberg pegmatite (Fig. 4.23a). Five oxygen isotope analyses of beryl from the pegmatite in NE Bavaria fall in the $\delta\,^{18}O$ range 11.9–16.8 ‰ (SMOW) and gave a mean $\delta\,^{18}O$ value of 13.4 ‰ (SMOW). These data do not coincide with those of true pegmatites but overlap with the data obtained from beryl called type-III or shear zone-hosted-type by Giuliani et al. (1998) which are characterized by $\delta\,^{18}O$ values of >+12 ‰. The minor element contents of beryl-bearing pegmatites of different type were listed in Table 4.10 (Dill et al. 2011c). Beryl is significantly enriched in Li, Zn, Rb and Cs in the Hagendorf pegmatite when compared with the pegmatite at Plössberg. Vice versa, the shear-zone hosted beryl at Plössberg is abundant in Cr, Mn, Fe and Ni. The REE contents of beryl from different pegmatites provide a similar picture. Beryl from intrusive and shear zone-hosted pegmatites gave flat but very much distinct graphs in the REE plots normalized to PAAS, although their REE_{tot} contents are very low (McLennan 1989). Intrusive pegmatites are depleted especially in Gd, Tb and Ho. There is almost no difference among the LREE and only little among the HREE. The REE contents of beryl from Plössberg are by some orders of magnitude lower in their REE contents relative to the REE

Table 4.10 Minor elements of pegmatites from different environments (Dill et al. 2011c)

	Shear-zone-hosted pegmatites	Intrusive pegmatites
Li	289	1291
Ti	61.5	58
V	14	9.66
Cr	54.7	47.25
Mn	48.8	28.03
Fe	2256	1118
Co	0.34	0.37
Ni	584	337.96
Cu	8.63	30.99
Zn	1218	7013
Y	0.78	0.06
Zr	2.04	1.03
Nb	18	13.77
Mo	0.26	0.35
Cd	0.71	2.17
Sn	2.15	3.69
Sb	0.32	0.13
Rb	71.5	363.75
Cs	1664	6078
Ba	6.36	6.24
Hf	0.17	0.34
Ta	20.8	19.16
W	0.65	0.52
Tl	0.37	1.77
Pb	19.4	23.57
Bi	0.21	0.09
Th	0.57	0.09
U	2.55	1.23

contents of the feldspar pegmatite and its alteration zone. Shear-zone-hosted/tabular pegmatites developed post-kinematically. Their beryl crystals accommodated a lot of those elements preconcentrated by the surrounding metasediments mainly contained there in the biotite and chlorites. These chemical exchange reactions stress a strong interaction between the siliceous melt and the country rocks. According to the oxygen isotopes, the beryl mineralization from Plössberg was disconnected from granites and pegmatites. The high ^{18}O-enrichment indicates an extensive isotopic exchange between the mineralizing fluids and the metamorphic ^{18}O-rich reservoir during beryl formation. The REE pattern of the Plössberg pegmatite shows a marked positive Eu anomaly and strong LREE/HREE fractionation, which is interpreted in terms of a strong wall rock – pegmatite interaction and hydrothermal alteration of the entire pegmatite system.

Fig. 4.23b A bundle of white beryl from Hagendorf-South

In the pegmatite stocks the interaction took place mainly between the early-formed Li phosphates and the Fe-bearing sphalerite, which supposedly predated the precipitation of beryl. At Püllersreuth a similar green beryl can be detected very much different from the pale bundles of the Hagendorf-South beryl (Fig. 4.23b). Its formation is debated in terms of shear-zone hosted process, involving a strong interaction between the felsic melts and the country rocks. Beryl in the pegmatite stocks formed separately from the phosphate concentration, otherwise herderite-series minerals should have developed. The resulting beryllophosphates highly dependent upon the pH, the temperature, and the availability of specific alkali cations in the solution (Kampf 1992). The formation of beryl can be envisaged to have taken place in strongly peraluminous parts where minerals of the typomorphic lazulite-scorzalite s.s.s were also emplaced. Beryl is closely associated with albite, K feldspar, quartz and muscovite suggesting an early precipitation, in parts even older than the silicates quoted above.

In conclusion, the tabular concordant beryl pegmatites are the result of a(dynamo) metamorphic mobilization. They have chemical signatures typical of the LCT (Li, Cs, Ta) pegmatites manifested by the high B and P contents and the low-P Abukuma amphibolite-facies conditions (Černý 1991). Yet there are also chemical characteristics that would place these pegmatites in the field of the NYF (Nb, Y, F) pegmatites, especially its subclass "REL-REE-MI-REE" (Černý and Ercit 2005). The reference types contain anomalously high values of Sc, Ti, Nb, U, Zr, Y and REE.

The conclusion may be drawn that these pegmatites have not been derived from any of the granitic plutons around. Both, the NNW-NW-striking granites and the tabular pegmatites are only different expressions of the same dynamo-metamorphic mobilization along zones of structural weakness in the crust.

A subtle investigation of pegmatites and aplites, different in structure, mineralogical and chemical composition of the same mineral province reveal that the present classification schemes are hardly applicable in the field. Hagendorf-South pegmatite stock and the external sheet-like Plössberg pegmatite had to be denominated as beryl-columbite-phosphate subtype pegmatites, whereas the large Kreuzberg and Hagendorf -North stocks would fall outside this subclass. Small-scale variations may place these felsic mobilizates in different compartments of the classification schemes so that it becomes meaningless for a practical use in the field, or in other words, for finding mineral deposits.

At the end, let us take again a brief look at Table 4.1, to try and find an explanation for the striking pattern of the distribution of Be minerals, which are widespread in the granitic pegmatites and granites to the NNW and SSE of the HPPP, while being almost absent from the HPPP pegmatites, proper. The Fichtelgebirge granites, enriched in Be minerals are aged 298 ± 4 Ma and 290 ± 4 Ma, the Hagendorf-South pegmatite yielded an age of 299.6 ± 1.9 Ma while the Matzersdorf pegmatite (Tittling Granite) gave an age of 337 ± 15 Ma. Beryl mineralization is obviously not a function of the age of host rocks. The great variety of unusual breakdown products of beryl in the Tittling area can be accounted for by a lithological variance. Novák and Filip (2010) recorded a series of Be minerals rather uncommon for the Bohemian Massif which they related to the orogenic ultrapotassic Třebíč Pluton, Czech Republic. The chemical composition of beryl shows a wide spread. The Be silicate is present from metapegmatites, through pegmatoids such as the albitic pegmatites of the Münchberg Gneiss Complex, Germany, or those known from the Bayan nuur-Lun region, Mongolia, to the pegmatites, proper. They are bound to peraluminous enclaves or parts of the selvage (Dill et al. 2006c). It is yet a hypothesis that has not been backed by a substantial number of samples, but not out of the ordinary to cast beryl in a similar role as the columbite-tantalite s.s.s. Both minerals, beryl and columbite s.s.s. can bridge the gap between the metamorphic and the magmatic environment of formation, relevant for the generation of pegmatitic rocks. Some of the beryl in the NE Bavarian basement, even predates the accumulation of the alumosilicates testifying a metamorphic origin.

4.24 Boron Silicates

Boron silicates are common all across the NE Bavarian Basement, particularly in aplitic and pegmatitic rocks but from the chemical point of view rather monotonous with only two different groups of minerals representing this the element boron. It is dumortierite only known from the Marchaney pegmatite and the tourmaline s.s.s. The boron silicates of the tourmaline series present in almost all felsic mobilizates

Fig. 4.24a Schorl disseminated in an aplite from the contact zone of the Brünst pegmatite

Fig. 4.24b Schorl arranged along the selvage of the Trutzhofmühle aploid

in the study area are either enriched in Mg dominated by dravite or abundant in Fe with schorl prevailing over dravite (Fig. 4.24a, b, c, d, e, f, g). Dravite seldom occurs within the felsic mobilizates themselves and mainly observed as an accessory minerals of the paragneisses (Fig. 4.24f). Schorl is the dominant tourmaline mineral in

Fig. 4.24c Schorl rosettes ("Tourmaline Sun") growing in the feldspar of a pegmatite. Pegmatite Mine Reil at Zessmannsrieth

Fig. 4.24d Schorl disseminated in translucent quartz of the pegmatite I. It is cut by a 2nd generation quartz vein causing an "en echelon" orientation of the tourmaline prisms. Quartz pegmatite Zottbachtal near Pleystein. *Red-blue arrowheads* denote the direction of movement during shearing

the aplite and pegmatites (Fig. 4.24g). Its dark slender prisms can easily be identified with the unarmed eye in the field, and from their texture and orientation within the groundmass of feldspar and quartz we can learn a lot about the evolution of the host aplites and pegmatites, how they were emplaced and what the relation between

Fig. 4.24e Tightly arranged schorl prisms within the shear-zone hosted aplite near the contact between aplite and biotite gneiss. Ploessberg aplite

Fig. 4.24f Well-shaped chestnut brown dravite in a solution cavity of the Trutzhofmühle aploid

formation and deformation is like. In aplites, such as at Brünst, at the northernmost edge of the HPPP, schorl is disseminated in a feldspar matrix. In these most-fine-grained aplites, frequently also (mis)interpreted as metaaplite, no preferred orientation of the prisms of schorl may be recognized it formed more or less simultaneously with the other rock-forming alumosilicates. A prekinematic structural relation can hardly be seen (Fig. 4.24a). Aploids that originated from a late to postkinematic process reproduce the temporal relation between crystallization and structural

Fig. 4.24g Black schorl in the quartz of the pegmatite at Beidl

deformation the better the more contrasting the grain size difference is between the felsic matrix and the tourmaline phenocrysts. This is exemplified by the elongated crystals of schorl floating in a feldspar matrix of the Trutzhofmühle aploid (Fig. 4.24b). Black tourmaline shows a preferred three-dimensional orientation of inequant grains parallel to the walls of the aploid, contributing to a faint zonation of the aploid into three different zones, a tourmaline- enriched border facies, grading into a feldspar-enriched facies away from the gneiss-aploid contact and a quartz-rich central zone. The different facies are not so pronounced as to speak of a zoned aplite or pegmatite similar to what is known from the "Big Three", Hagendorf-South, Hagendorf-North and Kreuzberg that show a clear separation of a quartz core and a feldspar rim with a small aplitic at the contact between the gneiss and the feldspar body. Aligning tourmaline prisms parallel to the contact plane has not been reported from the endocontact zone of these pegmatite stocks.

Lineation of elongate mineral aggregates such as tourmaline are not inevitably provoked by tectonic transport and deformation. At Trutzhofmühle, under the microscope, no crenulation or post-crystalline displacement or rotation of minerals is observed. Preferred orientation of some minerals is interpreted as a product of strain-related post-kinematic growth of boron minerals near the contact with the metamorphic country rocks. Different theories have been put forward to account for the layering next to the gneiss-pegmatite contact. Kleck (1996) has explained layering and aligned minerals in the rocks formed as a result of "soft sediment" deformation of a crystal-liquid mush by the gravitational settling of e.g., tourmaline from the melt.

Where the strain rate was lower than e.g. in the pegmatite stocks, a random orientation of elongated minerals or even rosettes of schorl as in the Zessmannsrieth pegmatite are the rule (Fig. 4.24e). The feldspar-quartz assemblage is evidence of a leucosome invading the metamorphic country rocks in the western part of the study area along shear planes together with the numerous quartz veins. These structural and textural patterns are consistent with the idea that the leucosome developed in the waning stages of metamorphic and structural overprinting of the psammopelitic country rocks in the northern Oberpfalz. Consequently, the aplitic dike rock cannot be considered metaaplite. It is beyond doubt that the slender prisms of black tourmaline from the Quartz pegmatite in the Zottbachtal near Pleystein got their *en echelon* structure or feathered structural pattern by a structural distortion along with the formation of the 2nd generation quartz veins (Fig. 4.24d). The schorl prisms were most tightly arranged within the shear-zone hosted aplite near the contact between aplite and biotite gneiss in the Ploessberg aplite (Fig. 4.24e). Tourmaline, although of minor value to characterize the chemical diversity of pegmatites, is a first-order marker unravel the stress-strain relationship in felsic mobilizates.

4.25 Uranium Minerals

Rare metal pegmatites containing Sn, Ta, Nb and Li mineralization, generally have also variable U, Th and rare earth element contents, see e.g., the Greenbushes and Wodgina pegmatites, Western Australia. The Greenbushes pegmatites commonly have 6–20 ppm U and 3–25 ppm Th, but rarely these pegmatites qualify as a U deposits like Bancroft, Canada, because their content of "uranium black ore minerals" is very low and renders these pegmatites subeconomic.

This is also the case with the pegmatites and aplites along the western edge of the Bohemian Massif. Hagendorf-South, Hagendorf-North and Trutzhofmühle have "uranium black ore" (uraninite) as well as "uranium yellow ore" minerals (uranyl-bearing phosphates and silicates), while at Miesbrunn coffinite is the only "uranium black ore mineral". Nevertheless, the "uranium black ore" as well as "uranium yellow ore" minerals can do a good job to decipher the complex history of the pegmatites and aplites during their emplacement as well as hypogene and supergene alteration due to their inherent clock (Sects. 3.1 and 4.15). Moreover the susceptibility of these U minerals to redox changes ($U^{++++} \Leftrightarrow UO_2^{++}$) and the great diversity of anion complexes accommodated in the structure of the uranyl compounds provides an excellent to constraining the physical and chemical conditions during which U minerals came into existence in the HPPP pegmatites and also elsewhere in pegmatites.

Uranium oxide is the most common host of tetravalent U in the different U deposits scattered across the NE Bavarian Basement (Dill 1986). Almost always in the vein-type deposits, the uranium oxide is the so-called pitchblende recognizable by its globular outward appearance and its dull black luster. Strunz (1962), who dealt with the uranium mineral assemblages in NE Bavaria, has devoted a great deal of his compilation, to the well-crystallized uraninite which was recorded for the first

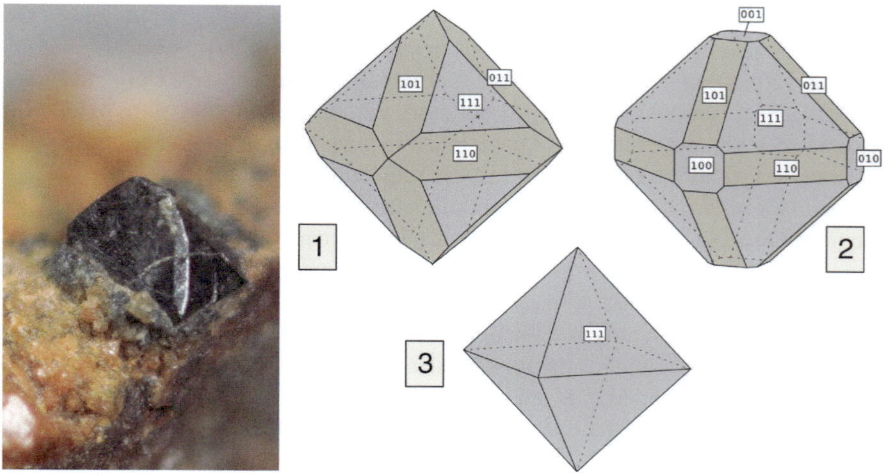

Fig. 4.25a Uraninite crystal at Hagendorf-South added up with a set of three cartoons illustrating the octahedral crystals and their intergrowth with the hexahedron and the rhomb dodecahedron. The crystal aggregate on display in the micro mount image is the natural analogue to the type 1 in the series of cartoons

time from Hagendorf-South by Scholz in 1924 (Fig. 4.25a). Uraninite from Hagendorf stands out from the numerous uranium oxides determined in the vein-type U deposits and is worth mentioning as to its crystal morphology, particularly, as to its intergrowth with columbite-(Fe) (Fig. 4.8d). These crystal aggregates may be quoted as an example of epitactical or law-like overgrowth of octahedral and cubic uraninite onto Nb-Ta oxides. Strunz (1962) has described in great detail the regular intergrowth based upon the preparatory investigations by C. Tennyson whose results may be presented as follows. The lattice parameters of columbite-(Fe) are $a_0 = 5.71$ Å, $b_0 = 14.15$ Å, and $c_0 = 5.08$ Å, while the lattice constant of uraninite measures 5.50 Å equal to the mean value of a_0 and c_0 of 5.40 Å. The pyramids of the uraninite octahedra on display in Fig. 4.8d grew on the (010) face of the host columbite-(Fe). In addition to these crystallographic peculiarities, reflecting the relation between uraninite and its associated minerals, the lattice constant of uraninite can also be applied in a different way to shed some light on the origin of this well-shaped uranium oxide and its host rock the, Hagendorf-South pegmatite. The lattice constant of uraninite shows a linear increase proportional to the temperature of formation, changing from low sedimentary U deposits with $a_0 = 5.39$ Å through 5.46 Å in high-temperature environments represented by pegmatites (Brooker and Nuffield 1952). Pechmann von and Bianconi (1982) measured the lattice constant of the uraninite from Tiraun and Trun in Graubünden, in the Swiss Alps and showed it to range from 5.439 to 5.454 Å which indicates a moderate to high temperature of formation for the uranium oxide. The uraninite in the Swiss Alps formed synmetamorphically to the surrounding metamorphic country rocks and was subjected to green schist metamorphic conditions of 400 °C/3 Kb. Considering the lattice constant of the Hagendorf uraninite, a considerable higher temperature of formation of greater than 400 °C may be invoked. The lattice constants of uraninite met in the mesothermal vein-type deposits are considerably lower than those from Hagendorf: Their a_0 values range from 5.428 to 5.434 Å which is equivalent to a $UO_{2.28}$ to $UO_{2.34}$ (Dill 1982). It is more or less self-explanatory that this primary uranium oxide be it associated with columbite or not, developed in a reducing environment during the incipient stages of the emplacement of the Hagendorf pegmatite.

Taking a brief look at what happened in terms of uranium outside the HPPP, in the northern Saxo-Thuringian Zone, reveals that primary and secondary U minerals in the granites and granitic pegmatites are not at variance with those from the HPPP, with one conspicuous exception, relating to the "uranium yellow ore minerals". In the Fichtelgebirge granites pitchblende and uraninite decomposed to uranyl phosphate and minor amounts of uranyl arsenates (zeunerite). Despite its elevated contents of arsenic contained in arsenopyrite-see Sect. 4.16-, and a significant amount of coffinite and allanite among the primary minerals, the Miesbrunn pegmatite-aplite swarm did not give rise to uranyl arsenates during supergene alteration.

To get an idea of physical-chemical regime of the U-Si mineralization containing uranophane, soddyite and coffinite the dissolved species were assumed to be log $_{aU4+}=-7$, log $_{aSiO2(aq)}=-3$ and log $_{a\ ca2+}=-3$. No uranium species developed at Miesbrunn under near- ambient conditions of 25 °C and the above dissolved species. Coffinite and arsenopyrite, although belonging to the same stage of mineralization behaved in a different way. At a low pH (<pH 4) under these conditions coffinite was still preserved in the infiltration zone, while arsenopyrite underwent decomposition so that the resultant trivalent As has been transported away to get converted into its pentavalent arsenic in the vadose zone and precipitated as scorodite. The case-history reflects a depth-related mineral formation in the supergene alteration zone of the HPPP. A close-up view of the Miesbrunn and Reinhardsrieth deposits, which is located to the North of the Miesbrunn deposit, reveals a striking paucity of both sites in terms of supergene U mineral assemblages at the northern edge of the HPPP. It forces one to the conclusion that the marginal part of the HPPP was more strongly uplifted relative to the central part of the HPPP and As and U responded in a different way to this uplift (Fig. 4.6a). Uranium was completely washed out from all hydraulic zones including the vadose and phreatic zones, whereas arsenic was preserved in the vadose zone, only.

A decisive question, frequently raised in context with the formation of ore deposits, is related to the chronology of uranium mineralizations and has not yet been answered (Sect. 3.1). In this section, the U/Pb isotope data obtained in the HPPP are discussed in view of the isotope data obtained during radiometric dating of "uranium black ore minerals" in vein-type deposits which are scattered across the NE Bavarian Basement and some of which are being located proximal to the pegmatitic study area. Pitchblende samples from the Nabburg-Wölsendorf fluorite vein-type deposits were dated using the U/Pb method (Carl and Dill 1984) (Fig. 1.1a). The ages yielded a minimum age of 295 ± 14 Ma. The Nabburg-Wölsendorf fluorite mining district is approximately 20 km away from the HPPP located in southwesterly direction. The U/Pb age of the uraninite-columbite paragenesis have already frequently been cited in the book and only summarized here for better handling:

Hagendorf-South 299.6 ± 1.9 Ma, Trutzhofmühle: 302.1 ± 3.3 Ma, Silbergrube/Waidhaus: 302.8 ± 1.9 Ma (Fig. 1.1b). On the opposite side, about 25 km north of the HPPP, two uranium vein-type deposits are located near Poppenreuth and Mähring (Fig. 1.1a). Two different U/Pb ages have been published by Carl et al. (1985). The oldest pitchblende generation gave an age of 336 ± 17 Ma, the younger pitchblende yielded age around 298 ± 4 Ma. The age of brannerite which is, in places, or more widespread occurrence in the vein-type mineral assemblage than pitchblende can only be reported at a lower level of precision with 288 ± 78 Ma due to low uranium concentration and a high common lead presence. If we ignore in this discussion the oldest U/Pb age of pitchblende from the Poppenreuth-Mähring U district, which bridges the gap between regional metamorphism, the intrusion of the oldest post-kinematic granites and the incipient stages of U mineralization, the remaining data enable us to an assessment of the chronological evolution of vein- and pegmatite deposits in the basement. Both, vein-type U mineralization and pegmatite-hosted U mineralization, although strikingly different from the mineralogical and physical-chemical point of view belong without any doubt to the same late Variscan heat event. In other words, granite-related (Sn-W-U-Pb-Zn-Li) and pegmatite-related (U-Nb-Ta-Zn-P-Li) mineralization at depth and the oldest stages of unconformity-related vein mineralization (U-F) emplaced near the Late Paleozoic paleo-surface formed more or less contemporaneously in this ensialic orogen (Fig. 2.2a). Although different in their mode of formation, the various U mineralizations extended only over a narrow depth range between 600 and 400 a.m.s.l. The HPPP presents a sort of pegmatite intrusion at rather shallow level.

In the aplitic rocks of the Miesbrunn pegmatite-aplite swarm, zircon enriched in Sc and Hf was found intergrown with coffinite which is isostructural to zircon and present in tetragonal – ditetragonal dipyramidal crystals. Similar zircon-coffinite aggregates were recorded also from the biotite-sillimanite gneisses country rocks of the uranium deposit near Poppenreuth north of the study area, but have not been met in the HPPP so far (Dill 1982) (Fig. 4.25b, c). Experimental studies showed a pressure-induced phase transition in the zircon-coffinite lattice (Zhang et al. 2009). The zircon–coffinite solid solution tends to exsolve the coffinite component in order to reduce the strain energy created at low temperatures by the substitution of the U^{4+} cation for the smaller Zr^{4+} ion (Cherniak and Watson 2003; Ferriss et al. 2010). Mina et al. (2005) measured homogenization temperatures for fluid inclusions of fluorite and calcite associated with coffinite range from 182 to 210 °C and from 126 to 178 °C, respectively. This stage of mineralization has all the hallmarks common to an intermediate- to low temperature hydrothermal mineralization.

Reddish to brown uranyl compounds which formed proximal to pitchblende or even pseudomorphose uraninite in U deposits were grouped together under the term "gummites" when the analytical devices had not yet reached the level of modern-day highly sophisticated methods or too many mineral species are intimately intergrown with each other. These "uranium yellow ores" present close to their parent material and, hence, abundant in tetravalent uranium have a rather monotonous spectrum of elements. The above general term is applied to mixtures of various secondary hydrated uranium oxides with impurities like "limonite" or "leucoxene"

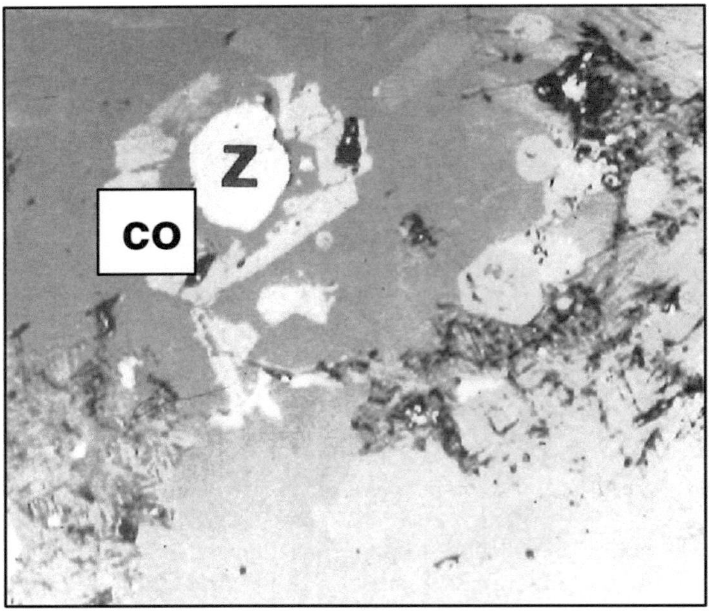

Fig. 4.25b Strongly chloritized biotite-sillimanite gneisses with zircon (z). Detrital zircon is overgrown with euhedral coffinite (co), a situation also observed in the Miessbrunn pegmatite-aplite swarm. Locality: Höhensteinweg fault-bound uranium mineralization, polished section, plain polarized light (Dill 1982)

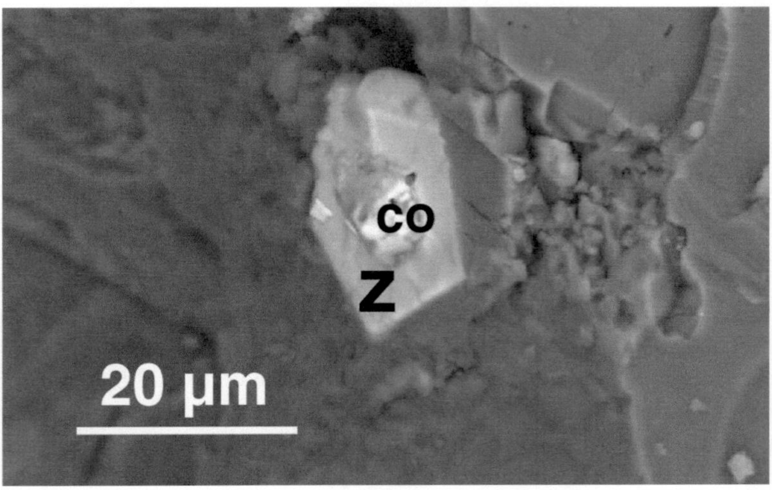

Fig. 4.25c A reversal of mineral growth shown in Fig. 4.25b. Tetragonal coffinite forms the nucleus while surrounded by a rim of zircon ($USiO_4 \Rightarrow ZrSiO_4$). Miessbrunn pegmatite-aplite swarm

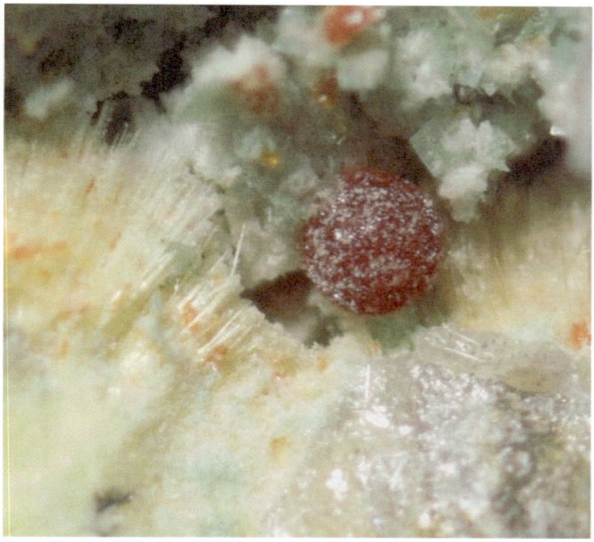

Fig. 4.25d Uranosphaerite globules from Hagendorf-South

Fig. 4.25e Uraninite octahedron on columbite-(Fe) pseudomorphosed by "gummite". The *brown*, resinous luster has derived from ianthinite, the *yellow spot* on the vertex from schoepite. The determination of the mineral has only been made by the author based on visual inspection and is not proved by XRD. Hagendorf-South

used for Fe and Mn oxide-hydrates. In the HPPP, this category of secondary U minerals is represented by uranosphaerite and vandendriesscheite, both of which are exclusive to Hagendorf-South (Fig. 4.25d). The "gummite" minerals pseudomorphosing octahedra which are epitactically intergrown with columbite are interpreted as a mixture of brown resinous ianthinite with minor amounts of schoepite (Fig. 4.25e). A similar outward appearance of "gummites" was also observed associated with pitchblende from the Johannes-Schacht Mine at Wölsendorf. This mineral determination is based only on visual examination comparing the mineral colors with other "gummite" minerals, the author became aware of during uranium exploration in NE Bavaria, and, consequently, these "gummite" minerals have not been listed in Table 4.1. Commonly, schoepite is an early weathering product of uraninite. It does not go alone very long as it has a strong tendency to incorporate cations into its interlayer spacings. Bismuth and lead which guaranteed the charge balance in the hydrated uranyl compounds have derived from the sulfide minerals present among the primary minerals at Hagendorf-South and from the decay of uraninite (Sowder et al. 1996, 1999). These hydrated uranyl minerals, the basis of which is schoepite, formed under oxidizing conditions around pH 7 dependent upon the exposure to different cations such as Bi or Pb, as exemplified by the Hagendorf pegmatite.

As far as the variability of the chemical composition among the uranyl compounds is concerned, uranyl phosphates gain an almost unrivalled significance among the secondary U host in the HPPP. Including their metaforms, 11 uranyl phosphates can be listed in Table 4.1: Torbernite, metatorbernite, autunite, metaautunite, bassetite, chernikovite, uranocircite, lehnerite, vochtenite, vyacheslavite, lermontovite.

Uranyl phosphates, which derived from the afore-mentioned uranyl oxide hydrates are among the most insoluble uranyl compounds, one reason why these minerals prevail over uranyl oxide hydrates uranosphaerite and vandendriesscheite (Sowder 1998).

Torbernite and its metaform rank first among the uranyl phosphates as to their morphological variability. Both minerals may easily be identified in the field with the naked eye by its vivid emerald green color and its characteristic crystal habits (Fig. 4.25f, g, h). It is not a whim of nature that the different morphological types of uranyl phosphate came into existence. A succession composed of five morphological types from stubby through thin tabular torbernite crystals has been established

Fig. 4.25f A stacked pattern of autunite (*yellowish tint*) and torbernite (*greenish tint*) in an oriented intergrowth ("uranyl mica mixed-layers"). The overall crystal morphology is built up of individual platy crystals leading to a stubby bipyramidal "crystal" bounded by the basal pinacoid {001}. This intergrowth occurred on strengite (*st*) in an solution cavity in rockbridgeite (*ro*). Hagendorf-South

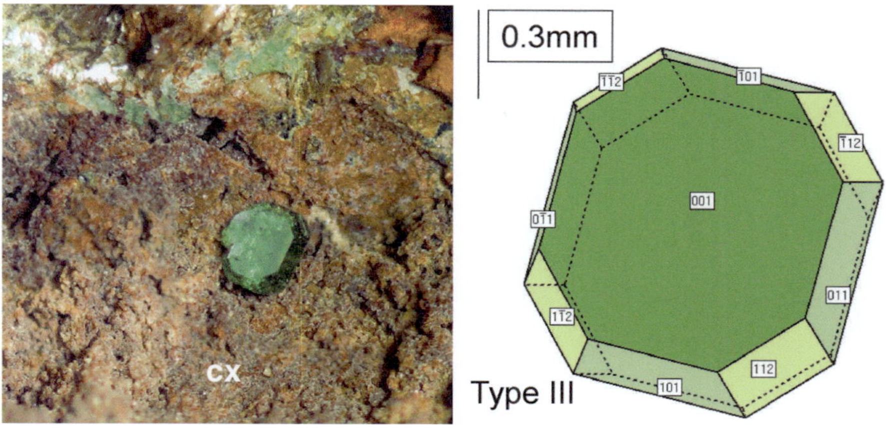

Fig. 4.25g *Thin plates* of emerald-green torbernite caused by a retarded growth of the tetragonal pyramids and a predominance of the basal pinacoid {001}. The substrate is brown cacoxenite (*cx*). Hagendorf-South

Fig. 4.25h Moderately prismatic dark green torbernite showing an almost equal share in the prisms {100} and bipyramid {103}. Torbernite grew onto ferrocolumbite (*co*). Hagendorf-South

for the NE Bavarian Basement; in the HPPP three isometric to stubby morphological types of torbernite – I, II, III- have been recognized (Fig. 4.25f, g, h) (Dill et al. 2007c). Only metatorbernite, the youngest member in this succession developed the ultimate stage, which is represented by thin plates (V) in cavities on strengite (Fig. 4.25i). Thick tabular crystals of type IV are known from miaroles in the Flossenbürg Granite and from vugs in the granite-hosted F-U veins at Nabburg-Wölsendorf. Pressure and temperature cannot have any impact on this mineralization

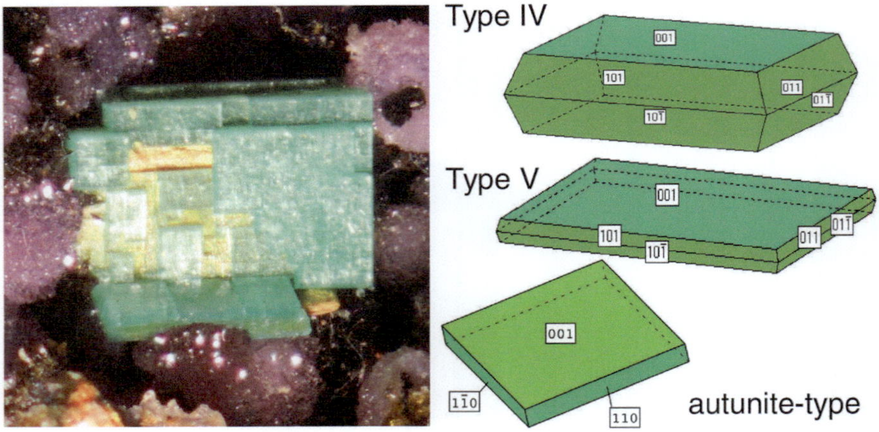

Fig. 4.25i Thin plates of metatorbernite in a cavity lined by strengite. Hagendorf-South. The corresponding morphological type in a fivefold succession is type V. The thick tabular crystals of type IV have no representative in the HPPP. They occur among others in cavities of the Flossenbürg Granite which is the nearest exposure of this type of torbernite to the HPPP

under near-ambient conditions during supergene alteration. It is mainly the variation of the meteoric solutions through time and the parent material which torbernite and metatorbernite have derived from.

Uranyl phosphates which immediately reside on primary minerals of the pegmatite, e.g. columbite and uraninite epitactically intergrown with this Nb-Ta oxides, used to develop type I and type II morphologies. The torbernite age obtained by LA-ICP-MS is 4.550 ± 0.022 Ma (Dill et al. 2007c). A "limonitic" core zone within this torbernite yielded an upper intercept age of 549 ± 12 Ma that lies far from the Late Variscan ages known for the HPPP pegmatites but comes close to apparent ages of between 473 ± 5 Ma and 494 ± 5 Ma reported by Glodny et al. (1998) for the metapegmatites Domažlice Crystalline Complex, Czech Republic. Fiala et al. (1995) gave age limits for the deposition of Moldanubian metasediments in the area. The older zircon population from the metasediments yielded somewhat heterogeneous ages in the range 1.6–2.0 Ga. The younger prismatic zircons only went through one sedimentary cycle and displayed a morphology typical of granitic rocks. Their age of formation was determined to 549 ± 5 Ma corresponding to the Cadomian intrusive activity in the source area. Therefore one can assume that these old ages measured in the "limonitic" core of torbernite may be traced back to a thermal event near the Precambrian-Cambrian boundary.

In the second mineral assemblage, torbernite of morphological type III and V are found next to phosphates with bivalent and/or trivalent iron in their crystal lattice such as strengite, beraunite, and cacoxenite lining cavities of rockbridgeite. The Fe concentration in the mineralizing fluids and the redox conditions of the underlying parent material are responsible for the various type of torbernite and its associated minerals.

The reaction path during weathering began with dissolution of Fe phosphates such as rockbridgeite (bivalent and trivalent Fe) and strengite (trivalent iron). If Cu

Fig. 4.25j Thin plates of bassetite in rockbridgeite at Hagendorf-South

solution had not entered the scene the reaction would have ended with bassetite being precipitated in solution cavities of the precursor Fe phosphate (Fig. 4.25j). Acidic Cu solution favored the evolution of torbernite under oxidizing conditions. Depending upon the concentrations of Mn, Fe, Ca and Cu in the meteoric fluids, solid solutions series of Fe-Mn-Cu phosphates may evolve or crystal aggregates of autunite and torbernite epitactically intergrown with each other may come into being (Fig. 4.25f). Autunite and torbernite are "mixed-layer of uranyl micas" giving rise to a sandwich-like crystal aggregate that at the very end looks like type I. Another structural feature known from phyllosilicates can also be observed in these "uranium micas". Torbernite of type V forms so-called integrates or core zones in uranyl phosphate aggregates. Towards younger stages the integrates got deprived of its and changed into autunite (Fig. 4.25k). Both morphological types, torbernite-type V and autunite ("autunite-type") resemble each other and therefore layered or integrated types of intergrowth are not anything out of the ordinary. Autunite is a common uranyl phosphate in the HPPP, whose formation during supergene alteration is preceded by another mineral chemically quite similar to it. Although chernikovite is not so widespread as autunite it is nevertheless an important basis for a series of uranyl phosphates to built upon, depending on the cations available in the meteoric solutions (Fig. 4.25l). Chernikovite, is also referred to as hydrogen autunite and transformed to autunite by reaction with calcium (Sowder 1998), and to lehnerite by reaction with bivalent manganese (Vochten 1990) (Fig. 4.25m). While the afore-mentioned uranyl phosphate are mainly found at

Fig. 4.25k Integrates of torbernite (*to*) forming the core of a uranyl aggregate whose rim is made up of autunite (*au*). Hagendorf-South. Thin plates of autunite are illustrated in a cartoon ("autunite-type")

Fig. 4.25l Chernikovite pseudomorphoses of uraninite octahedra intergrown with columbite-(Fe). Hagendorf-South

Hagendorf-South, lermontovite, vyacheslavite and vochtenite were only observed in the Trutzhofmühle Aploid under SEM. Uranocircite, the barium-bearing uranyl phosphate was found at Plößberg, where barite was concentrated along the contact in a way not matched by any other pegmatite in NE Bavaria. The quality of its crystals

Fig. 4.25m Lehnerite from Hagendorf with a cartoon to show its ideal crystal morphology (Image: C. Rewitzer)

renders this U mica from the pegmatite of HPPP less attractive than the U-F veins at Wölsendorf, where well-shaped tiny plates of this mineral are a common among the supergene alteration minerals. To complete the list of "uranium yellow ores" in pegmatites, uranophane has to be recorded from Hagendorf-South (Fig. 4.25n).

Uranyl phosphates (torbernite, autunite, uranocircite) and the hydrated uranyl silicates (uranophane) found at various erosion levels in the pegmatites and the Late Variscan granites at the western edge of the Bohemian Massif, Germany. Hydrated uranium oxides can be sidelined during this discussion, because of their position proximal to the source of tetravalent uranium. Supergene U minerals have an edge over rock-forming minerals, because of their inherent 'clock' and their swift response to chemical and physical environmental changes on different scales. It enables the student of these felsic rocks to discuss their embedding into the most recent parts of the geological history and the evolution of the geomorphology and hydraulic systems during the late Cenozoic. The impact of supergene alteration can be traced down to a depth of approximately 200 m below ground in the basement at the western edge of the Bohemian Massif by the evidence of uranyl phosphates and hydrated silicates. They are grouped into five depth-related compartments: (1) Fresh felsic rocks, (2) infiltration zone, (3) percolation zone, (4) saprock, (5) saprolite. Zones 2 and 3 are representative of the lowermost part of the alteration zone and called the hydraulith. Overlying zones 4 and 5 are indicative of that what has been described in the textbook as regolith. The vadose zone extends down to the

Fig. 4.25n Needle-shaped uranophane from Hagendorf-South

percolation zone, while the infiltration zone and the fresh rock are located within the phreatic hydraulic zone. Uranyl phosphates prevail in the zones 2, 3, and 4, whereas in zone 5 uraniferous silicacretes (e.g. uranophane) predominate. The various zones are physico-chemically characterized as follows: (1) $Eh<0 - pH>7$, (2) $Eh \leq 0 - pH>6$, (3) $Eh>0$ – pH 5–7.5, (4) $Eh>0$ – pH 5–7.5, (5) $Eh>0 - pH<6$. Uranium redeposition provoking the precipitation of uraniferous phoscretes and silicacretes may be traced back by means of radiometric age dating to 10 Ma from now.

4.26 Barium, Lead, Silver and Antimony Sulfur Minerals

The literature about epigenetic Ba-, Pb-, Ag and Sb- deposits in Central Europe amounts to an outstanding number of publications, although none of these deposits described is still operated today for any of the above raw materials (Dill et al. 2008a – further literature cited there and in Sect. 4.2). Some are given the status of a cultural heritage, like the base metal mine Rammelsberg, near Goslar, while others are run as adventure or visitor mines, to give interested people an insight into the underground world. In the pegmatites under study as well as elsewhere in the world, sulfides, sulfates and carbonates accommodating Ba, Pb, Ag, and Sb as major component in their lattice are rare minerals. Those elements neither saw some enrichment during the final stages of granitic fractionation nor in strongly differentiated zoned pegmatites.

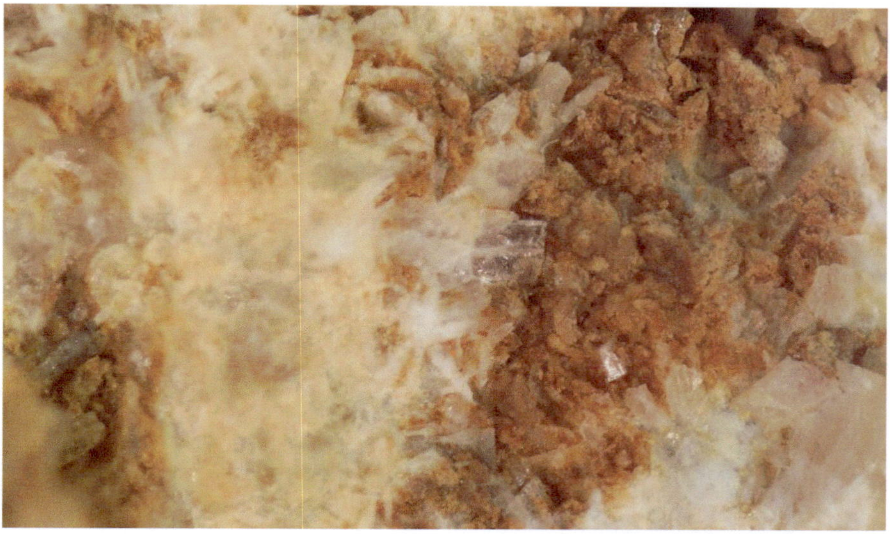

Fig. 4.26a Thin plates of transparent "white barite" associated with hematite in a micro fissure intersecting the contact zone of the Trutzhofmühle Aploid

Barite has only been met sporadically in the contact zone of the tabular pegmatites and aplites (Table 4.1). It is a typical "white barite", transparent and well-crystallized attesting to a low Eh, although the most common minerals associated with it, hematite and goethite, stress strongly oxidizing conditions. They precipitated subsequently to its emplacement and are meaningless as the redox regime of barite in pegmatites (Fig. 4.26a). In the adjacent Nabburg-Wölsendorf F-(Ba) veins barite is very widespread but due to its Fe contents, called "red barite", not liked very much by the mining engineers and entrepreneurs. Finest particles of hematites are disseminated among the plates of barite and hamper its economic use even if the concentration of barter may exceed the break even in some deposits. Barite attained its strongest accumulation among all the pegmatites in the NE Bavarian Basement, in the barite-bearing beryl-phosphate pegmatite at Plössberg (Fig. 4.26b). It is considered as a missing link between barite in shear zone-hosted pegmatites and vein-type barite deposits lined up along the western boundary of the Bohemian Massif- see e.g. Nabburg-Wölsendorf, Donaustauf, Warmensteinach, Nittenau and Issigau-Lichtenberg (Fig. 1.1a). Barium minerals such as barite are rare among pegmatitic minerals and thus little is known about the behavior of Ba minerals in igneous feldspar-quartz rocks (Černý et al. 1985). There are rather conflicting views if the barium contents are related to the pegmatite melt fractionation or were triggered mainly by transfer processes among pegmatitic melt and the host rocks of the crystalline basement. The crystalline basement rocks are enriched in Ba incorporated into K feldspar and mica, where Ba 2+ may proxy for K+ in the various rock-forming tectosilicates and phyllosilicates.

Fig. 4.26b Aggregates of barite, made up of plates of barite, randomly grown along the selvage of the Plössberg Pegmatite (SEM)

In the following sections, the chemistry, mineralogy and geology of the Plössberg pegmatite is described. Barium got only once enriched during the initial stages of the emplacement of the Trutzhofmühle aploid together with scandium (Sect. 4.22). At Plössberg, barite is a very common sulfur mineral besides pyrite and arsenopyrite. It is intimately intergrown with chloritized biotite and has Sr contents of 0.11 wt. % $SrSO_4$. Barium in the late Variscan pegmatites from NE Bavaria exhibits quantities similar to those reported from the Variscan pegmatites in Spain (Alfonso et al. 2003). Well-preserved rosettes of barite developed subsequently to the emplacement of the pegmatite and its alteration zone (with Ba being enriched by the factor 1–10 in this zone). Those elements enriched at a level greater than 10 are supposed to have been derived from the enclosing country rocks. Barium was re-mobilized at the transition from the ductile to the brittle deformation. Cambro-Ordovician (meta) sedimentary host rocks have vein-type barite with $SrSO_4$ contents in the range 1–2 wt. % $SrSO_4$, whereas barite cementing siliciclasts in upper Triassic arkoses and sandstones of the foreland of the NE-Bavarian Basement have $SrSO_4$ contents around 7 wt. %.

The Plössberg barite shows exceptionally low Sr contents and hence fits into the metallogenetic model depicted in Fig. 2.2a. It is the depth of where the barite mineralization of the vein-type and pegmatite-hosted barium mineralization developed that accounts for the variation of Sr contents. The vein-type barite mineralization under consideration is part of the late-Variscan unconformity-related barite mineralization that evolved at rather shallow depth. The barite mineralization associated with the shear-zone-related pegmatites, however, is part of a late-Variscan barite mineralization that formed at greater depth than the afore-mentioned veins. Granites and pegmatites are two sides of the same coin. Both were controlled among others by deep-seated NNW-SSE trending structural zones along the western edge of the Bohemian Massif. Barium does not form part of the differentiation of these magmatic pegmatites but is associated with late magmatic/ hydrothermal processes. There is also striking trend of barite in the HPPP getting more abundant to the north while sulfides and arsenides diminish in the same direction. On a local scale barite is a true facies minerals underscoring the presence of a more oxidizing facies at the margin of the HPPP.

The Ag-bearing sulfides acanthite, schapbachite, sternbergite, stromeyerite and native silver have been reported by Strunz et al. (1975) and almost all of them were encountered as tiny inclusions either in base metal sulfides or as discrete grains. None attained economic significance in the pegmatites. As I have not got any personal impression of these minerals their siting within the mineral association of the Hagendorf pegmatite is provisional and based only upon literature data. Keeping the acanthite apart, which might have been formed at a similar hydraulic level as, e.g., covellite in the phreatic zone, the overall Ag-Cu-Bi sulfides are likely to have formed together with the Bi-Cu sulfides (Sect. 4.17).

A similar scenario may also be envisaged for galena, that is only known from Hagendorf-South, and whose product of supergene alteration, cerussite, was only described from the southern Bayerischer Wald (Table 4.1).

Stibnite was identified under the SEM in one sample from the Miesbrunn pegmatite aplite swarm. No idiomorphic crystals or aggregates were encountered. Antimony sulfides are common to the Saxo-Thuringian zones only.

4.27 Zinc Minerals

Zinc is ubiquitous in the Pb-Zn veins across Central Europe, so that it is commonly treated under the header base metals together with Pb, which used to be present as galena. Not so in the HPPP pegmatites, where sulfidic and non-sulfidic Zn minerals occur from the initial stages of pegmatite emplacement through the hypogene alteration and even supergene alteration, without any genetic link to Pb. Locally it is present within extraordinary Zn phosphates or in the most common sulfide sphalerite yet with an uncommon trace element spectrum, which is decisive for the discussion of the origin of the pegmatites and aplites of the HPPP.

Spinel-group minerals are rather uncommon among the rock-forming minerals of the crystalline basement in NE Bavaria and gahnite is without any doubt the rarest constituent among the non-sulfidic Zn minerals of the pegmatites under study. It has only been encountered as a dissemination of tiny grains in degraded biotite of the tourmaline-bearing aplite at Miesbrunn (Fig. 4.27a). Zincian spinel developed as a consequence of the breakdown of biotite into chlorite in the biotite-sillimanite

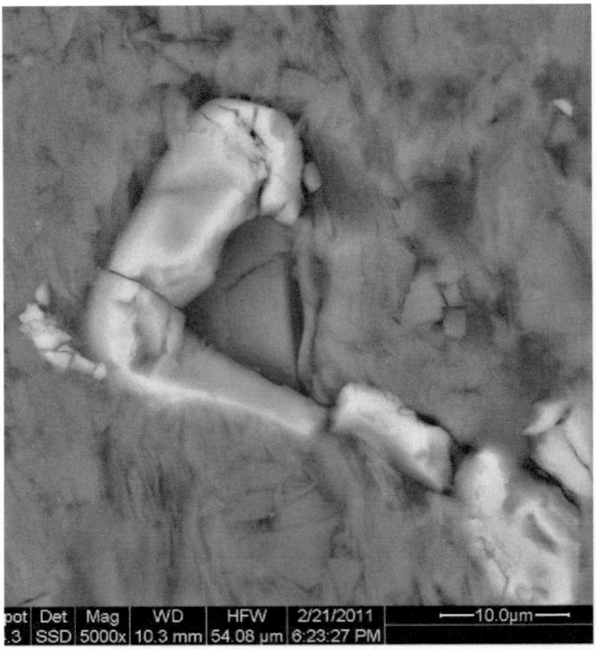

Fig. 4.27a Gahnite disseminated in degraded biotite of the tourmaline-bearing aplite at Miesbrunn

gneisses, where bivalent Zn was camouflaged by bivalent Fe and Mg in the micaceous phyllosilicates. Gahnite behaved as a restite in the pegmatite-aplite swarm at Miesbrunn, while most of the Zn was removed from the melting zone into the younger stock-like pegmatites, such as at Pleystein or Hagendorf and precipitated there as sulfidic Zn mineral. Zincian spinel is not only indicative of highly fractionated pegmatites but also a good tool in constraining the temperature of formation (Soares et al. 2007). Skeletal gahnite was obtained during laboratory trials in water of a mixture of zinc oxide and hydroxide of aluminum at 180–400 °C under a pressure of water vapor 1–26 Mpa (Ivakin et al. 2006). Batchelor and Kinnaird (1984) claimed that the an increase in the Zn content in gahnite is characteristic of a more fractionated magmas and associated with Li-rich pegmatites. Field evidence challenges this assessment because in the HPPP it is the other way round. Gahnite is absent from all those pegmatites enriched in Zn and are abundant in Li-see also Fig. 5.5d-B. There is a clear differentiation between tabular and stock-like pegmatites as to the presence of sulfidic and non-sulfidic zinc mineralization.

Hemimorphite is as seldom as gahnite in the pegmatites of NE Bavaria, but unlike its siliceous counterpart among the primary Zn minerals, the zincian spinel, it forms part of the mineral assemblage of the supergene alteration zone at Hagendorf-South which is barren as to Zn spinel (Fig. 4.27b). Hemimorphite is stable at slightly acidic to alkaline pH values at quartz saturation and dissolved zinc concentration of 10^{-5} m (McPhail et al. 2003). Zinc mobility is enhanced by low pH, oxidized conditions (sulfide-poor) and high salinity (depending on the importance

Fig. 4.27b Globular aggregates composed of tiny plates of hemimorphite. Hagendorf-South

of sorption). The source of Zn in hemimorphite has to be looked for among the Zn phosphates that precipitated in the aftermaths of sphalerite.

The major host mineral of Zn in the HPPP is sphalerite, which occurs in the pegmatite stocks and in some tabular aplites, pegmatites and aplite granites (Table 4.1). The Zn sulfide is present in at least two generations, strikingly different from each other as to their minor elements and the group of minerals accompanying each Zn sulfide.

The most conspicuous Zn sulfide is the Fe-enriched sphalerite (marmatite, christophite). Many mineralogists have paid attention to it for its black luster since opening up the Hagendorf pegmatites by the beginning of the last century (Fig. 4.27c). Strunz et al. (1975) reported the chemical composition of this Fe-enriched sphalerite as a function of depth and host silicates. With the distance decreasing from the quartz core and the depth increases more elevated Fe contents have been reported by the above authors. Differentiation played a vital role in the enrichment of Fe in sphalerite.

Element	109 m- level	76 m-level	K-feldspar	Na-feldspar	Quartz
Zn wt. %	49.15	54.51	55.8	51.5	46.5
Fe wt. %	16.80	11.40	7.4	10.2	14.6
S wt. %	33.75	34.40	34.2	33.3	32.8

Two samples one from the Kreuzberg Pegmatite at Pleystein and one from the Burkhardsrieth pegmatite, yielded 12.9 wt. % Fe and 10.9 wt. Fe. This type of

Fig. 4.27c Fe-enriched black sphalerite in the outer part of the quartzose core of the Kreuzberg Pegmatite at Pleystein

Fe-enriched sphalerite precipitated in the Hagendorf pegmatites, at Pleystein and at Burkhardsrieth during the initial stages of the primary mineralization attesting to a high temperature of formation in contrast to another Fe-poor sphalerite which was encountered among others at Miesbrunn, in a swarm of tabular aplites and pegmatites. The high-T sphalerite also contains exsolutions lamellae of chalcopyrite and pyrrhotite, another mineralogical evidence for higher temperatures of formation. The sample from Hagendorf -South which formed close to the quartz core is associated with younger ore minerals such as pyrite, chalcopyrite, covellite, native bismuth, bismuthinite and Cu-Bi sulfides (emplectite, cuprobismutite). At Pleystein, Fe-enriched sphalerite goes along with cassiterite, quartz and pyrite while at Burkhartsrieth the mineral association is mineralogically poorer, with sphalerite being the only ore mineral besides feldspar. The Fe content in sphalerite was applied by Lusk and Ford (1978), Lusk et al. (1993), and by Martin and Gil (2005) as a geobarometer. The variation of the Fe contents of sphalerite within the Hagendorf pegmatite dependent upon the host silicates rules out the usage of it as a geobarometer in the Fe-S-Zn system. While the Fe content can only be used for a rough temperature assessment, high, medium, low and for minero stratigraphic correlation, a closer look at the minor elements spectrum of the pegmatite-hosted sphalerite gives a more different picture that may also help disentangle the succession of the emplacement of pegmatite within the HPPP.

The trace element contents of indium in sphalerite from Hagendorf stand at 65 ppm In, while those of cadmium are at 5837 ppm Cd and of gallium at 648 ppm. At Burkhardsrieth the contents are higher for indium with 122 ppm In, the Cd is almost the same with 6070 ppm Cd and significantly lower for gallium, attaining only 26 ppm Ga. The highest In values have been analyzed in samples from Pleystein

(221 ppm In). Its gallium contents are at 264 ppm Ga and the Cd contents amount to 5261 ppm Cd. Gallium used to diminish with temperature of formation, whereas indium shows the reverse trend and shows a striking increase. Keeping in mind these relations, the Kreuzberg Pegmatite formed at the highest temperature level and Hagendorf-South at the lowest level. Burkhardsrieth lies between the two as far as these minor element contents are concerned. For comparison an Fe poor sphalerite from the Wendersreuth metapegmatite was discussed in context with these pegmatite from the HPPP (39 ppm In, 2300 ppm Cd 2100 ppm Ga). The trace element contents demonstrate that this pale brown sphalerite is genetically unrelated to the emplacement of the metapegmatites but has to be attributed to a late-stage fissure-bound mineralization. This type of low-Fe sphalerite is analogous to the type of sphalerite recorded from the Miesbrunn pegmatite-aplite swarm and cannot be correlated with the Fe- and In-enriched sphalerite from the HPPP.

For those taking into consideration the amount of cadmium present in the pegmatite-hosted black sphalerite, the presence of earthy greenockite is not anything unexpected either (Fig. 4.27d). The Cd sulfide marks a fluctuating Eh value still moving into negative territory and terminates the succession of Zn phosphates which began with the alteration of primary sphalerite. Greenockite characterizes the phreatic hydraulic level within the pegmatites, closely related to the spade of supergene minerals that developed at shallower depth.

Zinc phosphates are very common constituents besides sphalerite in the mineral assemblages of the Hagendorf-North, Hagendorf -South and Pleystein pegmatite stocks. It is hopeite, parahopeite, jungite, scholzite, parascholzite, phosphophyllite and schoonerite that have been found by various students of the pegmatites in the HPPP (Table 4.1). These Zn compounds are absent from the tabular pegmatites and

Fig. 4.27d Canary yellow earthy greenockite covers a joint within the quartz core of Hagendorf-North

Fig. 4.27e Acicular parahopeite in massive aggregates. Hagendorf-North

aplites and the Burkhardsrieth pegmatite. In the study area, there are two polymorphs of $Zn_3(PO_4)_2 \cdot 4H_2O$, hopeite (orthorhombic) and parahopeite (triclinic) present as void-filling crystal aggregates and as massive mineral concentration (Fig. 4.27e, f, g). The zincian phosphates are medium- to low-temperature alteration products which formed as a result of hydrothermal corrosion of primary Zn sulfide in the HPPP. The temperature can only be estimated by means of the minerals which they are associated with, since no detailed information is available on the temperature of formation of these Zn phosphates in the pertinent literature. Both polymorphs are rare Zn minerals indeed, yet it would be misleading to confine their presence to the pegmatites only as a result of some kind of autohydrothermal or retrograde process of the remaining fluids, held inside the pegmatite body upon its consolidation. Hopeite and parahopeite occur sporadically in non-pegmatitic deposits, e.g., the Kabwe Pb-Zn Mine, Namibia (Cairney and Kerr 1998). They show up in the upper levels of the mine, and may have been derived from the bones and other organic remains found in caustic cavities (Cairney and Kerr 1998). Based upon its structure parahopeite which appears as massive Zn ore and as void-filling phosphate may cover the entire alteration process from autohydrothermal at the beginning to low-temperature hydrothermal during the waning stages.

Scholzite and parascholzite follow suite with some of the Fe-Mn phosphates encountered in the same interval of alteration, when alkaline and earth alkaline elements were released during corrosion of primary silicates as well as the ubiquitous

Fig. 4.27f Concertina-type aggregates of parahopeite in a vug at Pleystein. The inset shows a thin platy ideal crystal characteristic of this Zn phosphate

Fig. 4.27g Hopeite single crystal at Hagendorf-South

Fig. 4.27h Scholzite at Hagendorf-South

Fig. 4.27i Parascholzite bundle at Hagendorf-South

apatite (Fig. 4.27h, i) (Sturman et al. 1981). The phosphates took up Ca^{++} to build up of Ca-Zn phosphates. Scholzite and parascholzite grew into vugs and cavities in the same way as many other Fe-Mn phosphate precipitating contemporaneously with these rare Zn-Ca phosphates from low-T hydrothermal solutions. Jungite is the only zincian phosphate which needs calcium and iron to develop during the latest stages of alteration (Fig. 4.27k). The prerequisite of Fe, Zn and Ca to be present in

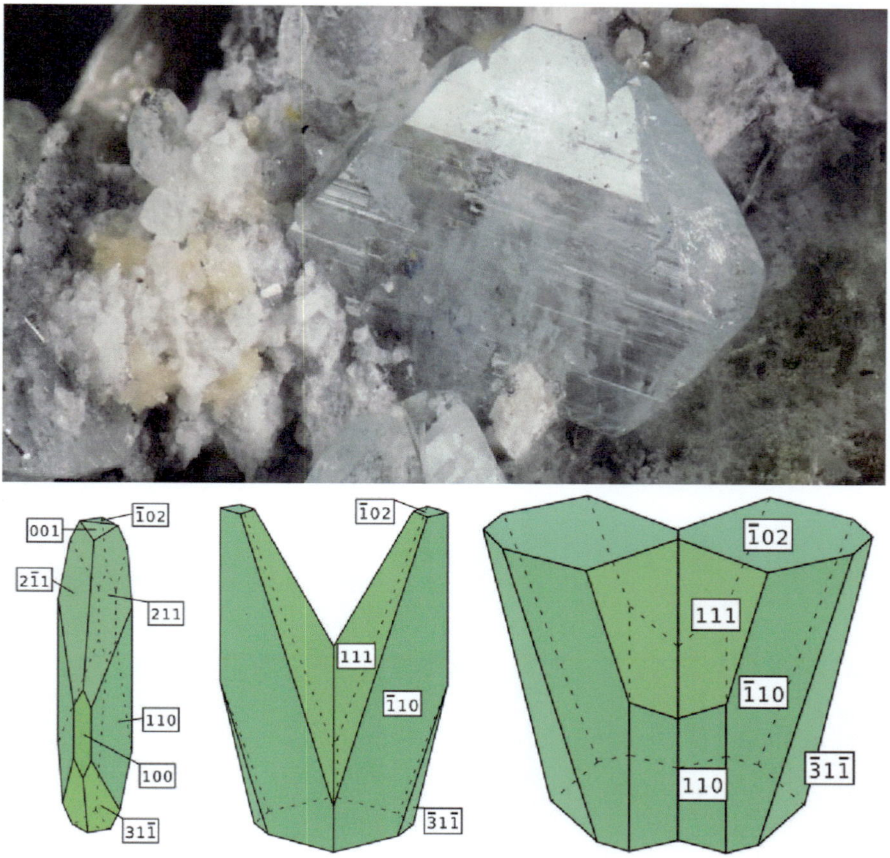

Fig. 4.27j Phosphophyllite at Hagendorf-South with a series of cartoons showing the crystal habits found at Hagendorf

sufficient amounts to yield Zn-Fe-Ca phosphate is only rarely fulfilled in a strongly oxidizing regime where iron is present only in its trivalent state. Schoonerite and phosphophyllite do not demand chemical conditions as rigorous as for jungite and therefore have a broader field of occurrence during the late hydrothermal stages (Moore and Kampf 1977) (Fig. 4.27j, l).

In conclusion, the high-T Fe-Cd-enriched black sphalerite pertains to one of the early minerals in the pegmatite system, while the Fe-Cd poor sphalerite is part of a late-stage low-temperature mineralization similar to the vein mineralization found at various sites in the NE-Bavarian Basement. The string of Zn phosphates is genetically linked to the Fe-Cd sphalerite only. Brown sphalerite poor in Fe did not bring about any Zn phosphates. Based upon this spatial relationship Zn phosphates may be indicative of the "pegmatite's kitchen", even if the primary Zn sulfides have been gone as a result of hydrothermal alteration or in the course of a pervasive supergene alteration. Zinc forms minerals of its own during the regional metamorphism of medium and high grade in the Moldanubian Zone, resultant among others in Zn spinel and Zn

Fig. 4.27k Jungite at Hagendorf-South

Fig. 4.27l Schoonerite at Hagendorf-South

staurolite, but below the level of what might be called a metamorphic non-sulfidic Zn deposit (Dill 1990; Tajčmanová et al. 2007, 2009). On the other hand, these Zn anomalies in the metamorphic country rocks, attaining higher than low-grade regional metamorphism, furnish additional evidence that the exceptional concentration of Zn without Pb and Ag in pegmatites is genetically related to the metamorphic processes.

Chapter 5
The Geological Setting of the HPPP

Abstract Contact metasomatism and contact metamorphism is closely related with the pegmatites of the HPPP and places the thermal "hot spot" for the early mineralization in the area between Pleystein and Gsteinach. Wollastonite is the diagnostic mineral found in the skarn and calcsilicates next to the Kreuzberg pegmatite. The intimate intertonguing of scheelite with calcsilicate rocks near Pleystein, accompanied by silicates such as grossularite (hessonite), vesuvianite, zoisite, and minerals of the diopside-hedenbergite s.s.s., in context within a well-defined zone of minerals typical of a high-T contact metamorphism lends support to a W mineralization genetically related to the pegmatites. The presence of W-bearing "nigrine" supports this idea. Neither the large granite massifs, e.g., the Flossenbürg Granit Massif at the northern rim of the HPPP, nor the shallow granitic stocks intruded along deep-seated lineamentary fault zones (e.g., Křížový kámen granite, Czech Republic) had the thermal potential to provoke the crystallization of wollastonite within their calcareous country rocks. Consequently, magmatic rocks of dioritic to even gabbroic composition area around Pleystein possess the thermal potential to raise the temperature in the wall rocks of the HPPP so that minerals of the uppermost hornfels facies can develop in the pertinent host rock lithology. The intimate spatial relationship between the pegmatites and a wide variety of ultrabasic, basic magmatic rocks of subcrustal origin and even metacarbonates has to be paid more attention to during the study of pegmatites. The stock-like pegmatites have a characteristic aureole with stockwork-like veins and veinlets in the hanging wall zone. Within this zone the amount of pegmatitic veinlets gradually increases at the expense of aplites when approaching the pegmatite stock. It is accompanied by a strong kaolinization representing a supergene alteration onto a hypogene one. By contrast, the footwall sequence of the pegmatite consists of a layered aplite. The intensity of the kaolinization gradually changes towards the pegmatite reached its maximum in zone underneath and to the east of the Kreuzberg Pegmatite where feldspar was no longer stable under the existing physical-chemical regime and completely eradicated from the aplites, aploids and pegmatoids. The hypogene kaolinization which has neither an equivalent in the remaining pegmatites of the HPPP nor elsewhere in the granitic complexes in NE Bavaria has eaten away the feldspar rim of the Pleystein pegmatite and left behind the "quartz pegmatite ruin". The timing of the process may be seen in context with the proposed epithermal processes (shallow hydrothermal mineralization) at the end of the Variscan orogeny. Corresponding to their structural setting,

H.G. Dill, *The Hagendorf-Pleystein Province: the Center of Pegmatites in an Ensialic Orogen*, Modern Approaches in Solid Earth Sciences 15,
DOI 10.1007/978-3-319-18806-5_5

the complete suite from pre-kinematic meta-lamprophyres to post-kinematic lamprophyres could be mapped in the environs of the Kreuzberg Pegmatite. They point to a subcrustal source. High Nb, Ta, P and Ti contents underscore the impact of these mafic rocks on the felsic intrusive rocks around Pleystein. From the structural point of view the lamprophyres are part of a dyke swarm, whose individual dykes are arranged NNE-SSW being an integral part of a shear zone or strike-slip fault together with the Kreuzberg Pegmatite and the newly discovered aplite. The pegmatite at Pleystein developed in an open cylindrical fold whose axis gently plunged towards the WSW. The fold axis has been calculated to dip away at angle between 10° and 15° towards the WSW at a distance of approximately 100 m west off the quartz core. All tabular pegmatites and aplites strike NW to NNW. Their stock-like counterparts, such as Hagendorf-South reside on fold structures running almost in the same direction NW-SE (2nd order fold). The pegmatite bodies of the stocks pinch out towards the WSW to SW. The central facies of the pegmatite system composed of the Hagendorf-South, Hagendorf-North and Pleystein stocks plunges towards the E and SE. Mimetic or facsimile crystallization played the most decisive role from the micro- to the larger scales when it came to the emplacement of the stock-like and tabular pegmatites and aplites in the HPPP. Preexisting fold structures created during the Variscan Orogeny behaved in two different ways as conduit and as traps for the mineralizing solutions and melt, all in one.

5.1 Lithology and Regional Economic Geology

5.1.1 Contact Metasomatism and Contact Metamorphism and Pegmatites

In the previous Chap. 4, emphasis has been placed upon the minerals found in the rare metal pegmatites and aplites of the HPPP (Tables 4.1 and 4.2). The physical-chemical regime under which the minerals evolved is debated there. This discussion in section 4 might, however, may lead the reader to suppose that the "Big Three" (Hagendorf-S, Hagendorf-N and Pleystein), together with their satellite aplites and pegmatites, all rife with phosphates, are the only felsic mobilizates in the region. This is not the case as the reader may have learnt from Fig. 5.1. Numerous feldspar-, quartz- and mica pegmatites devoid of rare metals such as Li or Nb crop out mainly to the West and South of the HPPP. The chemical symbols used in the map of Fig. 5.1 reveal the ratio of K feldspar to Na feldspar while the locality of aplites and pegmatites relative to each other is given. The zone of contact metamorphism is marked by a shaded areas taking on an elliptical shape in the map and extending in NNE-SSW direction from the Kupferholz towards the Galgenberg, to the W of Pleystein. This yellow area delineates a zone exclusive to wollastonite in the region, the index mineral of high-temperature contact metamorphism (Fig. 5.2a). The zone went unnoticed, as the region was mapped for the official geological map on scale

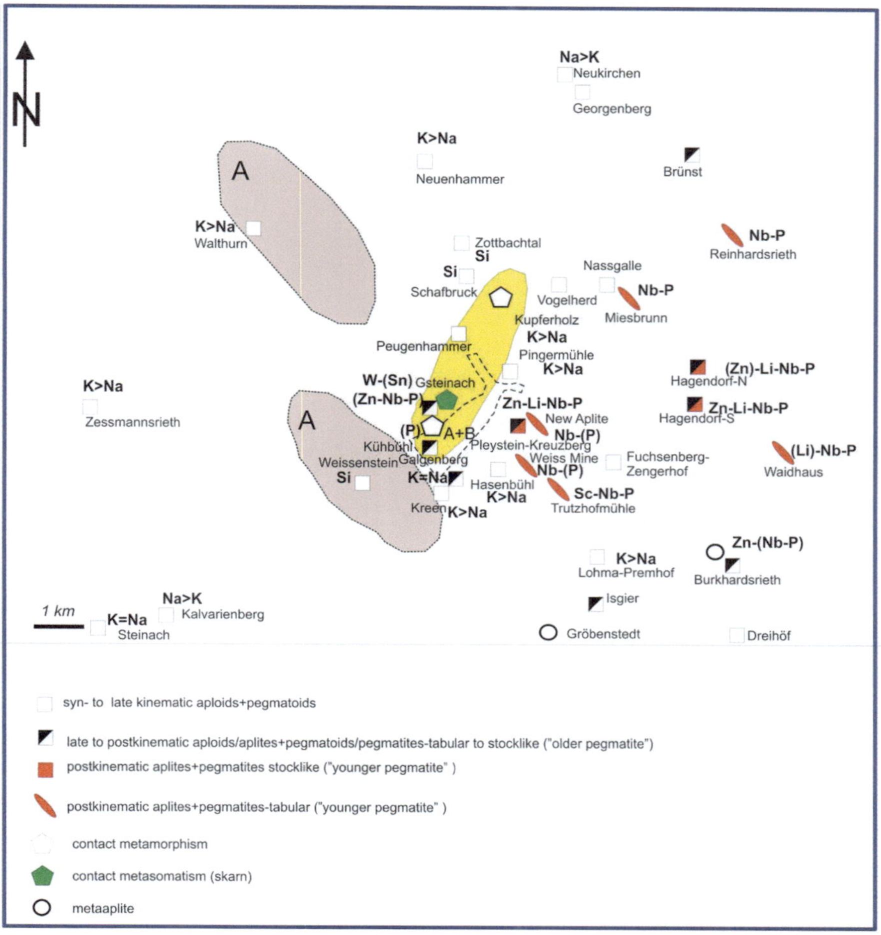

Fig. 5.1 Pegmatitic and aplitic rocks at outcrop within and along the external contact of the HPPP. The chemical symbols denote the specialization of rare metal pegmatites. In the pegmatites and aplites barren as to those rare elements the ratio of K feldspar to Na feldspar is indicated by the chemical symbols K and Na. The elliptically-shaped areas in *grey shades* (A) outline the area abundant in quartz dikes striking NW-SE. The area shaded in *yellow marks* the "hot spot" with skarn and wollastonite-stage contact metamorphism to the west of the HPPP

1:25,000. The process leading to wollastonite when CaO and SiO_2 react with each other has already been discussed b V.M Goldschmidt and can be found in almost each textbook of petrology. And as simple it may appear, the temperature of formation is difficult to deduce from the reaction if the partial pressure of CO_2 (X_{CO2}) is unknown. Observations in the field suggest that the X_{CO2} was rather high and a temperatures of 600 °C existed at a depth of 2 km below ground, corresponding to 500 bar. Doubling the depth needs temperatures of between 650 and 670 ° C to produce wollastonite. Whatever temperature of formation has been invoked in relation to

Fig. 5.2a Bundles of pale white wollastonite make up the hornfels at the Galgenberg area to the W of Pleystein

depth, one thing is for sure. There is a thermal anomaly to the west of Pleystein at the external contact of the HPPP. At the internal contact of the HPPP, the chemical composition of the black Fe-enriched sphalerite from the Kreuzberg Pegmatite is a backing to the idea. The thermal center or the "hot spot" of the rare metal-bearing pegmatites of the HPPP lies in the environs of Pleystein rather than in the Hagendorf area (Sect. 4.27).

During the mid 1970s a German exploration company drilled at Gsteinach, immediately at the outskirts of Pleystein, two DDH into calcsilicate rocks in search of tungsten deposits along the western edge of the Bohemian Massif (Fig. 5.1) (Schmid and Weinelt 1978). The geologists of the company were able to delineate a small but uneconomic mineralization of scheelite, which then was interpreted as stratabound in view of the Hg-Sb-W Formation newly designed by Maucher and published in 1974. The theory was successfully applied in the Felbertal Region, Austria, near Mittersill, where one of the most significant W deposits were discovered (Figs. 1.2d and 5.2b).

The intimate intertonguing of scheelite with calcsilicate rocks near Pleystein, accompanied by silicates such as grossularite (hessonite), vesuvianite, zoisite, and minerals of the diopside-hedenbergite s.s.s., in context within a well-defined zone of minerals typical of a high-T contact metamorphism lends support to a W skarn deposit and discards this former idea of a stratabound scheelite mineralization timebound to the early Paleozoic rocks in NE Bavaria (Fig. 5.2c, d). The development of the HPPP is closely linked with contact metamorphic and contact-metasomatic processes of late Paleozoic age at its edge. While the spatial relation between the heat source immediately to the West of Pleystein and the emplacement of the rare-metal pegmatites in the HPPP is beyond doubt, the question how the thermal event is precisely related in time to the formation of the pegmatites and aplites of the HPPP is a question difficult to be answered.

Fig. 5.2b Anhedral aggregates of scheelite in the garnet-diopside-vesuvianite skarn west of Pleystein at Gsteinach. SEM

Fig. 5.2c Vesuvianite fels with acicular vesuvianite rosettes (*red arrowheads*) at Gsteinach

Neither the large granite massifs, e.g., the Flossenbürg Granit Massif at the northern rim of the HPPP, nor the shallow granitic stocks intruded along deep-seated lineamentary fault zones, such as the afore-mentioned Sparnberg-Pottiga Granite in the Saxo-Thuringian zone or the Křížový kámen (German: Kreuzstein) Granite, Czech republic, an equivalent of the latter, had the thermal potential to provoke the crystallization of wollastonite within their calcareous country rocks. Granitic

Fig. 5.2d Vesuvianite-zoisite fels peppered with well-shaped pinkish porphyroblasts of hessonite Gsteinach

magmas of stocklike or sheet like intrusions with temperatures of as much as 700–800 °C do not reach the level capable of generating stable phases of the high-T hornfels facies, such as wollastonite. Consequently, the area around Pleystein was carefully scouted for more basic magmatic rocks, of dioritic to even gabbroic composition, which possess the thermal potential to raise the temperature in the wall rocks of the HPPP so that minerals of the uppermost hornfels facies can develop in the pertinent host rock lithology.

Three different types of lithology could be mapped proximal to or within the wollastonite zone. Type I has been denominated during field work as biotite-quartz pegmatite *sensu lato* (the structural relation to the wall rocks could not properly determined) (Fig. 5.2e). Biotite is to a great extent converted into clinochlore 1MIIb. The lithology of this pegmatitic rock has a dioritic to gabbroic chemical composition and is not exclusive to the outcrop at Gsteinach but has also attracted notice at Burkhardsrieth, Isgier, Galgenberg and Kühbühl (Fig. 5.1).

Type II magmatic rocks has a gabbroic outward appearance that has been corroborated by the chemical and to some extent also by the mineralogical composition, which encompasses the common biotite, Mg hornblende, clinochlore, albite and Plagioclase. The field-based classification ought to be refined petrographically in favor of a metagabbro (Fig. 5.2f).

To classify the type-III lithology is a hard nut to crack, because it displays a heterolithic composition, as one can see already with the naked eye (Fig. 5.2g). A more detailed mineralogical examination reveals predominantly biotite with some vermiculite, calcite and clinopyroxene whose chemical composition comes close to that of diopside. Texturally it has much in common with a microbreccia (pipe) and looks like a micaceous lamprophyre devoid of feldspar, the source of which is unknown. Even if some readers may rise the eye brows as they hear the result of the petrological diagnosis, there is only one lithology that fits best: kimberlite.

Fig. 5.2e Mafic rock type 1 biotite-quartz pegmatite *sensu lato* in the contact metamorphic and contact metasomatic zone to the West of the Pleystein pegmatite

No	SiO_2	TiO_2	Al_2O_3	Fe_2O_3	MnO	MgO	CaO	Na_2O	K_2O	P_2O_5	Ce	Cr	Cs	La	Nb	Nd	Ni	Sm
1	54.70	1.537	15.47	13.30	0.117	6.41	0.203	0.21	5.927	0.050	57	75	69	42	32	41	108	50
2	50.32	1.404	15.00	9.45	0.161	8.99	7.860	2.08	2.074	0.358	73	726	72	45	9	63	16	48
3	31.46	2.443	5.04	9.85	0.138	16.17	16.210	0.28	1.560	0.520	138	1381	73	91	96	61	776	49

SiO_2 to P_2O_5 in wt. %
Ce to Sm in ppm

The igneous lithologies have a considerable amount of P and Nb while type III is also enriched in Ti, Cr and REE. Even if the enigma of their derivation has not yet fully been solved, one can draw the conclusion that a local (ultra)mafic intrusion preceded the emplacement of the pegmatites in the HPPP. The temperature event was most pronounced at the western edge and faded out towards the SE along the southwestern boundary of the HPPP (Fig. 5.1). An influence of the mantle onto the pegmatite emplacement in this region is more than a mere speculation.

In conclusion, the lamprophyres irrespective of the temporal relation they formed to the pegmatites are the least affected mantle derivates. Type I to III are related to

Fig. 5.2f Mafic rock type 2 metagabbro

the pegmatites but they did not sweep their way through the crust unaffected. On one side they gave rise to the basic pegmatites of type I. On the side, type II here called megagabbro based on texture and lithology is an endocontact reaction zone of the dioritic magma from a subcrustal source when getting in touch with the metacarbonates/skarn suffering slight increase of Ca. Near Žulova, Czech Republic, a similar lithological setting may be recognized along the contact of a dioritic magma and metacarbonates in an area rife with pegmatitic dikes. The intimate spatial relationship between the pegmatites and a wide variety of ultrabasic, basic magmatic rocks and even metacarbonates, which has been shown in numerous publications on pegmatites from across the globe, has been a bit suppressed or it has not been paid the attention to by the students in their striving to find the next parental granitic pluton for the pegmatite.

5.1.2 The Hanging Wall Stockworks and Footwall Layers of Pegmatites in the HPPP

Mining of pegmatites in the HPPP looks back on a very long history – see Sect. 1.3.2 – using open-cut and underground-mining techniques which both resulted in a good coverage of the various pegmatitic and aplitic stocks and sheet like bodies,

Fig. 5.2g Mafic rock type brecciated micaceous lamprophyre

e.g., Hagendorf-South. This is not true for the exocontact zone of these pegmatites, because it is common practice to grind drifting to a halt when the quantity of raw material falls below cut-of-grade. On the other hand, sinking an exploration drill hole rarely hits the target by the first attempt while offering an insight into wall rocks instead. Reading the alteration zones around a rare metal pegmatites tells the mining and exploration geologists what the spatial position relative to the target of exploration is like. In my personal opinion, these subcrops of pegmatites and aplites are under-investigated compared to the huge number of studies devoted to the mineralogy of the pegmatites, proper. As an economic geologist, it simply means: "Do as you would be done by".

Currently, underground operation on pegmatites does not exist anymore and information on the wall rock lithology in the HPPP is poor. In the town center of Pleystein, the relic quartz core of the Kreuzberg pegmatite stands out as a picturesque quartz reef or "aesthetic deposit" while its marginal facies is covered by residential areas. What may be considered as a weak point at first glance, rendering impossible

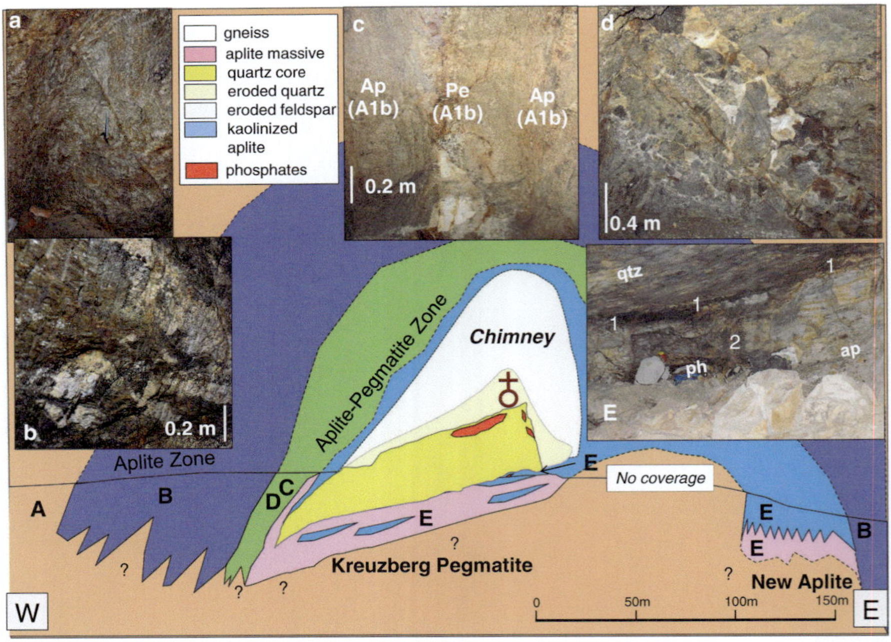

Fig. 5.3a The hanging wall and the footwall zones of the Kreuzberg Pegmatite in the city of Pleystein along a W-E cross section including the New Aplite. *A*: Intensively folded biotite gneiss intersected by a fault in the cellar room 1. Hammerhead for scale. *B*: Disharmonically folded aploid layer within the biotite gneiss. Cellar room 2. *C*: Zoned aplite-pegmatite affected by pervasive kaolinitisation. It dips almost vertically towards the SW. This structural feature may be correlated with the Kreuzberg Pegmatite the plunge of which goes towards the SW. *Ap A1b* aplite tectonic stage 1b, *PeA1b* pegmatite tectonic stage 1b. For more information see section on 5.1.3 Cellar room 4. *D*: Rotated breccia with gneiss clasts floating within an aplitic matrix made up of alkaline feldspar and quartz. It is supposed to be the youngest deformation indentified underneath the housing area of the city of Pleystein. Cellar room 4. *E*: The quartz core (*qz*) of the Kreuzberg Pegmatite exposed by weathering lies in sharp contact with the footwall aplite (*ap*). The fault zone gently dips towards the west and is host to fault gouge of kaolinite (1). Another subvertical joint underneath this subhorizontal fault is also filled with kaolinite (2). The zone marked with "ph" in the aplite is rife with phosphate minerals. The symbol of the *cross marks* the position of the Salesian Monastery on *top* of the Kreuzberg -see also Fig. 1.3. For the "chimney" see also Fig. 5.3b and for discussion Sect. 6.4.2

any surface mapping, is compensated by a number of underground storage rooms which unlike their modern analogues are non-stone built but excavated by the owner from the soft bedrock. Mapping these subcrops enables us to determine the lithological and structural variation around the pegmatite. Along a W-E transect perpendicular to the strike of the quartz pegmatite the different zones are lithologically described in this section and interpreted as to their origin (Fig. 5.3a, Table 5.2).

Zone A In the underground storage rooms belonging to the most distal zone A only tightly folded biotite-sillimanite gneisses are exposed (Fig. 5.3a-A). They are intersected by some normal faults, while aplitic and pegmatitic bodies are absent. Passing

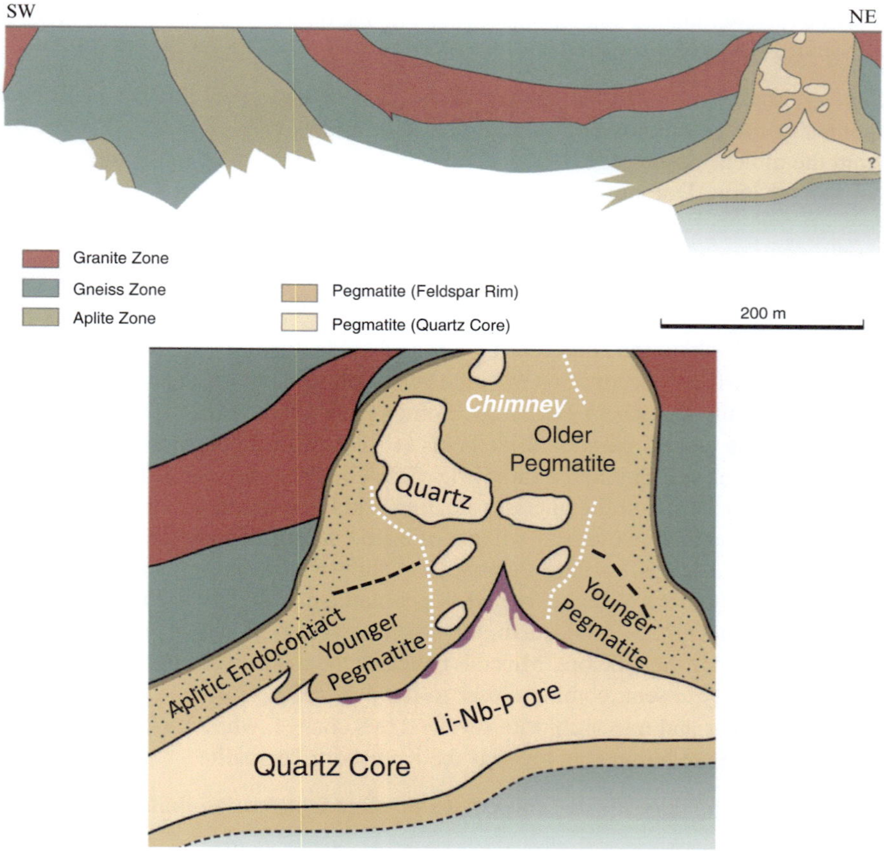

Fig. 5.3b The cross section on *top* illustrates the external zonation within the hanging wall rocks of the Hagendorf-South Pegmatite based on examination of DDH drilling records (simplified) and the structural position of the pegmatite within a fold structure. The Aplite Zone features the same structures as shown in the underground photographs of Fig. 5.3a. Below the Hagendorf-South pegmatite is on display in a close-up view (Modified after Teuscher and Weinelt 1972). Internal subdivision into the Older Pegmatite and Younger Pegmatite corresponding to Uebel (1975). The chimney is a structure of ore control for the rare elements and similar in essence to the greisen structures (For discussion see Sect. 6.4.2)

towards the Kreuzberg Pegmatite cm-sized aplitic mobilizates developed along the cleavage planes of the gneisses. The gneisses were overprinted by a retrograde metamorphism resultant in the conversion of biotite into chlorite and considerable albitization of the preexisting plagioclase. It has to be noted that the fine-grained aplitic margin of the various stocklike pegmatites is also enriched in albite. Diopsidic clinopyroxene and clinozoisite hallmark the presence of calcsilicate rocks among the biotite gneisses (Sect. 5.1.1). Although no coherent masses of calcsilicate rocks were encountered in the cellars, there is a conspicuous accumulation of these calcsilicate rocks in the biotite gneisses of the HPPP relative to the those being further afield.

Mineral assemblage: quartz, muscovite, alkaline feldspar plagioclase, chlorite, ± clinopyroxene, ± clinozoisite

Zone B Zone B is called the aplite zone because of a growing number of disharmonically folded aploids and joints filled with aplites (Fig. 5.3a-B). Concluding from the mineral assemblage below there is only a slight change in the mineralogical composition. Dolomite was found in some of the aplites and K feldspar still prevails over Na feldspar.

Mineral assemblage: quartz, muscovite, alkaline feldspar plagioclase, chlorite, ± clinopyroxene, ± clinozoisite, ± dolomite, ± vermiculite

Zone CD This alteration zone adds up a considerable amount of pegmatite veins to the stockwork-like concentration of aplite veins and veinlets in the gneissic country rocks around the Kreuzberg Pegmatite. Some of the aplite veins increase in thickness while the central parts of these veins is filled with a coarse-grained ground mass of alkaline feldspar and quartz (Fig. 5.3a-C). The presence of breccia zones with rotated clasts of biotite gneisses floating within an aplitic matrix made up of alkaline feldspar and quartz bear witness of brittle deformation in the course of the emplacement of the Kreuzberg Pegmatite. Aplitic melts were still intruding into gneissic country rocks as the main part of pegmatite has already been consolidated to a great extent (Fig. 5.3a-D). As to the mineral association, a conspicuous change may be observed in the subcrops. Minerals typical of the skarn and contactmetamorphic origin are totally absent, coming closer to the pegmatite. Only quartz, muscovite, alkaline feldspar and minor chlorite survived this change, while kaolinite becomes more and more predominating towards the Kreuzberg Pegmatite.

Mineral assemblage: quartz, muscovite, alkaline feldspar, kaolinite, ± chlorite

Zone E Compared to previous zones, in zone E, the chemical alteration has been taken to the extreme. In the monographic study, Wilk (1975) described this zone as "eutectic zone" of a muscovite aplite and kaolinite as a minor component of the Pleystein pegmatite. Both statements have to be corrected, because there is no zonation like that recorded by Wilk (1975) and kaolinite, previously downgraded to minor component, is the most widespread mineral in the aplite, which no longer deserves this technical lithological term on account of the pervasive argillitisation this aplite was subjected to. This can be inferred from the mineral assemblage and from the chemical composition listed in the table below (Table 5.1). There is a remarkable zonation along a profile perpendicular to the quartz core for Zn, Ba, Fe, and Mn, whereas Nb and Ta in columbite-(Fe) are left untouched. Phosphate is contained in Fe- and Mn phosphates associated with goethite and cryptomelane, the latter oxides are supergene and were precipitated from meteoric waters percolating along a subhorizontal contact fault zone between the footwall aplite and the quartz core (Fig. 5.3a-E).

Mineral assemblage: quartz, muscovite, kaolinite, ± goethite.

The hanging wall and footwall rocks of the Kreuzberg Pegmatite, exposed in the city of Pleystein, were subjected to the same type of regional metamorphism- (Sect. 2.1.5). Calcium silicates, excluding wollastonite typical of a medium-grade contact

Table 5.1 Chemical profiling underneath the quartz core in the so-called "aplite". See zone "E" in Fig. 5.3a

Site (above 0 in m)	Major	Minor	SiO_2 wt. %	TiO_2 wt. %	Al_2O_3 wt. %	Fe_2O_3 wt. %	MnO wt. %	MgO wt. %	CaO wt. %	Na_2O wt. %	K_2O wt. %	P_2O_5 wt. %	Ba wt. %	Bi wt. %	Nb wt. %	Ta wt. %	U wt. %	Zn wt. %	Nb/Ta	Quartz core
1.40	Quartz, Muscovite	Kaolinite, goethite	23.71	0.054	19.42	26.15	10.880	0.15	0.234	0.09	4.595	1.773	1787	8	108	113	42	1974	0.96	Zn-P-Mn-Fe ore zone
1.30	Quartz, Muscovite	Kaolinite	69.42	0.022	18.63	2.43	0.462	0.08	0.062	0.11	3.976	0.142	81	13	152	197	8	155	0.77	Muscovite Zone
1.20	Quartz, Muscovite	Kaolinite	76.66	0.015	14.27	1.60	0.142	0.05	0.082	0.06	2.297	0.275	226	14	204	313	12	135	0.65	Muscovite Zone
1.10	Quartz, Muscovite	Kaolinite	78.88	0.021	12.91	1.76	0.035	0.04	0.091	0.09	3.323	0.109	63	88	154	201	8	82	0.77	Muscovite Zone
1.00	Quartz	±Muscovite	98.79	0.001	0.36	0.39	0.019	0.01	0.015	0.01	0.042	0.034	61	133	5	10	8	14	0.50	Fault zone with quartz
0.80	Quartz, Muscovite	Kaolinite	74.95	0.020	15.60	1.65	0.045	0.05	0.145	0.09	3.351	0.051	63	52	173	216	8	115	0.80	Muscovite Zone
0.40	Quartz, Muscovite, Kaolinite		69.92	0.010	19.53	1.22	0.161	0.05	0.087	0.08	2.718	0.070	62	5	152	242	8	144	0.63	Kaolinite Zone
0.20	Quartz, Muscovite, Kaolinite		67.03	0.015	21.57	1.59	0.089	0.06	0.077	0.09	3.134	0.079	62	20	149	191	8	130	0.78	Kaolinite Zone
0.00	Quartz, Muscovite, Kaolinite		68.82	0.014	20.66	1.31	0.047	0.06	0.072	0.09	3.100	0.056	62	34	155	241	8	104	0.64	Kaolinite Zone

Table 5.2 Chemical composition of aplites and pegmatites (mean) in the alteration zones around the Kreuzberg Pegmatite at Pleystein

Zone	SiO_2	TiO_2	Al_2O_3	Fe_2O_3	MnO	MgO	CaO	Na_2O	K_2O	P_2O_5	Ba	Bi	Ce	Nb	Nd	Rb	Sr	Ta	Zn	Zr	Rb/K2O *100	Nb/Ta
NA	63.16	0.02	24.06	1.18	0.11	0.02	0.04	0.10	3.62	0.13	80	10	17	105	12	890	29	33	156	30	3.05	3.44
E	69.80	0.02	15.88	4.23	1.32	0.06	0.10	0.08	2.95	0.29	274	41	85	139	65	747	91	192	317	18	3.16	0.72
CD	70.41	0.42	16.76	3.10	0.06	0.62	0.06	0.63	3.02	0.12	370	6	80	23	37	374	64	12	64	145	1.48	1.85
B	68.22	0.36	16.40	3.00	0.09	0.73	1.89	2.01	3.33	0.33	572	4	71	11	30	168	190	8	55	104	0.56	1.92
A	75.82	0.39	12.72	3.02	0.04	1.06	0.24	1.83	1.97	0.07	786	2	49	8	13	61	121	4	43	168	0.37	2.00

NA **New Aplite**
Kreuzberg Pegmatite: Zone E, zone CD, zone B, zone A
SiO_2 to P_2O_5 given in wt. %, Ba to Zr given in ppm
For location see Fig. 5.3a

metamorphism of between 500 and 550 °C have been observed in these metamorphic rocks which are exposed in zone A (Fig. 5.3a). Calcsilicate minerals are arranged in a characteristic zonation around the Kreuzberg Pegmatite. They do not only act as a marker of the temperature regime but are also indicative of chemical reaction. There is a depletion in Ca along with a gradual increase in Na as approaching the present-day outcrop of the stock-like pegmatite and last but not least a concentration in K and Si. The same chemical scenario may be reported from the sheet-like aplite- and pegmatite swarms. The mineralogical scenario is slightly different, particularly in the exocontact zone of the pegmatites. In the exocontact of the stock-like pegmatites, a calcsilicate mineralization is the source of Ca, whereas in the exocontact of the tabular pegmatites, plagioclase behaves as source of Ca. Plagioclase is inherent to the medium- to high-grade regional metamorphism of the gneissic country rocks, whereas the calcsilicates minerals vesuvianite, grossularite etc. are the mineralogical expression of an additional heat pulse.

Conclusively, a strong differentiation, confined to the zone pegmatite stocks is characterized by the succession calcsilicate minerals ⇒ Na feldspar ⇒ K feldspar ⇒ quartz. A moderate differentiation, confined to the tabular aplitic or pegmatitic rocks is characterized by plagioclase ⇒ Na feldspar ⇒ K feldspar (Fig. 5.5d-B).

Compared with zone A that is scarcely endowed with some aplitic gneisses, in zone B true aplites and aploids occur in otherwise similar physical-chemical regime. The onset of minor quantities of kaolinite, while feldspar is still preserved, can be recorded from zone CD onward. The intensity of kaolinization reached its maximum in zone E, where feldspar was no longer stable under the existing physical-chemical regime and completely eradicated from the aplites, aploids and pegmatoids. There is no doubt that a supergene kaolinization affected the granites and pegmatites alike in the region. Quite in contrast to this supergene kaolinization which has a region wide extension with subhorizontal boundaries in the weathering mantle, the kaolinization at Pleystein is of hypogene origin with its boundaries more or less arranged vertically in an onion-shell pattern around the pegmatite. The hypogene kaolinization which has neither an equivalent in the remaining pegmatites of the HPPP nor elsewhere in the granitic complexes in NE Bavaria has eaten away the feldspar rim of the Pleystein pegmatite (Fig. 5.3a). An assessment of the size of the original pegmatite, using the size of the still existing quartz core as a yardstick, lead to the conclusion that the Kreuzberg Pegmatite was almost of the same size as Hagendorf-South or even larger. Why did the "two brethren", only a few kilometers away from each other, not share the same fate with respect to this hypogene argillitisation and when did their evolution parted ways?

The intensity of kaolinization which has no match in or outside the HPPP in NE Bavaria may be accounted for by the position of the Kreuzberg Pegmatite right on top of the "hot spot" of the pegmatite system. Acidic hydrothermal fluids (<5 pH) at temperature below 390 °C (kaolinite in – pyrophyllite out) eradicated the feldspar rim of the Kreuzberg Pegmatite. The present-day landscape of a bowl-shaped depression with the pinnacle of the quartz core in the midst of it, is intersected by the Zottbach-Pflaumbach river drainage system. Its linear erosion was kickstarted and speeded up by the hypogene kaolinization when its thalweg cut into the soft alteration zone underneath the metamorphic roof rocks (Fig. 6.1a). As the intensity of the kaolinization was more intensive at the eastern side of the Kreuzberg Pegmatite, the drainage system, following the path of least resistance shifted its meander bend towards the E (Figs. 5.3a and 6.1a). The fluvial morphology around Pleystein is controlled in its evolution by the spatial distribution of the aureole of kaolinization around the pegmatite (Fig. 5.5d-E).

The timing of the process may be seen in context with the proposed epithermal processes at the end of the Variscan orogeny. During this period of time kaolin only formed from hydrothermal solutions at depth, see also the stock-shaped intrusion in the granite porphyry at Altenberg, Germany, with its extensive greisen zones (lithium-albite granite in which feldspar and biotite are converted to a disseminated assemblage of quartz, topaz, muscovite, zinnwaldite and protolithionite (both Li-micas), cassiterite, sericite, fluorite, dickite, kaolinite, wolframite and scheelite) but also under supergene conditions (Baumann et al. 1986). Further examples from western Europe can be reported from Cornwall, Great Britain (Dominy and Camm 1998).

Late Variscan uplift led to extensive erosion as a result of pervasive chemical weathering and denudation of the basement rocks under (sub)tropical climatic conditions as it is currently the case in central Africa, where a thick regolith of kaolin is being formed and felsic intrusive rocks either were exposed or brought into the reaches of the hydraulic zone of the late Paleozoic peneplain. The essential geomorphological and hydraulic prerequisites for shallow epithermal deposits to form were provided. Kaolin quarried at Podbořany, Czech Republic, immediately overlies this unconformity in the form of a palaeoregolith/saprolite or is found redeposited within arkoses and conglomerates of Carboniferous age (Kužvart 1968).

The timing of the kaolinization at Pleystein can only be constrained by a closer look at the phosphate minerals. Primary phosphates are present but frankly spoken are very rare in terms of quantity and as to the diversity are no real match to Hagendorf-South. In the kaolinized zone strengite, phosphosiderite, vivianite and rockbridgeite have been identified. Their presence well agrees with the physical-chemical regime supposed to have been during the kaolinization. The phosphate minerals belong to mineral assemblages that formed during the hydrothermal stage.

The results obtained by the underground mapping of the Kreuzberg Pegmatite at Pleystein paved the way to an in-depth study of the wall rocks of the nearby Hagendorf-South pegmatite. Drill records provided to the local mining authorities formed the basis for the simplified SW-NE cross section through the Hagendorf-South Structure. The aplite zone in this cross section does not represent a homogeneous or coherent layer but a strongly deformed section of the metamorphic country rocks intersected by several criss-crossing aplitic veinlets (stockwork-like) prevailing over pegmatitic veins. The same is true for the granite zone that consists of a series of mobilizates, different in their thickness and intercalated into the folded gneisses.

Both pegmatites from Hagendorf-South, shown in Fig. 5.3b, and at Pleystein depicted in the cross section of Fig. 5.3a, found enough accommodation space to develop an economic ore body within a complex fold structure. Both pegmatite stocks have the same pattern of alteration within the hanging wall rocks and the footwall rocks.

The distinct features of the hanging wall and footwall rocks of pegmatites can be explained in two different ways. While the overall tectonic pattern around the pegmatites is controlled in its 3-D representation by the stress-strain relations in the region, the textural and structural outward appearance, such as multiple aplitic veinlets and brecciation is clear evidence for a forceful emplacement of the pegmatitic bodies. High contents of volatiles in the siliceous melt shattered the roof rocks but left the footwall rocks unharmed. It is not a centripetal growth of minerals filling an open space but a unidirectional top-down precipitation of minerals during the initial stages of the emplacement of pegmatites. The position of the planar architectural elements and the intensity of fracturation as well as brecciation in the hanging wall rocks of pegmatite system can be used as exploration tools and proximity indicators. While these structural and textural phenomena positively correlate with the nearness to the pegmatite system, mineralogical features, such as skarn mineral assemblages negatively correlate with the proximity to the pegmatite system. Given the chemical composition of the country rocks allows for the formation of such minerals, the metasomatic processes wane towards the pegmatite system.

5.1.3 *The Lamprophyres and Pegmatites in the HPPP – The Mantle Impact*

Who will be concerned about rainy days and cloudbursts while spending the holidays at the sea sun tanning ? Who will take a look at thin dykes of black dull rocks, being in front of a huge wall of white felsic rocks with colorful phosphates? The question is self-explanatory. This is to describe how lamprophyres are treated in the HPPP. When Forster (1965) issued the official geological map, he also reported on thin dykes of massive black rocks cutting through the Silbergrube Aplite at Waidhaus. He failed to determine the mineralogical composition of the igneous rocks due to the strong alteration he claimed to have affected the original mineral assemblage. Under the petrographic microscope the magmatic dikes investigated during the current study display a micro-porphyritic texture. In the thin section Mg-enriched hornblende, plagioclase, biotite, quartz, Fe-Mg chlorite and muscovite can be identified (Fig. 5.4a). These rock-forming minerals are arranged in a random granular texture, despite being deformed by the late Variscan orogeny, and cut by step-wise faults. This fracturing in the intensively folded gneisses is conducive to a noticeable boudinage of the lamprophyres and, in places, led to isolated lens-shaped lamprophyres floating in a gneissic matrix (Fig. 5.4b). Aplite veins, themselves disharmonically folded, cut through these meta-lamprophyres which may be assigned a late-kinematic position within the Variscan structural distortion. Lamprophyres were only spotted within the "Aplite Zone B" – see Fig. 5.3a. Corresponding to their structural setting, the complete suite from pre-kinematic meta-lamprophyres to

Fig. 5.4a Meta-lamprophyres from Pleystein under the petrographic microscope (crossed polars)

plagioclase (white) +ilmenite (opaque)

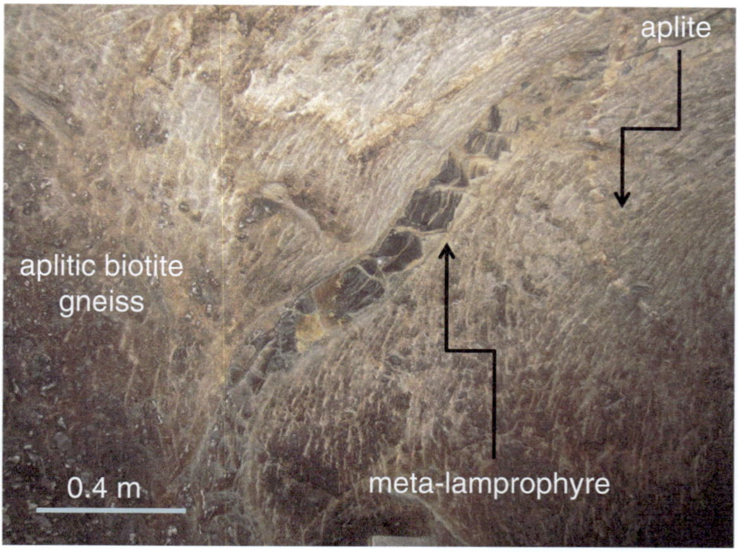

Fig. 5.4b Meta-lamprophyres at outcrop in a cellar next to the Kreuzberg Pegmatite. The dyke rocks formed late-kinematically relative to the Late Variscan structural disturbances, they are intersected by aplites which themselves were involved in the fold processes affecting the aplitic biotite gneisses. Beginning boudinage of the lamprophyre during on-going local extension

post-kinematic lamprophyres could be mapped in the environs of the Kreuzberg Pegmatite. The meta-lamprophyres are by far the most homogeneous magmatic rocks in the region and have the most elevated contents of Ti which is contained in ilmenite grains (Fig. 5.4a, Table 5.3). The lamprophyric rocks have a basaltic chemical composition and have rather high P and Nb contents. A significantly high susceptibility of 0.43 S.I. measured with a handheld kappameter attests to a considerable concentration of magnetite, too. In the common classification scheme these lamprophyres would plot into the field of kersantites *sensu lato* or camptospessartites. The classification of these dark fine-grained magmatic dyke rocks is often fraught with difficulties and the issue of classification not solved to everybody's satisfaction (Wooley et al. 1996; Le Bas 2007). From the Sn-W-Mo mining district of Krupka, Novák et al. (2001b) reported lamprophyres which were found in a similar position as the ore mineralization and the granites as at Pleystein. Förster et al. (1999) published a lower limit for the age of formation of kersantites of 320 Ma. These dykes were intersected by S-type granites at Ehrenfriedersdorf, Germany. These basaltic dyke rocks pre-date and post-date the pegmatite emplacement at Pleystein and imply a deep-seated magmatic process of mantle derivation framing the emplacement of the pegmatites. Even if the number of outcrops where lamprophyric rocks are exposed is scarce, a preliminary statement can be made. Meta-lamprophyres were discovered only in gneisses adjacent to pegmatite stocks, while lamprophyres have been found intersecting aplites such as at Silbergrube.

Table 5.3 The chemical composition of meta-lamprophyres from the "Aplite Zone B" (see Fig. 5.3a) at Pleystein

SiO_2	TiO_2	Al_2O_3	Fe_2O_3	MnO	MgO	CaO	Na_2O	K_2O	P_2O_5	Ce	Cr	La	Nb	Nd	Ni	Rb	Sm	Sn	Ta	Y	Zn	Zr
50.31	4.210	12.81	15.01	0.22	5.79	6.455	0.43	0.713	0.51	56	108	44	40	42	152	74	52	38	15	43	231	293
50.23	4.167	12.60	15.17	0.21	5.94	7.476	0.39	0.490	0.54	58	116	22	39	27	97	22	22	13	6	46	171	302
49.93	4.151	13.09	15.01	0.20	5.83	8.057	0.43	0.383	0.55	68	112	27	38	26	75	15	18	13	7	46	149	292
50.50	4.218	12.86	15.40	0.22	6.03	7.985	0.40	0.222	0.50	66	114	45	41	43	81	15	53	38	16	47	219	303
49.43	3.680	13.58	14.11	0.22	5.99	7.454	0.49	1.531	1.02	57	160	45	36	42	97	100	52	39	15	42	260	268

SiO_2 to P_2O_5 are given in wt. %
Ce to Zr are given in ppm

High Nb, Ta, P and Ti contents underscore the impact of these mafic rocks on the felsic intrusive rocks around Pleystein. With the aid of these mafic dykes a more precise picture of the structural evolution in the HPPP can be presented in the following Sect. 5.1.4. Van Lichtervelde et al. (2006) came to a different view as to the role of metabasic igneous rocks within the pegmatite at Tanco, Canada. The Tanco granitic pegmatite, in SE Manitoba, is not only emplaced in metagabbros, but also encloses rafts of the host metagabbro, associated with abnormally high Ta concentrations close to these rafts. The authors concluded that the rafts do not appear to have any chemical influence on the crystallization of columbite group minerals but acted as a physical barrier that separated distinct pegmatite cells that evolved independently from the whole body.

5.1.4 *The Structural Geology of the Pegmatites and Aplites in the HPPP*

The storage rooms underneath the residential sites in the city of Pleystein provided "lithological documents" indispensible to map the variegated lithologies around the Kreuzberg Pegmatite and to establish a schematic picture of its alteration zones (Sect. 5.1.2). Moreover, the cellars proved to be a good occasion to collect data on the strike, dip, and thickness of the various felsic mobilizates so as to cast a bit more light on the tectonic setting of the Kreuzberg Pegmatite that is essential for a better understanding of the overall structural evolution of the HPPP.

To achieve this goal of fine-tuning the structural evolution of the felsic mobilizates and to show the 3-D arrangement of aplites and pegmatites in the HPPP, their planar elements (contact zones, joints) and the plunge of their axes have been plotted into a stereo net diagram (Schmidt Net equal area presentation) to generate the great circles and calculate the density contour intervals of the different 2-D and 1-D structural elements. Stereo net diagrams of sites outside the town of Pleystein of relevance for the understanding of the genesis of the Kreuzberg Pegmatite, such as those at Finkenhammer at outskirts of Pleystein were also integrated into this structural analysis. They were equiangularly shifted in accordance with their correct position and fitted into the city map in order to assist the correlating of the various planar structural elements (Fig. 5.5a).

A handheld high-sensitive gamma ratemeter has proved to be a useful supplement to measure the intensity of the gamma radiation in the surroundings of the Kreuzberg Pegmatite in microsievert per hour for several station points (Fig. 5.5b). This large-scale radiometric survey compared to the regional geophysical survey discussed in Sect. 3.3 is considered in this section because of its intimate relation to the structural geology of the pegmatites. A large-scale coverage of the mapping area around Pleystein with gamma-readings gives a clearer picture of how the pegmatite evolves towards depth. The pegmatite has a higher gamma rate for their elevated potassium and uranium contents, and even if the felsic rocks are not exposed along the walls or

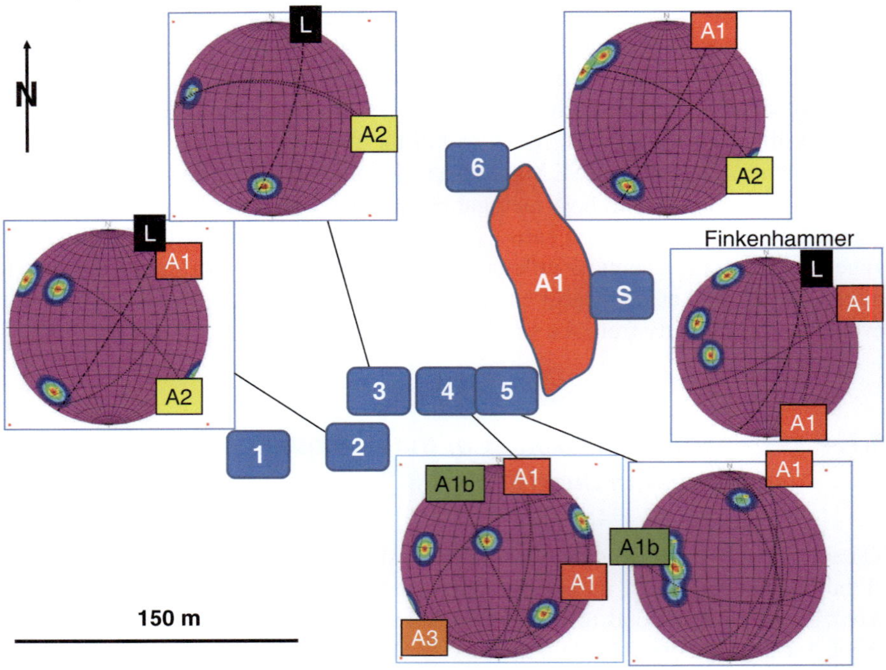

Fig. 5.5a Structural geology of the environs of the Kreuzberg Pegmatite. The planar structural elements are plotted into a stereo net diagram (Schmidt Net equal area presentation) to show the great circles and density contour intervals of the various features. The stereo net diagram of the site Finkenhammer is located outside the city map. It was shifted equiangularly towards the left and fit into city map to assist in correlating the various planar structural elements. L = great circle of lamprophyre dikes. A1 = great circle of the pegmatites and aplites that are genetically linked to the structural processes inherent to the emplacement of the Kreuzberg Pegmatite. A2+A3 = great circles of the aplites pertaining to younger structural processes. *Blue symbols* with *Arabic Numerals* denote sites studied by underground mapping at Pleystein. The symbol *S* shows the adit of the Kreuzberg trial mining operation

roofs of the non-stone-built underground storage sites used for the gamma survey, the gamma read-outs are helpful in delineating the shape and dip of the pegmatite system, which are shown in the Fig. 5.5b and discussed in more detail in the next paragraph.

Taking into consideration all the results that were presented in the previous sections on mineralogy, chemistry and geophysics, one can rule out a simple emplacement of the Kreuzberg Pegmatite as a result of an in-situ differentiation of one of the large Oberpfalz granites adjacent to the HPPP. Structural geology has played a significant part during the formation of the pegmatite but for quite a long time not been paid the attention to it deserves (Fig. 5.5d). According to Weber and Vollbrecht (1989), the latest structural disturbances took place during the Variscan Orogeny between 450 and 330 Ma. The early aplites and lamprophyres were distorted by these processes and assigned a late-kinematic position within the

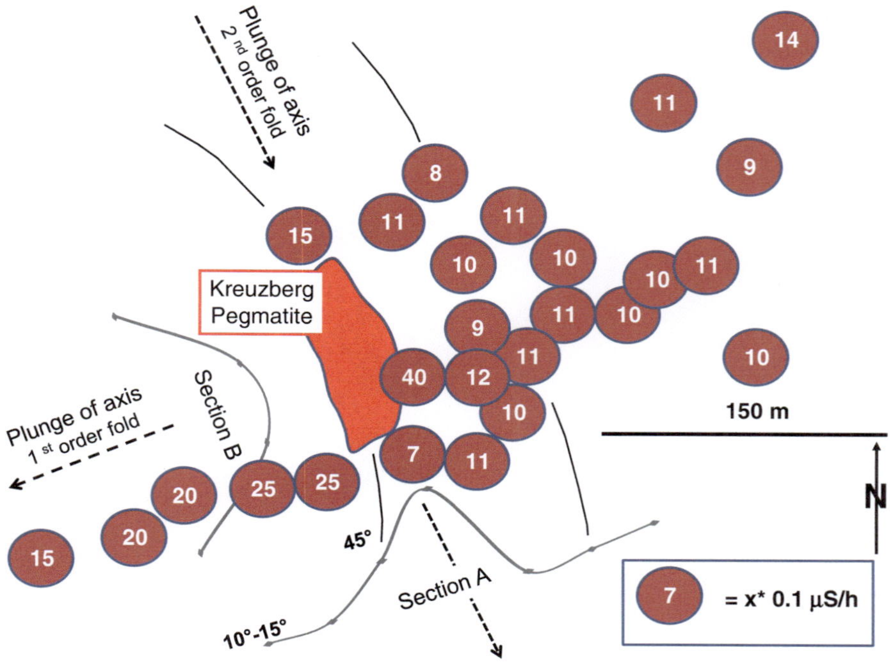

Fig. 5.5b Intensity of the gamma radiation in the surroundings of the Kreuzberg Pegmatite given in micro sievert per hour for each station point. The *curves* denote the gradient of the gamma radiation which is equal to the dip and shape of the pegmatite body. The different folds are marked by the plunge of the fold axis. The first-order fold is a symmetrical fold, the second-order fold an asymmetrical fold (Dill and Weber 2014b)

structural setting. From the structural point of view the lamprophyres are part of a dyke swarm, whose individual dykes are arranged NNE-SSW being an integral part of a shear zone or strike-slip fault together with the Kreuzberg Pegmatite and the newly discovered aplite (Fig. 5.5a, c). The A 2 deformation gave rise to a structural pattern with NW-SE-striking structures that is representative of a more advanced level during the lateral movement. The opening of the structure providing the accommodation space of the "New Aplite" took place during this younger stage of deformation. Deformation stage A 3 can be used to contribute to the shaping of the pegmatite-see next paragraph. For a regional outline see also the block diagram of Fig. 3.12, where the lineamentary structure zones formed the guideline to the structural deformation initiated by deep-seated mantle-related processes.

Aided by the structural analysis and the data obtained from the radiometric survey the morphology of the pegmatite body can be approximated at depth and displayed in two cross sections intersecting each other at a right angle (Fig. 5.5b). The pegmatite at Pleystein developed in an open cylindrical fold – Fig. 5.5a section B- whose axis gently plunged towards the WSW (Fig. 5.5b). This is also indicated by the great circles representing the deformation of stage A 3 in the stereo net

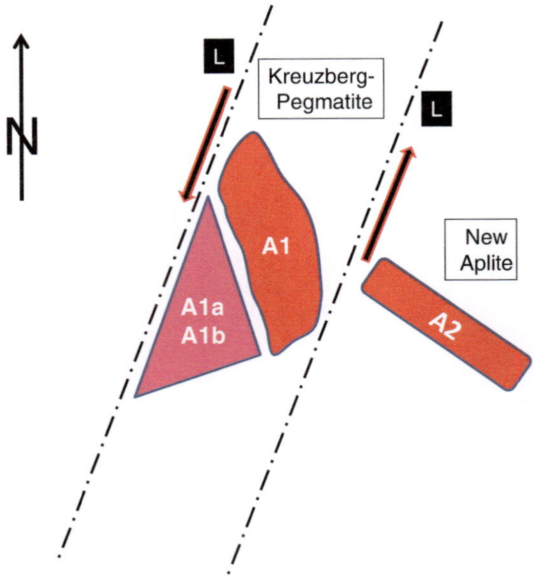

Fig. 5.5c Interpretation of the structural environment of the Kreuzberg Pegmatite in Pleystein as a strike slip fault system. For legend see Fig. 5.5a

diagram at station point 4 (Fig. 5.5a). Based upon the readings in this stereo net diagram the fold axis has been calculated to dip away at angle between 10° and 15° towards the WSW at a distance of approximately 100 m west off the quartz core (Fig. 5.5a, b-section A). The dip angle of the pegmatite abruptly rises to approximately 45° as approaching the quartz core of the Kreuzberg (Fig. 5.5b-section A). The orientation of this pegmatite is controlled by a second-order fold showing a NW-SE trend and running perpendicular to the master or first-order fold with its fold axis gently dipping towards the SE (Fig. 5.5b).

In general, what was the structural setting like that governed the emplacement of the various tabular and stocklike pegmatites and aplites, respectively, in the HPPP. All tabular pegmatites and aplites strike NW to NNW. Their stock-like counterparts, such as Hagendorf-South reside on fold structures running almost in the same direction NW-SE (2nd order fold) as it was determined for the Pleystein host structure. The pegmatite bodies of the stocks pinch out towards the WSW to SW as shown by the gamma map (Fig. 5.5b). The central facies of the pegmatite system composed of the Hagendorf-South, Hagendorf-North and Pleystein stocks plunges towards the E and SE.

The most recent structural analysis of the NE Bavarian Basement encompassing the country rocks of HPPP was performed by Stein (1988), who subdivided the Variscan deformation of the Moldanubian basement rocks into six different stages, each characterized by folds and faults different as to their morphology, dip and plunge of their fold axes. Petrographic results mentioned in previous sections stress a late to post-kinematic emplacement of the pegmatites and aplites of the HPPP. Pre-kinematic metapegmatites have been mapped in the area and displayed in the official geological map. Concluding from our most recent investigations, their significance among the felsic mobilizates has been overestimated and the majority of felsic rocks shown in Fig. 5.1 needs to be downgraded from the level of a metapegmatite to the level of a pegmatoid. Sheared marginal zones and a faint alignment of rock-forming minerals near the contact with the gneissic country rocks do not qualify these felsic mobilizates as metapegmatites because the overall textural features attest to an unstrained growth of the rock-forming minerals (Fig. 4.24a). Mimetic or facsimile crystallization played the most decisive role from the micro- to the larger scales when it came to the emplacement of the stock-like and tabular pegmatites and aplites in the HPPP. Preexisting fold structures created during the Variscan Orogeny behaved in two different ways as conduit and as traps for the mineralizing solutions and melt, all in one. In the cartoons of Fig. 5.5d, the different steps of the structural built-up of pegmatites and aplites and their destruction covering the interval from the Late Paleozoic through the Cenozoic are depicted in an evolution scheme.

Stage A is represented by an ideally similar fold of Late Paleozoic age that was still in the making when the pegmatites were emplaced. The two opposing arrowheads indicate the pressure exerted on the limbs of this fold structure. The ideas put forward by Uebel (1975) who showed in a cartoon the structural differentiation into an upper “Older Pegmatite” and a “Younger Pegmatite” can be agreed upon – see also (Fig. 5.3b). As to the mode of emplacement of both pegmatites, modifications to the model are necessary, because in describing the emplacement of the pegmatites the author sticked too strongly to a one-sided, mineralogically-minded view, underestimating the structural framework and ignoring satellite pegmatites and aplites adjacent to Hagendorf. Many of these felsic mobilizates were either not yet discovered during the mid-1970s or they were held to be of lesser importance, an argument certainly that is very much convincing as far as the wealth and beauty of phosphates is concerned. Hagendorf-South is still today second to none. The aplitic and pegmatitic layers on the limbs underwent strong deformation with their initial thickness being drastically reduced. Little accommodation space was produced during this late-kinematic movements along the limps and tabular bodies are to be expected only near the hinge area. During the emplacement of the older pegmatites which have region-wide distribution from Zessmannsrieth to Georgenberg and from Burkhardsrieth to Neuenhammer, fold hinges were the preferred *loci* for the concentration of pegmatitic melts (Fig. 5.1). This is especially true for the Pleystein and Hagendorf anticlines (Figs. 5.3a, b and 5.5d-A). While at Hagendorf-South the original arrangement of older and younger pegmatites has well been preserved it can only be assumed to have existed at Pleystein. Corresponding to the cooling ages of muscovite from Brünst yielding 316±3 Ma and of Hagendorf-South yielding 317±3 Ma the age of formation of the older pegmatite is definitely older than 317±3 Ma, and hence predates the intrusion of the Flossenbürg Granite at 311.9±2.7 Ma (Wendt et al. 1994; Glodny et al. 1995b). There have been various ideas on how such a zoning in pegmatites might have taken place, put forward among others by Munoz (1971) and Stewart (1978). The theories encompass monophase processes such as sequential crystallization of a single melt, and alteration of a parent siliceous melt by later aqueous solutions or a multi-stage process. As shown in the cartoons and in the text the situation is a bit more complex and the result of a wide range of processes.

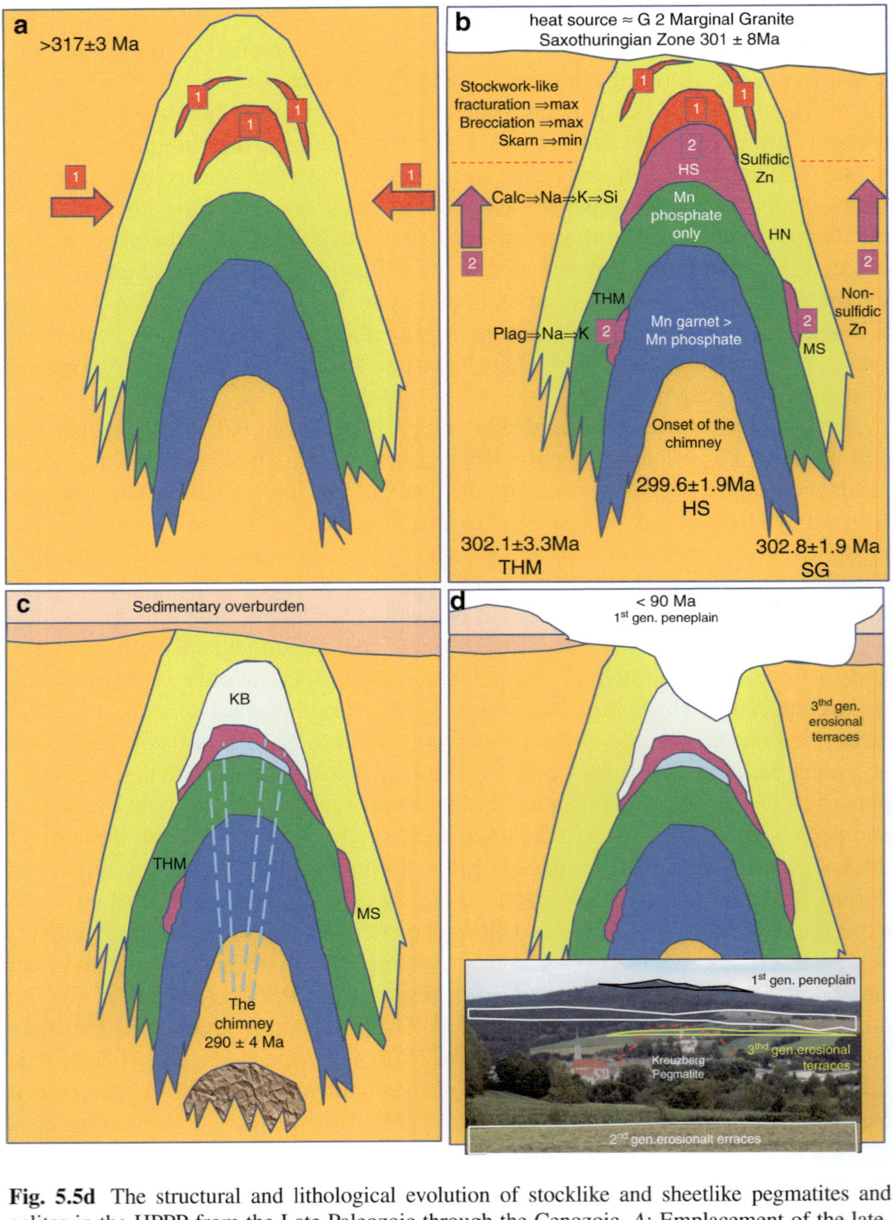

Fig. 5.5d The structural and lithological evolution of stocklike and sheetlike pegmatites and aplites in the HPPP from the Late Paleozoic through the Cenozoic. *A*: Emplacement of the late-kinematic older pegmatites "1" – see Fig. 5.1 – during the late Variscan structural disturbances. The older pegmatites are concentrated in the hinge area of anticlinal folds, e.g., Burkhardsrieth, Older Pegmatite at Hagendorf-South with little tabular pegmatites such as at Brünst. *B*: Emplacement of the post-kinematic Younger Pegmatites in the hinge area underneath the Older Pegmatite and some tabular aplites and pegmatites squeezed into the limb area. Lateral compression was substituted for by a vertical movement. The ensuing uplift lead to the emersion of the basement rocks. While the late Paleozoic peneplain truncated the folded basement rocks small

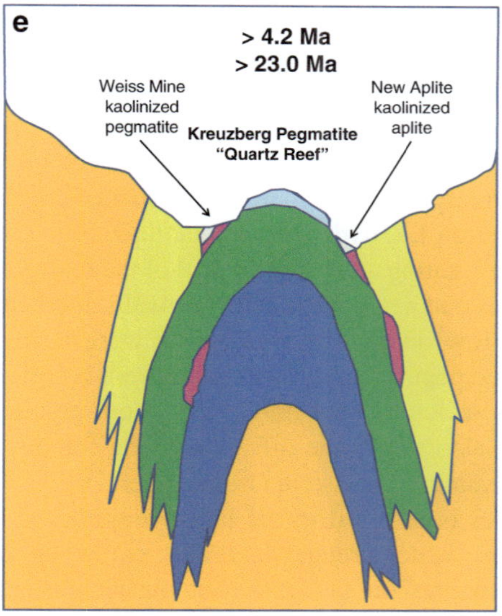

Fig. 5.5.d (continued) granite stocks were intruded along deep-seated fault zone into the host anticlines already providing host to the older pegmatites. Pegmatites were emplaced in a way quite similar to what has been known from oil and gas traps in so-called anticlinal or caprock traps (related to the preceding older pegmatites). THM = Trutzhofmühle, MS = Miesbrunn Pegmatite Aplite Swarm, H-S = Hagendorf-South, HM = Hagendorf North. Mn is shown in apatite and garnet as function of depth. The age of formation of pegmatites and aplites is given for some sites. The thermal regime is equivalent to that at the intrusion of the G 2 Granite. Calc⇒Na⇒K⇒Si = strong differentiation calcsilicate minerals = e.g. zoisite, vesuvianite, hessonite…, Na feldspar (albite), K feldspar (microcline), Si (quartz). Plag⇒Na⇒K = moderate differentiation plagioclase, …, Na feldspar (albite), K feldspar (microcline). *C*: At the passage from the primary to the secondary phosphate mineralization, the epithermal mineralization I involving a hydrothermal kaolinisation affected the pegmatites, particularly at Pleystein eating away a great deal of the feldspar rim of the pegmatite and altering the aplites and pegmatites in its footwalls and hanging walls. The age of the potential heat source is given together with the age of formation. KB = Kreuzberg Pegmatite at Pleystein. *D*: A new uplift, provoked the rivers to incise into the early Cretaceous peneplain (<100 Ma). Hydrothermal alteration produced another round of phosphates, while denudation transported away the decomposed feldspar rim at Pleystein. The photograph gives on overview of the ancient peneplain and the various erosional terraces carved into the crystalline rocks by the Pflaumbach-Zottbach drainage system on fluvial incision. *E*: Neogene uplift, chemical weathering and erosion unroofed the quartz core of the Pleystein pegmatite and shaped the landscape according to the vulnerability of the crystalline basement rocks, leaving behind an inversion of the landscape

Stage B illustrates an ideal host anticline at the passage from lateral compression towards uplift. Granite stocks are supposed to have been intruded still at depth, while on the dryland created by the continuous vertical movement, peneplanation truncated the folded basement rocks. In the hinge area the younger pegmatites were accommodated, e.g. Hagendorf-South at 299.6 ± 1.9 Ma. Due to the verge of the fold planes towards the SW, some of them show an asymmetrical shape, the most conspicuous of which is the Kreuzberg Pegmatite. Along the limps of the folds aplites and aplite granites formed among others at Trutzhofmühle (dipping west)-

302.1 ± 3.3 Ma and at Waidhaus-Silbergrube (dipping east)- 302.8 ± 1.9 Ma bearing witness of the large-scale doming up. The entire structural deformation was accompanied by a remarkable mineralogical differentiation affecting mainly the Ca-Na-K-Si-, the Mn-(Fe)-Si-P and Zn-S-O systems (Fig. 5.5d-B).

Stage C reflects rather staple conditions from the tectonic point of view. At the passage from the primary to the secondary phosphate mineralization, during the epithermal mineralization I, a hydrothermal kaolinization affected the pegmatites, particularly at Pleystein eating away a great deal of the feldspar rim of the pegmatite and altering the aplites and pegmatites in its footwalls and hanging walls. Although taking place at shallow depth, the roof rocks were still tight and sealed the hydrothermally altered pegmatites. This thermal event is correlated with the youngest granitic activities which did not only bring about granite stocks in the Saxo-Thuringian Zone but also sent some offshoots along fracture zones further south, such as the Křížový Kámen/Kreuzstein Granite aged 297 ± 2 Ma (Fig. 5.5d-C). For a 3-D view the reader is referred to the block diagram of Fig. 3.12, where the lineamentary structure zones created favorable pathways to vent these isolated granitic stocks at Sparnberg-Pottiga in the North outside the area of nappe movement and in the South near the root zone of the allochthonous units at Křížový Kámen. The concave thrust plane dipping towards the SE is the planar architectural element controlling the evolution of the ore mineralization. The lithium mineralization is getting younger from the SE towards the NW while the accompanying rare elements change by quality and by quantity along with it in time and space (Fig. 3.12).

Stage D saw a new uplift, provoking the rivers to incise into the early Cretaceous peneplain capping the basement and platform sediments in the foreland alike (<100 Ma). Hydrothermal alteration produced another round of phosphates, while denudation shaping the landscape transported away the decomposed feldspar from the rim of the Kreuzberg Pegmatite at Pleystein (Fig. 5.5d-D).

During stage E uplift continued and resulted in an inversion of the landscape. After the pervasively altered feldspar rim was eroded, the underlying quartz core, more resistant to weathering than the kaolinized feldspar, came into the reaches of supergene alteration. It stands now out as a pinnacle from the center of a bowl shaped depression with some relic pegmatites that still have some pockets of kaolinization all of which are surrounded by the metamorphic country rocks of the Moldanubian Zone (Fig. 5.5d-E).

5.2 Epigenetic Mineralization in NE Bavaria and Beyond the Border – Minero-Stratigraphy of Pegmatites and Vein-Type Deposits

A great variety of secondary phosphates, and sulfides appeared in the aftermath of the primary mineralization in the pegmatites. They are often difficult to grasp and elude the chronological access, since U/Pb or other methods of radiometric dating failed to contribute to the timing of these mineral assemblages. In this case, one can only resort to what might called "minero-stratigraphy" of different mineral assemblages in the region. A careful examination and selection of particular phosphate minerals from the nearby foreland sediments may help to reduce the field of speculation as to the age of formation of the secondary mineralization in the HPPP. Given the sound paleontological and stratigraphic data available in these platform sediments, the age of mineralization hosted by these Mesozoic sediments can precisely be constrained. The nearest of these ore deposits suitable for this minero stratigraphic correlation in the foreland sediments and also abundant in phosphate minerals, is located in the Auerbach Fe mining district (Gudden 1975; Ruppert 1984; Dill et al. 2009b) (Fig. 1.1a).

The Auerbach siderite-goethite ironstone deposit evolved during the Cenomanian in a fluvial-lacustrine depositional environment on the Jurassic karstified carbonate platform. Long-lasting normal faulting controlled the subsidence of the Fe sinkholes, protected the ironstone deposit from erosion and was responsible for the precipitation of Fe phosphates in fissures and vugs. Five mineral assemblages were established: (1) sedimentary-diagenetic, (2) early hydrothermal, (3) late hydrothermal/epithermal-reducing to oxidizing, (4) late hydrothermal/epithermal- oxidizing, (5) supergene/weathering. Mineral assemblage 1 evolved during the Cenomanian, mineral assemblages, assemblages 2 through 4 from the Coniacian through the Neogene and mineral assemblage 5 formed during the Pliocene. The Auerbach Fe-P mineralization within well-dated Cretaceous platform sediments is taken reference to chronologically constrain similar Fe-P mineralizations in pegmatites and vein-type deposits in the basement. "Telescoping" of phosphate minerals is typical of these mineral assemblages, characterized by a small-scale fluctuation of the redox conditions in the phreatic-vadose hydrological system at temperatures <100 °C. This epithermal phosphate mineralization is genetically bound to NW to NNW trending faults, common also to the HPPP, and related to the Alpine unconformity as a horizontal reference plane.

The following phosphate minerals can be identified in the Auerbach Fe deposit and correlated with equivalent phosphates from the HPPP: vivianite, rockbridgeite, phosphosiderite, ferroan variscite, cacoxenite, oxi-beraunite, wavellite, crandallite s.s.s. and churchite-(Y). Equivalent mineral assemblages hosting these phosphate minerals in the pegmatites and aplites of the HPPP supposedly precipitated during Post-Coniacian time (<90 Ma) (Fig. 5.5d-D). After a period of relative tectonic quiescence, renewed uplifted triggered an incision into the Jurassic carbonate platform in the foreland and speeded up erosion into the strongly kaolinized pegmatite at Pleystein.

Another deposit, although rather unknown among pegmatologists, plays a key role in linking the pegmatite deposits of the HPPP in the Moldanubian Zone with pegmatites in the Saxo-Thuringian Zone. It is the Weisser Stein Pegmatite near Lázně Kynžvart, located just beyond the border on Czech territory (Figs. 1.1a and 3.12). In an unpublished project report by F. Veselovský, P. Ondruš, A. Gabašová, R. Škoda, J. Hloušek, P. Bouše and J. Malec (Komplexní mineralogická studie pegmatitu, "Weisser Stein" u Lázní Kynžvartu, 2007) a list of minerals has been given which is very much of assistance in correlating the various granitic pegmatites and pegmatites *sensu stricto* in the two geodynamic realms of the Variscan Orogen. The pegmatite is a representative example, half-way between the Erzgebirge-Fichtelgebirge Anticline of the Saxo-Thuringian Zone and the Moldanubian HPPP, that shows impressively the chemical transition from the Moldanubian rare metal pegmatites *sensu stricto* (P > Nb > B > Li) into the granitic pegmatites from the Saxo-Thuringian Zone (F ≈ Be > Li > B > P)- see also Fig. 2.5a, b. Lithium in the Weisser Stein Pegmatite is bound to phosphates like at Hagendorf but was accommodated into the lattice of montebrasite [$LiAl(PO_4)(OH,F)$] rather than triphylite, which is the protagonist in this role in the HPPP. The Weisser Stein Pegmatite contains a great variety of Ca-Be phosphates such as greifensteinite and hurlbutite. It has a complex As-Bi-Cu-Sn-Zn-Mo mineralization and uraninite-coffinite association resembling that of the HPPP. The pegmatite bridges the gap between the low-F/high-Li pegmatites *sensu stricto* of the HPPP and the peraluminous granitic pegmatites and greisen zones in the apex of the highly differentiated S-type granites of the Erzgebirge-Fichtelgebirge Anticline, where fluorine found its mineralogical outlet in topaz, an Al-F-Si compound completely absent from the HPPP. There is a geodynamic differentiation as a function of the proximity to the northern collision zone. The chemical transition from S to N is governed by deep-seated structural elements some of which acted as conduits for small granite stocks and also were also responsible for the localization of the HPPP (Fig. 3.12).

Chapter 6
Synopsis and Conclusions

Abstract Pegmatites are representatives of a unique lithology, in view of their mega-crystals and colorful minerals spread across in the felsic groundmass, but on the other hand they are not unique in terms of the lithological processes, involving magmatic, metamorphic, hydrothermal and sedimentary processes, if by common consensus we accept weathering and denudation as part of the sedimentary processes. Metapegmatites are restricted to the allochthonous units of the ensialic orogen. A transitional facies with an extraordinary concentration of Nb-Ta together with Be can be correlated with two crustal highs of a high-velocity zone indicating a lower crustal thermal high underneath. Unlike the metapegmatites, which are prekinematic, the younger pegmatitic mobilizates, called pegmatoids, were emplaced late kinematically with only minor or no textural adjustment to the stress-strain relations in the NE Bavarian basement rocks. One can find this type of felsic mobilizates in the allochthonous units. The built-up, alteration and destruction of pegmatites and aplites in the HPPP is a complex process.

Stage 1: Metamorpho-tectonic syn- to late kinematic mobilization is similar to the processes known from the nappe emplacement with its numerous pegmatoids. The transition from lens-shaped and schlieren of pegmatoids to pegmatites s.str. suggests that these melts were very mobile, and separated gradually from their site of formation through selective separation such as filter-pressing and seismic-pumping along with thrustal motion. Intracrustal and subcrustal processes of element mobilization have acted in the same direction by providing heat and mobilizing elements.

Stage 2: Late kinematic (to postkinematic) subcrustal magmatic mobilization was in full swing as the deformational processes of Variscan Orogeny waned. Metagabbros and lamprophyres are the only magmatic rocks that can be held accountable for this early mobilization. Accumulation of rare elements started off with beryllium and continued into zinc.

Stage 3: Postkinematic subcrustal magmatic mobilization prevailing over late kinematic one culminated in the dissemination of "nigrine" confined to the country rocks of the pegmatites.

Stage 4a and b: At the end of stage-4b postkinematic mobilization all mineral assemblages and element association had been completely emplaced in the HPPP that enable us to categorize the majority of pegmatites and aplites of the central zone of the HPPP as (Be-Zn)-Li-Nb-P pegmatites (stock-like and tabular) and aplites

H.G. Dill, *The Hagendorf-Pleystein Province: the Center of Pegmatites in an Ensialic Orogen*, Modern Approaches in Solid Earth Sciences 15,
DOI 10.1007/978-3-319-18806-5_6

(tabular). A characteristic "chimney-like" thermal cell within the pegmatite system was of ore control and fostered the fluid migration and accumulation of the rare metals which used to be concentrated at the contact between the massive quartz underneath and the zone of quartz fragments above. The "chimney" has much in common with the Sn-bearing siliceous greisen deposits in the Saxo-Thuringian Zone. This structural similarities stimulate the idea that between siliceous greisen deposits at rather shallow depth and quartzose pegmatites at a deeper level is no real difference by quality but only by quantity. The accommodation space was provided by a fundamental change in the strain vectors from lateral to vertical. From the SW-NE trending Luhe Line to the granites along the SW-NE striking Fichtelgebirge-Erzgebirge Anticline in the Saxo-Thuringian Zone a gradual change in the mineralogy can be observed.

Stages 4c and 4d: Postkinematic mobilization and contact metasomatic processes along the margin with the metasedimentary rocks have mainly to do with the close interdigitating of metamorphic country rocks and felsic mobilizates.

Stages 4e, 4f and 4g: The stages are representative of the postkinematic mobilization and autometasomatism with autometasomatic retrograde reactions.

Stage 4h: Postkinematic mobilization-retrograde process took place along the margin with the metasedimentary rocks.

Stages 5a and 5b: Epithermal mineralization I is related to shallow intrusions of two-mica granites. A decisive criterion for its denomination in this case has been the level of mineral deposition, which is assumed to be shallower than the level of formation of stage 4 taking into account that the paleohydraulic conditions controlled by the morphology of the late Paleozoic landscape.

Stage 5c: Epithermal mineralization I reflects a transition into the post-Variscan unconformity-related vein-type mineralization. The shift from the phreatic into the vadose hydraulic zone that developed in line with the shaping of the paleo-landscape is a common phenomenon as one deals with unconformity-related mineralization.

Stage 5d and 5e: Epithermal autohydrothermal alteration I gave rise to a true hypogene kaolinization which affected the apical parts of the "Older" and "Younger Pegmatites", particularly of the pegmatites and aplites of the Pleystein Trend. The ensuing secondary mineralization, that contains a lot of phosphate minerals stable only under reducing conditions, corroborates the idea of the roof rocks to have been still intact prior to the onset of the secondary mineralization.

Stages 6: Early (auto)hydrothermal mineralization gave rise to a wide range of phosphate minerals which were precipitated in vugs and cavities, while others line fracture planes. Mn-bearing phosphates point to Mn as a marker element of stage 6.

Stages 7: Late hydrothermal mineralization (autohydrothermal to externally driven) appeared during the uplift of the pegmatite system as two fluid systems started to mix and another so-called epithermal phase (II) is looming. It is the period of time where secondary minerals absolutely unknown from outside the pegmatite system, such as schoonerite appear together with rockbridgeite. Correlation with mineral associations in the foreland hosting equivalent mineral assemblages lead to the conclusion that the secondary phosphate minerals precipitated during Post-Coniacian time.

Stage 8: Epithermal mineralization II is a topomineralic alteration. The persisting impact of the primary pegmatite mineralization conduced to a mineral suite, behaving as an alien element in the "home-grown" epigenetic mineralogy of the NE Bavarian basement and its foreland. It warrants to be called a "topomineralic" effect.

Stage 9: Supergene alteration by nature in the pegmatites of the HPPP is not kickstarted by any sudden effect but is a continuum which is hard to grasp in time due to the imminent destruction of its results by concurrent denudation. The preservation potential for the pertinent minerals was raised as the uplift was moderate during periods of relative tectonic quiescence.

Stage 10: Supergene alteration by man need not be mentioned in this treatment of pegmatites. Some of these chemical compounds may still form today in some cavities or along the open pit which is flooded up to the present-day groundwater level.

The granitic pegmatites require a direct link with the host granite so as to prove their genetic relation to a parental magma. In the Fichtelgebirge, the younger suite of granites have miarolitic cavities and produced pegmatitic schlieren with rare elements such as Sn, W, Be, F, Li, Nb-Ta, B, REE, As and U contained in a variegated group of minerals.

The pseudopegmatites belong to the Variscan metallotect in the Bohemian Massif which was reactivated in the course of the Alpine Orogeny. The mineral association differs from what we have heard about along the western edge of the Bohemian Massif but the element composition is not very much at variance with the pegmatite districts in the Saxo-Thuringian and the Moldanubian Zones.

6.1 The Ensialic Orogen

The conceptual model of ensialic orogeny is not new and its evolution within the scope of global tectonics has already been discussed for the Proterozoic ensialic mobile belts of the Damara Orogen (Martin and Porada 1978; Kröner 1980; Tankard et al. 1982). An initial extensional phase, creating all the sedimentary and magmatic rocks of common to a rift basin, is followed by a compressional stage, causing a thickening of the crust, accompanied by metamorphism, magmatic activity, and last but not least strong uplift. The processes required to create and alter the pegmatitic deposits in Central Europe between Laurussia and Gondwana correspond to the inherent nature of an ensialic orogen (Figs. 2.1a, 2.1b, 2.1c, and 2.1d). The Variscan-type metallogenetic belts encompass a wide variety of mineral deposits (Fig. 2.2a) among others the entire succession of pegmatitic rocks from metapegmatites, through pegmatoids, pegmatites *sensu stricto* to granitic pegmatites (Fig. 2.2a). Pseudopegmatites fall outside this metallogenetic belt and were only created in

Alpine-type orogens where part of the different felsic rocks got incorporated in the newly formed orogen and their mineral assemblage reactivated. It has to be noted that a lot of Precambrian pegmatites have to be classified as pseudopegmatites rather than pegmatites according to the newly introduced classification scheme. Compare, e.g., the Greenbushes and Koralpe lithium pegmatites. Moving towards the left in the cartoon of Fig. 2.2b, into the Alpine-type orogen the ensialic orogen gradual turns into an ensimatic one, with spreading of the lithospheric plates and subsequent closure of an ocean. Even a quick look at the cartoon unveils that the Alpine-type orogen is not correlative to subduction-related magmatic rocks, of the Andean and island-arc type metallogenetic belts which are truly ensimatic and barren as to the pegmatitic rocks under consideration in this book. The rift type being placed on the opposite side of the Variscan type metallogenetic belt in this idealistic transect in the cartoon of Fig. 2.2b has a strong vertical component during which a bulge formed in the crust. the crust underwent considerable thinning, became more and more ensimatic and as consequence of that delamination alkaline magmatic rocks formed instead of calc-alkaline rocks, a fact which has also a strong impact on the mineralogy of pegmatites encountered in this geodynamic setting. Although pegmatites of this reference geodynamic type have not been treated in this book, they need to be mentioned here for the sake of completeness. As an example for this metallogenetic realm, the Zomba-Malosa Pegmatite Province, Malawi, is quoted here. Agpaitic pegmatites and nepheline syenites brought about a spade of rare REE and Be minerals, aegirite and arfvedsonite which attract not only mineral collectors and dealers.

The current interpretation of what an ensialic orogen is like may slightly differ from what it was meant by the beginning as the technical term was coined and discussed by the above authors mentioned at the beginning of this section. Nonetheless it offers the most suitable approach to explain the various types of pegmatitic rocks on a global and geodynamic (small scale-first order). In the succeeding paragraphs the NE Bavarian Basement with the HPPP as the centerpiece is discussed on a larger scale (second order). In context with the Alpine Orogen the synoptical overview shows what a complex and multi-facetted process called “pegmatitization” is. Pegmatites are representatives of a unique lithology, there is no doubt about in view of their mega-crystals and colorful minerals spread across in the felsic groundmass, but on the other hand they are not unique in terms of the lithological processes, involving magmatic, metamorphic, hydrothermal and sedimentary processes, if by common consensus we accept weathering and denudation as part of the sedimentary processes. It goes without saying that such a complex lithology resultant from a polyphase geological evolution cannot be addressed with experiments behind closed doors only, which demonstrate their high value in some other ways, but needs the full-blown spectrum of geosciences to successfully create a model for exploration and exploitation. Such models in my opinion ought to be the target of a study on mineral deposits. It requires a holistic approach to disentangle the origin of pegmatites.

6.2 The Metapegmatites

Metapegmatites are restricted to the allochthonous units of the ensialic orogen, the Zone of Erbendorf-Vohenstrauß, immediately west of the HPPP, and, to a lesser extent, to the Münchberg Gneiss Complex, 80 km north-west of the HPPP (Figs. 2.1c and 3.1a). True metapegmatites formed in the period from 470 to 440 Ma. They are intercalated as lens-shaped bodies among paragneisses and metabasic rocks and without any doubt older than the orthogneisses, representing original granites (Fig. 3.1b). Their mineral assemblage is rather monotonous devoid of any rare elements, excluding the Püllersreuth (meta) pegmatite. Nevertheless their feldspar and quartz contents, render mining feasible, provided that thickness and striking length allow for an economic operation. The issue why we find rare elements only in the Püllersreuth metapegmatite and not in the other metapegmatites cannot be brought to a conclusion satisfying all readers. In columbite from the Trutzhofmühle aploid a relic age of 376 ± 14 Ma has been recognized besides its true age of 302.1 ± 3.3 Ma (Fig. 3.1d). This "inherited age" data are similar to the age date obtained for the formation of the pegmatoids in the nappe units (Sect. 6.3). On Czech territory, analyses of metapegmatite-hosted columbite at Domazlice yielded 482.2 ± 13 Ma, an age which has never been equaled within the district of metapegmatites of the ZEV and can only be correlated with a magmatic event manifested lithologically by the orthogneisses in the Münchberg Gneiss Complex. Niobium and tantalum accumulations seem to have the highest preservation potential in (meta)pegmatitic lithologies.

Another interpretation of this extraordinary concentration of Nb-Ta oxides in metapegmatites together with Be can be deduced from the block diagram in Fig. 3.12 which was designed based upon surface mapping and an extensive seismic survey. The diagrammatic three-dimensional representation reveals two crustal highs of a high-velocity zone indicating a lower crustal thermal high underneath. One of these highs or thermal anomalies is shown along the NE-SW cross section underneath the HPPP, the other less well expressed, because of truncation of its northwestern limp by shear zones, is located immediately underneath the ZEV, where the metapegmatites came to rest after their horizontal thrusting from the south. The mineralization with Be-Nb-Ta in the Püllersreuth metapegmatite, which chemically resembles the Be-Nb-Ta mineralization in the older pegmatites of the HPPP was caused by a lower crustal reactivation casting the Püllersreuth-Type metapegmatite in the position of transitional pegmatite between metapegmatites *sensu lato* and the older pegmatoids, which will discussed below at the beginning of the section on the HPPP pegmatites (Sect. 6.4.2).

6.3 The Pegmatoids

Unlike the metapegmatites, which are pre-kinematic as far as the deformational history is concerned, the younger pegmatitic mobilizates, called pegmatoids, were emplaced late kinematically with only minor or no textural adjustment to the stress-strain relations in the NE Bavarian basement rocks. One can find this type of felsic mobilizates in the allochthonous, e.g., Münchberg Gneiss Complex, and in the autochthonous units of the Moldanubian Zone. Some of the pegmatoids have a rather monotonous mineral assemblage, totally unattractive to mineral enthusiasts, while others bridge the gap into the rare metal pegmatites, for which the HPPP has been taken reference in this book. Therefore, the latter group will be discussed in the next section as the incipient stage of pegmatite evolution in the HPPP (Table 4.2). In many places in this book, pegmatoids have been discussed and demonstrated to be metamorphic in origin, be it of prograde or retrograde style. It is mainly a feldspar-quartz assemblage with minor muscovite, totaling up to 100,000 t in a favorable structural setting where no granite was close by. The hinge and trough areas of fold structures have shown to be the most successful target areas- see also Fig. 5.5d.

6.4 The Pegmatites

Those who want to broaden their knowledge on how pegmatites formed and thus going to browse the current literature in an unbiased way might read a lot about fractional crystallization of siliceous melts, learn a lot about the cooling rate of these melts or can follow the discussion of the degree of their undercooling. The evolution of the graphic quartz-feldspar textures which plays only a minor part in the pegmatites in NE Bavaria and the growth of megacrystals which are less widespread in the HPPP than expected may be a significant phenomena and, without any doubt, need experimental work. But to put in a nutshell, each pegmatite evolves as a three-dimensional body with characteristic geological features to describe the relationship with its country rocks. In this sense, there is much need in the current studies on pegmatites to catch up and to avoid the risk that "giant crystals are grown in a test tube" while leaving behind an unrealistic copy of the emplacement of pegmatites in nature. As long as it comes close to a dogma to look for a parental granite for each pegmatite, the hopes to a solution of the conundrum of "pegmatitization" in harmony with nature are fading.

6.4.1 Geophysical Surveys and Siting of the HPPP at a Glance

There is an unspoken question: Why was the HPPP emplaced just in this place of the Moldanubian Region in what is called today the northern Oberpfälzer Wald ? Some aspects have already been referred to in different sections of this book. The answer of this key question can only be given after a careful examination of the various data which were obtained during seismic, gravimetric, magnetic and to a lesser extent geoelectric surveys.

The small-scale contour map of Fig. 3.9a presenting the results of the gravimetric survey with a 2 mgal-spacing demarcates two zones of steep geophysical gradients from areas where the change is rather moderate. One of these linear elements is oriented NW-SE, stretching between Erbendorf (ERB) and Tirschenreuth (TIR), the other one runs NE-SW, by-passes south of Vohenstrauß (VOH) and ends up near the Czech-German border in a gravity low marked by the circular −38 contour line. The HPPP sits right at the intersection of two deep-seated lineamentary basement structures.

Another contour map of Fig. 3.10 illustrates the magnetic anomalies. Even a quick look at the densely packed contour lines reveals these magnetic data to be processed and plotted at a larger scale than equivalent data in the gravity map giving us the opportunity to insert the sites of the main pegmatites and aplites of the HPPP and thus enhance any correlation between geophysical and geological data. The NE-SW trend already known from the gravimetric survey becomes consequently more conspicuous and the elongated positive anomalies attest to a swarm of basic to ultrabasic deep-seated magmatic rocks delimitating the felsic bodies of the HPPP towards the NW and also towards the NE yet less well expressed. Between the magnetic and gravimetric fields there are some striking similarities but also conspicuous differences being apparent also for the reader not very well acquainted with interpreting geophysical maps. The gravimetric plots visualize an extensive segmentation of NE Bavaria, taking into account that these different blocks are not only confined to the basement, proper, but are to be encountered also under the foreland basin. The HPPP resides on the intersection of two of these prominent structures, as mentioned above.

The most positive anomaly of the magnetic field, reaching its maximum at +745, forms the south-western branch of the ENE-WSW striking structure which is highlightened in Fig. 3.10 by the dashed line. The map of Fig. 3.10 fails to deliver on an equivalent magnetic anomaly in NNW-SSE direction parallel to the structure detected during the gravimetric survey between the Flossenbürg and Bärnau Granites. The ore control of the HPPP is exerted by two deep-seated structures strongly contrasting in the lithologies (basic vs. felsic).

The seismic survey completes the picture outlined so far by the gravimetric and magnetic data. Figure 3.12, a block diagram mainly based upon the seismic data sets of DEKORP encompasses the interim results of Figs. 3.9a, 3.9b, and 3.10 and provides the structural model for the HPPP emplacement as far the physical approach is concerned. The focus lies on the two culminations of the high-velocity layers observed in the NE-SW and in the NW-SE sections, both of which are located underneath basement lithologies hosting pegmatitic rocks (see also Sect. 6.2).

The anticlinal structure exposed in the NW-SE cross section of Fig. 3.12 is an asymmetrical one, northern limb has been truncated by a listric shear fault, gently dipping towards the N. The high-velocity layer and the various minor reflectors dip towards the South into an ENE-striking zone, called the Luhe Line. This ENE-to NE trending zone is marked by ultrabasic and basic rocks (Fig. 3.10). The Luhe Line has been taken as the root zone of the allochthonous units that gave host to the pegmatoids and metapegmatites, including those transitional into the pegmatites *sensu stricto* at Püllersreuth.

Another anticlinal structure can be recognized along the NE-SW cross section in Fig. 3.12, accentuated by the updoming of the high-velocity layer. In contrast to the anticlinal structure depicted in the previous paragraph, this anticline is symmetrical in shape and coincides with the eastern branch of the Luhe Line in Fig. 3.10, where the elongated strong positive magnetic anomaly became perforated like a "Swiss Cheese" by a set of circular and elliptically-shaped negative magnetic anomalies. It is accounted for by felsic intrusives in the HPPP. The depth of the MOHO is 11–12 s. TWT equal to 33–36 km according to the geophysicists calculations during their reprocessing of the seismic data (Sect. 3.3.4). The Luhe Line was intersected by another deep-seated structure demarcating a lower crustal thermal zone. Subhorizontal through gently dipping planar elements were active along with thrusting and contingent upon the Luhe Line with its eastward extension (ENE-NE).

A series of vertical through steeply dipping NW- striking structures pertinent to the late Variscan deformation, intersected the Luhe Line together with NNE-trending structures – see also (Fig. 5.5c) – where the HPPP is located. In the geophysical contour maps only the master or first-order structures can be fixed, subordinate tectonic structures of higher order escape this approach and need to be mapped on the surface as demonstrated in Sect. 5.1.4. This is valid also for the ore traps. Stock- and sheet-like pegmatites and aplites were controlled in their emplacement by those late Variscan fold structures, providing at the time of mobilization the maximum accommodation space (Fig. 5.5d A-B).

The geoelectric measurements carried out in and adjacent to those areas in NE abundant in pegmatites were not in vain as to the interpretation of the genesis of the pegmatites and the data collected are not superfluous. Yes indeed, these geophysical results do not add very much to enhance the regional picture visualized in the block diagram of Fig. 3.12 and help localize still undiscovered pegmatites or aplites. Their strong points lie in a totally different field of geosciences as these data on the apparent resistivity or conductivity may shed some light on the derivation of chemical compounds relevant for the formation of the pegmatites. Two zones of high conductivity can be outlined in the study area, one at 10 km, the other between 0.3 and 1.0 km depth. Both of them play into the hands of those geoscientists in search of a source of carbonaceous matter (Sect. 3.3.3) and who try and find an explanation for the intensive hydrothermal alteration of the pegmatites (Sect. 3.3.3). Saline brines, nothing unusual in a foreland basin abundant in chemical sediments, used these zones of very-low resistance in the basement as pathway for lateral migration down under. Being hit by one of those cross-cutting structures discussed in the previous paragraphs, these brine reservoirs could have been tapped. Basement brines have been traced several 1,000 m deep into the basement by the samplers of the Continental Deep Drilling Program. Electrical methods as reported in Sect. 3.3.3 are among those geophysical techniques able to track down these (hydrothermal) fluids. A correlation of epigenetic mineralization in the foreland and the NE Bavarian basement in the scope of the minero-stratigraphy of pegmatites receives additional support by those geoelectric measurements providing illuminating insight into the situation at depth where normally petrography and mineralogy have no access without superdeep drill holes (Sect. 5.2).

6.4.2 Built-Up, Alteration and Destruction of Pegmatites and Aplites in the HPPP

Heart and soul of this Sect. 6.4.2 are the Tables 4.1 and 4.2, listing and interpreting the minerals discussed in detail as to their physical-chemical environment in section 4.

Stage 1: Metamorpho-Tectonic syn- to Late Kinematic Mobilization The incipient lithological processes in the HPPP are similar to those leading at the very end of nappe emplacement of the Münchberg Gneiss Complex to the numerous pegmatoids in this allochthonous unit (Table 4.2). This is no longer a surprise after a quick look at the block diagram of Fig. 3.12 where these felsic mobilizates have one planar architectural element in common, the southeastward dipping reflectors and thrust zones plunging into the Luhe Line. The felsic mobilizates are syn- to late kinematic ≈ 600°C with a rather monotonous alkaline feldspar-quartz-muscovite assemblage. The majority of deposits worked in the past for feldspar at the margin of the HPPP belong to this type of pegmatitic deposits which came into existence without any granite intrusion close by (Fig. 5.5d-A). The transition from lens-shaped and schlieren of pegmatoids to pegmatites s.str.suggests that these melts were very mobile, and separated gradually from their site of formation through selective separation such as filter-pressing and seismic-pumping along with thrustal motion. Intracrustal and subcrustal (see succeeding sections) processes of element mobilization have acted in the same direction by providing heat and mobilizing elements from different sources to create in a multistage process a complex rock called pegmatite s.str.

Stage 2: Late Kinematic (to Postkinematic) Subcrustal Magmatic Mobilization Felsic melts of stage 2 were produced as the deformational processes of Variscan Orogeny waned in the region which today bounded in the W by Vohenstrauß and the E by the Czech-German border. Subcrustal thermal processes underwent temperature conditions which considered on a regional scale were sufficient to create calcsilicate rocks of medium-grade contact metamorphism (500–550 °C), on a local scale, e.g., W of Pleystein, the thermal regime attained a higher T-level able to bring about wollastonite. Metagabbros and lamprophyres are the only magmatic rocks that can be held accountable for this early mobilization of the "Older" pegmatitic rocks of stage 2a due to their higher temperature of intrusion (Figs. 5.2f, 5.2g, and 5.5d-A, Table 4.2). The afore-mentioned basic magmatic rocks are attributed to the "redwitzites", which have been described in Sects. 2.1.4.3, 3.2.1, and 5.1.1. The "Older Pegmatite" followed as guideline during its emplacement NW-SE trending fold structures and predated the intrusion of the Flossenbürg Granite.

Upon transition from the "Older" into the "Younger Pegmatite" those elements diagnostic to the pegmatites of the HPPP were concentrated and listed in Table 4.2 for the stages 2b, 2c and 2d. Accumulation of rare elements started off with beryllium, already known from the Püllersreuth transitional metapegmatite and continued into zinc which shows up in two distinct facies, a sulfidic facies prevailing over a non-sulfidic one, with gahnite as the sole representative (Table 4.2-stages 2c and 2d, Fig. 5.5d-B).

Stage 3: Postkinematic Subcrustal Magmatic Mobilization Prevailing Over Late Kinematic One The more advanced level of subcrustal mobilization culminated in the dissemination of "nigrine" confined to the country rocks of the pegmatites, while it is absent from the felsic mobilizates, proper. Its type-B Ti mineral aggregates contain a spectrum of Nb-U oxide minerals similar to that known from the "Older Pegmatites", where fractionation of Nb-Ta in columbites went on into the "Younger Pegmatites" (Table 4.2, stages 3a and 3b, Fig. 4.8i).

Stage 4a and b: Postkinematic Mobilization At the end of stage-4b mobilization all those mineral assemblages and element association had been completely emplaced in the HPPP so as to enable us to categorize the majority of pegmatites and aplites of the central zone of the HPPP (Tables 1.1a and 4.2, Fig. 5.5d-B):

(Be-Zn)-Li-Nb-P pegmatites (stock-like and tabular)
and aplites (tabular)

If somebody wants it shorter, feel free and abbreviate the term to **BZLNP pegmatite**. It is without any doubt closer to the natural state and better to use in practice than any pigeonholing using the triplets LCT- or NYT pegmatite.

It is difficult to hand down a final judgment on whether there was a slight hiatus or not between the two pegmatitic rocks. I interpret a second aplitic zone within the Hagendorf-South stock found by Uebel (1975) as a slight hiatus and the emplacement of the "Younger Pegmatite" to be done in an unbalanced physical regime, in other words, the younger melt came in contact with a cooler and semi consolidated preexisting pegmatitic rock. Textural and structural descriptions of previous workers are interpreted this way (I have been underground at Hagendorf-South by the end of the 1970s and beginning of the 1980s in search of uranium anomalies. During this time the younger pegmatite was worked). Differentiation and separation of a quartz core and feldspar rim was at full swing only in the "Young Pegmatite", where it evolved in the hinge zone of one of the NW-striking anticlines (Table 4.2 – stage 4a). I am in doubt about the "Older Pegmatite" to have developed a similar siliceous core as its successor (Fig. 5.3b). Neither the relic quartz pegmatite at Pleystein, nor the numerous older pegmatites in the immediate vicinity of Hagendorf-South show any hint of such an early zonation in the HPPP pegmatite system. It is more realistic to speak of a funnel-shaped vertical fading out of the quartz core at Hagendorf South in the Older Pegmatite. At Hagendorf-North the situation is not very much different from that (Fig. 1.4e). Although this neighboring pegmatite does not show a zonation into an older and a younger pegmatite similar to Hagendorf-South, the siliceous zone fades out in the hanging wall feldspar rim. There is only one silicification by the end of the precipitation of the feldspar. In both pegmatites the isolated quartz bodies "floating" in the massive feldspar rim at Hagendorf-South and Hagendorf-North represent a reaction front parallel to the top of the quartz core underneath. It reflects a "chimney-like" thermal cell within the pegmatite system that is also of ore control fluid migration and on accumulation of the rare metals which used to be concentrated at the contact between the massive quartz underneath and the zone of quartz fragments above (Figs. 1.4e and 5.3b). Even in the Kreuzberg Pegmatite, where all feldspar is gone, the phosphates and Nb-Ta oxides are concentrated at the apex of the quartz core (Fig. 5.3a). We might expect a similar zone of isolated quartz fragments within the feldspar rim at Pleystein that fell victim to the intensive kaolinization. Mapping the shape of the silicification front in a pegmatite *sensu stricto* is the key to the primary rare element concentration. The "chimney" has much in common with the Sn-bearing siliceous greisen deposits at Altenberg and Sadisdorf, located in the Saxo-Thuringian Zone. This structural similarities stimulate the idea that between siliceous greisen deposits at rather shallow depth and quartzose pegmatites at a deeper level is no real difference by quality but only by quantity (Figs. 1.4e, 5.3a, and 5.3b). In the broadest sense, this structure may be explained with the so-called up-dip-effect of ascending hyperfusible-rich fluids *sensu* Černý (1991). The 3-D representation of Fig. 3.12 offers the architectural planar element that geodynamically connects both types and governs which type of ore-bearing structure prevails in the deposit.

The accommodation space was provided by a fundamental change in the strain vectors from lateral to vertical (Fig. 5.5d-B). Stock-like and tabular pegmatites as well as aplites developed more or less synchronolously along with the rock-mechanical and chemical changes. The heat source can neither be looked for in the Oberpfalz granites around nor in the granitic mobilizates intercalated into the gneissic rocks. These granitic rocks which were mapped and hit by diamond drilling operations as many aplitic and pegmatitic veinlets are not the "reason" but the "result" of a process sparking among others the rare-metal pegmatites- see also Sect. 5.1.2.

A common saying among exploration geologists is: If you look for a mine, look near a mine. In search of the "kitchen" of these rare-metal pegmatites, this phrase would end up in a cul-de-sac. The key figure in this case it the 3-D representation of Fig. 3.12., because the geodynamic setting guides our view from down-under the Moldanubian Region along the SW-NE trending Luhe Line to the granites along the SW-NE striking Fichtelgebirge-Erzgebirge Anticline in the Saxo-Thuringian Zone. The G 2 Marginal Granite in the northwestern Fichtelgebirge was intruded at shallow depth into a cooler environment and as such can be referred to as the "chilled margin" – facies equivalent to the aplitic zones found proximal to the "Younger Pegmatite". The age of the G 2 Marginal Granite stands at 301 ± 8 Ma (Besang et al. 1976) (Table 3.1). These data reported from the northern Fichtelgebirge well accord with those given in Fig. 5.5d-B. The thermal contrast between the G 1 and G 2 Granites in the Fichtelgebirge is analogous to the aplitic rime facies in the pegmatite stocks of the Oberpfälzer Wald. In some tabular bodies of moderate thickness the aplitic facies has become the sole representative of this felsic mobilization with neither a pegmatitic rim nor quartz core to develop. It formed in a thermally unbalanced environment.

Stages 4c and 4d: Postkinematic Mobilization and Contact Metasomatic Processes Along the Margin with the Metasedimentary Rocks Contact metasomatic or topo-mineralic reactions-dependent upon the mineralogical composition of the country/wall rocks- are confined to tabular aplites and pegmatites. In stocks like Hagendorf-South these minerals are either absent or of minor importance (Table 4.2 stages 4c and 4d). It has mainly to do with the close interdigitation of metamorphic country rocks and felsic mobilizates. Large bodies tend to respond in a different way as shown by the processes operative during stages 4e, 4f and 4g. A similar zircon mineralization was recorded also from the (Sn-As/Zn-Zr)-Tb/Nb-U-B-Be-P pegmatites Mount Mica, USA by Brownfield et al. (1993). Textural observations suggest that the K–Ba–Zr phosphate and the Sc–Zr phosphate(−silicate) found at Trutzhofmühle are magmatic high-temperature phases. Although the pegmatite from the USA and the aplite from Germany are not located in a peralkaline intrusion which used to have originated from mantle intrusions a subcrustal impact on the pegmatite system during the initial stages of its evolution cannot be ruled out for the phosphate pegmatites in Germany as well as northeastern USA.

Stages 4e, 4f and 4g: Postkinematic Mobilization and Autometasomatism with Retrograde Reaction The mineralization in the tripartite series of stages 4e through 4g resulted from processes exclusive to stock-like intrusions, whereas being absent from tabular ones. Opposed to the stages 4c and 4d, where, on account of the reduced thickness exchange reactions with the inclosing country rocks are favored, large pegmatite bodies keep the fluids after cooling down and solidification so that they can hardly escape to react with the surrounding rocks but attack pre-existing minerals. They spark a series of minerals during processes most suitably be called autometasomatic retrograde reactions (Table 4.2 – stages 4e–4g).

Stage 4h: Postkinematic Mobilization-Retrograde Process Along the Margin with the Metasedimentary Rocks Stage 4h is not the direct consequence of a further lowering of the temperature or a hydration of minerals mentioned under stage 4f or 4g. It is more closely related to those minerals grouped under header contact metasomatic process along the margin with metasedimentary rocks (Table 4.2 – stage 4c). Lazulite was hydrated and ended up as gordonite or manganogordonite, dependent upon the availability of Mg and Mn.

Stages 5a and 5b: Epithermal Mineralization I Related to Shallow Intrusions of Two-Mica Granites The technical term epithermal ore mineralization as entrenched into the present-day literature on economic geology is used for a wide range of mainly Cu-Au deposits. Going into further detail and taking this wording literally shows this term to be a misnomer, because the temperature of epithermal Cu-Au deposits is not significantly lower than that of mesothermal vein-type deposits. In the current study the temperature of this mineralizing stage in the pegmatite is lower than the T values recorded for stage 4, witnessing a general decline in temperature along the pegmatites emplacement and retrograde autometasomatism. A decisive criterion for its denomination in this case has been the level of mineral deposition, which is assumed to be shallower than the level of formation of stage 4 taking into account that the paleohydraulic conditions controlled by the morphology of the late Paleozoic landscape had an ever increasing impact also on the alteration of the pegmatite mineral assemblages at least during the waning substages of this epithermal mineralization. Uranium-lead dating in the region proved this synchronous evolution of pegmatites, U vein-type deposits and supergene kaolinization, as the pegmatite got uplifted and the fluid regime was no longer able to be kept separate from the surrounding country rocks.

The various substages of stage 5 bridge the gap between the primary and secondary mineralization. While in the primary mineralization massive, granular textures, in places, with some alignment of its minerals predominate, in the secondary mineralization of the pegmatite cut-and-fill structures, vugs and cavities with minerals growing in the open space as well as earthy, friable mineral aggregates prevail.

During stages 5a and 5b a variegated sulfide mineral assemblage and U black ore mineralization evolved at a subcritical fluid level (Table 4.2). Small stock-like two-mica granites were hit by drill holes immediately north of the Saxo-Thuringian granite belt, e.g., the Sn-W-bearing Sparnberg, Pottiga and Eichigt Granites. The westernmost representative of these granites is located near Weitisberga in the Thüringer Wald, hosting a subeconomic molybdenite mineral assemblage. Along deep-seated structure zones, striking predominantly NNW-SSE, these latest stock-like intrusions of the late Variscan felsic magmatic activity may also be traced down to the S into the Moldanubian Region (Behr et al. 1989). Around 290 ± 4 Ma, the G 4 Tin Granite concluded the granitic activity in the Fichtelgebirge region (Besang et al. 1976) (Fig. 5.5d- C). It is even younger than the youngest intrusions within then Moldanubian Region, represented by the Křížový Kámen/Kreuzstein Granite at 297 ± 2 Ma (Siebel et al. 1995a, b, 1997; Müller et al. 1998).

Stages 5c: Epithermal Mineralization I Transition into Post-variscan Unconformity-Related Vein-Type Mineralization While during stage 5a and 5b the mineralization developed under reducing conditions, a fact which has not explicitly been mentioned for the predecessors among the primary pegmatite mineralization, in the stage under consideration the gradual change from reducing into oxidizing conditions at the edge of the HPPP has to be emphasized. The appearance of barite does not allow for a different explanation (Table 4.2). The shift from the phreatic into the vadose hydraulic zone that developed in line with the shaping of the paleo-landscape is a common phenomenon as one deals with unconformity-related mineralization of whatever element assemblage may be concerned. The paleo-landscape was not a simple erosional plane narrowing the gap between the level of pegmatite emplacement and the near-surface hydraulic system but is has already encroached during the late Paleozoic at least in parts upon by transgressions leaving behind a clastic overburden on the basement of the Bohemian Massif (Fig. 3.7).

Stage 5d and 5e: Epithermal Mineralization I Autohydrothermal Alteration The phyllosilicate assemblages of stages 5d and 5e in combination with the aluminum phosphates are a stark message in favor of a gradual acidification of the relic fluids in the waning epithermal mineralization 1. It is a true hypogene kaolinization which affected the apical parts of the “Older” and “Younger Pegmatites”, particularly of the pegmatites and aplites of the Pleystein Trend (Fig. 5.5d–C). The site where it all began with a high-temperature event, the argillitisation was most pervasively among the HPPP mineralized sites to the W of Pleystein (Fig. 5.1). In the hinge area of the host anticline, sealed off completely towards the topstratum, the hydrothermal fluids were concentrated and focused on the most vulnerable part of the pegmatite, its feldspar rim. It is a matter of conjecture to tell anything about to what extent erosion into the roof rocks by paleo-drainage systems capitalized on the presence of friable argillisation zone underneath the gneissic roof support. Objective evidence for a denudation of the soft alteration rim cannot be provided for the time span before the Early Cretaceous. The ensuing secondary mineralization, that contains a lot of phosphate minerals stable only under reducing conditions, corroborates the idea of the roof rocks to have been still intact prior to the onset of the secondary mineralization (Table 4.2, stage 6).

Stages 6: Early (Auto)Hydrothermal Mineralization During the initial stages of hydrothermal mineralization a wide range of phosphate minerals were precipitated in vugs and cavities, while others line fracture planes. It has to be noted that the secondary Fe phosphates encountered during this early stage of hydrothermal alteration are still enriched in Mn, e.g., Mn-bearing vivianite (Fig. 4.1.2e). The close proximity to the primary Mn-bearing phosphates simply account for the presence of Mn as a marker element of stage 6. Other elements following suit in this way are Ca-bearing phosphoferrite and Zn-bearing ludlamite. They incorporated these elements and accommodated them into their lattice as minor elements. Phosphates are present today at different altitudes above mean sea level (a.m.s.l.) from 525 m through 630 m a.m.s.l. No alteration front swept through the primary mineralization and each deposit in the HPPP irrespective of its level of emplacement was affected under the same physical-chemical conditions but dependent upon its individual primary mineral association. This is especially evident among the group of Zn minerals. The chemical composition of the pegmatitic parent material has still a strong say on the composition of the secondary phosphates and a homogenizing effect – see, e.g.,. supergene alteration- on the pegmatite system cannot be identified during this period of time. It is difficult to chronologically constrain this early (auto)hydrothermal alteration, because no independent age data making use of one or the other secondary Fe phosphate exist for this stage.

Stages 7: Late Hydrothermal Mineralization (Autohydrothermal to Externally Driven) While autohydrothermal alteration processes continued under decreasing temperatures and increasing Eh there are processes overlapping with this intrinsic alteration, often causing a relapse in the Eh or even an oscillation of the physical-chemical parameters conducive to a sandwich-structured mineralization. Upon uplift of the pegmatite system two fluid systems started to mix and another so-called epithermal phase (II) is looming. It is the period of time where secondary minerals absolutely unknown from outside the pegmatite system, such as schoonerite appear together with rockbridgeite (Sect. 5.2). Correlation with mineral associations in the foreland hosting equivalent mineral assemblages lead to the conclusion that the secondary phosphate minerals precipitated during Post-Coniacian time (<90 Ma)

(Fig. 5.5d-E). It has to be noted that as to the youngest hypogene alteration processes affecting the pegmatites and aplites of the HPPP we enter a reservation on the chronological constraining. The lack of elements amenable to radiometric age dating, excluding potassium in leucophosphite, is one of the big stumbling blocks. Dating K-bearing sulfates and phosphates does not contribute to making friends among the geochronologists who want to use their measuring chamber for more than one sample.

Stage 8: Epithermal Mineralization II Topomineralic Alteration The epithermal alteration II took place much closer to the paleosurface than during equivalent alteration processes at the end of the primary mineralization (epithermal mineralization I). The massive granular texture known from the stage 5 mineralization was substituted for by an open texture in the stage-8 mineralization. In both mineralizing processes the pegmatitic parent material plays an leading role in the precipitation of phosphate minerals, be as to the Zn, F or the P contents. This persisting impact of the primary pegmatite mineralization conduced to a mineral suite in stage 8, behaving as an alien element in the "home-grown" epigenetic mineralogy of the NE Bavarian basement and its foreland. It warrants to be called a "topomineralic" effect. Opposed to those effects in stages 4c and 4d, where an external effect by the fluids derived from the wall rocks caused a typical mineralization within the pegmatite, in stage 8 fluids of regional extent suffered from mixing with fluids percolating through the pegmatite's primary mineralization (Table 4.2, Fig. 5.5d–E).

Stage 9: Supergene Alteration by Nature In the current and previous studies on supergene alteration, the boundary between the hypogene and supergene mineralization has been drawn based on datable minerals, e.g., hydrated Mn-K oxides or uranyl minerals which are amenable to K/Ar, Ar/Ar or U/Pb dating, respectively. The supergene alteration in the pegmatites of the HPPP is not kickstarted by any sudden effect but is a continuum which is hard to grasp in time due to the imminent destruction of its results by concurrent denudation. Supergene alteration becomes evident as an economically destructive process within the mineral succession as the afore-mentioned minerals amenable to age dating become more and more widespread. The preservation potential for these minerals was raised a s the uplift was moderate and erosion retarded, in other words during periods of relative tectonic quiescence (Figs. 5.5d–F and 6.1a). The true boundary overlaps with the hypogene processes at depth.

Stage 10: Supergene Alteration by Man Anthropogenic minerals were discussed in the pertinent sections and reasonably need not be mentioned in this treatment of pegmatites. Some of these chemical compounds may still form today in some cavities or along the open pit which is flooded up to the present-day groundwater level. In the HPPP, the Hagendorf-South pegmatite offers the most complete mineral succession with almost all stages mentioned in Table 4.2 developed. The Kreuzberg Pegmatite could not follow suit and developed all stages to completeness as a consequence of the pervasive kaolinization affecting the pegmatites and aplites in and around Pleystein (Table 4.2, stage 5e). The Reinhardsrieth aplite, the northernmost

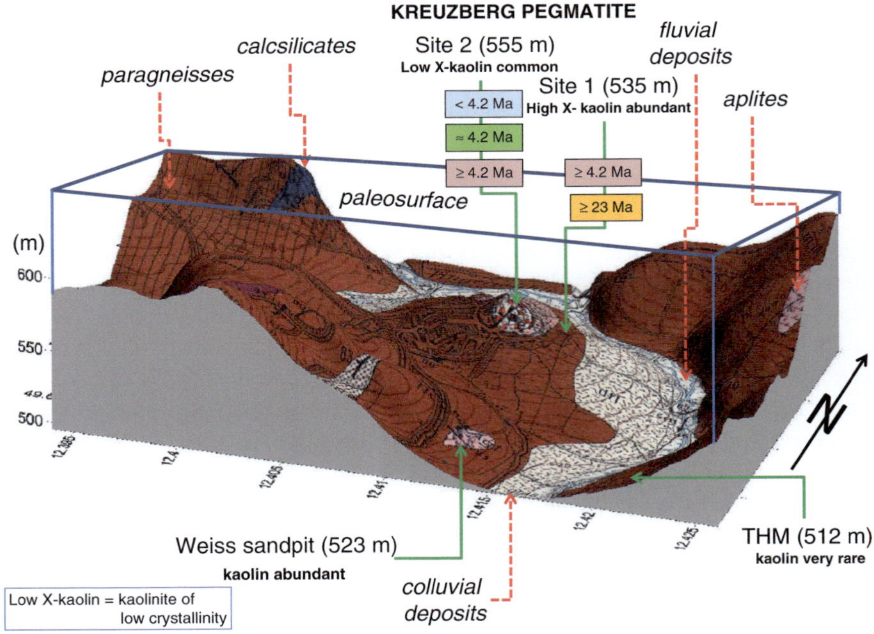

Fig. 6.1a Three dimensional view of the kaolinization at outcrop in the present-day geomorphology around Pleystein. The lithologies and sites mentioned in the text are given: site 1: New Aplite, site 2: Kreuzberg Quartz Pegmatite, Weiss Pit: aplite, *THM*: Trutzhofmühle Aplite. The numbers in the *rectangular boxes* denote the age of kaolinization

rare-element aplite of the HPPP strongly contrasts with the remaining felsic mobilizates. It has a rather poor mineral assemblage, but nevertheless all stages listed in Table 4.2 can be found, too. In the New Aplite, on account of its proximity to the hot spot and vent system for the hydrothermal solutions suffered most strongly from this epithermal mineralization and exhibit the widest hiatus in the minerostratigraphy. All primary and secondary phosphates excluding those of stage 9 were completely wiped out by the hypogene and supergene kaolinization. Many pegmatites and aplites, being situated N and S of the HPPP have a rather moderate mineral assemblage as to the primary and secondary phosphates. The onset of secondary phosphates can be observed within stage 7 when a fault-bound mineralization during the late Cretaceous affected pegmatitic and non-pegmatitic deposits, alike. The pegmatitic mineralization in the Fichtelgebirge and in the adjacent Erzgebirge can no longer be denominated as pegmatitic *sensu stricto* but need to be classified as granitic-pegmatitic, or even episyenitic, and greisen-type dependent upon the driving processes and the resultant minerals in the altered granites. Although strikingly different in structure and grain size, the processes leading to pegmatitic or greisen-type alteration in the granites are inherent to the physical-chemical processes of the host granite system and resemble those processes discussed for the stages 4 and 5 for a true pegmatite/aplite system, exemplified by the HPPP.

6.5 The Granitic Pegmatites

Granitic pegmatites require a direct link with the host granite so as to prove their genetic relation to a parental magma. Even the Hopfau and Grünberg pegmatites both of which formed within the exocontact zone of the Steinwald and Fichtelgebirge granites do not qualify as granite pegmatites s.st. (Fig. 1.1a). In case of the Grünberg Sn-Nb-Li pegmatite (lepidolite-amblygonite) it is a tabular pegmatite cutting through the wall rocks-see also Figs. 2.4a and 2.4b. It would be too premature and lacks any evidence to attribute both pegmatites to the nearby granites. It meant to fall back on the geoscientific level and strategy of uranium exploration attained in the 1950s when every epigenetic U mineralization in a granitic terrain was attributed to the nearest granite possible. There is no doubt that granites have higher U and Li background values than many of the crystalline basement rocks around, but it is absolutely not necessary to genetically link these deposits to granites related with them in space but not temporal particularly as any genetic link is missing.

In the Fichtelgebirge, the younger suite of granites encompassing the G 3 Epprechtstein, Kornberg and Waldstein granites and the G 4 Rudolfstein and Fuchsbau granites have miarolitic cavities and produced pegmatitic schlieren with rare elements such as Sn, W, Be, F, Li, Nb-Ta, B, REE, As and U contained in a variegated group of minerals (Table 4.1). Černý (2000) addressed these open space fillings in pegmatites and claimed three basic processes to be responsible for these special textures: (1) decompression of a magma during its ascent, (2) fractional crystallization during isobaric solidification, (3) depletion of a magma in fluxing agents during precipitation of minerals accommodating mainly B, F, and P in their lattice.

According to Simmons (2007), the crystallization sweeps its way from the contact towards the center while the residual melt gradually becomes more and more enriched with volatiles including the rare elements above. One fundamental issue in the formation of the mineralized miaroles and druses is the pressure which according to the above authors must not exceed 3 kbar. High-pressure conditions are detrimental to the built-up of such cavities filled with well-shaped minerals such as topaz, tourmaline or be minerals (Simmons et al. 2003). It goes without saying that moving southward, away from the granitic pegmatites into the area of the pegmatites and aplites sensu stricto, this structural type disappears. Upon these textural variation a pressure estimation may be made for the granitic pegmatites in the Saxo-Thuringian Zone.

The granitic pegmatites mainly belong to the Sn-W-, Be-, U-, F- and Li-bearing pegmatites. Dependent upon the main elements, the felsic mobilizates have to be denominated F granite pegmatites (fluorite-topaz) or Li granite pegmatites (lepidolite-zinnwaldite) according to Table 1.1. The G 2 Granite in Fichtelgebirge was intruded at shallow depth into a cooler environment of the G1 Granite as a some kind of a "chilled margin" – facies for the ensuing Granites G 3 and G4. The Granite G 2 is barren as to granite pegmatites and rare metal accumulations are absent (Sect. 6.4.2). From the textural and chemical point of view the scenario of the

Fichtelgebirge granites resembles that described from the pegmatites of the HPPP – see, e.g., Fig. 5.3b. The reason for this why there exists such a striking contrast in the presence or absence of economic rare metal pegmatites is the striking difference in the geodynamic setting summarized in the various sections of Figs. 2.1c, 2.2a, and 3.12 and the diminishing impact of the mantle towards the collision zone (Fig. 6.1b). The miarolitic granitic pegmatites in the Saxo-Thuringian Zone were dealt with by Thomas et al. (2009) and Thomas and Davidson (2012) as to their physical-chemical regime. The peralkaline melt containing as much as 30 mass % water started to crystallize at about 700 °C in a low-pressure regime of 1 to 3 kbar (see also succeeding paragraphs as to the formation of miaroles). The graphic zone resulted from magmatic-metasomatic reactions (see in the pegmatites *sensu stricto*, the autometasomatic processes). Quartz precipitated in two generations around 573 °C which is the crucial temperature of the phase transition in quartz ($\beta - \alpha$).

6.6 The Pseudopegmatites

Following the classification scheme outlined in Table 1.1 a great deal of the larger bodies of pegmatites especially those in Precambrian terrains can hardly be pigeon-holed either as pegmatites s.st. or granitic pegmatites. They reside in crystalline basement rocks, intercalated into biotite schists and gneisses, in metapelitic through metapsephitic host rocks, and metabasic magmatic rocks (Sect. 1.2.3). Unlike the metapegmatites whose quartz-feldspar-mica association is pre-kinematic in origin, the pseudopegmatites rarely suffered from any superimposition of tectonic textures on to their granular pegmatitic or aplitic groundmass by structural disturbances. There is a small-scale alignment of aplitic and pegmatitic bodies but no large-scale orientation of the rock-forming minerals. In the wider study area the reworking of the pegmatites s.str. belonging to the Variscan metallotect in the Bohemian Massif took place in the course of the Alpine Orogeny when, e.g., lithium-bearing aplitic schlieren of economic importance formed in southern Austria at Koralpe and several other sites of Nb Be-Li pseudopegmatites (spodumene-holmquisite) nearby (Figs. 1.2a, 1.2b, and 1.2c). The mineral association differs from what we have heard about along the western edge of the Bohemian Massif-see Table 4.1- but the element composition is not very much at variance with the pegmatite districts in the Saxo-Thuringian and the Moldanubian Zones (Fig. 6.1b).

6.7 Learning from Nature

Pegmatitic rocks are very similar and, not to say, they are identical to granitic rocks in their major components, being composed of feldspar, quartz and mica. Those pegmatites, called rare-metal pegmatites due to their abundance in niobium, tantalum, lithium, beryllium, cesium or phosphorus, make up only a small proportion of

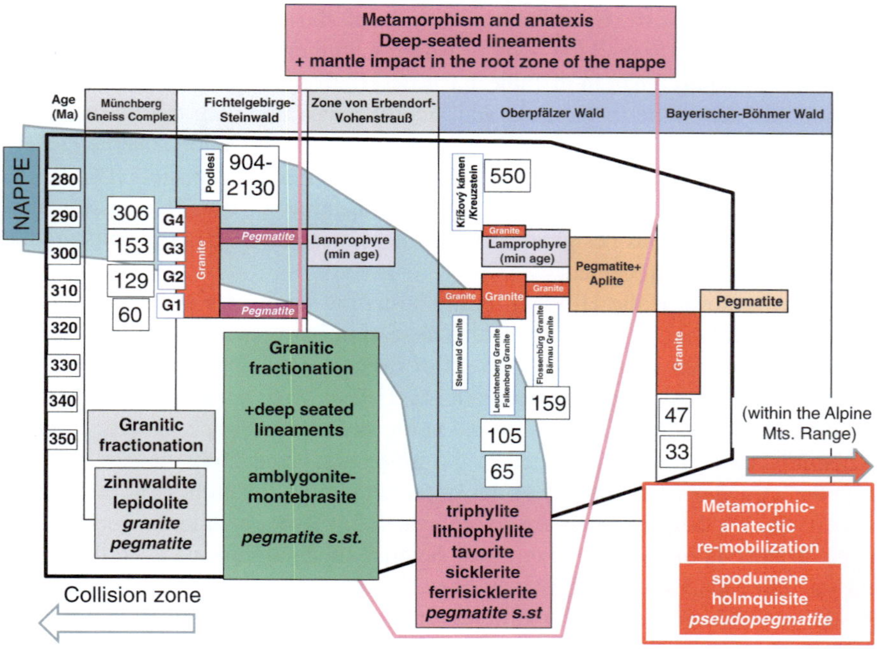

Fig. 6.1b The evolution of pegmatites, granitic pegmatites and pseudopegmatites resultant from intra- and subcrustal mobilization, as an example portrayed by the chemical distribution of lithium and its minerals in the different pegmatitic and associated igneous rocks

less than 1 % among the pegmatites but have been paid an enormous amount of attention to by geoscientists, who will be listed by the name but not by their publications which would go far beyond what might be acceptable to a list of references for this book: Fersman, Niggli, Sederholm, Holmquist, Scheumann, Mehnert, Barth, Wegmann, Ramberg, De Figueroia, Lafitte, Roubalt, Perrin, Valemoff, Wlassow, Termier, Cameron, McNair, Page, Wickmann, Eskola, Volborth, Ginsburg, Gevers, Quensel, Andersen, Goldschmidt, Landes, Lindgren, Schaller, Hess, Kemp, Lacroix, Schneiderhöhn, Strunz, Björlykke, Vogt, Rosenbusch, Brögger, Oelsner, Brotzen, Eitel, Kittel, Rosenbuch, Spurr.

Frequently, people who try and find an introduction into pegmatites and thus skimming the literature on pegmatites might come to the conclusion that the world of pegmatites started off with Jahns and Burnham in the USA as they published in 1969 their experimental work on pegmatites. They used the adjective "granitic" to describe their experimental objects, and showed the right path to Rome. We need experiments, and also must use highly sophisticated methods, no doubt about this, but there is no geology without field geology. The model and laboratory results must follow suit and adapted to nature. Granitic pegmatites only play o minor role in the wide spectrum of pegmatitic and aplitic rocks. The trends are illustrated for some rare elements and element couples in an ensialic orogen, with the HPPP being at the heart of this orogen, described by Weber and Behr (1983). A holistic approach offers ways out of the many roads leading to Rome for all those who like it less dogmatic and who are willing to read the book of nature with their own eyes. The classification scheme proposed herein in Table 1.1 is descriptive solely based upon what the students can determine in the field and identify when applying the routine techniques at hand in his or her study. It is in essence a classification scheme closely following "The chessboard classification scheme of mineral deposits" (Dill 2010) which is open for modification and adaptation to the practical use. The classification scheme in Table 1.1 was designed with the intension to be used in the field and the study and open for supplements.

The ore control of the HPPP from the structural point of view can simply be described as follows. It is the area of intersection between a southward-dipping thrust plane near its root zone with a set of deep-seated lineamentary fault zones tapping the subcrustal parts underneath the HPPP. Motion along these faults is mainly that of transform faults, mirrored at the surface by *en echelon* felsic dykes. Accommodation space is provided in the hinge areas and limbs of late Variscan anticlinal structures following the rules of mimetic crystallization. How do the rare elements which the pegmatites are so renowned, for behave within the geodynamic subenvironments of an ensialic orogen ? Similar subenvironments as in the HPPP and its immediate surroundings can also be found elsewhere in the Bohemian Massif, yet often obliterated to a different degree or not developed to the full completeness as along the western edge of the Bohemian Massif. A synoptical overview of the rare elements diagnostic for the pegmatites in this province is given in the concluding paragraphs of this study- see also Fig. 3.12.

Lithium Lithium is a marker element which enables us to track down the pathway of rare elements within pegmatites. Metamorphism, including the regional and contact types and the ascent of basic to intermediate magmas lead to the formation of Li phosphates of the triphylite series in the pegmatites s.st. (Fig. 6.1b). In the root zone the mobilizing impact of the mantle is evidenced. It is not a new idea that in addition to huge felsic bodies of pegmatites more mafic magmatic rocks contributed to the built-up of pegmatites. The so-called NYF pegmatites are said to have chemical affinities to A-type granites (Martin and De Vito 2005). Their source is a twofold one, from the lower crust and the mantle (Christiansen et al. 2007) or they might have entirely originated from the mantle (Haapala et al. 2007). While the differentiation from a more mafic igneous source rock well accords with a subcrustal or even mantle impact in this region (Sect. 5.1.1) the chemical composition and the mineral suite of the so-called NYF pegmatites does not correspond to the element composition of the HPPP pegmatites.

Towards the collision zone, granitic pegmatites came into existence, abundant in Li mica such as zinnwaldite and lepidolite instead of triphylite. Differentiation evolves perfectly well according to what has been written in many textbooks. The only dark spot on the white shirt is the absence of pegmatites coming at least close to some of those of the HPPP in economic terms or with regard to the variety in rare minerals. Granitic fractionation did not straightforward worked its way towards the W but also spawned some highly-fractionated granites stocks which were intruded along lineamentary fracture zones. These stocks contain in their apical parts an amblygonite-montebrasite type Li mineralization. The only place of lithium enrichment of economic interest is currently re-assessed south of the border of the Bohemian Massif. Reworking of lithium mineralization within the Bohemian Massif took place outside the ensialic orogen in the Alpine Mountains giving rise to pseudopegmatites with spodumene and holmquisite.

Beryllium Beryllium is modestly enriched in the root zone, where beryl is a rare constituent among the primary minerals of the pegmatite. This also true for the pegmatites found along the lineamentary structure zones fading out towards the south within the Moldanubian Region. The upper mantle has a temperature effect in the root zone, upgrading the content of elements pre-concentrated in the crust such as lithium but does not feed beryllium from the mantle to a large extent into this crustal system. Beryllium and the zinc described in the following paragraph have some striking similarities. Both start off from subcrustal sources, where they occur in rather exotic associations such as genthelvite [$Zn_4Be_3(SiO_4)_3S$] and willemite [$Zn_2(SiO_4)$]. Both minerals were recorded from the alkaline magmatic rocks of the Motzfeld Alkaline Complex, South Greenland (Finch 1990). The protolith of both metals is held to be of (ultra)basic origin. On their way up interaction with crustal rocks by, e.g., metasomatism affected beryllium much more than zinc. A decisive mineral association for the positioning of pegmatitic rocks which needs further investigation and can only be touched in this book for lack of the Be-bearing spinel is the assemblage made up of chrysoberyl, gahnite, beryl and sphalerite.

Zinc Zinc which is common to many oxidic and silicate minerals in the metamorphic units of the Moldanubian Zone, is not unexpectedly, a major and exclusive element to the root zone, represented by the HPPP. Unlike, beryllium, discussed above, it forms only a few minerals of its own relevant for the formation of pegmatites. It is sphalerite, gahnite s.s.s. and to a lesser extent Zn staurolite that bear witness of an intermediate repository of zinc in crustal rocks such as metasediments. This intermediate crustal repository can almost neglected for the metamorphic rocks in and around the HPPP, in the northern Moldanubian Zone, where gahnite is only sporadic. It is totally absent in the frontal parts of the collision zone of the Saxo-Thuringian Zone, where granitic pegmatites devoid of Zn developed at a minor extent. The non-sulfidic Zn compounds are of widespread occurrence, however, in the southern Moldanubian Zone, in the core zone of the ensialic orogen. Zinc sulfide is concentrated near the “hot spot” in the HPPP and does not show up anywhere else in the ensialic orogen of the Central European Variscides. Sphalerite occurs in two different mineral associations and bears witness of two distinct processes. On one hand, sphalerite enriched in Fe, is associated with pegmatitic minerals in the three stock-like pegmatites, it marks the heat center in the pegmatite field and based upon its trace element contents (indium) shows the heat gradient. On the other hand, sphalerite in most primitive “nigrine” accompanied by uraniferous pyrochlore-group minerals and columbite-(Fe) is representative of the primary subcrustal source of Zn (Fig. 3.12). Both parameters play hand in hand when it comes to account for why the HPPP is localized in this place in the basement, where deep-seated lineamentary fault zone tapping the mantle and subhorizontal thrust planes steepened towards the root zone intersect each other (Fig. 3.12).

Niobium-Tantalum-Iron-Manganese The oxidic Nb-Ta-Fe-Mn system like no other chemical system in the pegmatitic provinces underscores the transition from the metamorphic into the magmatic realm and, hence, was extensively discussed in Sect. 4.8. The metamorphic impact on the COLTAN mineralization within a pegmatites diminishes with decreasing Mn content as a function of the Ta/Nb ratio. It is most suitable described by an anticlock-wise trend. The Ta/Nb ratio is controlled by the magmatic fractionation in a consanguineous system. The most strongly fractionated magmatic system occurs in the granitic pegmatites of the Fichtelgebirge granites, where tapiolite developed besides columbite. The marker role of columbite-(Fe), together with pyrochlore and sphalerite in "nigrine" has been already emphasized in the foregoing section on zinc.

Phosphate-Calcium-Manganese Manganese is preferably taken up by garnet (spessartite) at a more abyssal level of intrusion, whereas shallow intrusions used to take up bivalent Mn into the lattice of apatite. The pegmatites located further to the S have garnet s.s.s. more strongly endowed with spessartite.

Tin-Tungsten Both elements represented mainly by cassiterite, wolframite s.s.s. and scheelite tend to be enriched in granitic pegmatites rather than in pegmatites s.st. There they bridge the gap in the state, what formerly was called the "pneumatolitic state" and featured by the greisen zones.

Uranium Felsic systems average the highest uranium contents in the crust but neither intragranitic (e.g. Rössing, Namibia) nor pegmatitic rocks (e.g. Bancroft, Canada) are U high-grade deposits. In the pegmatites s.st., uranium black ore minerals developed very early associated with Nb-Ta oxides, in the granite pegmatites they appear rather late together with Sn and W minerals.

The essence of this comprehensive outline of trends and enrichments along with geodynamic variations in an ensialic orogen contains one pragmatic point. Granites are not the parents of pegmatites and aplites. If we try and find any family relationship between the two rocks under consideration, they behave like brother and sister, there may be younger or older ones. Both, granites and pegmatites are two sides of the same coin- a heat event within the crust or sparked in the crust. Economic geology is a "*mixtum compositum*" of all geoscientific disciplines focused on one goal, finding new mineral deposits and enhancing their exploitation. The keystones of this "*mixtum compositum*" are geology and mineralogy whose studies are centered around the emplacement of the ore body and the development of its minerals and rocks. The lesser insight we have into the crustal interior, the more geophysical methods can demonstrate their value in reducing the field of speculation (Fig. 3.12).

Why granites are so frequently distributed within the Earth's crust and rare-metal pegmatites are comparatively seldom, albeit both rock series share almost the same composition with feldspar, quartz and mica as the major constituents ? May I counter with another question: "Why form the colored gemstones such as emerald and ruby that belong to a group of minerals rather seldom in the mineral kingdom while their hosts beryl and corundum are more widespread ?" Both chemical compounds, the rare-metal pegmatites and colored gemstones, do not simply develop from a parent rock or a source mineral by some kind of a breading reaction but they are the result of distinct multi-phase processes that bring together elements and chemical compounds from sources very much different as to their physical-chemical regime. A beryl from the Plössberg or Hagendorf-South Pegmatite never will end up in an aquamarine or emerald known from Zimbabwe or Malawi. On the other hand a Flossenbürg Granite will never spawn a Hagendorf-Type (Be-Zn)-Li-Nb-P pegmatite (stock-like and tabular) or aplite (tabular). Host and chromophores have derived from rather distinct lithologies. The physical-chemical regime and the geological setting of granites and pegmatites *sensu stricto* are two totally different entities. The degree of crustal to subcrustal impact on both felsic systems or in other words the geodynamic setting has a major say in which way the rather simple mineral association made up of feldspar, quartz and mica develop, whether it comes close to a normal granite like the majority of pegmatites and aplites or turns into one of the most-looked-for rare element pegmatites such as the Hagendorf-Pleystein Pegmatite Province.

Acknowledgment

A lot of papers have been published together with colleagues of mine – see References-who I need to express my gratitude for their support. I would like to include all the technical staff members who contributed to the success of these publications. In addition to these colleagues I would like to mention some, who were especially engaged in the run up to this book. Burkhart, B., Forster J., Göd, R., Amberger Kaolinwerke AKW Hirschau, Buchtal AG. Schwarzenfeld, Gottfried Feldspat GmbH, Max Schmidt Company, Heimatkundlicher Arbeitskreis Waidhaus e.V., Museumsverein Pleystein, Novak, M. /Masaryk University Brno, Reger G., T. Schubert (BGR Hannover). H.-J. Sturm performed the drafting of some of the maps and the cross sections (BGR Hannover).

My sincere thanks I would like to express to Franco Pirajno editor of this series of books who kickstarted this project and was the driving force behind it. The book was realized by SPRINGER under the technical guidance by C. van der Giessen and P. van Steenbergen who I would express my gratitude. I express my gratitude also to Mr. Prakash Marudhu and his team for their technical assistance.

H.G. Dill, *The Hagendorf-Pleystein Province: the Center of Pegmatites in an Ensialic Orogen*, Modern Approaches in Solid Earth Sciences 15,
DOI 10.1007/978-3-319-18806-5

About the Author

The author with students from the Czech Republic, Finland, Germany, The Netherlands and Slovakia at Hagendorf-North Pegmatite, 2015

Harald G. Dill After 2 years active service in an artillery battalion with the German Army (his rank when resigning from the armed forces in 2006 was colonel of the German Armed Forces Reserve) he began studying geology in 1971 at Würzburg University (minor: geography, mineralogy) followed by economic geology at the Technical University at Aachen. He received his M.Sc. degree in geology in 1975 after having submitted his master thesis on stratigraphy and paleoenvironmental studies. In 1978 he was graduated from Erlangen University with a PhD thesis on pyritiferous Pb-Cu-Zn deposits in Tuscany, Italy, submitted to the Dept. of Mineralogy.

Subsequently, he entered upon a 1-year research work at the Dept. of Soil Sciences and Soil Geography of Bayreuth University, where he was mainly engaged in shallow geophysical sounding and the study of duricrusts. Since 1979 he has been with the Federal Institute for Geosciences and Natural Resources (BGR), Hannover-Dept. of Geophysics/Radiometric age dating. He was mainly involved in the study

H.G. Dill, *The Hagendorf-Pleystein Province: the Center of Pegmatites in an Ensialic Orogen*, Modern Approaches in Solid Earth Sciences 15,
DOI 10.1007/978-3-319-18806-5

of uranium concentration processes, a joint research projects carried out in close cooperation with international agencies and exploration companies in France, Italy and Australia. From 1986 through 1991 he was a staff member of the project management group of the "Continental Deep Drilling Program of the F.R. Germany", being responsible for economic geology, mineralogy and geochemistry. In 2014 he retired from BGR continuing his work at the university as lecturer and researcher.

In 1982 he became lecturer for applied geology at Mainz University, where he obtained his Dr. habil. degree in 1985 after submission of his thesis entitled "Ore Mineralization at the Western Edge of the Bohemian Massif" (Assist. Prof.). In 1991 he was appointed Associated Professor at Hannover University, in 2008 honorary Professor at Mainz University and in 2010 profesor invitado de la Universidad Nacional del Sur – Bahia Blanca, Argentine and in 2013 he was awarded a doctorate *honoris causa* from Iasi University, Romania.

He gives lectures in economic geology (metallic and non-metallic deposits), applied sedimentology and lecture courses in petrographic microscopy and is mentoring young colleagues (Ms. students) during more than 35 years. Aside of his regular teaching posts he was/is involved in teaching at universities in Bangkok (Thailand), Cottbus (Germany), Doha (Qatar), Vilnius (Lithuania), Ulan Bataar (Mongolia), Zomba, Malawi, Tashkent (Uzbekistan), Riga (Latvia), Muscat (Oman), Amman (Jordan), Tunis (Tunisia), Hanoi (Vietnam), Athens (Greece), Bahia Blanca (Argentina), Izmir (Turkey) and Iasi (Romania), Brno (Czech Republic). In BGR he was senior research scientist in "Geophysical exploration and technical mineralogy". His main interest lies in the field of chemistry and mineralogy of ancient and modern depositional systems and related fossil fuel, metallic and non-metallic deposits. His work on this subject matter has led to 320 publications, more than 100 abstracts, 1 patent and several open file reports. His scientific work led also to the discovery of a smectite-bearing clay deposit at the edge of the Gobi Desert, Mongolia, helped delineating a gypsum-celestite deposit in the desert on the Qatar Peninsula, and contributed to the finds of oil in the Permo-Carboniferous basins in SE Germany as well as three pegmatite occurrences. Furthermore he conducts studies in the field of archeometallurgy, mining history and only recently has written his second three-volume work on the history of aerial warfare over Germany.

References

Abdakhmanov K, Zhautikov T, Kulishkin H, Poletaev A (1997) Field trip in Kazakhstan. In: Bekzanov GR, Dudich E, Gaál G, Jenchuraeva RJ (eds) Paleozoic granite-related Au, Cu, Mo, W, REE deposits and epithermal gold deposits, IUGS/UNESCO Deposit Modeling Program, workshop, Kazakhstan and Kyrgyzstan, Budapest, pp 65–80, 31 August–14 September 1997

Achstetter M (2007) Aus Hornberg im Schwarzwald: Aquamarin in Edelsteinqualität und weitere Neufunde. Lapis 32:13–21

Ackerman L, Zachariáš J, Pudilová M (2007) P–T and fluid evolution of barren and lithium pegmatites from Vlastějovice, Bohemian Massif, Czech Republic. Int J Earth Sci 96:623–638

Adiwidjana G, Friese K, Klaska KH, Schlüter J (1999) The crystal structure of kastningite (Mn, Fe, Mg)(H2O)4[Al2(OH)2(H2O)2(PO4)2]·2H2O – a new hydroxyl aquated orthophosphate hydrate mineral. Z Krist 214:465–468

Adusumilli MS, Castro C, Bhaskara Rao A (1994) Blue and green tourmalines from Greogrio pegmatite, Brazil. Int Mineral Assoc Meet Abstr 16:3–4

Aichler J, Fojt B, Vaněček M (1995) Metallogenesis of the [Moravo Silesian zone]. In: Dallmeyer RD, Franke W, Weber K (eds) Tectono-stratigraphic evolution of the Central and East European orogens. Springer, Heidelberg, pp 512–517

Aldrich LT, Davis GL, Tilton GR, Wetherill GW (1956) Radioactive ages of minerals from the Brown Derby Mine and the Quartz Creek Granite near Gunnison, Colorado. J Geophys Res 61:215–232

Aleksandrov VB (1963) Isomorphism of cations in titaniferous tantalo-niobates of composition AB2X6. Doklady Akademia Nauk 153:672–675 (English translation 129-131)

Aleksandrowski P, Mazur S (2002) Collage tectonics in the northeasternmost part of the Variscan Belt: the Sudetes, the Bohemian massif. In: Winchester JA, Pharaoh TC, Verniers J (eds) Palaeozoic amalgamation of Central Europe, Geological Society special publication, 201. Geological Society, London, pp 237–277

Aleksandrowski P, Kryza R, Mazur S, Zaba J (1997) Kinematic data on major Variscan strike-slip faults and shear zones in the Polish Sudetes, northeast Bohemian Massif. Geol Mag 133:727–739

Alfonso P, Melgarejo JC, Yusta I, Velasco F (2003) Geochemistry of feldspars and muscovite in granitic pegmatite from the Cap de Creus Field, Catalonia, Spain. Can Mineral 41:103–116

Altherr R, Holl A, Hegner E, Langer C, Kreuzer H (2000) High-potassium, calc alkaline I-type plutonism in the European Variscides: northern Vosges (France) and northern Schwarzwald (Germany). Lithos 50:51–73

Amacher P, Schüpbach T (2011) NEAT-Mineralien: Kristallschätze tief im Berg. 230 S. Verlag GEO-Uri, Amsteg, pp 52–59

H.G. Dill, *The Hagendorf-Pleystein Province: the Center of Pegmatites in an Ensialic Orogen*, Modern Approaches in Solid Earth Sciences 15,
DOI 10.1007/978-3-319-18806-5

Anthes G, Reischmann T (1996) Geochronologie und Isotopengeochemie der NE Mitteldeutschen Kristallinschwelle. Terra Nostra 96:9–10

Armbruster T, Bonazzi P, Akasak M, Bermanec V, Chopin C, Giere R, Heuss-Assbichler S, Liebscher A, Menchetti S, Pan Y, Pasero M (2006) Recommended nomenclature of epidote-group minerals. Eur J Mineral 18:551–567

Awdankiewicz M, Kryza R, Szczepara N (2013) Timing of post-collisional volcanism in the eastern part of the Variscan Belt: constraints from SHRIMP zircon dating of Permian rhyolites in the North-Sudetic Basin (SW Poland). Geol Mag 151:611–628

Baldwin JR, Hill PG, von Knorring O, Oliver GJH (2000) Exotic aluminum phosphates, natromontebrasite, brazilianite, goyazite, gorceixite, and crandallite from rare-element pegmatites in Namibia. Mineral Mag 67:1147–1164

Bariand P, Poullen JF (1978) The Pegmatites of Laghman, Nuristan, Afghanistan. Miner Rec 301–308

Bartholomew MJ, Whitaker AE (2010) The Alleghanian deformational sequence at the foreland junction of the Central and Southern Appalachians. In: Tollo RP, Bartholomew MJ, Hibbard JP, Karabinos PM (eds) From Rodinia to Pangea: the lithotectonic record of the Appalachian Region, vol GSA Memoir, 206. Geological Society of America, Boulder, pp 431–454

Barton MD (1986) Phase equilibria and thermodynamic properties of minerals in the BeO-AI, O,-SiO,-H, O (BASH) system, with petrological applications. Am Mineral 71:277–300

Batchelor R, Kinnaird J (1984) Gahnite compositions compared. Mineral Mag 48:425–429

Bates RL, Jackson JA (1987) Glossary of geology. American Geological Institute, Alexandria, 788 pp

Bauberger W (1957) Über die "Albitpegmatite" der Münchberger Gneissmasse und ihre Nebengesteine. Geologica Bavarica 36:5–77

Baumann L (1979) Some aspects of mineral deposit formation and the metallogeny of Central Europe. Verhandlungen der Bundesanstalt Wien 1978:205–220

Baumann L, Kölbel B, Kraft S, Lächelt J, Rentzsch J, Schmidt K (1986) German democratic republic. In: Dunning FW, Evans AM (eds) Mineral deposits of Europe, vol. 3 Central Europe. IMM & Mineralogical Society, London, pp 303–329

Baumann L, Kuschka E, Seifert TH (2000) Lagerstätten des Erzgebirges. Enke im Thieme Verlag, Stuttgart, pp 1–300

Baumgärtner J, Zoback MD (1989) Interpretation of hydraulic fracturing pressure-time records using interactive analysis methods. Int J Rock Mech Min Sci Geomech Abstr 26:461–469

Bayerisches Oberbergamt (1924) Die nutzbaren Mineralien, Gesteine und Erden Bayerns. R. Oldenburg, Piloty & Loehle, München, 210 pp

Bayerisches Staatsministerium für Wirtschaft und Verkehr (1978) Rohstoffprogramm für Bayern. Bayerisches geologisches Landesamt, München, 129 pp

Bederke E (1924) Das Devon in Schlesien und das Alter Sudetenfaltung. Fortschritte der Geologie und Paläontologie 7:1–55

Behr H-J, Engel W, Franke W, Giese P, Weber K (1984) The Variscan belt in Central Europe: main structures, geodynamic implications, open questions. Tectonophys 109:15–40

Behr HJ, Grosse S, Heinrichs T, Wolf U (1989) A reinterpretation of the gravity field in the surroundings of the KTB drill site-implications for the granite plutonism and terrane tectonic in the Variscan. In: Emmermann R, Wohlenberg J (eds) The German continental deep drilling program (KTB). Springer, Berlin, pp 501–525

Benisek A, Finger F (1993) Factors controlling the development of prism faces in granite zircons: a microprobe study. Contrib Mineral Petrol 114:441–451

Beran P, Sekora J (2006) The Krásno Sn-W ore district, near Horní Slavko: mining history, geological and mineralogical characteristics. J Czech Geol Soc 51:3–187

Berger EL, Lauretta DS, Keller LP (2012) The thermodynamic properties of cubanite 75th annual Meteoritical Society meeting, 5008

Bernard JH (1980) Paragenetic units of the European Variscan megazone. Freiberger Forschungshefte C 354:55–117

Bernhard F, Walter F, Ettinger K, Taucher J, Merreiter K (1998) Pretulite, Sc(PO4), a new scandium mineral from Styrian and Lower Austrian lazulite occurrences, Austria. Am Mineral 83:625–630

Besang C, Harre W, Kreuzer H, Lenz H, Müller P, Wendt I (1976) Radiometrische Datierung, geochemische und petrographische Untersuchungen der Fichtelgebirgsgranite. Geol Jahrb E 8:3–71

Bierlein FP, Groves DI, Cawood PA (2009) Metallogeny of accretionary orogens-the connection between lithospheric processes and metal endowment. Ore Geol Rev 36:282–292

Bigham JM, Schwertmann U, Traina SJ, Winland RL, Wolf M (1996) Schwertmannite and the chemical modeling of iron in acid sulfate waters. Geochim Cosmochim Acta 60:185–195

Bilal E, Horn AH, de Mello FM (2012) P-Li-Be Bearing Pegmatites of the South East Brazil. Int J Geosci 3:281–288

Bingen B, Davis WJ, Austrheim H (2001) Zircon U-Pb geochronology in the Bergen arc eclogites and their Proterozoic protoliths, and implications for the pre-Scandian evolution of the Caledonides in western Norway. Geol Soc Am Bull 113:640–649

Birch WD, Grey IE, Mills SJ, Pring A, Bougero C, Ribaldi-Tunnicliffe A, Wilson NC, Keck E (2011) Nordgauite, MnAl2(PO4)2(F, OH)2. 5H2O, a new mineral from the Hagendorf-Süd pegmatite, Bavaria, Germany: description and crystal structure. Mineral Mag 75:269–278

Bjørlykke H (1937) Mineral parageneses of some granite pegmatites near Kragerø, Southern Norway. Nor Geol Tidsskr 17:1–16

Blümel P (1986) Metamoprhic processes in the Variscan crust of the central segment. In: Freeman R, Mueller S, Giese P (eds) Proceedings of the 3rd EGT workshop. European Science Foundation, Bad Honef, pp 149–155

Bohlen SR, Montana A, Kerrick DM (1991) Precise determinations of the equilibria kyanite ¼ sillimanite and kyanite ¼ andalusite and a revised triple point for Al_2SiO_5 polymorphs. Am Mineral 76:677–680

Boiron MC, Barakat A, Cathelineau M, Banks DA, Durisova J, Moravek P (2001) Geometry and P-V-T-X conditions of microfissural ore fluid migration: the Mokrsko gold deposit (Bohemia). Chem Geol 173:207–225

Borisova AY, Thomas R, Salvi S, Candaudap F, Lanzanova A, Chmeleff J (2012) Tin and associated metal and metalloid geochemistry by femtosecond LA-ICP-QMS microanalysis of pegmatite–leucogranite melt and fluid inclusions: new evidence for melt–melt–fluid immiscibility. Mineral Mag 76:91–113

Bossart PJ, Meier M, Oberli F, Steiger RH (1986) Morphology versus U-Pb systematics in zircon: a high-resolution isotopic study of a zircon population from a Variscan dyke in the Central Alps. Earth Planet Sci Lett 78:339–354

Bottke H (1963) Zur Kenntnis der dichten Roteisenerz aus Eisenerzlagerstätten des Lahn-Dill-Typs und deren Bildungsbedingungen. Erzmetall 16:437–443

Braithwaite RSW, Cooper BV (1982) Childrenite in south-west England. Mineral Mag 46:119–126

Breiter K (1998a) Granites of the Krušne Hory/ Erzgebirge Mts. Slavkovský Les area. In: Breiter K (ed) Excursion guide: genetic significance of phosphorus in fractionated granites, International geological correlation program, IGCP 373. Czech Geological Survey, Peršlák, pp 21–31

Breiter K (1998b) Phosphorus- and fluorine-rich granite system at Podlesi. In: Breiter K (ed) Excursion guide: genetic significance of phosphorus in fractionated granites, International geological correlation program, IGCP 373. Czech Geological Survey, Peršlák, pp 59–76

Breiter K (1998c) P-rich muscovite granites-latest product of two-mica granites fractionation in the South Bohemian Pluton. In: Breiter K (ed) Excursion guide: genetic significance of phosphorus in fractionated granites, International geological correlation program, IGCP 373. Czech Geological Survey, Peršlák, pp 107–126

Breiter K, Siebel W (1995) Granitoids in the Rozvadov pluton. Geol Rundsch 84:506–519

Breiter K, Fryda J, Seltmann R, Thomas R (1997) Mineralogical evidence for two magmatic stages in the evolution of an extremely fractionated P-rich rare-metal granite: the Podlesı stock, Krušne Hory, Czech Republic. J Petrol 38:1723–1739

Breiter K, Novák M, Koller F, Cempírek J (2005) Phosphorus a omnipresent minor element in garnet of diverse textural types of leucocratic granitic rocks. Mineral Petrol 85:205–221

Brindley GW, Kao C-C, Harrison JL, Lipsicas M, Raythatha R (1986) Relation between structural disorder and other characteristics of kaolinites and dickites. Clays Clay Miner 34:239–249

Bröcker M, Żelaźniewicz A, Enders M (1998) Rb–Sr and U–Pb geochronology of migmatitic gneisses from the Góry Sowie (West Sudetes, Poland): the importance of Mid–Late Devonian metamorphism. J Geol Soc Lond 155:1025–1036

Bromiley G, Hilaret N (2005) Hydrogen and minor element incorporation in synthetic rutile. Mineral Mag 69:345–358

Brooker EJ, Nuffield EW (1952) Studies of radioactive compounds IV, Pitchblende from lake Athabaska. Am Mineral 37:363–385

Broska I, Williams CT, Uher P, Koneçný P, Leichmann J (2004) The geochemistry of phosphorus in different granite suites of the Western Carpathians, Slovakia: the role of apatite and P-bearing feldspar. Chem Geol 205:1–15

Brownfield ME, Foord EE, Sutley SJ, Botinelly T (1993) Kosnarite, $KZr_2(PO_4)_3$, a new mineral from Mount Mica and Black Mountain, Oxford County. Maine Am Mineral 78:653–656

Brunet F, Morineau D, Schmid-Beurmann P (2004) Heat-capacity of lazulite $MgAl_2(PO_4(OH)_2$, from 35 to 300K and a (S-V) value for P_2O_5 to estimate phosphate entropy. Min Mag 68:123–134

Buder W, Schuppan W, Seltmann R, Wolf M (1993) The Poehla Haemmerlein deposit. In: Seltmann R, Breiter K (eds) Hercynian tin granites and associated mineralisation from the Saxonian and Bohemian parts of the Erzgebirge: excursion guide, IAGOD working group on Tin and Tungsten. GeoForschungsZentrum, Potsdam, pp 67–71

Bültemann W (1954) Mineralien aus den Wölsendorfer Flussspatgruben. Aufschluss 5:211–212

Cairney T, Kerr CD (1998) The geology of the Kabwe area: explanation of degree sheet 1428, NW quarter. Geol Surv Zamb Rep 47:1–40

Cameron EN, Jahns RH, McNair AH, Page LR (1949) Internal structure of pegmatites. Econ Geol Monogr 2:1–115

Carl C, Dill HG (1983) Uranium disequilibria and modern redistribution phenomena in alteration zones in the Höhensteinweg uranium occurrence. Uranium 1:113–125

Carl C, Dill HG (1984) U-Pb Datierungen an Pechblenden aus dem Nabburg-Wölsendorfer Flussspatrevier. Geol Jahrb D 63:59–76

Carl C, Dill HG (1985) Age of secondary uranium mineralization in the basement rocks of northeastern Bavaria, F.R.G. Chem Geol 52:295–316

Carl C, Wendt I (1993) Radiometrische Datierungen der Fichtelgbirgsgranite. Zeitschrift Geologische Wissenschaften 21:49–72

Carl C, Dill HG, Kreuzer H, Wendt I (1985) U/ Pb- und K/ Ar-Datierungen des Uranvorkommens Höhenstein Oberpfalz. Geol Rundsch/Int J Earth Sci 74:483–504

Čech F, Mísa Z, Povondra P (1971) A green lead-containing orthoclase. Miner Pet 15:213–231

Čech F, Rieder M, Novák F, Novotný J (1978) Accessory nigerite in a granite from central Bohemia, Czechoslovakia. Neues Jb Mineral Monat 8:337–346

Cempírek J, Novák M (2006a) Hydroxylherderit a sdružené berylofosfáty z pegmatitu Rožná-Borovina. Acta Musei Moraviae, Scientiae Geologicae 91:79–88 (in Czech, with English summary)

Cempírek J, Novák M (2006b) Mineralogy of dumortierite-bearing abyssal pegmatites at Starkoč and Běstvina, Kutná Hora Crystalline Complex. J Czech Geol Soc 51:259–270

Černý P (1972) The Tanco pegmatite at Bernic Lake Manitoba W, Eucryptite. Can Mineral 11:708–713

Černý P (1982) Mineralogy of rubidium and cesium. In: Černý P (ed) Granitic pegmatites, science and industry, Short course handbook, 8. Mineralogical Association of Canada, Toronto, pp 149–162

Černý P (1989) Exploration strategy and methods for pegmatite deposits of tantalum. In: Möller P, Cerný P, Saupé F (eds) Lanthanides, tantalum and niobium. Springer, Heidelberg, pp 274–310

Černý P (1991) Rare-element granitic pegmatites: Part I: Anatomy and internal evolution of pegmatite deposits. Part II: Regional and global environments and petrogenesis. Geosci Can 18:49–81

Černý P (1992) Geochemical and petrogenetic features of mineralization in rare-element granitic pegmatites in the light of current research. Appl Geochem 7:393–416

Černý P (2000) Constitution, petrology, affiliations and categories of miarolitic pegmatites. In: Pezzotta F (ed) Mineralogy and petrology of shallow depth pegmatites, Memorie della Società Italiana di Scienze Naturali e del Museo Civico di Storia naturale di Milano, 30. Società Italiana di Scienze Naturali, Milano, pp 5–12

Černý P (2002) Mineralogy of beryllium in granitic pegmatites. Rev Mineral Geochem 50:405–444

Černý P, Ercit TS (2005) The classification of granitic pegmatites revisited. Can Mineral 43:2005–2026

Černý P, Ferguson RB (1972) The Tanco pegmatite at Bernic Lake Manitoba. VI, Petalite and spodumene relations. Can Mineral 11:690–707

Černý P, Povondra P (1966) Beryllian cordierite from Věžná: (Na+K)+Be ⇒ Al. Neues Jb Mineral Monat 1966:36–44

Černý P, Simpson FM (1977) The Tanco pegmatite at Bernic Lake Manitoba: the Beryl. Can Mineral 15:489–499

Černý P, Simpson FM (1978) The Tanco pegmatite at Bernic Lake Manitoba X, Pollucite. Can Mineral 16:325–333

Černý P, Meintzer RE, Anderson AJ (1985) Extreme fractionation in rare-element granitic pegmatites: selected examples of data and mechanisms. Can Mineral 23:381–421

Černý P, Goad BE, Hawthorne FC, Chapman R (1986) Fractionation trends of the Nb-and Ta-bearing oxide minerals in the Greer Lake pegmatitic granite and its pegmatite aureole, southeastern Manitoba. Am Mineral 71:501–517

Černý P, Novák M, Chapman R (1992) Effects of sillimanite-grade metamorphism and shearing on Nb-Ta oxide minerals in granitic pegmatites: Maršíkov. Northern Moravia, Czechoslovakia. Can Mineral 30:699–718

Černý P, Staněk J, Novák M, Baadsgaard H, Rieder M, Ottolini L, Kavalová M, Chapman R (1995) Chemical and structural evolution of micas at the Rožná and Dobrá Voda pegmatites, Czech Republic. Mineral Petrol 55:177–202

Černý P, Chapman R, Simmons WB, Chackowsky E (1999) Niobian rutile from the McGuire granitic pegmatite, Park County, Colorado: solid solution, exsolution, and oxidation. Am Mineral 84:754–763

Černý P, Fryer BJ, Chapman R (2001) Apatite from granitic pegmatite exocontacts in Moldanubian serpentinites. J Czech Geol Soc 46:15–20

Chakhmouradian AR, Mitchell RH (1997) Compositional variation of perovskite-group minerals from the carbonatite complexes of the Kola alkaline province, Russia. Can Mineral 35:1293–1310

Charalampides G, Ericsson T, Nord AG, Khangi F (1988) Studies of hydrothermally prepared (Fe, M),(PO4)2-sarcopsides. Neues Jb Mineral Monat 1988:324–336

Chatterjee ND (1976) Margarite stability and compatibility relation in the system CaO-Al_2O_3-SiO2-H_2O as a pressure-temperature indicator. Am Mineral 61:699–709

Cherniak DJ, Watson EB (2003) Diffusion in zircon. In: Hanchar JM, Hoskin PWO (eds) Zircon, Reviews in mineralogy & geochemistry, 53. Mineralogical Society of American, Washington, DC, pp 113–143

Chopin C, Oberti R, Camara F (2006) The arrojadite enigma: II. Compositional space, new members, and nomenclature of the group. Am Mineral 91:1260–1270

Chovan M, Rojkovič I, Andráš P, Hanas P (1992) Ore mineralization of the Malé Karpaty Mts. (Western Carpathians). Geol Carpath 43:275–286

Christiansen EH, Haapla I, Garret GL (2007) Are Cenozoic topaz rhyolites the erupted equivalent s of Proterozoic Rapakivi granites? Examples from the western United States and Finland. Lithos 97:219–246

Chrt J (1959) Current result of geological research at the skarn-type deposits in the Krušne Hory Mts. and the Krkonoše Mts. Geological Průk 180–185. (in Czech)
Chukanov NV (2005) Minerals of the Kerch iron-ore basin in Eastern Crimea, Mineralogical almanac, 8. Ocean Pictures Ltd, Moscow, pp 1–109
Ciesielczuk J, Domańska-Siuda J, Szuszkiewicz A, Turniak K (2008) Strzegom-Sobótka massif (Sudetes, SW Poland) – an example of a complex late-Variscan granitic intrusion and its pegmatitic mineralization. Mineralogia – Special Papers 32: 181–187
Claus G (1936) Schwerminerale aus kristallinen Gesteinen des Gebietes zwischen Passau und Cham. Neues Jahrbuch Mineralogie Beilage 71:1–58
Collins AS, Kryza R, Żelaźniewicz J (2000) Macrofabric fingerprints of late Devonian-early Carboniferous subduction in the Polish Variscides, the Kaczawa complex, Sudetes. J Geol Soc Lond 157:283–288
Cota AC, Wise MA, Owens BE (2010) Textural and chemical constraints on the origin of epidote in granitic pegmatites. Northeastern Section (45th annual) and Southeastern Section (59th annual) Joint Meeting (13–16 March 2010), Session No. 65 – Booth 17, Igneous/Metamorphic Petrology and Geochemistry (Posters)
Crane MJ, Sharpe JL, Williams PA (2001) Formation of chrysocolla and secondary copper phosphate in the highly weathered supergene zones of some Australian deposits. Rec Aust Mus 53:49–56
Cruft EF (1966) Minor elements in igneous and metamorphic apatite. Geochim Cosmochim Acta 30:375–398
Cuney M, Barbey P (2014) Uranium, rare metals and granulite facies metamorphism. Geosci Front 5:729–745
Dabinett TR, Humberstone D, Leverett P, Williams PA (2008) Synthesis and stability of wroewolfeite, $Cu_4SO_4(OH)_6.2H_2O$. Pure Appl Chem 80:1317–1323
Dachille F, Simons PY, Roy R (1968) Pressure-temperature studies of anatase, brookite, rutile, and TiO_2 (II). Am Mineral 53:1929–1939
Dallmeyer RD, Urban M (1998) Variscan vs. Cadomian tectonothermal activity in northwestern sectors of the Teplá-Barrandian zone, Czech Republic: constraints from $^{40}Ar/^{39}Ar$ ages. Geol Rundsch 87:94–106
Dallmeyer RD, Franke W, Weber K (1995a) Pre-permian geology of Central and Eastern Europe. Springer, Berlin/Heidelberg, 604 pp
Dallmeyer RD, Fallick AE, Koller F, Slapansky P (1995b) The Nebelstein complex: a Variscan mineralized granite intrusion in the Bohemian Massif (Austria). Beiheft I. Eur J Mineral 7:52
Davis H, Reynolds SJ (1996) The structural geology of rocks and regions, 2nd edn. Wiley, New York
De Jong G, Rotherham J, Phillips GN, Williams PJ (1997) Mobility of rare-earth elements and copper during shear-zone-related retrograde metamorphism. Geol Mijnb 76:311–319
DEKORP Research Group (1990) Results of deep-seismic reflection investigations in the Rhenish Massif. Tectonophysics 173:507–515
Delaney PJV (1992) Gemstones of Brazil: geology and occurrences, Revista Escola de Minas, Praca Tiradentes 20. Editora REM-Revista Escola de Minas, Ouro Preto, 125 pp
Demartin F, Gramaccioli CM, Pilati T, Sciesa E (1996) Sigismundite, (Ba, K, Pb) Na3 (Ca, Sr) (Fe, Mg, Mn)14 (OH)2(PO4)12, a new Ba-rich member of the arrojadite group from Spluga Valley, Italy. Can Mineral 34:827–834
Demartin F, Gay HD, Gramaccioli CM, Pilati T (1997) Benyacarite, a new titanium-bearing phosphate mineral species from Cerro Blanco, Argentina. Can Mineral 35:707–712
Dickin AP (2005) Radiogenic isotope geology. Cambridge University Press, Cambridge, 471 pp
Dill HG (1979) Der Feldspat-Bergbau in der Münchberger Gneismasse. Bergbau 30:301–304
Dill HG (1981) Zwei selenid-führende Uran-Mineralisationen aus dem ostbayerischen Moldanubikum und ihre mögliche Bedeutung für die Klärung der Lagerstättengenese. Geol Jahrb D 48:37–57
Dill HG (1982) Geologie und Mineralogie des Uranvorkommens am Höhensteinweg bei Poppenreuth (NE Bayern) – Ein Lagerstättenmodell. Geol Jahrb D 50:3–83

Dill HG (1983a) On the formation of the vein-type uranium "yellow ores" from the Schwarzach-Area (NE-Bavaria, Germany) and on the behavior of P, As, V, and Se during supergene processes. Geol Rundsch/Int J Earth Sci 72:955–980

Dill HG (1983b) Vein- and metasedimentary-hosted carbonaceous matter and phosphorus from NE Bavaria (F.R. Germany) and their implication on syngenetic and epigenetic uranium concentration. Neues Jb Mineral Abh 148:1–21

Dill HG (1983c) Plutonic mobilization, sodium metasomatism, propylitic, wall rock alteration and element partitioning from Höhensteinweg uranium occurrence (Northeast Bavaria). Uranium 1:139–166

Dill HG (1983d) Lagerstättengenetische Untersuchungen im Bereich der Uranerz-Struktur Wäldel/Mähring (NE Bayern). Geol Rundsch/Int J Earth Sci 72:229–252

Dill HG (1985a) Die Vererzung am Westrand der Böhmischen Masse: Metallogenese in einer ensialischen Orogenzone. Geol Jahrb D 73:3–461

Dill HG (1985b) The polymetallic and monotonous uranium parageneses- a contribution to the position of endogenous uranium mineralization at the western edge of the Bohemian Massif. Neues Jb Mineral Monat 1985:184–192

Dill HG (1986) Fault-controlled uranium black ore mineralization from the western edge of the Bohemian Massif (NE Bavaria/F.R. Germany). In: Fuchs HD (ed) Vein-type uranium deposits, International Atomic Energy Agency. The Agency, Vienna, pp 275–291

Dill HG (1989) Metallogenetic and geodynamic evolution in the central European Variscides: a pre-well site study for the German Continental Deep Drilling Programme. Ore Geol Rev 4:279–304

Dill HG (1990) Chemical basin analysis of the metalliferous "Variegated Metamorphics" of the Bodenmais ore district (F.R. of Germany). Ore Geol Rev 5:151–173

Dill HG (2001) The geology of aluminium phosphates and sulphates of the alunite supergoup: a review. Earth-Sci Rev 53:35–93

Dill HG (2007) A review of mineral resources in Malawi: with special reference to aluminium variation in mineral deposits. J Afr Earth Sci 47:153–173

Dill HG (2010) The "chessboard" classification scheme of mineral deposits: mineralogy and geology from aluminum to zirconium. Earth-Sci Rev 100:1–420

Dill HG (2015) Pegmatites and aplites: their genetic and applied ore geology. Ore Geol Rev 69:417–561

Dill HG, Klosa D (2011) Heavy-mineral-based provenance analysis of Mesozoic continental-marine sediments at the western edge of the Bohemian Massif, SE Germany: with special reference to Fe-Ti minerals and the crystal morphology of heavy minerals. Int J Earth Sci 100:1497–1513

Dill HG, Kolb SG (1986) The Grossschloppen-Hebanz uranium occurrences: a prototype of mineralized structure zones characterized by desilification and silification. In: Fuchs HD (ed) Vein-type uranium deposits, International Atomic Energy Agency. The Agency, Vienna, pp 261–274

Dill HG, Lohmann D (2011) Die Zeit der billigen Rohstoffe ist vorbei – Prof. Harald G. Dill im Interview. In: Lohmann D, und Podbregar N (Editoren und Autoren) Im Fokus: Bodenschätze-Auf der Suche nach Rohstoffen. Springer, Heidelberg/Dordrecht/London/New York, pp 29–39

Dill HG, Röhling S (2007) Bodenschätze der Bundesrepublik Deutschland 1: 1 000 000 (unter Mitarbeit der Geologischen Dienste der Bundesländer). Bundesanstalt für Geowissenschaften und Rohstoffe, Hannover

Dill HG, Weber B (2009) Eine ungewöhnliche zink- und wismutführende Benyacarit-Paragenese von Pleystein/Oberpfalz. Geologische Blätter Nordost-Bayern 59:69–78

Dill HG, Weber B (2010a) Eine störungsgebundene Zinnober-Molybdänglanz-Remobilisation in Amphiboliten der ZEV westlich von Vohenstrauss. Geologische Blätter NO-Bayern 60:127–140

Dill HG, Weber B (2010b) Accessory minerals of fluorite and their implication concerning the environment of formation (Nabburg-Wölsendorf fluorite district, SE Germany): with special reference to fetid fluorite ("Stinkspat"). Ore Geol Rev 37:65–86

Dill HG, Weber B (2010c) Variation of color, structure and morphology of fluorite and the origin of the hydrothermal F-Ba deposits at Nabburg-Wölsendorf, SE Germany. Neues Jb Mineral Abh 187:113–132

Dill HG, Weber B (2011) Die Oberpfälzer Flussspat-Anthologie – "Bunte Steine" prägen die Region und ihre Menschen um den Wölsenberg. Verlag Druckkultur Späthling, Weissenstadt, 311 pp

Dill HG, Weber B (2014a) Die Entstehung von Blei-Zink-Kupfer-Mineralisationen in der nördlichen Oberpfalz (Nordostbayern) unter besonderer Berücksichtigung ihrer Spurenelemente. Geologische Blätter Nordost-Bayern 63:55–79

Dill HG, Weber B (2014b) Die Stadtgeologie von Pleystein – Ein Beitrag zur lithologisch-strukturgeologischen Entwicklung des Kreuzberg-Pegmatits in der nördlichen Oberpfalz. Geologische Blätter Nordost-Bayern 63:81–105

Dill HG, Weiser T (1981) Eine Molybdän-Sulfid-Impsonit – Mineralisation aus dem Uran-Vorkommen Wäldel/Mähring (Oberpfalz). Neues Jb Mineral Monat 1981:452–458

Dill HG, Skoda R (2015) The new Nb-P aplite at Reinhardsrieth: a keystone in the lateral and depth zonations of the Hagendorf-Pleystein Pegmatite field, SE Germany. Ore Geol Rev 70:208–227

Dill HG, Bosse H-R, Henning K-H, Fricke A, Ahrend H (1997) Mineralogical and chemical variations in hypogene and supergene kaolin deposits in a mobile fold belt–The Central Andes of northwestern Peru. Mineral Deposita 32:149–163

Dill HG, Weber B, Fuessl M, Melcher F (2006a) The origin of the hydrous scandium phosphate kolbeckite from the Hagendorf – Pleystein pegmatite province, Germany. Mineral Mag 70:281–290

Dill HG, Melcher F, Fuessl M, Weber B (2006b) Accessory minerals in cassiterite: a tool for provenance and environmental analyses of colluvial-fluvial placer deposits (NE Bavaria, Germany). Sediment Geol 191:171–189

Dill HG, Khishigsuren S, Majigsuren Y, Myagmarsuren S, Bulgamaa J, Hongor O (2006c) A review of industrial minerals of Mongolia: the impact of geological and geographical factors on their formation and use. Int Geol Rev 48:129–170

Dill HG, Fuessl M, Botz R (2007a) Mineralogy and (economic) geology of zeolite-carbonate mineralization in basic igneous rocks of the Troodos Complex, Cyprus. Neues Jb Mineral Abh 183:251–268

Dill HG, Melcher F, Fuessl M, Weber B (2007b) The origin of rutile-ilmenite aggregates ("nigrine") in alluvial-fluvial placers of the Hagendorf pegmatite province, NE Bavaria, Germany. Mineral Petrol 89:133–158

Dill HG, Gerdes A, Weber B (2007c) Cu-Fe-U phosphate mineralization of the Hagendorf-Pleystein pegmatite province, Germany: with special reference to Laser-Ablation-Inductive-Coupled-Plasma Mass Spectrometry (LA-ICP-MS) of iron-cored torbernite. Mineral Mag 71:371–387

Dill HG, Sachsenhofer RF, Grecula P, Sasvári T, Palinkaš LA, Borojević-Šoštarić S, Strmić-Palinkaš S, Prochaska W, Garuti G, Zaccarini F, Arbouille D, Schulz H-M (2008a) Fossil fuels, ore – and industrial minerals. In: McCann T (ed) Geology of Central Europe, Special publication. Geological Society of London, London, pp 1341–1449

Dill HG, Sachsenhofer RF, Grecula P, Sasvári T, Palinkaš LA, Borojević-Šoštarić S, Strmić-Palinkaš S, Prochaska W, Garuti G, Zaccarini F, Arbouille D, Schulz H-M, Locmelis B (2008b) The origin of mineral and energy resources of Central Europe (map 1: 2500000). Geological Society of London, London (on CD ROM)

Dill HG, Melcher F, Gerdes A, Weber B (2008c) The origin and zoning of hypogene and supergene Fe-Mn-Mg-Sc-U-REE-Zn phosphate mineralization from the newly discovered Trutzhofmühle aplite (Hagendorf pegmatite province, Germany). Can Mineral 46:1131–1157

Dill HG, Weber B, Gerdes A, Melcher F (2009a) The Fe-Mn phosphate aplite "Silbergrube" near Waidhaus, Germany: epithermal phosphate mineralization in the Hagendorf-Pleystein pegmatite province. Mineral Mag 72:1143–1168

Dill HG, Weber B, Kaufhold S (2009b) The origin of siderite-goethite-phosphate mineralization in the karst-related faultbound iron ore deposit Auerbach, Germany, a clue to the timing of hypogene and supergene Fe-Al phosphates in NE Bavaria. Neues Jb Mineral Abh 186:283–307

Dill HG, Gerdes A, Weber B (2010a) Age and mineralogy of supergene uranium minerals – tools to unravel geomorphological and palaeohydrological processes in granitic terrains (Bohemian Massif, SE Germany). Geomorphology 117:44–65

Dill HG, Hansen B, Keck E, Weber B (2010b) Cryptomelane a tool to determine the age and the physical-chemical regime of a Plio-Pleistocene weathering zone in a granitic terrain (Hagendorf, SE Germany). Geomorphology 121:370–377

Dill HG, Kaufhold S, Weber B, Gerdes A (2010c) Clay mineralogy and LA-ICP-MS dating of a supergene U-Cu- mineralization bearing nontronite at Nabburg-Wölsendorf, SE Germany. Can Mineral 48:497–511

Dill HG, Škoda R, Weber B (2011a) Preliminary results of a newly-discovered lazulite-scorzalite pegmatite-aplite in the Hagendorf-Pleystein Pegmatite Province, SE Germany. Asociación Geológica Argentina, Serie D, Publicación Especial N° 14. Pegmatite, Mendoza, Argentina, pp 79–81

Dill HG, Hansen BT, Weber B (2011b) REE contents, REE minerals and Sm/Nd isotopes of granite- and unconformity-related fluorite mineralization at the western edge of the Bohemian Massif: with special reference to the Nabburg-Wölsendorf District, SE Germany. Ore Geol Rev 40:132–148

Dill HG, Weber B, Botz R (2011c) The barite-bearing beryl-phosphate pegmatite Plössberg – a missing link between pegmatitic and vein-type barium mineralization in NE Bavaria, Germany. Geochemistry 71:377–387

Dill HG, Skoda R, Weber B, Berner Z, Müller A, Bakker RJ (2012a) A newly-discovered swarm of shearzone-hosted Bi-As-Fe-Mg-P aplites and pegmatites in the Hagendorf-Pleystein pegmatite province, SE Germany: a step closer to the metamorphic root of pegmatites. Can Mineral, Special Volume dedicated to Petr Černý 50: 943–947

Dill HG, Weber B, Klosa D (2012b) Crystal morphology and mineral chemistry of monazite–zircon mineral assemblages in continental placer deposits (SE Germany): ore guide and provenance marker. J Geochem Explor 112:322–346

Dill HG, Skoda R, Weber B, Müller A, Berner ZA, Wemmer K, Balaban S-I (2013a) Mineralogical and chemical composition of the Hagendorf-North Pegmatite, SE Germany – a monographic study. Neues Jb Mineral Abh 190:281–318

Dill HG, Garrido MM, Melcher F, Gomez MC, Weber B, Luna LI, Bahr A (2013b) Sulfidic and non-sulfidic indium mineralization of the epithermal Au-Cu-Zn-Pb-Ag deposit San Roque (Provincia Rio Negro, SE Argentina) – with special reference to the "indium window" in zinc sulfide. Ore Geol Rev 51:103–128

Dill HG, Weber B, Botz R (2013c) Metalliferous duricrusts ("orecretes") – markers of weathering: a mineralogical and climatic-geomorphological approach to supergene Pb-Zn-Cu-Sb-P mineralization on different parent materials. Neues Jb Mineral Abh 190:123–195

Dill HG, Weber B, Melcher F, Wiesner W, Müller A (2014) Titaniferous heavy mineral aggregates as a tool in exploration for pegmatitic and aplitic rare-metal deposits (SE Germany). Ore Geol Rev 57:29–52

Dill H, Dohrmann R, Kaufhold S, Balaban S-I (2015) Kaolinization – a tool to unravel the formation and unroofing of the Pleystein pegmatite-aplite system (SE Germany). Ore Geol Rev 69:33–56

Dill HG, Kaufhold S, Ehling A, Bowitz J (2016) Oxidized and reduced kaolin fan deposits: their sedimentological-mineralogical facies and physical-chemical regime (North-Bavarian Kaolin Mining District, Germany). Ore Geol Rev 72: 459–484

Dombrowski A, Okrusch M, Henjes-Kunst F (1994) Geothermobarometry and geochronology on mineral assemblages of orthogneisses and related metapelites of the Spessart Crystalline Complex, NW Bavaria, Germany. Chem Erde 54:85–101

Dominy SC, Camm GS (1998) Geology and hydrothermal development of Bostraze-Balleswidden kaolin deposit, Cornwall, United Kingdom. Inst Min Metall Trans Sect B Appl Earth Sci 107:148–157

Don J (1990) The difference in the Paleozoic facies-structural evolution of the West Sudetes. Neues Jb Geol Paläontol Abh 179:307–328

Dörr W, Zulauf G, Fiala J, Franke W, Vejnar Z (2002) Neoproterozoic to Early Cambrian history of an active plate margin in the Teplá-Barrandian unit – a correlation of U-Pb isotopic dilution TIMS ages (Bohemia, Czech Republic). Tectonophysics 352:65–85

Dosbaba M, Novák M (2012) Quartz replacement by "kerolite" in graphic quartz–feldspar intergrowths from the Věžná I pegmatite, Czech Republic: a complex desilication process related to episyenitization. Can Mineral 50:1609–1622

Drost K, Linnemann U, McNaughton N, Fatka O, Kraft P, Gehmlich M, Tonk C, Marek J (2004) New data on the Neoproterozoic – Cambrian geotectonic setting of the Teplá-Barrandian volcano-sedimentary successions: geochemistry, U-Pb zircon ages, and provenance (Bohemian Massif, Czech Republic). Int J Earth Sci 93:742–757

Drozdzewski G (1993) The Ruhr coal basin (Germany): structural evolution of an autochthonous foreland basin. Int J Coal Geol 23:231–250

DuBois C, Walsh J (1970) Minerals of Kenya. Geol Surv Kenya Bull 11:1–82

Duthou JL, Couturie JP, Mierzejewski MP, Pin CH (1991) Age determination of the Karkonosze granite using isochrone Rb–Sr whole rock method. Prz Geol 39:75–79

Duvall AR, Clark MK, van der Pluijm BA, Li C (2011) Direct dating of Eocene reverse faulting in northeastern Tibet using Ar-dating of fault clays and low-temperature thermochronometry. Earth Planet Sci Lett 304:520–526

Ebner F, Cerny I, Eichhorn R, Götzinger MA, Paar WH, Prochaska W, Weber L (2000) Mineral resources in the Eastern Alps and adjoining areas. In: Neubauer F, Höck V (eds) Aspects of geology in Austria, vol 92, Mitteilungen der Österreichischen Geologischen Gesellschaft. Österreichische Geologische Gesellschaft, Vienna, pp 157–184

Ecrit TS (2005) REE-enriched granitic pegmatites. In: Linnen RL, Sampson IM (eds) Rare-element geochemistry and mineral deposits. Geological Association of Canada, St. John's, pp 175–199

Eichhorn R, Höll R, Loth G, Kennedy A (1999) Implication of U-Pb SHRIMP zircon data on the age and evolution of the Felberthal tungsten deposit (Tauern Window, Austria). Int J Earth Sci 88:496–512

Eichhorn R, Loth G, Höll R, Finger F, Schermaier A, Kennedy A (2000) Multistage Variscan magmatism in the Central Tauern Window (Austria) unveiled by U/Pb SHRIMP zircon data. Contrib Mineral Petrol 139:418–435

Emmermann R, Lauterjung J (1997) German continental deep drilling program KTB: overview and major results. J Geophys Res 102:18179–18201

Emmert U, Stettner G (1968) Erläuterung zur geologischen Karte von Bayern 1.25000, Sheet No. 5737, Schwarzenbach a.d. sächsischen Saale, Geological Survey of Bavaria, München, 236 pp

Emmert U, Weinelt Wi (1962) Erläuterung zur geologischen Karte von Bayern 1.25000, Sheet No. 5935, Marktschorgast, Geological Survey of Bavaria, München, 279 pp

Emmert U, Horstig von G, Weinelt Wi (1960) Erläuterung zur geologischen Karte von Bayern 1.25000, Sheet No. 5835, Stadtsteinach, Geological Survey of Bavaria, München, 279 pp

Ericsson T, Nord AG, Aberg G (1986) Cation partitioning in hydrothermally prepared olivine-related (Fe, Mn)-sarcopsides. Am Mineral 71:136–141

Evensen JM, London D (2002) Experimental silicate mineral/melt partition coefficients for beryllium and the crustal Be cycle from migmatite to pegmatite. Geochim Cosmochim Acta 66:2239–2265

Falk F, Franke W, Kurze M (1995) Stratigraphy. In: Dallmeyer RD, Franke W, Weber K (eds) Tectono-stratigraphic evolution of the Central and East European orogens. Springer, Heidelberg, pp 221–234

Falster A, Simmons WB, Moore PB (1988) Fillowite, Lithiophilite, Heterosite/Purpurite and Allvaudite–Varulite group minerals from a Pegmatite in Florence County, WI. In: Abstract 15th Rochester Mineralogical Symposium, Rocks and Minerals 63, p. 455

Fauth H, Hindel R, Siewers U, Zinner J (1985) Geochemischer Atlas der Bundesrepublik Deutschland: Verteilung von Schwermetallen in Wässern und Bachsedimenten. Bundesanstalt für Geowissenschaften und Rohstoffe, Hannover

Fernandez-Caliani JC, Crespo E, Rodas M, Barrenechea JF, Luque FJ (2004) Formation of nontronite from oxidative dissolution of pyrite disseminated in Precambrian felsic metavolcanics of the Southern Iberian Massif (Spain). Clay Clay Miner 52:106–114

Ferriss EDA, Ewing RC, Becker U (2010) Simulation of thermodynamic mixing properties of actinide-containing zircon solid solutions. Am Mineral 95:229–241

Fersman A (1930) O geokhimicheskoi geneticheskoi klassifikatsii granitnykh pegmatitov (A geochemical genetic classification of pegmatites). Monograph Akademiia Nauk SSSR. (in Russian)

Fettel M (1971) Mineralienfunde bei Hornberg im Schwarzwald. Aufschluss 22:216–217

Fiala J, Fuchs G, Wendt JI (1995) Stratigraphy. In: Dallmeyer RD, Franke W, Weber W (eds) Prepermian geology of Central and Eastern Europe. Springer, Berlin/Heidelberg, pp 417–428

Finch A (1990) Genthelvite and willemite, zinc minerals associated with alkaline magmatism from the Motzfeld centre, South Greenland. Mineral Mag 54:407–412

Finger F, Roberts MP, Haunschmid B, Schermaier A, Steyrer HP (1997) Variscan granitoids of central Europe: their typology, potential sources and tectonothermal relations. Mineral Petrol 61:67–96

Forster A (1965) Erläuterungen zur Geologischen Karte von Bayern 1:25000 Blatt Vohenstrauss/ Frankenreuth. GLA, München, 174 pp

Forster A, Kummer R (1974) The pegmatites in the area of Pleystein-Hagendorf/North Eastern Bavaria. Fortschritte Mineralogie 52:89–99

Forster A, Strunz H, Tennsyson C (1967) Die Pegmatite des Oberpfälzer Waldes, insbesondere der Pegmatit von Hagendorf-Süd. Aufschluss Spec Pub 16:137–198

Förster HJ, Seltmann R, Tischendorf G (1995) High-fluorine, low phosphorous A-type (post collision) silicic magmatism in Erzgebirge. In: 2nd symposium on Permo-Carboniferous igneous rocks, Terra Nostra, 7, Bonn, pp 32–35

Förster HJ, Tischendorf G, Trumbull RB, Gottesmann B (1999) Late-collisional granites in the Variscan Erzgebirge, Germany. J Petrol 40:1613–1645

Fossen H (2010) Structural geology. Cambridge University Press, Cambridge, 480 pp

Frank W, Kralik M, Schabert S, Thöni M (1987) Geochronological data from the Eastern Alps. In: Flügel HW, Faupl P (eds) Geodynamics of the eastern Alps. F. Deuticke, Vienna, pp 272–281

Franke W (1995) Introduction-Saxo-Thuringian Basin. In: Dallmeyer D, Franke W, Weber K (eds) Pre-permian geology of Central and Western Europe. Springer, Berlin, pp 33–49

Franke W (2000) The mid-European segment of the Variscides: Tectonostratigraphic units, terrane boundaries and plate tectonic evolution. In: Franke W, Haak V, Oncken O, Tanner D (eds) Orogenic processes, quantification and modeling in the Variscan belt, Geological Society special publication, 179. Geological Society Publication, London, pp 35–62

Franke W, Bram K (1988) Geowissenschaftliche Umfelduntersuchungen in der KTB-Arge 2, Ziele, Probleme, Projekte.- 1. KTB-Schwerpunkt-Kolloquium Giessen 1988. Zusammenfassung der Beiträge 16

Franke W, Stein E (2000) Exhumation of high –pressure rocks in the Saxo-Thuringian belt: geological constraints and alternative models. In: Franke W, Haak V, Oncken O, Tanner D (eds) Orogenic processes: quantification and modeling in the Variscan Belt, Geological Society special publications, 179. Geological Society Publications, London, pp 337–354

Franke W, Kreuzer H, Okrusch M, Schüssler U, Seidel E (1995) Saxothuringian Basin: exotic metamorphic nappes, stratigraphy, structure and igneous activity. In: Dallmeyer D, Franke W, Weber K (eds) Pre-permian geology of Central and Western Europe. Springer, Berlin, pp 277–294

Fransolet A-M (1980) The eosphorite-childrenite series associated with the Li-Mn-Fe phosphate minerals from the Buranga pegmatite, Rwanda. Mineral Mag 43:1015–1023

Fransolet A-M, Hatert F, Fontan F (2004) Petrographic evidence for primary hagendorfite in an unusual assemblage of phosphate minerals, Kibingo Granitic Pegmatite, Rwanda. Can Mineral 42:697–704

Frey M, De Capitani C, Liou JG (1991) A new petrogenetic grid for low-grade metabasites. J Metamorph Geol 9:497–509

Frisch W, Neubauer F (1989) Pre-Alpine terranes and tectonic zoning in the eastern Alps. In: Dallmeyer RD (ed) Terranes in the circum-Atlantic Paleozoic orogens, Geological Society of America, special publications, 230. Geological Society of America, Boulder, pp 91–100

Frizzo P, Mills J, Visoná D (1982) Ore petrology and metamorphic history of Zn-Pb ores, Monteneve, Tyrol, N. Italy. Mineral Deposita 17:333–347

Froitzheim N, Plašienka D, Schuster R (2008) Alpine tectonics of the Alps and Western Carpathians. In: McCann T (ed) Geology of Central Europe, Geological Society special publication. Geological Society, London, pp 1141–1232

Frondel C (1958) Strunzite, a new mineral. Naturwissenschaften 45:37–38

Fry N (1991) The field description of metamorphic rocks, Geological Society of London handbook series, Book 14. Wiley, London, 112 pp

Frýda J, Breiter K (1995) Alkali feldspar as a main phosphorus reservoir in rare metal granites: three examples from the Bohemian Massif (Czech Republic). Terra Nova 7:315–320

Fuchs G, Matura A (1976) Zur Geologie des Kristallins der südlichen Böhmischen Masse. Jahrb Geol Bundesanst 119:1–43

Gajda E (1960a) Pegmatite veins of the region of Szklarska Poręba (Karkonosze Mts.). Kwartalnik Geologiczny 4:545–564

Gajda E (1960b) Minerals of pegmatite veins in the vicinity of Szklarska Poręba region in Karkonosze Mountains (Riesengebirge). Kwartalnik Geologiczny 4:565–587

Galadí-Enríquez E, Dörr W, Zulauf G, Galindo-Zaldívar J, Heidelbach F, Rohrmüller J (2010) Variscan deformation phases in the southwestern Bohemian Massif: new constraints from sheared granitoids. Z Dtsch Ges Geowiss 161:1–23

Galliski MA, Marquez Zavalia MF, Lomniczi de Upton I, Oyarzabal JC (1998) Mitridatite from the San Luis granitic pegmatite, La Florida, Argentina. Can Mineral 36:395–397

Galliski MA, Lira R, Dorais MJ (2004) Low-pressure differentiation of melanephelinitic magma and the origin of ijolite pegmatites at La Madera, Córdoba, Argentina. Can Mineral 42:1799–1823

García-Lorenzo ML, Pérez-Sirvent C, Martínez-Sánchez MJ, Molina-Ruiz J (2012) Trace elements contamination in an abandoned mining site in a semiarid zone. J Geochem Explor 113:23–35

Gaschnitz R (2001) Gasgenese und Gasspeicherung im flözführenden Oberkarbon des Ruhr-Beckens, Berichte des Forschungszentrums Jülich, 3859. Forschungszentrums, Jülich, pp 1–342

Gebauer D (1993) Geochronologische Übersicht. In: Bauberger W (ed) Geologische Karte von Bayern 1.25000, Erläuterungen zum Blatt Nr. 6439 Tännesberg. Geologisches Landesamt Bayern, München, pp 10–22

Gebauer D, Grünenfelder M (1979) U-Pb zircon and Rb-Sr mineral dating of´eclogites and their country rocks; example: Münchberger Gneiss Massif, northeast Bavaria. Earth Planet Sci Lett 42:35–44

Gebauer D, Grünenfelder M (1982) Geological development of the Hercynian Belt of Europe based on age and origin of high-grade and high-pressure mafic and ultramafic rocks. In: 5th international conference on geochronology, cosmochronology, and isotope geology, Nikko National Park, Japan, Abstracts 111–112, 27 June–2 July

Gerstenberger H, Haase G, Wemmer K (1995) Isotope systematics of the Variscan postkinematic granites in the Erzgebirge (E Germany). In: 2nd symposium on Permocarboniferous igneous rocks, Terra Nostra 7, Bonn, pp 36–41

Ginsburg AI, Rodionov GG (1960) On the depths of the granitic pegmatite formation. Geologiya Rudnykh Mestorozhdenyi, Izd. Nauka, Moskva l:45–54

Ginsburg AI, Timofeyev IN, Feldman LG (1979) Principles of geology of the granitic pegmatites. Nedra, Moscow, 296 pp. (in Russian)

Giuliani G, France-Lanord C, Coget P, Schwarz D, Cheilletz A, Branquet Y, Giard D, Martin-Izard A, Alexandrov P, Piat DH (1998) Oxygen isotope systematics of emerald: relevance for its origin and geological significance. Mineral Deposita 33:513–519

Glodny J, Grauert B, Krohe A, Vejnar Z, Fiala J (1995a) Altersinformation aus Pegmatiten der Westlichen Böhmischen Masse: ZEV, Teplá-Barrandium und Moldanubikum. 8. DFG-Kolloquium KTB, Giessen (abstract)

Glodny J, Grauert B, Krohe A (1995b) Ordovizische Pegmatite in variszischen HT-Metamorphiten des KTB-Umfeldes: Hinweis auf hohe Stabilität des Rb–Sr-Systems in Muskoviten. Terra Nostra 95:98

Glodny J, Grauert B, Fiala J, Vejnar Z, Krohe A (1998) Metapegmatites in the western Bohemian massif: ages of crystallization and metamorphic overprint, as constrained by U–Pb zircon,

monazite, garnet, columbite and Rb–Sr muscovite data. Geol Rundsch/Int J Earth Sci 87:124–134
Göd R (1978) Vorläufige Mitteilung über einen Spodumen-Holmquistit führenden Pegmatit aus Kärnten. Anzeiger der Akademie der Wissenschaften, mathematisch-naturwissenschaftliche Klasse 115:161–165
Göd R (1989) The spodumene deposit at "Weinebene" Koralpe, Austria. Mineral Deposita 24:270–278
Grauby O, Petit S, Decarreau A, Baronnet A (1994) The nontronite-saponite series; an experimental approach. Eur J Mineral 6:99–112
Grecula P (1982) Gemericum – segment of the Paleotethyan riftogenous basin, Mineralia slovaca monograph. AFLA, Bratislava. 263 pp. (in Slovak with English abstract)
Grecula P, Roth Z (1978) Kinematický model Západnich Karpat v souborĕm řezu. Zb Geol Vĕd 32:49–73
Grecula P, Abonyi A, Abonyiová M, Antaš J, Bartalský B, Bartalský J, Dianiška I, Drnzík E, Ďuďa R, Gargulák M, Gazdačko Ľ, Hudáček J, Kobulský J, Lörincz L, Macko J, Návesňák D, Németh Z, Novotný L, Radvanec M, Rojkovič L, Rozložník O, Varček C, Zlocha J (1995) Mineral deposit of the Slovak Ore Mountains. Mineralia Slovaca Monogr 1:1–834
Grew ES (1988) Kornerupine at the Sar-e-Sang, Afghanistan, whiteschist locality: implications for tourmaline-kornerupine distribution in metamorphic rocks. Am Mineral 73:345–357
Grew ES (2002) Mineralogy, petrology and geochemistry of beryllium: an introduction and list of beryllium minerals. In: Beryllium, mineralogy, petrology, and geochemistry, Reviews in mineralogy and geochemistry 50. Mineralogical Society of America, Washington, DC, pp 1–76
Grey IE, Mumme WG, Neville SM, Wilson NC, Birch WD (2010) Jahnsite-whiteite solid solutions and associated minerals in the phosphate pegmatite at Hagendorf-Süd, Bavaria, Germany. Mineral Mag 74:969–978
Grey IE, Macrae CM, Keck E, Birch WD (2012) Aluminium-bearing strunzite derived from jahnsite at the Hagendorf-Süd pegmatite, Germany. Mineral Mag 76:1165–1174
Gudden H (1975) Die Kreide-Eisenerzlagerstätten in Nordost-Bayern. Geol Jahrb D 10:201–238
Haak V (1989) Electrical resistivity studies in the vicinity of the KTB drill site, Oberpfalz. In: Emmermann R, Wohlenberg J (eds) The continental deep drilling program (KTB). Springer, Heidelberg, pp 224–241
Haapala I, Frindt S, Kandara I (2007) Cretaceous Gross Spitzkoppe and Klein Spitzkoppe Stocks in Namibia: Topaz-bearing A-type granites related to continental rifting and mantle plume. Lithos 97:174–192
Habel M (2003) Neufunde aus dem östlichen Bayerischen Wald. Mineralienwelt 14:24–29
Habel A, Habel M (1991) Die Granitbrüche von Tittling im Bayerischen Wald (Matzersdorf). Emser Hefte 91:37
Habel M, Reinhardt M (2011) Neufunde aus dem O Bayerischen Wald. Mineralienwelt 21:81–82
Habler G, Thöni M (2001) Preservation of Permo–Triassic low-pressure assemblages in the Cretaceous high-pressure metamorphic Saualpe crystalline basement (Eastern Alps, Austria). J Metamorph Geol 19:679–697
Hann HP, Chen F, Zedler H, Frisch W, Loeschke J (2003) The Rand Granite in the southern Schwarzwald and its geodynamic significance in the Variscan belt of SW Germany. Int J Earth Sci 92:821–842
Hansen BT, Teufel S, Ahrendt H (1989) Geochronology of the Moldanubian-Saxothuringian Transition Zone, Northeast Bavaria. In: Emmermann R, Wohlenberg J (eds) The German continental deep drilling program (KTB). Springer, Berlin, pp 55–66
Harben PW, Kužvart M (1996) Industrial minerals: a global geology. London Industrial Minerals Information Ltd., London, 462 pp
Harder H (1970) Boron content of sediments as a tool in facies analysis. Sediment Geol 4:153–175
Harrison TM, McDougall I (1980) Investigations of an intrusive contact, northwest Neslon New Zealand-I: thermal chronological and isotope constraints. Geochim Cosmochim Acta 44:1985–2003
Harrison TM, Célérier J, Aikman AB, Hermann J, Heizler MT (2009) Diffusion of ^{40}Ar in muscovite. Geochim Cosmochim Acta 73:1039–1051

Hatert F, Roda-Robles E, Keller P, Fontan F, Fransolet A-M (2007) Petrogenetic significance of the triphylite + sarcopside intergrowths in granitic pegmatites: an experimental investigation of the $Li(Fe,Mn)(PO_4)$-$(Fe,Mn)_3(PO_4)_2$ system. Granitic pegmatites: the state of the art, Book of Abstracts, 44

Hatert F, Ottolini L, Schmid-Beurmann P (2011) Experimental investigation of the alluaudite + triphylite assemblage, and development of the Na-in-triphylite geothermometer: applications to natural pegmatite phosphates. Contrib Mineral Petrol 161:531–546

Hauzenberger CA, Häger T, Sutthirat C, Bojar A-V, Kienzel N (2005) Geochemical characterization of corundum from different gem deposits: a stable isotope and trace element study. Gem-materials and modern analytical methods. GEM.MAT.MAM, Hanoi, 3rd international workshop, pp 55–62

Hein UF (1993) Synmetamorphic Variscan siderite mineralization of the Rhenish Massif, Central Europe. Mineral Mag 57:451–467

Heimann A (1997) The chemical composition of gahnite and garnet as exploration guides to and indicators of rare element (Li) granitic pegmatites. Final technical report U.S.G.S. mineral resources external research program U.S. Department of energy, Award number g10ap00051, 24 pp

Heinrich EW (1951) Mineralogy of triplite. Am Mineral 36:256–271

Henk A, von Blanckenburg F, Finger F, Schaltegger U, Zulauf G (2000) Syn-convergent high-temperature metamorphism and magmatism in the Variscides: a discussion of potential heat sources. In: Franke W, Haak V, Oncken O, Tanner D (eds) Orogenic processes: quantification and modeling in the Variscan Belt, Geological Society special publications, 179. Geological Society Publications, London, pp 387–399

Herzog T, Lehrberger G, Stettner G (1997) Goldvererzungen bei Neualbenreuth im Saxothuringikum des Waldsassener Schiefergebirge, Oberpfalz. Geologica Bavarica 102:173–206

Hladíková J, Aichler J, Mixa P (1992) Source of sulfur of base metal sulfide deposits at the NE margin of the Bohemian Massif (Czechoslovakia), Abstract 29th IGC Kyoto, 3:757

Hochleitner R, Fehr KT (2010) The keckite problem and its bearing on the crystal chemistry of the jahnsite group: Mössbauer and electron-microprobe studies. Can Mineral 84:1445–1453

Hoffmann V, Krupka F, TrdliČka Z (1965) Tapiolite and columbotantalite in cassiterites from the Krušne Hory. Véstnik Ústav Geologizki 49:295–296, Prag (in Czech)

Hogarth D (1977) Classification and nomenclature of pyrochlore group. Am Mineral 62:403–410

Hohl J-L (1994) Minéraux et Mines du Massif Vosgien. Editions du Rhin, Mulhouse, 271 pp

Höll R (1975) Die Scheelitlagerstätte Felbertal und der Vergleich mit anderen Scheelitlagerstätten in den Ostalpen, Abhandlungen Bayerischen Akademie der Wissenschaften N.F., 157A. Verl. d. Bayer. Akad. d. Wiss, München, pp 1–114

Hölle H (1964) Ein Graphitpegmatit vom Hirnkogel bei Pusterwald/Steiermark. Mitteilungen des naturwissenschaftlichen Vereins für Steiermark 94:86–88

Höller H (1959) Ein Spodumen-Beryll-Pegmatit und ein mineralreicher Marmor im Wildbachgraben bei Deutschlandsberg. Mitteilungsblatt Abteilung Mineralogie Landesmuseum Joanneum 1:19

Holub FV (1997) Ultrapotassic plutonic rocks of the durbachite series in the Bohemian Massif: petrology, geochemistry and petrogenetic interpretation. Sbor Geol Véd LoŽisková Geol Mineral 31:5–25

Holub FV, Klečka M, Matějka D (1995) Igneous activity. In: Dallmeyer RD, Franke W, Weber K (eds) Tectono-stratigraphic evolution of the Central and East European orogens. Springer, Heidelberg, pp 444–452

Holub FV, Cocherie A, Rossi P (1997) Radiometric dating of granitic rocks from the Central Bohemian Plutonic Complex (Czech Republic): constraints on the chronology of the thermal and tectonic events along the Moldanubian-Barrandian boundary. CR Acad Sci Ser IIA 325:19–26

Hölzl S, Hofmann B, Köhler H (1993) U-Pb and Sm-Nd dating on a metabasite from the KTB main bore hole.-KTB-Report 93–2:392–392

Horstwood MSA, Foster GL, Parrish RR, Noble SR, Nowell GM (2003) Common-Pb corrected in situ U-Pb accessory mineral geochronology by LA-MC-ICP-MS. J Anal At Spectrom 18:837–846

Hovorka D, Meres Š, Ivan P (1992) Pre-Alpine Western Carpathians Mts basement complexes: geochemistry, petrology, geodynamic setting. Terra Abstract 4:32

Hughes RW (1990) Corundum. Butterworth-Heinmann, London, 314 pp

Iiishi K, Tomisaka T, Kato T, Takeno S (1970) Syntheses of valleriite. Am Mineral 55:2107–2110

Ioannou SE, Spooner ETC (2000) Miocene epithermal Au-Ag vein mineralization, dixie claims, Midas District, north-central Nevada; characteristics and controls. Explor Min Geol 9:233–252

Ivakin Yu D, Danchevskaya MN, Ovchinnikova OG, Muravieva GP (2006) Thermo vaporous synthesis of fine crystalline gahnite ($ZnAl_2O_4$). J Mater Sci 41:1377–1383

Jacob H (1967) Petrologie von Asphaltiten und asphaltischen Pyrobitumen. Erdöl, Kohle, Erdgas, Petrochemie 20:393–400

Jahns RH, Burnham CW (1969) Experimental studies of pegmatite genesis: I: a model for the derivation and crystallization of granitic pegmatites. Econ Geol 64:843–864

Jakes P, Waldhauserova J (1987) Orogenic sequences in the upper Proterozoic of the Bohemian Massif. Proterozoic geochemistry (IGCP 217) Lund, 1987:14

Janeczek J, Sachanbiński M (1989) Beryl pegmatites in two-micas granite at the Eastern part of Strzegom-Sobótka Massif. Arch Mineral 54:57–79 (in Polish)

Janousek V, Finger F (2003) Whole-rock geochemistry of felsic granulites from the Gföhl Unit (Moldanubian Zone, Austria and Czech Republic): petrogenetic implications. Mitteilungen der Österreichischen Mineralogischen Gesellschaft 148:174–176

Janoušek V, Rogers G, Bowes DR (1995) Sr-Nd isotopic constraints on the petrogenesis of the Central Bohemian Pluton, Czech Republic. Geol Rundsch 84:520–534

Jaszczak A, Dimovski S, Hackney SA, Robinson GW, Bosio P, Gogotsi Y (2007) Micro- and nanoscale graphite cones and tubes from Hackman Valley, Kola Peninsula, Russia. Can Mineral 45:379–389

Jęczmyk M, Juskowiakowa M (1989) Geology and geochemical characteristics of the crystalline rocks of the Bogatynia area (Western Sudetes Mts.). Bull Inst Geol 360:5–38 (in Polish)

Jiang S-Y, Palmer MR, Slack JF, Shaw DR (1998) Paragenesis and chemistry of multistage tourmaline formation in the Sullivan Pb–Zn–Ag deposit, British Columbia. Econ Geol 93:47–67

Jura D, Jurecka J, Krieger W, Kuzak R, Lesnik D, Perski Z (2000) Conditions of hard coal exploitation and its environmental impact in the western part of the Upper Silesian Coal Basin. In: Guide to field trips, 4th European Coal conference, Ustron, Polish Geological Institute, Warsaw, pp 5–39

Kalenda F, Vaněček M (1989) Base metal deposits of the Zlaté Hory District. Excursion Guide 5A Gold Districts of the Bohemian Massif, Gold 89 in Europe, pp 39–43

Kalt A (1995) Petrologie cordieritführender Metapelite das Bayerischen Waldes und des Schwarzwaldes. Terra Nostra 95:108

Kalt A, Altherr R, Hanel M (2000) The Variscan basement of the Schwarzwald. Beiheft zum Europäischen J Mineral 12:1–43

Kampf AR (1992) Beryllophosphate chains in the structures of fransoletite, parafransoletite and erleite and some general comments on beryllophosphate linkages. Am Mineral 77:848–856

Kastl E, Tonika J (1984) The Marianske Lazne metaophiolite complex (West Bohemia). Krystalinikum 17:59–76

Kastning J, Schlüter J (1994) Die Mineralien von Hagendorf und ihre Bestimmung, vol 2, Schriften des Mineralogischen Museums der Universität Hamburg. C. Weise Verlag, Munich, 95 pp

Keck E (1990) Hagendorf-Sued-Ein kurzer historischer Überblick. Aufschluss 41:54–66

Keck E (2001) Carlhintzeit $Ca_2AlF_7.H_2O$ vom Kreuzberg in Pleystein, Oberpfalz. Aufschluss 52:219–222

Keller P, Von Knorring O (1989) Pegmatites at the Okatjimuku farm, Karibib, Namibia Part I: phosphate mineral associations of the Clementine II pegmatite. Eur J Mineral 1:567–593

Keller P, Fontan F, Fransolet AM (1994) Intercrystalline cation partitioning between minerals of the triplite-zwieselite-magniotriplite and the triphylite-lithiophilite series in granitic pegmatites. Contrib Mineral Petrol 118:239–248

Kempe U, Wolf D (2006) Anomalously high Sc contents in ore minerals from Sn-W deposits: possible economic significance and genetic implications. Ore Geol Rev 28:103–122

Kesler SE, Gruber PW, Medina PA, Keoleian GA, Everson MP, Wallington TJ (2012) Global lithium resources: relative importance of pegmatite, brine and other deposits. Ore Geol Rev 48:55–69

Kettner R (1918) Mineralogický výlet na Rožnou. Vědy přírodni 2:1–10

Kievlenko EY (2003) Geology of gems. Ocean Publications Ltd., Littleton. 432. pp

Kippenberger C, Krauss U, Kruzona M, Schmidt H, Thormann A, Priem J, Wettig E (1988) Lithium. Untersuchungen über Angebot und Nachfrage mineralischer Rohstoffe 21:1–212

Kirsch H, Kober B, Lippolt HJ (1988) Age of intrusion and rapid cooling of the Frankenstein gabbro (Odenwald, SW Germany) evidenced by 40Ar/39Ar and single zircon 207Pb/206Pb measurements. Geologische Rundschau 77:693–711

Kleck W (1996) Crystal settling in pegmatite magma. Abstracts and Program. Geological Association of Canada and Mineralogical Association of Canada, A-50

Kleinschmidt G, Neugebauer J, Schönenberg R (1975) Gesteinsinhalt und Stratigraphie der Phyllitgruppe in der Saualpe. Clausthaler Geol Abh, Special Volume 1:11–44

Knanna YK (1977) Note on the unusually high concentration of rubidium in a lithium mica from Govindpal area of Bastar district. J Geol Soc India 18:500–502

Kodym O (1966) Moldanubicum. In: Svoboda J (ed) Regional geology of Czechoslovakia I. Geological Survey of Czechoslovakia, Prague, pp 40–98

Köhler H, Müller-Sohnius D (1976) Ergänzende Rb/Sr-Altersbestimmungen an Mineral- und Gesamtgesteinsproben des Leuchtenberger und Flossenbürger Granits, NE-Bayern. Neues Jb Mineral Monat 8:354–365

Köhler H, Müller-Sohnius D, Cammann KC (1974) Rb/Sr-Altersbestimmungen an Mineral- und Gesamtgesteinsproben des Leuchtenberger und Flossenbürger Granits, NE-Bayern. Neues Jb Mineral Abh 123:63–85

Köhler H, Propach G, Troll G (2008) Isotopische (Sr, Nd) Charakterisierung und Datierung Varistischer Granitoide der Moldanubischen Kruste Nordbayerns. Geologica Bavarica 110:170–203

Kohút M (2013) How many events occurred during the Variscan orogeny in the Western Carpathians. Crustal evolution and geodynamic processes in Central Europe. In: Proceedings of the Joint Conference of the Czech and German geological societies held in Plzeň (Pilsen), volume 82, 63, 16–19 September 2013

Koller F (1996) Plutonische Gestein. In: Steininger FF (ed) Erdgeschichte des Waldviertels, Das Waldviertel 45. Waldviertel Heimatbund, Horn, Austria, pp 25–36

Konopásek J, Schulmann K (2005) Contrasting early carboniferous field geotherms: evidence for accretion of a thickened orogenic root and subducted Saxothuringian crust (Central European Variscides). J Geol Soc Lond 162:463–470

Kopp R (1987) Kurze Betriebsschronik der Feldspatgrube Gertrud by Weiden Opf. Aufschluss 37:75–76

Kossmat F (1927) Gliederung des varistischen Gebirgsbaus. Abhandlungen Sächsische Geologische Landes-Anstalt 1:1–39

Kovách Á, Svingor É, Grecula P (1986) Rb/Sr isotopic ages of granitoid rocks from the Spiš-Gemer metalliferous Mts., West Carpathians, Eastern Slovakia. Mineralia Slovaca 18:1–14

Kozłowski A, Sachanbiński M (2007) Karkonosze intragranitic pegmatites and their minerals. In: Kozłowski A, Wiszniewska J (eds) Granitoids in Poland, Archiwum Mineralogiczne monograph, 1. Faculty of Geology of The Warsaw University, Warszawa, pp 155–178

Kraus F (2014) Geheimnis des stinkenden Minerals – Von "Stinkspat" und elementarem Fluor. Fluor More 2(14):38–42

Kreuzer H, Harre W (1975) K/Ar-Altersbestimmungen an Hornblenden und Biotiten des Kristallinen Odenwaldes. Aufschluss Spec Publ 27:71–77

Kreuzer H, Seidel E, Schüssler U, Okrusch M, Lenz K-L, Raschka H (1989) K-Ar geochronology of different tectonic units at the northwestern margin of the Bohemian Massif. In: Meisner R, Gebauer D (eds) The evolution of European continental crust: deep drilling, geophysics, geology and geochemistry, vol 157, Tectonophysics. Elsevier, New York, pp 149–178

Kreuzer H, Henjes-Kunst F, Seidel E, Schüssler U, Bühn B (1993) Ar-Ar spectra on minerals from KTB and related medium pressure units. KTB-Report 93–2. Project Management Group KTB, Hannover, pp 133–136
Kröner A (1980) Pan African crustal evolution. Episodes 2:3–8
Kroner U, Hahn T (2004) Sedimentation, deformation und Metamorphose im Saxothuringikum während der variszischen Orogenese: Die komplexe Entwicklung von Nord. Gondwana während kontinentaler Subduktion und schiefer Kollision. In: Linnemann U (ed) Das Saxothurigikum, Geologica Saxonia 48/49. Staatliche Naturhistorische Sammlungen, Dresden, pp 133–146
Kryza R, Mazur S, Oberc-Dziedzic T (2004) The Sudetic geological mosaic: insights into the root of the Variscan orogen. Prz Geol 52:761–773
Kryza R, Crowley QG, Larionov A, Pin C, Oberc-Dziedzic T, Mochnacka K (2012) Chemical abrasion applied to SHRIMP zircon geochronology: an example from the Variscan Karkonosze Granite (Sudetes, SW Poland). Gondwana Res 21:757–767
Kucha H (1980) Continuity in the monazite-huttonite series. Mineral Mag 43:1031–1034
Küster D (1995) Rb-Sr isotope systematics of muscovite from the Pan-African granitic pegmatites of Western and Northeastern Africa. Mineral Petrol 55:71–83
Kuznetsova L (2013) Kystaryssky granite complex: tectonic setting, geochemical peculiarities and relations with rare-element pegmatites of the south Sangilen Belt (Russia, Tyva Republic). In: PEG 2013, the 6th international symposium on granitic pegmatites. Rubellite Press, New Orleans, pp 73–74
Kužvart M (1968) Kaolin deposits of Czechoslovakia. International Geological Congress, 23th Prague, 1968, Proceedings 15:47–73
Lammerer B, Weger M (1998) Footwall uplift in an orogenic wedge of the Tauern Window in the Eastern Alps of Europe. Tectonophysics 285:213–230
Landes KK (1933) Origin and classification of pegmatites. Am Mineral 18:95–103
Langenaeler V (2000) The Campine Basin: stratigraphy, structural geology, coalification and hydrocarbon potential of the Devonian to Jurassic. Aardkundige Mededelingen 10:1–142
Laubmann H (1924) Die Minerallagerstätten der Pegmatite. Aus: Die Minerallagerstätten von Bayern. Piloty und Loehle, München, pp 39–65
Laubmann H, Steinmetz H (1920) Phosphatführende Pegmatite des Oberpfälzer und des Bayerischen Waldes. Z Krist 55:523–586
Laurs BM, Simmons WB, Rossman GR, Quinn EP, McClure SF, Peretti A, Armbruster T, Hawthorne FC, Falster AU, Günther D, Cooper MA, Grobéty B (2003) Pezzottaite from Ambatovita, Madagascar: a new gem mineral. Gems Gemol 2003:284–301
Le Bas M (2007) Igneous classification revisited 4: Lamprophyres. Geol Today 23:167–168
Lee SM, Holdaway MJ (1977) Significance of Fe-Mg cordierite stability relations on temperature pressure and water pressure in cordierite granulites. AGU monography 20 (The Earth's Crust). American Geophysical Union, Washington, pp 79–94
Lehrberger G, Preinfalk C, Morteani G, Lahusen L (1990) Stratiforme Au-As-Bi-Vererzung in Cordierit-Sillimanit-Gneisen des Moldanubikums bei Oberviechtach-Unterlangau, Oberpfälzer Wald (NE-Bayern). Geologica Bavarica 95:133–176
Lenz H (1986) Rb/Sr-Gesamtgesteins-Altersbestimmung am Weissenstadt-Markleuthener Porphyrgranit des Fichtelgebirges. Geol Jahrb E 34:67–76
Linhardt E (2000) Der Beryllpegmatit der Feldspatgrube Maier, Püllersreuth. Lapis 25:13–23
Linnemann U (2003) Die Struktureinheiten des Saxothurigikums. In: Linnemann U (ed) Das Saxothuringikum, Gelogica Saxonica 48/49. Staatliche Naturhistorische Sammlungen, Dresden, pp 19–28
Linnemann U, Nance RD, Kraft P, Zulauf G (2007) The evolution of the Rheic Ocean: from Avalonian-Cadomian Active Margin to Alleghanian-Variscan collision. The Geological Society of America Special Paper, 423, Boulder, pp 1–630
Linnen RL, Cuney M (2005) Granite-related rare-element deposits and experimental constraints on Ta-Nb-W-Sn-Zr-Hf mineralization. In: Linnen RL, Samson IM (eds) Rare element geochemistry and mineral deposits, vol 17, Geological Association of Canada short course notes. Geological Association of Canada, St. John's, pp 45–68

Linnen RL, Keppler H (1997) Columbite solubility in granitic melts: consequences for the enrichment and fractionation of Nb and Ta in the Earth's crust. Contrib Mineral Petrol 128:213–227

Liou JG, De Capitani C, Frey M (1991) Zeolite equilibria in the system $CaAl_2Si_2O_8$–$NaAlSi_3O_8$–SiO_2–H_2O. N Z J Geol Geophys 34:293–301

Lippolt HJ (1986) Nachweis altpaläozoischer Primäralter (Rb-Sr) und karbonischer Abkühlungsalter (K-Ar) der Muskowit-Biotit-Gneise des Spessart und der Biotit-Gneise des Böllsteiner Odenwald. Geol Rundsch 75:569–583

Lis J, Sylwestrzak H (1979) Episyenites and perspectives of occurrences of intragranite uranium deposits in the Karkonosze Massif. Przegl Geol 27:223–229 (in Polish)

Lisle RJ, Brabham P, Barnes JW (2011) Basic geological mapping (Geological Field Guide) paperback, 5th edn. Wiley, Oxford, 230 pp

London D (2008) Pegmatites, Canadian mineralogist, special publication, 10. Mineralogical Association of Canada, Ottawa, pp 1–347

Lorenz W (1991) Criteria for the assessment of non-metallic mineral deposits. Geol Jahrb A 27:299–326

Lorenz W (1995) Planvolle Ressourcennutzung von Industriemineralien, Steinen und Erden als Teil des Umweltschutzes. Z Angew Geol 41:98–105

LOTEM-Group (1986a) Report on the LOTEM measurements Oberpfalz

LOTEM-Group (1986b) Poster 2nd conference of the KTB, Seeheim

Lottermoser BG, Lu J (1997) Petrogenesis of rare-element pegmatites in the Olary Block, South Australia, part 1. Mineralogy and chemical evolution. Mineral Petrol 59:1–19

Lowenstern JB (1995) Applications of silicate-melt inclusions to the study of magmatic volatiles. In: Thompson JFH (ed) Magmas, fluids, and ore deposits, vol 23, Mineralogical Association of Canada, short course. Mineralogical Association of Canada, Nepean, pp 71–99

Ludhová L, Janák M (1999) Phase relations and P-T path of cordierite-bearing migmatites, Western Tatra Mts, Western Carpathians. Geol Carpath 50:283–293

Lusk J, Ford CE (1978) Experimental extension of the sphalerite geobarometer to 10 kbar. Am Mineral 63:516–519

Lusk J, Scott SD, Ford CE (1993) Phase relations in the Fe–Zn–S system to 5 Kbars and temperatures between 325 and 150 °C. Econ Geol 88:1880–1903

Madel J (1967) Die Umgebung von Rabenstein bei Zwiesel. Geologica Bavarica 58:67–76

Malitch KN, Thalhammer OAR, Knauf VV, Melcher F (2003) Diversity of platinum-group mineral assemblages in banded and podiform chromatite from the Kraubath ultramafic massif, Austria: evidence for an ophiolitic transition zone? Mineral Deposita 38:282–297

Malkovský M (1979) Tektogenese der Plattformbedeckung des Böhmischen Massivs. Knihovna Ústř Úst Geol Praha 53:1–176

Marakushev AA, Gramenitskiy YN (1983) Problem of the origin of pegmatites. Int Geol Rev 25:1179–1186

Markl G (1995) Bertrandit vom Leuchtenberg ein weiteres Berylliumsilikat aus dem Gebiet von Hornberg. Erzgräber 9:96–98

Markl G, Schumacher JC (1996) Cassiterite-bearing greisens and late-magmatic mineralization in the Variscan Triberg Granite Complex, Schwarzwald Germany. Econ Geol 91:576–589

Markl G, Schumacher C (1997) Beryl stability in local hydrothermal and chemical environments in a mineralized granite. Am Mineral 82:194–202

Marschall HR, Altherr R, Gmeling K, Kasztovszky Z (2009) Lithium, boron and chlorine as tracers for metasomatism in high-pressure metamorphic rocks: a case study from Syros (Greece). Mineral Petrol 95:291–302

Martin RF, De Vito C (2005) The patterns of enrichment in felsic pegmatites ultimately depend on tectonic setting. Can Mineral 43:2027–2048

Martin JD, Gil ASI (2005) An integrated thermo-dynamic mixing model for sphalerite geobarometry from 300 to 850 °C and up to 1 GPa. Geochim Cosmochim Acta 69:995–1006

Martin H, Porada H (1978) The intracratonic branch of the Damara Orogen in Southwest Africa: discussion of geodynamic models. Precambrian Res 5:311–338

Marzoni Fecia Di Cossato Y, Orlandi P, Vezzalini G (1989) Rittmannite, a new mineral species of the whiteite group from the Mangualde Granitic Pegmatite, Portugal. Can Mineral 27:447–449

Máška M, Zoubek V (1960) The principal division of the West Carpathians and their pre-Neodic basement. In: Buday T, Koym O, Mahel M et al (eds) Tectonic development of Czechoslovakia. Nakladatelstvi Československé Akademie Věd, Praha, pp 139–151

Matte P (2001) The Variscan collage and orogeny (480±290 Ma) and the tectonic definition of the Armorica microplate: a review. Terra Nova 13:122–128

Matte P, Maluski H, Rajlich R, Franke W (1990) Terrane boundaries in the Bohemian Massif: results of large-scale Variscan thrusting. Tectonophysics 177:151–170

Matthes S (1961) Ergebnisse zur Granatsynthese und ihre Beziehungen zur natürlichen Granatbildung innerhalb der Pyralspit-Gruppe. Geochim Cosmochim Acta 23:233–246

Maucher A (1974) Zeitgebundene Erzlägerstätten. Geol Rundschau 63:263–275

Mazur S, Aleksandrowski P, Szczepański J (2005) The presumed Teplá-Barrandian/Moldanubian terrane boundary in the Orlica Mountains (Sudetes, Bohemian Massif): structural and petrological characteristics. Lithos 82:85–112

Mazur S, Aleksandrowski P, Kryza R, Oberc-Dziedzic T (2006) The Variscan orogen in Poland. Geol Q 50:89–118

McCann T (2008a) The geology of Central Europe: precambrian and palaeozoic: 1. The Geological Society of London, London, Special volume, 748 pp

McCann T (2008b) The geology of Central Europe: mesozoic and cenozoic: 2. The Geological Society of London, London, Special volume, 700 pp

McKerrow WS, Ziegler AM (1972) Palaeozoic oceans. Nature Phys Sci 240:92–94

McKerrow WS, MacNiocaill C, Ahlberg PE, Clayton G, Cleal CJ, Eagar RMC (2000) The late Paleozoic relations between Gondwana and Laurussia. In: Franke W, Haak V, Oncken O, Tanner D (eds) Orogenic processes, quantification and modeling in the Variscan belt, Geological Society special publication, 179. Geological Society Publishing House, London, pp 9–20

McLennan SM (1989) Rare earth elements in sedimentary rocks: influence of provenance and sedimentary processes. In: Lipin BR, McKay GA (eds) Geochemistry and mineralogy of rare earth elements, Reviews in mineralogy. Mineralogical Society of America, Washington, DC, pp 169–200

McPhail DC, Summerhayes E, Welch S, Brugger J (2003) The geochemistry and mobility of zinc in the regolith. Adv Regolith 2003:287–291

Medrano MD, Piper DZ (1997) Fe-Ca-phosphate, Fe-silicate, and Mn-oxide minerals in concretions from the Monterey Formation. Chem Geol 138:9–23

Meixner H (1957) Kurzbericht über neue Kärntner Minerale und Mineralfundorte. Karinthin 17:119–122

Meixner H (1966) Neue Mineralfunde in den österreichischen Ostalpen XXI. Carinthia 156/76. Fachgruppe für Mineralogie und Geologie des Naturwissenschaftlichen Vereines für Kärnten, Hüttenberg, Austria, pp 97–108

Melleton J, Gloaguen E, Frei D, Novák M, Breiter K (2012) How are the emplacement of rare-element pegmatites, regional metamorphism and magmatism interrelated in the Moldanubian domain of the Variscan Bohemian Massif, Czech Republic? Can Mineral 50:1751–1773

Métais D, Chayes F (1963) Varieties of lamprophyres. Carnegie Inst Wash Year B 62:156–157

Métais D, Chayes F (1964) Kersantites and vogesites: a possible example of group heteromorphism. Carnegie Inst Wash Year B 63:196–199

Miethig A, von Drach V, Köhler H (2008) Sr- und Nd-Isotopensystematik an Gesteinen der Gabbroamphibolitmasse von Neukirchen b. Hl. Blut (Nordostbayern)-Kdyne (Tschechische Republik). Geologica Bavarica 110:129–169

Mina M, Fanga C, Fayek M (2005) Petrography and genetic history of coffinite and uraninite from the Liueryiqi granite-hosted uranium deposit, SE China. Ore Geol Rev 26:187–197

Misař Z, Urban M (1996) Introduction [in to the Moravo-Silesian zone]. In: Dallmeyer RD, Franke W, Weber K (eds) Tectono-stratigraphic evolution of the Central and East European orogens. Springer, Heidelberg, pp 469–473

Misař Z, Dudek A, Havlena V, Weiss J (1983) Geologie ČSSR I, Český masív. Státni pedagogické nakladatestvi, Praha, 333 pp

Mochnacka K, Banas M (2000) Occurrence and genetic relationships of uranium and thorium mineralization in the Karkonosze Izera Block (the Sudety Mts., SW Poland). Ann Soc Geol Pol 70:137–150

Mochnacka K, Banaś M, Kramer W, Pošmourný K (1995) Metallogenesis of the Lugicum. In: Dallmeyer RD, Franke W, Weber K (eds) Tectono-stratigraphic evolution of the Central and East European orogens. Springer, Heidelberg, pp 357–372

Moore PB (1971) Crystal chemistry of the alluaudite structure type: contribution to the, paragenesis of pegmatite phosphate giant crystals. Am Mineral 56:1955–1975

Moore PB, Ito J (1980) Jungit und Matulait: zwei neue tafelige Phosphatmineralien. Aufschluss 31:55–61

Moore PB, Kampf AR (1977) Schoonerite, a new zinc-manganese-iron phosphate mineral. Am Mineral 62:246–249

Morávek P, Lehrberger G (1997) Die genetische und geotektonische Klassifikation der Goldvererzungen in der Böhmischen Masse. Geologica Bavarica 102:7–31

Morávek P, Pouba Z (1990) L ór dans la métallogénie du massif de Bohème. Mineral Deposita (Suppl) 25:90–98

Moretz L, Heimann A, Bitner J, Wise M, Rodrigues Soares D, Mousinho Ferreira A (2013) The composition of garnet as indicator of rare metal (Li) mineralization in granitic pegmatites. In: PEG 2013- the 6th international symposium on granitic pegmatites. Rubellite Press, New Orleans, pp 94–95

Mrázek P (1986) Metallogeny of the West Bohemian Upper Proterozoic. In: Proceedings of conference on metallogeny of the precambrian, IGCP, Proj. 91:49–53

Mrázek P, Pouba Z (1995) Metallogenesis. In: Dallmeyer RD, Franke W, Weber K (eds) Tectono-stratigraphic evolution of the Central and East European orogens. Springer, Heidelberg, pp 411–414

Mücke A (1977) Mineralien aus dem Pegmatit von Hagendorf. Aufschluss 28:353–358

Mücke A (1978) Sekundäre Phosphatmineralien aus dem Pegmatit von Hagendorf/Opf. und deren Paragenesen. Aufschluss 29:211–217

Mücke A (1980) Über einige Mineralien aus dem Pegmatit von Hagendorf/Opf. und deren Paragenesen. Aufschluss 32:85–93

Mücke A (1983) Wilhelmvierlingit, (Ca, Zn)MnFe 3+[OHl(PO4)2]2H2O, a new mineral from Hagendorf/Oberpfalz. Aufschluss 34:267–274

Mücke A (1987) Sekundäre Phosphatmineralien (Perloffit, Brasilianit, Mineralien der Kingsmountit-Gruppe) sowie Brochantit und die Zwieselit-Muschketoffit-Stilpnomelan-Pyrosmalith-Paragenese der 115-m-Sohle des Hagendorfer Pegmatits. Aufschluss 38:5–28

Mücke A (1988) Lehnerit $Mn[UO_2|PO_4]_2$ $8H_2O$, ein neues Mineral aus dem Pegmatit von Hagendorf/Oberpfalz. Aufschluss 39:209–217

Mücke A (2000) Die Erzmineralien und deren Paragenesen im Pegmatit von Hagendorf-Süd, Oberpfalz. Aufschluss 51:11–24

Mücke A, Keck E (2008) Untersuchungen an Columbiten (Fe, Mn)(Nb, Ta)2O6 und dem mit Columbit verwachsenen Neufund Petscheckit U(Fe, Mn)(Nb, Ta)2O8 aus dem Pegmatit von Hagendorf-Süd/Oberpfalz. Aufschluss 59:372–392

Mücke A, Keck E (2011) Karbonate aus dem Pegmatit von Hagendorf-Süd/Opf.: Zusammensetzung, Verbreitung und begleitende Phosphat-Mineralien (Apatit, Hagendorfit und Eosphorit-Gruppe) – darunter einige Neufunde (Triplit, Mineralien der Arrojadit-Dickinsonit Reihe, Goyazit und Variscit). Aufschluss 62:87–117

Mücke A, Keck E, Haase J (1990) Die genetische Entwicklung des Pegmatits von Hagendorf-Süd/Oberpfalz. Aufschluss 41:33–51

Müllbauer F (1925) Die Phosphatpegmatite i. Bayern (Neue Beoabachtungen.). Zeitschrift für Kristallographie, Zeitschrift f Kristallographie 64:319–336

Müller A, Breiter K, Novák JK (1998) Phosphorus-rich granites and pegmatites of Northern Oberpfalz and Western Bohemia. In: Breiter K (ed) Excursion guide: genetic significance of phosphorus in fractionated granites, International geological correlation program, IGCP 373. Peršlák. Český geologický ústav, Praha, pp 93–105

Munoz JL (1971) Hydrothermal stability relations of synthetic lepidolite. Am Mineral 56:2069–2087

Murowchick JB, Barnes HL (1986) Marcasite precipitation from hydrothermal solutions. Geochim Cosmochim Acta 50:2615–2629

Nägele M (1982) Ein bemerkenswerter Eigenfund: Orientierte Verwachsung von Torbernit und Autunit von Hagendorf-Süd. Lapis 6:38

Neiva AMR, Silva MMVG, Antunes IMHR, Ramos JMF (2001) Phosphate minerals of some granitic rocks and associated quartz veins from northern and central Portugal. J Czech Geol Soc 46:35–43

Němec D (1973) Das Vorkommen der Zn-Spinelle in der Böhmischen Masse. Tschermaks Mineralogische und Petrographische Mitteilungen 19:95–109

Němec D (1978) Zink im Staurolith. Chem Erde 37:307–314

Němec D (1998) The Rožná pegmatite field, western Moravia (Czech Republic). Chem Erde 58:233–246

Neuroth H (1997) K-Ar-Datierungen an detritischen Muskowiten-"Sicherungskopien" orogener Prozesse am Beispiel der Varisziden. Göttinger Arbeiten Geologie und Paläontologie 72:1–143

Nickel E, Fettel M (1985) Odenwald, vol 65, 2nd edn, Sammlung geologischer Führer. Bornträger, Berlin/Stuttgart, 231 pp

Niedermayr G (1990) Die Mineralien der Kor- und Saualpe in Kärnten, Österreich. Mineralien-Welt 1:58–67

Niggli P (1920) Die leichtflüchtigen Bestandteile im Magma. Preisschriften der Fürstlich Jablonowskischen Gesellschaft. B.G. Teubner, Leipzig, 272 pp

Nizamoff JW, Simmons WB, Falster AU (2004) Geology and geophysics, Univ of New Orleans, New Orleans, Phosphate mineralogy and paragenesis of the Palermo #2 Pegmatite, North Groton, New Hampshire. Denver annual meeting, 7–10 November 2004

Novák M (1992) Locality no 1: Rožná near Bystřice nad Pernštejnem, a large pegmatite dike of the lepidolite subtype, type locality of the lepidolite. In: Novák M, erný P (eds) Field trip guidebook, Lepidolite 200, pp 21–26

Novák M (2005) Granitické pegmatity Českého masivu (Česká Republika); mineralogická, geochemická a regionální klasifikace a geologický význam. Acta Musei Moraviae, Scientiae geologicae 90:3–74 (in Czech, with an abstract in English)

Novák M, Černý P (2001) Distinctive compositional trends in columbite-tantalite from two segments of the lepidolite pegmatite at Rožná, western Moravia, Czech Republic. J Czech Geol Soc 46:1–8

Novák M, Filip J (2010) Unusual (Na, Mg)-enriched beryl and its breakdown products (beryl II, bazzite, bavenite) from euxenite-type NYF pegmatite related to the orogenic ultrapotassic Třebíč pluton, Czech Republic. Can Mineral 48:615–628

Novák M, Gadas P (2010) Internal structure and mineralogy of a zoned anorthite- and grossular-bearing leucotonalitic pegmatite in serpentinized lherzolite at Ruda Nad Moravou, Staré Mešto unit, Czech Republic. Can Mineral 48:629–650

Novák M, Hyršl J (1992) Locality no.3: Vlastějovice near Zruč nad Sázavou, pegmatites with fluorite penetrating skarn. In: Novák M, erný P (eds) International symposium on mineralogy, petrology and geochemistry of granitic pegmatites Lepidolite 200, Nové Město na Moravě, Czech Republic. Field trip guidebook, pp 33–37

Novák M, Povondra P (1995) Elbaite pegmatites in the Moldanubicum: a new subtype of the rare-element class. Mineral Petrol 55:159–176

Novák M, Selway JB (1997) Locality no. 1: Rožná near Bystřice nad Pernštejnem, Hradisko hill, a large lepidolite subtype pegmatite dike. In: Novák M, Selway JB (eds) International symposium Tourmaline 1997, Nové Město na Moravě, Czech Republic. Field trip guidebook, pp 23–38

Novák M, erný P, ech F, Staněk J (1992) Granitic pegmatites in the territory of the Bohemian and Moravian Moldanubicum. In: Novák M, erný P (eds) International symposium on mineralogy, petrology and geochemistry of granitic pegmatites, Lepidolite 200, Nové Město na Moravě, Czech Republic. Field trip guidebook, pp 11–20

Novák M, Klečka M, Šrein V (1994) Compositional evolution of Nb-Ta oxide minerals from alkali-feldspar muscovite granites Homolka and Šejby, southern Bohemia, and its comparison with other rare-element granites. Mitteilungen der Österreichischen Mineralogischen Gesellschaft 139:353–354

Novák JK, Pivec E, Štemprok M (1996) Hydrated iron phosphates in muscovite-albite granite from Waidhaus (Oberpfalz, Germany). J Czech Geol Soc 41:201–207

Novák M, Selway JB, Staněk J (1997) Locality no. 10: Myšenec near Protivín, barren pegmatite with giant tourmaline aggregates. In: Novák M, Selway JB (eds) International symposium Tourmaline 1997, Nové Město na Moravě, Czech republic. Field trip guidebook, pp 105–110

Novák M, Houzar S, Pfeiferová A (1998a) Přehled mineralogie, petrografie a historie klasické lokality lepidolitového pegmatitu v Rožné u Bystřice nad Pernštejnem, západní Morava. Acta Musei Moraviae, Scientiae Geologicae 83:3–48. (in Czech,with English summary)

Novák M, Černý P, Kimbrough DL, Taylor MC, Ercit TS (1998b) U-Pb Ages of monazite from granitic pegmatites in the Moldanubian Zone and their geological implications. Acta Univ Carol Geol 42:309–310

Novák M, Selway JB, Černý P, Hawthorne FC, Ottolini L (1999) Tourmaline of the elbaite–dravite series from an elbaite-subtype pegmatite at Bližná, southern Bohemia, Czech Republic. Eur J Mineral 11:557–568

Novák F, Pauliš P, Süsser C (2001a) Chemical composition of crandallite, goyazite and waylandite from Krásno near Horni Slavkov. Bull Mineral Petrol Odd Nar Muz 9:230–234 (in Czech)

Novák J, Pivec E, Holub FV, Štemprok M (2001b) Greisenization of lamprophyres in the Krupka Sn-W district in the eastern Krušné Hory/Erzgebirge, Czech Republic. In: Piestrzynski A et al (eds) Mineral deposits at the beginning of the 21st century. A. A. Balkema, Rotterdam, pp 465–467

Novák M, Cerný P, Uher P (2003) Extreme variation and apparent reversal of Nb-Ta fractionation in columbite-group minerals from the Scheibengraben beryl-columbite granite pegmatite, Marsikov, Czech Republic. Eur J Mineral 15:565–574

Novák M, Černý P, Cempírek J, Šrein V, Filip J (2004) Ferrotapiolite as a pseudomorph of stibiotantalite from the lastovicky lepidolite pegmatite, Czech Epublic; an example of hydrothermal alteration at constant Ta/(Ta + Nb). Can Mineral 42:1117–1128

Novák M, Škoda R, Gadas P, Krmíček L, Černý P (2012) Contrasting origins of the mixed (NYF + LCT) signature in granitic pegmatites, with examples from the Moldanubian Zone, Czech Republic. Can Mineral 50:1077–1094

Nriagu JO (1976) Phosphate-clay mineral relations in soils and sediments. Can J Earth Sci 13:717–736

Nriagu JO, Moore PB (1984) Phosphate minerals. Springer, Berlin/Heidelberg/New York 442 pp

O'Donoghue M (2006) Gems their sources, descriptions and identification. Elsevier, Amsterdam, 873 pp

Oberc-Dziedzic T, Kryza R (2012) Late stage Variscan magmatism in the Strzelin Massif (SW Poland): SHRIMP zircon ages of tonalite and Bt-Ms granite of the Gęsiniec intrusion. Geol Q 56:225–236

Oberc-Dziedzic T, Kryza R, Białek J (2010) Variscan multistage granitoid magmatism in Brunovistulicum: petrological and SHRIMP U-Pb zircon geochronological evidence from the southern part of the Strzelin Massif, SW Poland. Geol Q 54:301–324

Oncken O (1997) Transformation of a magmatic arc and an orogenic root during oblique collision and its consequence for the evolution of the European Variscides (Mid-German Crystalline Rise). Geol Rundsch 86:2–20

Oncken O, Massonne HJ, Schwab M (1995) Metamorphic evolution of the Rhenohercynian. In: Dallmeyer RD, Franke W, Weber K (eds) Tectono-stratigraphic evolution of the Central and East European orogens. Springer, Heidelberg, pp 512–517

Orville PM (1963) Alkali ion exchange between vapor and feldspar phases. Am J Sci 261:201–237

Osann A (1927) Die Mineralien Badens. Schweizerbart, Stuttgart, 239 pp

Oyarzábal J, Galliski MÁ, Perino E (2009) Geochemistry of K-feldspar and Muscovite in Rare-element Pegmatites and Granites from the Totoral Pegmatite Field, San Luis, Argentina. Resour Geol 59:315–329

Parc S, Nahon D, Tardy Y, Viellard P (1982) Estimated solubility products and fields of stability for cryptomelane, nsutite, birnessite, and lithiophorite based on natural lateritic weathering sequences. Am Mineral 74:466–475

Partington GA, Mc Naughton NJ, Williams IS (1995) A review of the geology, mineralization, and geochronology of the Greenbushes Pegmatite, Western Australia. Econ Geol 90:616–635

Pašava J, Hladikova J, Dobes P (1996) Origin of metal-rich black shales from the Bohemian Massif, Czech Republic. Econ Geol 91:63–79

Patočka F, Vrba J (1989) The comparison of stratabound massive sulfides deposits using the fuzzy-linguistic diagnosis of the Zlaté Hory deposits, Czechoslovakia, as an example. Mineral Deposita 4:192–198

Peacor DR, Rouse RC, Coskren TD, Essene EJ (1999) Destinezite ("diadochite"), $Fe_2(PO_4)(SO_4)(OH).6H_2O$: its crystal structure and role as a soil mineral at Alum Cave Bluff, Tennessee. Clay Clay Miner 47:1–11

Pedersen RB, Dunning GR, Robins B (1989) U-Pb ages of nepheline syenite pegmatites from the Seiland Magmatic Province, N. Norway. In: Gayer RA (ed) The Caledonide geology of Scandinavia. Graham and Trotman, London, pp 3–8

Pertold Z (1978) Prospects of the stratiform deposits of the Bohemian massif. In: Kužwart M (ed) Theoretical basic prognoses of the raw materials in Czechoslovakia. Czech Geological Survey, Praha, Czechoslovakia, pp 118–123. (in Czech)

Pertold Z, Chrt J, Budil V, Burda J, Burdová P, Kribek B, Pertoldová J, Gaskarth B (1994) The Tisova Cu- deposit: a Besshi-type mineralization in the Krusné Hory Mts., Bohemian massif, Czech Republic. Minerogr Ser Miner Depos 31:71–95

Petersen JS (1978) Structure of the larvikite – lardalite complex, Oslo-region, Norway, and its evolution international. J Earth Sci 67:330–342

Petrakakis K (1997) Evolution of Moldanubian rocks in Austria: review and synthesis. J Metamorph Geol 15:203–222

Pezzotta F (2001) Madagascar, a mineral and gemstone paradise. Ed, Extralapis English 1. Lapis International LLC, East Hampton, 100 pp

Pezzotta F, Guastoni A (1998) Rossmanit aus Rozna (CR) und Elba (I). Lapis 9:38–40

Pfaffl F (1966) Die Blötz bei Bodenmais/Bayrischer Wald. Aufschluss 17:207–208

Piantone P, Itard Y, Pillard F, Boulingui B (1995) Compositional variation in pyrochlores from the weathered Mabounié carbonatite (Gabon). In: Pasava J, Kríbek B, Zák K (eds) Mineral deposits: from their origin to their environmental impacts. Balkema, Rotterdam, pp 629–632

Pieczka A (2000) A rare mineral-bearing pegmatite from the serpentinite massif, the Fore-Sudetic Block, SW Poland. Geol Sudet 33:23–31

Pieczka A (2007) Beusite and an unusual Mn-rich apatite from the Szklary granitic pegmatite, Lower Silesia, southwestern Poland. Can Mineral 45:901–914

Pin C, Puziewicz J, Duthou JL (1989) Ages and origins of a composite granitic massif in the Variscan belt: a Rb-Sr study of the Strzegom-Sobótka Massif, W Sudetes (Poland). Neues Jb Mineral Abh 160:71–82

Pirard C, Hatert F, Fransolet A-M (2007) Alteration sequences of aluminium phosphates from Montebras Pegmatite, Massif Central, France. Granitic Pegmatites: The State of the Art – International Symposium, 6–12 May 2007, Porto

Pitcher WS (1979) The nature, ascent and emplacement of granitic magmas. J Geol Soc Lond 136:627–662

Pitcher WS (1982) Granite type and tectonic environment. In: Hsü KJ (ed) Mountain building processes. Academic, London, pp 19–40

Pitcher WS (1987) Granites and yet more granites forty years on. Geol Rundsch 76:51–79

Pitra P, Boulvais P, Antonoff V (2008) Wagnerite in a cordierite-gedrite gneiss: witness of long-term fluid-rock interaction in the continental crust (Ile d'Yeu, Armorican Massif, France). Am Mineral 93:315–326

Plaumann S (1976) Schweremessungen im Bereich Erbendorf. Niedersächsisches Landesamt für Bodenforschung, File No 75564, Hannover

Plaumann S (1986) Die Schwerekarte der Oberpfalz und ihre Bezüge zu Strukturen der oberen Erdkruste. Geol Jahrb E 33:5–13

Plaumann S (1995) Die Schwerekarte 1:500000 der Bundesrepublik Deutschland (Bouguer-Anomalien) Blatt Süd. Geol Jahrb E 53:3–13

Pöllmann H, Bäumler W, Meier S (2005) Sekundärphosphate aus dem Steinwaldgranit von Hopfau bei Erbendorf/Oberpfalz. Aufschluss 56:71–79

Pollner R, Schmidt C, Fischer G, Kuhn K, Poschl E, Dalla Lana IG, Fiedorow R, Gaworski B, Song X, Arnaud NO, Kelley SP (1997) Argon behaviour in gem-quality orthoclase from Madagascar: experiments and some consequences for 40Ar/39Ar geochronology. Geochim Cosmochim Acta 61:3227–3255

Popov VA, Popova VI (2006) Ilmeny mountains. Mineralogical Almanac 9, 156 pp

Postl W, Golob P (1979) Ilmenorutil (Nb-Rutil), Columbit und Zinnstein aus dem Spodumenpegmatit im Wildbachgraben, Koralpe (Steiermark). Mitt-Bl Abt Miner Landesmuseum Joanneum 47:27–35

Pouba Z, Ilavsky J (1986) Czechoslovakia. In: Dunning FW, Evans AM (eds) Mineral deposits of Europe, 3, Central Europe. IMM & Mineralogical Society, London, pp 117–173

Povondra P, Pivec E, Čech F, Lang M, Novák F, Prachař I, Ulrych J (1987) Přibyslavice peraluminous granite. Acta Univ Carol Geol 3:183–283

Povondra P, Staňková J, Staněk J (1992) CO_2 –bearing cordierite of Moldanubian leptynite rock series from Horní Bory, Czech Republic. Acta Univ Carol Geol 1992:331–349

Povondra P, Lang M, Pivec E, Ulrych J (1998) Tourmaline from the Přibyslavice peraluminous alkali-feldspar granite, Czech Republic. J Czech Geol Soc 43:3–8

Prachař I, Povondra P, Novák F (1983) Manganese-rich siderite from granite at Přibyslavice near Čáslav. Acta Univ Carol Geol 1983:13–25

Prentice JE (1990) Geology of construction materials. Chapman and Hall, London/New York/ Tokyo/Melbourne/Madras, 201 pp

Pucher R (1986) Interpretation der magnetischen Anomalie von Erbendorf (Oberpfalz) und dazugehörige gesteinsmagnetische Untersuchungen. Geol Jahrb E 33:31–52

Pucher R, Wonik T (1990) Eine Karte der Magnetfeldanomalien für die Umgebung der KTB-Bohrlokation in der Oberpfalz. Geol Jahrb E 44:3–13

Pucher R, Bader K, Stettner G, Wonik T (1990) Eine Karte der Magnetfeldanomalien für die Umgebung der KTB-Bohrlokation in der Oberpfalz. Geol Jahrb E 44:1–48

Pupin JP (1980) Zircon and granite petrology. Contrib Mineral Petrol 73:207–220

Pupin JP, Turco G (1981) Le zircon, minéral commun significatif des roches endogènes et exogènes. Bull Minéral 104:724–731

Raade G, Ferraris G, Gula A, Ivaldi G, Bernhard F (2002) Kristianesite a new calcium-scandium-tin sorosilicate from granite pegmatite from Tørdal, Telemark, Norway. Mineral Petrol 75:89–99

Rabe H (1975) Der Nachweis einer geochemischen Volumenregel am Beispiel von Diatexien aus dem Pegmatit von Hagendorf /Oberpfalz. PhD dissertation, Technical University, Berlin

Radvanec M, Grecula P, Žák K (2004) Siderite mineralization of the Gemericum Superunit (Western Carpathians, Slovakia): review and a revised genetic model. Ore Geol Rev 24:267–298

Rafailovich M (1997) Epithermal gold deposits of Kazakhstan. In: Bekzanov GR, Dudich E, Gaál G, Jenchuraeva RJ (eds) Paleozoic granite-related Au, Cu, Mo, W, REE deposits and epithermal gold deposits, IUGS/UNESCO Deposit Modeling Program, workshop, 31 August–14 September, 1997, Kazakhstan and Kyrgyzstan, Budapest, pp 54–55

Rakotondrazafy AFM, Giuliani G, Ohnenstetter D, Fallick AE, Rakotosamizanany S, Andriamamonjy A, Ralantoarison T, Razanatseheno M, Offant Y, Garnier V, Maluski H, Dunaigre C, Schwarz D, Ratrimo V (2008) Gem corundum deposits of Madagascar: a review. Ore Geol Rev 34:134–154

Rammelsberg Bergbaumuseum (2011/2012) Suche und Erkundung am Rammelsberg und in seiner Umgebung. Förderverein e.V., Goslar, 166 pp

Rasser MW, Harzhauser M (2008) Paleogene and neogene. In: McCann T (ed) Geology of Central Europe, Geological Society special publication. Geological Society, London, pp 1031–1139

Redhammer GJ, Tippelt G, Bernroider M, Lottermoser W, Amthauer G, Roth G (2005) Hagendorfite (Na, Ca)MnFe2(PO4)3 from type locality Hagendorf (Bavaria, Germany): crystal structure determination and 57Fe Mössbauer spectroscopy. Eur J Mineral 17:915–932

Reicherter K, Froitzheim N, Jarosinki M (co-ordinators), Badura J, Franzke H-J, Hansen M, Hübscher C, Müller R, Poprawa P, Reinecker J, Stackebrandt W, Voigt T, von Eynatten H, Zuchiewicz W (2008) Alpine tectonics north of the Alps. In: McCann T (ed) Geology of Central Europe. Geological Society special publication. Geological Society, London, pp 1233–1285

Richardson SW, Gilber MC, Bell PM (1969) Experimental determination of kyanite-andalusite and andalusite-sillimanite equilibria; the aluminum silicate triple point. Am J Sci 267:259–272

Richter P, Stettner G (1979) Geochemische und petrographische Untersuchungen der Fichtelgebirgsgranite. Geologica Bavarica 78:1–144

Richter P, Stettner G (1986) Untersuchungen zur Form und räumlichen Lage der Granitplutone im Projektgebiet der Lokation Oberpfalz. German Science Foundation- Continental Deep Drilling Program of the Federal Republic of Germany, Final report

Robbins LM, Falster AU, Simmons WB, Stark S (2008) Benyacarite; new data from the Hagendorf-Süd Pegmatite, Eastern Bavaria, Germany. Rocks Miner 83:340–347

Roberts RJ, Corfu F, Torsvik TH, Ramsay DM, Ashwal LD (2006) Short-lived mafic magmatism at 570 Ma in the northern Norwegian Caledonides- U/Pb zircon ages from the Seiland Igneous Province. Geol Mag 143:1–17

Robertson BT (1982) Occurrence of epigenetic phosphate minerals in a phosphatic iron formation Yulon Territory. Can Mineral 20:177–187

Robinson GW, Van Velthuizen J, Ansell HG, Sturman BD (1992) Mineralogy of the rapid creek and big fish river area, Yukon territory. Mineral Rec 23:1–47

Rock NMS (1991) Lamprophyres (with contributions by D. R. Bowes & A. E. Wright). Glasgow & London: Blackie, New York: Van Nostrand, Reinhold. Rock

Rodgers KA (1989) The thermochemical behavior of vivianite and vivianite/metavivianite admixtures from Borne (Netherlands) and Mangualde (Portugal). Geol Mijnb 68:257–262

Rodgers KA, Henderson GS (1986) The thermochemistry of some iron phosphate minerals: vivianite, metavivianite, bariéite, ludlamite and metavivianite/vivianite admixtures. Thermochim Acta 104:1–12

Rogers RJ, Brown FH (1979) Authigenic mitridatite from the Shungura formation, southwestern Ethiopia. Am Mineral 64:169–171

Rohrmüller J (1998) Geologische Karte von Bayern 1:25000 Erläuterungen zum Blatt Nr. 6240 Flossenbürg. Geological Survey of Bavaria, München, 95 pp

Rolland Y, Cox S, Boullie AM, Pennacchioni G, Mancktelow N (2003) Rare earth and trace element mobility in mid-crustal shear zones: insights from the Mont Blanc Massif (Western Alps). Earth Planet Sci Lett 214:203–219

Romeiro JCP, Pedrosa-Soares AC (2005) Controle do minério de espodumênio em pegmatitos da Mina da Cachoeira, Araçuai, MG. Geonomos 13:75–85

Romer RL, Smeds S-A (1994) Implications of U-Pb age of columbite-tantalites from granitic pegmatites for the Paleoproterozoic accretion of 1.90-1.85 Ga magmatic arcs to the Baltic Shield. Precambrian Res 67:141–158

Romer RL, Smeds S-A (1996) U-Pb columbite ages of pegmatites from Sveconorvegian terranes in southwestern Sweden. Precambrian Res 76:15–30

Romer RL, Smeds S-A (1997) U-Pb columbite chronology of post-kinematic Paleoproterozoic pegmatites in Sweden. Precambrian Res 82:85–99

Rose H (1845) Über die Zusammensetzung der Tantalite und über ein im Tantalit von Baiern entaltenes neues Mineral. J Prakt Chem 34:36–42

Rossovskiy LN, Konovalenko SI (1977) Corundum Plagioclasite of the Southwestern Pamirs. Doklady Academie Science. U.S.S.R., Earth Science Section 235:145–147

Ruppert H (1984) Physiko-chemische Betrachtungen zur Entstehung der Kreide-Eisenerz-Lagerstätten in Nordost-Bayern. Geol Jahrb D 66:51–75

Russell JD, Fraser AR (1994) Infrared methods. In: Wilson MJ (ed) Clay mineralogy: spectroscopic and chemical determinative methods. Chapman and Hall, London, pp 11–67

Ruzicka V (1971) Geological comparison between East European and Canadian uranium deposits. Can Geol Surv 70–48:1–196

Rykart R (1995) Quarz-Monographie – Die Eigenheiten von Bergkristall, Rauchquarz, Amethyst, Chalcedon, Achat, Opal und anderen Varietäten. Ott Verlag Thun, 2nd, 462 pp

Satish-Kumar M, Santosh MA (1998) Petrological and fluid inclusion study of calc-silicate–charnockite associations from southern Kerala, India: implications for CO2 influx. Geol Mag 135:27–45

Schaaf P, Sperling T, Müller-Sohnius D (2008) Pegmatites from the Bavarian Forest, SE Germany: geochronology, geochemistry and mineralogy. Geologica Bavarica 108:204–303

Scharbert S, Vesela M (1990) Rb-Sr systematic of intrusive rocks from the Moldanubicum around Jihlava. Thirty years of geological co-operation between Austria and Czechoslovakia. CGU and GBA Prague, pp 262–272

Schaltegger U (2000) U-Pb geochronology of the southern Black Forest Batholith (Central Variscan Belt): timing of exhumation and granite emplacement. Int J Earth Sci 88:814–828

Schaltegger U, Corfu F (1995) Late Variscan "Basin and Range" magmatism and tectonics in the Central Alps: evidence from U-Pb geochronology. Geodin Acta 8:82–98

Schleicher H (1994) Collision-type granitic melts in the context of thrust tectonics and uplift history (Triberg granite complex, Schwarzwald, Germany). Neues Jb Mineral Abh 166:211–237

Schlüter J, Klaska K-H, Friese K, Adiwidjaja G (1999) Kastningite, (Mn,Fe,Mg)AI2(PO4)2(OH)2. 8H20, a new phosphate mineral from Waidhaus, Bavaria, Germany. Neues Jahrb Mineralogie, Monatsh: 40–48

Schmid H, Weinelt W (1978) Lagerstätten in Bayern. Geologica Bavarica 77:1–160

Schmid-Beurmann P, Knitter ST, Cemicï L (1999) Crystal chemical properties of synthetic lazulite ± scorzalite solid-solution series. Phys Chem Miner 26:496–505

Schneiderhöhn H (1961) Die pegmatite [The pegmatites]. Gustav Fischer Verlag, Stuttgart, 720 pp. (in German)

Schneiderhöhn H (1962) Erzlagerstätten [Ore deposits]. Gustav Fischer Verlag, Stuttgart, 371 pp. (in German)

Schnorrer G, Kronz A, Pascher G (2003) Cheralith, Monazit und Xenotim drei neue Minerale der Monazit-Gruppe sowie Uranosphärit, ein sekundäres Bi-Uran-Mineral vom ehemaligen Phosphatpegmatit Hagendorf-Süd/Oberpfalz. Aufschluss 54:267–272

Scholz A (1925) Untersuchungen über Mineralführung und Mineralgenese der bayerischen Pegmatite, Bericht für das Jahr 1924 des Naturwissenschaftlichen Vereins Regensburg e.V. Regensburg 17:1–46

Schorr T (1984) Beryll aus dem Schwarzwald. Lapis 9:22

Schroecke H (1966) Solid-solution series in the columbite-tapiolite, the columbite-tapiolite-euxenite, and the columbite-tapiolite-FeNbO sub 4 groups. Neues Jb Mineral Abh 106:1–54

Schüssler U, Oppermann U, Kreuzer H, Seidel E, Okrusch M, Lenz K-L, Raschka H (1986) Zur Altersstellung des ostbayerischen Kristallins-Ergebnisse neuer K-Ar-Datierungen. Geologica Bavarica 89:21–47

Sebastian U (2013) Die Geologie des Erzgebirges. Springer Spektrum, Berlin/Heidelberg 268 pp

Seeland F (1876) Der Hüttenberger Erzberg und seine nächste Umgebung. Jahrbuch der Geologischen Reichsanstalt 26:49–112

Seeliger E, Mücke A (1970) Ernstit, ein neues Mn $^{2+}$Fe $^{3+}$ – Phosphat und seine Beziehungen zum Eosphorit. Neues Jb Mineral Monat 1970:289–298

Seifert T (2008) Metallogeny and petrogenesis of lamprophyres in the Mid-European Variscides – Postcollisional magmatism and its relationship to Late Variscan ore forming processes in the Erzgebirge (Bohemian Massif). IOS Press BV, Amsterdam

Sekjora J, Skoda R, Ondruš P (2006) New naturally occurring mineral phases from the Krásno-Horní Slavkov area, western Bohemia. Czech Republic. J Czech Geol Soc 51:159–187

Seltmann R, Faragher AE (1994) Collisional orogens and their related metallogeny-a preface. In: Seltmann R, Kämpf H, Möller P (eds) Metallogeny of collision orogen. Czech Geological Survey, Prague, pp 7–19

Seltmann R, Schilka W (1995) Late-Variscan crustal evolution in the Altenberg-Teplice caldera. Evidence from new geochemical and geochronological data. Terra Nostra 7(95):120–124

Seltmann R, Förster HJ, Gottesmann B, Sala M, Wolf D, Štemprok M (1998) The Zinnwald greisen deposit related to post-collisional A-type silicic magmatism in the Variscan eastern

Erzgebirge/Krušne Hory. In: Breiter K (ed) Excursion guide: genetic significance of phosphorus in fractionated granites, International geological correlation program, IGCP 373. Peršlák. Český geologický ústav, Praha, pp 33–50
Selway JB, Novák M, Hawthorne FC, Černý P, Ottolini L, Kyser TK (1998) Rossmanite LiA 2Al 6(BO3)3Si6O18(OH)4, a new alkali-deficient tourmaline: description and crystal structure. Am Mineral 83:894–900
Selway JB, Novák M, Černý P, Hawthorne FC (1999) Compositional evolution of tourmaline in lepidolite-subtype pegmatites. Eur J Mineral 12:569–584
Selway JB, Breaks FW, Tindle AG (2006) A review of rare-element (Li-Cs-Ta) pegmatite exploration techniques for the Superior province, Canada, and large worldwide tantalum deposits. Explor Min Geol 14:1–30
Shannon JR, Walker BM, Carten RB, Geraghty EP (1982) Unidirectional solidification textures and their significance in determining relative ages of intrusion at the Henderson Mine, Colorado. Geology 10:293–297
Siebel W, Höhndorf A, Wendt I (1995a) Origin of late Variscan granitoids from NE Bavaria: Germany, exemplified by REE and Nd isotope systematics. Chem Geol 125:249–270
Siebel W, Breiter K, Höhndorf A, Wendt I, Henjes-Kunst F (1995b) Preliminary note on age relationships and Nd isotope composition of granitoids from the Bärnau-Rozvadov pluton, Western Bohemia. In: 8th conference of the German Continental Deep Drilling Programme (KTB) Giessen, 25–26 May 1995
Siebel W, Trzebski R, Stettner G, Hecht L, Casten U, Höhndorf A, Müller P (1997) Granitoid magmatism of the NW Bohemian Massif related: gravity data composition, age relations and phase concept. Geol Rundsch 86:45–63
Siebel W, Thiel M, Chen F (2006) Zircon geochronology and compositional record of late- to post-kinematic granitoids associated with the Bavarian Pfahl zone (Bavarian Forest). Mineral Petrol 86:45–62
Simmat R, Rickers K (2000) Wagnerite in high-MgAl granulites of Anakapalle, Eastern Ghats Belt, India. Eur J Mineral 12:661–666
Simmons WB (2007) Gem-bearing pegmatites. In: Groat LA (ed) Geology of gem deposits, vol 37, Mineralogical Association Canada short course. Mineralogical Association Canada, Québec, pp 169–206
Simmons WB, Webber KL, Falster AU, Nizamoff JW (2003) Pegmatology: pegmatite mineralogy, petrology and petrogenesis. Rubellite Press, New Orleans, 176 pp
Škoda R, Novák M (2007) Y, REE, Nb, Ta, Ti-oxide (AB 2 O 6)minerals from REL-REE euxenite-subtype pegmatites of the Třebíč Pluton, Czech Republic; substitutions and fractionation trends. Lithos 95:43–99
Škoda R, Novák M, Houzar S (2006) Granitické NYF pegmatity třebíčského plutonu. Acta Musei Moraviae, Scientiae geologicae 91:129–176 (in Czech, with English summary)
Sláma J, Košler J, Condon DJ, Crowley JL, Gerdes A, Hanchar JM, Škoda R, Novák M, Houzar S (2006) Granitické NYF pegmatity třebíčského plutonu. Acta Musei Moraviae, Scientiae Geologicae 91:129–176 (in Czech, with English summary)
Soares DR, Beurlen H, Ferreira ACM, Da Silva MRR (2007) Chemical composition of gahnite and degree of pegmatitic fractionation in the Borborema Pegmatite province Northeastern Brazil. Anais da Academia Brasiliera de Ciencias 79:395–404
Söllner P, Köhler H, Müller-Sohnius D (1981) Rb/Sr Altersbestimmungen an Gesteinen der Münchberger Gneismasse (M.N:), NE-Bayern; Teil 1 Gesamtgesteinsdatierungen. Neues Jb Mineral Abh 141:90–112
Sowder AG (1998) The formation, transformation, and stability of environmentally relevant uranyl mineral phases. PhD dissertation, Clemson University, Clemson
Sowder AG, Clark SB, Fjeld RA (1996) The effect of silica and phosphate on the transformation of schoepite to becquerelite and other uranyl phases. Radiochim Acta 74:45–49
Sowder AG, Clark SB, Fjeld RA (1999) The transformation of uranyl oxide hydrates: the effect of dehydration on synthetic metaschoepite and its alteration to becquerelite. Environ Sci Technol 33:3552–355

Stacey JS, Kramers JD (1975) Approximation of terrestrial lead isotope evolution by a 2-stage model. Earth Planet Sci Lett 26(2):207–221

Stein E (1988) Die strukturgeologische Entwicklung im Übergangsbereich Saxothuringikum/Moldanubikum in NE-Bayern. Geologica Bavarica 92:5–131

Štemprok M, Seltmann R (1994) The metallogeny of the Erzgebirge (Krušné Hory). In: Seltmann R, Kämpf H, Möller P (eds) Metallogeny of collision orogen. Czech Geological Survey, Prague, pp 61–69

Stettner G (1960) Erläuterung zur geologischen Karte von Bayern 1.25000, Sheet No. 5836, Münchberg. Geological Survey of Bavaria, München, 163 pp

Stettner G (1964) Erläuterung zur geologischen Karte von Bayern 1.25000, Sheet No. 5837, Weissenstadt. Geological Survey of Bavaria, München, 194 pp

Stettner G (1990) KTB Umfeldgeologie. Bayerisches Geologisches Landesamt, München, 31 pp

Stewart DB (1978) Petrogenesis of lithium-rich pegmatites. Am Mineral 63:970–980

Stochs HG, Lugmair GW (1986) Geochemistry and evolution of eclogites from the Münchberg Gneiss Massif, W-Germany. Terra Cognita 6:254

Stochs HG, Lugmair GW (1987) Geochronology and geochemistry of eclogites from the Münchberg Gneiss Massif Germany. Terra Cognita 7:163

Strmić Palinkaš S, Bermanec V, Palinkaš LA, Boev B, Gault RA, Prochaska W, Bakker RA (2012) The evolution of the Čanište epidote-bearing pegmatite, Republic of Macedonia: evidence from mineralogical and geochemical features. Geologia Croatica 65:423–434

Strunz H (1948) Scholzit, ein neue Mineralart. Fortschr Mineral 27:31–32

Strunz H (1952) Mineralien und Lagerstätten in Ostbayern. Acta Albertina Ratsibonensia 20:81–203

Strunz H (1954a) Laueit, MnFe23+[OH|PO4]2 • 8H2O, ein neues Mineral. Naturwissenschaften 41:256–256

Strunz H (1954b) Hagendorfit, ein neues Mineral der Varulith-Hühnerkobelit-Reihe. Neues Jb Mineral Monat 1954:252–255

Strunz H (1956) Pseudolaueit, ein neues Mineral. Naturwissenschaften 6:128

Strunz H (1961) Epitaxie von Uraninit auf Columbit. Aufschluss 12:81–84

Strunz H (1962) Die Uranfunde in Bayern von 1804 bis 1962 (einschliesslich der radiometrischen Messergebnisse). Naturwissesnchaftlicher Verein zu Regensburg, Regensburg, 92 pp

Strunz H (1971) Mineralien und Lagerstätten des Bayerischen Waldes. Aufschluss 21:7–91

Strunz H (1974) Granites and pegmatites of in eastern Bavaria. Fortschritte der Mineralogie 52 Beihefte 1: 1–32

Strunz H, Fischer H (1957) Childro-Eosphorit, Tavorit und Fairfieldit von Hagendorf. Neues Jb Mineral Monat 1957:78–88

Strunz H, Tennyson C (1956) Kristallographie von Scholzit, $CaZn_2[PO_4]_2 \cdot 2H_2O$. Z Krist 107:318–324

Strunz H, Tennyson C (1975) Über den Columbit vom Hühnerkobel im Bayerischen Wald und seine Uran-Paragenese. Aufschluss 12:313–324

Strunz H, Forster A, Tennyson C (1975) Die Pegmatite der nördlichen Oberpfalz. Aufschluss 26:117–189

Strunz H, Tennyson C, Mücke A (1976) Mineralien von Hagendorf/Ostbayern – Fortschrittsbericht. Aufschluss 27:329–340

Sturman BD, Rouse RC, Dunn PJ (1981) Parascholzite, a new mineral from Hagendorf, Bavaria, and its relationship to scholzite. Am Mineral 66:843–851

Suess E (1888) Das Antlitz der Erde. Zweiter Band. Temsky/Freytag, Prag/Wien/Leipzig, 703 pp

Suess E (1912) Vorläufige Mitteilungen über die Münchberger Deckscholle. Sitzungsbericht der Akademie der Wissenschaften Wien, Mathematisch-Naturwissenschaftliche Klasse 121:1–253

Swaziland Natural Trust Commission (2007) Cultural resources – Malolotja archaeology. Lion Cavern

Szełęg E, Škoda R (2008) Y, REE-rich zirconolite from the Skalna Brama pegmatite near Szklarska Poręba (Karkonosze Massif, Lower Silesia, Poland). Mineralogia – Special Papers 32:160

Tajčmanová L, Konopásek J, Connolly JAD (2007) Diffusion-controlled development of silica-undersaturated domains in felsic granulites of the Bohemian Massif (Variscan belt of Central Europe). Contrib Mineral Petrol 153:237–250

Tajčmanová L, Konopásek J, Košler J (2009) Distribution of zinc and its role in stabilization of spinel-bearing mineral assemblages in high-grade felsic rocks of the Moldanubian domain (Bohemian Massif). Eur J Mineral 21:407–418

Talapatra AK (1968) Sulfur mineralisation associated with migmatization in the Southeastern part of the Singhbhum Shear Zone, Bihar. India Econ Geol 63:156–165

Tankard AJ, Jackson MPA, Eriksson KA, Hobday DK, Hunter DR, Minter WEL (1982) Crustal evolution of southern Africa. Springer, Heidelberg/New York/Berlin 523 pp

Taucher J, Walter F, Postl W (1992) Mineralparagenesen in Pegmatiten der Koralpe. Teil1: Die Lithium-Lagerstätte am Brandrücken, Weinebene, Koralpe, Kärnten. Die Minerale des feinkörnigen Spodumenpegmatits (MH-Pegmatit). Matrix 1:23–72

Taucher J, Walter F, Postl W (1994) Mineralparagenesen in Pegmatiten der Koralpe. Part 2: Die Lithium-Lagerstätte am Brandrücken, Weinebene, Kärnten. Die Minerale des grobkörnigen Spodumenpegmatits (AH-Pegmatit) sowie die Minerale der Pegmatitrandgesteine. Matrix 3:19–52

Tavakkoli B (1985) Das Granitmassiv von Flossenbürg-Erkundung seiner Lagerung und Genese auf Grund geochemischer Untersuchungen mit einem Beitrag zur Geochemie seiner Feldspäte und deren möglichen technologischen Abtrennung. PhD thesis, University of München, 184 pp

Taylor BE, Beaudoin G (2000) Sulphur stratigraphy of the Sullivan Pb–Zn–Ag deposit, B.C: evidence for hydrothermal sulphur, and bacterial and thermochemical sulphate reduction. In: Lydon JW, Höy T, Slack JF, Knapp M (eds) The Sullivan Deposit and its geological environment, Special publication Mineral Deposits Division of the Geological Association of Canada, 1. Geological Association of Canada, Mineral Deposits Division, St. John's, pp 696–719

Taylor SR, McLennan SM (1995) The geochemical evolution of the continental crust. Rev Geophys 33:241–265

Teipel U, Eichhorn R, Höll R, Kennedy A (2002) Neoproterozoic and lower Ordovician magmatic events in the western Bohemian Massif (Bayerischer Wald, Germany)-preliminary results from SHRIMP dating. In: Niebuhr B (ed) Geo 2002, Planet Erde, Vergangenheit, Entwicklung, Zukunft, October 1 through 5, 2002, Würzburg. Deutsche Geologische Gesellschaft, Hannover, pp 327–328

Teng F-Z, McDonough WF, Rudnick RL, Dalpé C, Tomascak PB, Chappell BW, Gao S (2004) Lithium isotopic composition and concentration of the upper continental crust. Geochim Cosmochim Acta 68:4167–4178

Tennyson C (1954) Phosphoferrit und Reddingit von Hagendorf. Neues Jb Mineral Abh 87:185–217

Tennyson C (1958) Columbitkristalle von Hagendorf, Bayern. Neues Jb Mineral Monat 1958:121–124

Tennyson C (1981) Zur Mineralogie der Pegmatite des Bayerischen Waldes. Aufschluss 31:49–73

Teuscher EO, Weinelt W (1972) Die Metallogenese im Raum Spessart – Fichtelgebirge-Oberpfälzer Wald-Bayerischer Wald. Geologica Bavarica 65:5–73

Thomas R, Davidson P (2012) Water in granite and pegmatite-forming melts. Ore Geol Rev 46:32–46

Thomas R, Davidson P, Rhede D, Leh M (2009) The miarolitic pegmatites from the Königshain: a contribution to understanding the genesis of pegmatites. Contrib Mineral Petrol 157:505–523

Thöni M, Miller C (2004) Ordovician meta-pegmatite garnet (N-W Ötztal basement, Tyrol, Eastern Alps): preservation of magmatic garnet chemistry and Sm–Nd age during mylonitization. Chem Geol 209:1–26

Thöni M, Miller C, Zanetti A, Habler G, Goessler W (2008) Sm–Nd isotope systematics of high-REE accessory minerals and major phases: ID-TIMS, LA-ICP-MS and EPMA data constrain multiple Permian–Triassic pegmatite emplacement in the Koralpe, Eastern Alps. Chem Geol 254:216–237

Timmermann H, Krenn E, Dörr W, Finger F, Zulauf G (2002) Cadomian and Variscan metamorphism in the Teplá-Barrandian unit, northwestern Bohemian Massif: implications for the resetting of the U-Pb systematics in monazites. In: Niebuhr B (ed) Geo 2002, Planet Erde, Vergangenheit, Entwicklung, Zukunft, October 1 through 5, 2002, Würzburg. Deutsche Geologische Gesellschaft, Hannover, pp 331–332

Tischendorf G, Förster HJ (1990) Acid magmatism and related metallogenesis in the Erzgebirge. Geol J 25:443–454

Tischendorf G, Geisler M, Gerstenberger H, Budzinski H, Vogler P (1987) Geochemistry of Variscan granites of the Westerzgebirge-Vogtland Region-an example of tin deposit-generating granites. Chemie der Erde/Geochem 46:213–235

Tischendorf G, Förster HJ, Frischbutter A, Kramer W, Schmidt W, Werner CD (1995a) Igneous activity. In: Dallmeyer D, Franke W, Weber K (eds) Pre-permian geology of Central and Western Europe. Springer, Berlin, pp 248–259

Tischendorf G, Dill HG, Förster HJ (1995b) Metallogenesis of the Saxothuringian Basins. In: Dallmeyer RD, Franke W, Weber K (eds) Tectono-stratigraphic evolution of the Central and East European orogens. Springer, Heidelberg, pp 266–273

Tkachev AV (2011) Evolution of metallogeny of granitic pegmatites associated with orogens throughout geologic time. In: Sial AN, Bettencourt JS, De Campos CP (eds) Granite-related ore deposits, Geological Society special publication, 350. Geological Society, London, pp 7–23

Todt W (1979) U-Pb-Datierungen an Zirkonen des kristallinen Odenwaldes. Fortschritte Mineralogie (Beiheft) 57:153–154

Tollari N, Toplis MJ, Barnes S-J (2006) Predicting phosphate saturation in silicate magmas: an experimental study of the effects of melt composition and temperature. Geochim Cosmochim Acta 70:1518–1536

Tollmann A (1977) Geologie von Österreich, Band 1. Die Zentralalpen. Deuticke-Verlag, Wien 766 pp

Tollmann A (1982) Grossräumiger variszischer Deckenbau im Moldaunubikum und neue Gedanken zum Variszikum Europas. Geotektonische Forschung 64:1–91

Trümpy R (1980) Geology of Switzerland. Wepf, Basel, 334 pp

Tschernich RW (1992) Zeolites of the world. Geoscience Press, Phoenix, 563 pp

Tuyet NN, Minh TNT, Ngoc AV, Van NN (2006) Gem minerals in rare metal Pegmatite from Lucyen mining area (North Vietnam). Asia Oceania Geosciences Society, Singapore AOGS 2006, 905

Ucik F (2005) Die Feldspatpegmatite des Millstättersee-Rückens. Arbeitstagung der Geologischen Bundesanstalt Blatt 182 Spittal an der Drau, Gmünd/ Kärnten, 12–16 September, p 135

Uebel P-J (1975) Platznahme und Genese des Pegmatit von Hagendorf-Süd. Neues Jb Mineral Monat 1975:318–322

Uher P, Broska I (1995) Pegmatites in two suites of Variscan orogenic granitic rocks (Western Carpathians, Slovakia). Mineral Petrol 55:27–36

Uher P, Černý P (1998) Zircon in Hercynian granitic pegmatites of the western Carpathians, Slovakia. Geol Carpath 49:261–270

Uher P, Černý P, Chapman R, Hatar J, Miko O (1998a) Evolution of Nb, Ta-oxide minerals in the Prašivá granite pegmatites, Slovakia. primary Fe, Ti-rich assemblages. Can Mineral 36:525–534

Uher P, Černý P, Chapman R, Hatar J, Miko O (1998b) Evolution of Nb, Ta-oxide minerals in the Prasiva granitic pegmatites, Slovakia; II, External hydrothermal Pb, Sb overprint. Can Mineral 36:535–545

Uher P, Dávidová Š, Vikár I (2001) Schorl composition from the barren granitic pegmatites in the Western Carpathians: two examples from the Nízké Tatry mountains. J Czech Geol Soc 46:21–26

Uher P, Bačík P, Ozdín D, Števko M (2012) Beryl in granitic pegmatites of the Western Carpathians (Slovakia): compositional variations, mineral inclusions and breakdown products. Acta Mineralogica-Petrographica, Abstract Series, Szeged, 7:144

Van Breemen O, Bowes DR, Aftalion M, Żelaźniewicz A (1988) Devonian tectonothermal activity in the Sowie Góry Gneissic Block, Sudetes, Southwestern Poland: evidence from Rb-Sr and U-Pb isotopic studies. Rocz Pol Towarz Geol 58:3–19

Van der Pluijm BA, Marshak S (2004) Earth structure – an introduction to structural geology and tectonics, 2nd edn. W. W. Norton, New York. 656pp

Van Houten FB, Hou HF (1990) Stratigraphic and palaeogeographic distribution of Palaeozoic oolitic ironstones. In: McKerrow WS, Scotese CR (eds) Palaeozoic palaeogeography and biogeography, Geological Society Memoire, 12. Geological Society, London, pp 87–93

Van Lichtervelde M, Linnen RL, Salvi S, Béziat D (2006) The role of metagabbro rafts on tantalum mineralization in the Tanco granitic pegmatite, Manitoba. Can Mineral 44:625–644

Vaněček M, Patočka F, Pošmourný K, Raijlich P (1985) The use of the isotope composition of ore lead in metallogenic analysis of the Bohemian Massif. Rozpravy Ceskoslovenské Akademie Ved, Rada Matematických a Prírodních Ved 95:1–114

Vasyukova EA, Izokh AE, Borisenko AS, Pavlova GG, Sukhorukov VP, Anh TT (2011) Early Mesozoic lamprophyres in Gorny Altai: petrology and age boundaries. Russ Geol Geophys 52:1574–1591

Velichkien VI, Chernyhov IV, Simonova LI, Yudintsev SV (1994) Geotectonic position, petrochemical and geochronological features of the Younger Granite Complex in the Krušne Hory (Erzgebirge) of the Bohemian Massif. J Czech Geol Soc 39:11

Veselovský F, Ondruš P, Gabašová A, Škoda R, Hloušek J, Bouše P, Malec J (2007) Komplexni mineralogická stduie pegmatite "Weisser Stein" u Lázni Kynžvartu.- unpublished report, Česká geologicka služba Klárov, Prag, 105 pp

Vignola P, Fransolet AM, Diella V, Ferrari ES (2011) Complex mechanisms of alteration in a graftonite + sarcopside + triphylite association from the Luna pegmatite, Piona, Lecco province, Italy. Can Mineral 49:765–776

Vochten R (1990) Transformation of chernikovite and sodium autunite into lehnerite. Am Mineral 75:221–225

Vochten R, De Grave E, Van Springel K, Van Haverbeke L (1995) Mineralogical and Moessbauer spectroscopic study of some strunzite varieties of the Silbergrube, Waidhaus, Oberpfalz, Germany. Neues Jb Mineral Monat 1995:11–25

von Gehlen K (1989) Ore and mineral deposits of the Schwarzwald. In: Emmermann R, Wohlenberg J (eds) The German continental deep drilling program (KTB). Springer, Berlin, pp 277–295

von Gümbel CW (1868) Geognostische Beschreibung des ostbayerischen Grenzgebirges oder des bayerischen und Oberpfälzer Waldgebirges. Justus Perthes, Gotha/Leipzig, 968 pp

von Gümbel CW (1879) Geognostische Beschreibung des Fichtelgebirges mit dem Frankenwald und westlichen Vorlande. Perthes, Gotha/Leipzig, 698 pp

von Horstig G (1972) Mineralabfolge und Tektonik in den flussspatführenden Mineralgängen des Frankenwaldes. Geologica Bavarica 65:160–184

von Knorring O, Mrose ME (1963) Westgrenite and waylandite, two new bismuth minerals from Uganda. Geol Soc Am Spec Pap 73:265A

von Knorring O, Sahama TG (1982) Some Fe-Mn phosphates from the Buranga Pegmatite, Rwanda. Schweizer Mineralogische-Petrographische Mitteilungen 62:343–352

Von Pechmann E, Bianconi F (1982) Synmetamorphic uranium mineralization from Tiraun, Graubfinden, Switzerland. Mineral Mag 46:173–178

von Raumer JF, Stampfli GM, Bussy F (2003) Gondwana-derived microcontinents – the constituents of the Variscan and Alpine collision orogens. Tectonophysics 365:7–22

Vozárová A, Vozár J (1988) Late Paleozoic in West Carpathians. Dionys Štúr Institute Geological, Bratislava, pp 1–314

Vozárová A, Vozár J (1996) Terranes of West Carpathian-North Panonian domain. Slovak Geol Mag 1:61–83

Vrána S, Blümel P, Petrakakis K (1995) Metamorphic evolution. In: Dallmeyer RD, Franke W, Weber K (eds) Pre-permian geology of central and eastern Europe. Springer, Heidelberg/Berlin/Tokyo, pp 453–466

Wagner T, Boyce AJ (2001) Sulphur isotope characteristics of recrystallisation, remobilisation and reaction processes: a case study from the Ramsbeck Pb-Zn deposit, Germany. Mineral Deposita 36:670–679

Wagner T, Boyce AJ (2003) Sulphur isotope geochemistry of black shale-hosted antimony mineralization, Arnsberg, northern Rhenish Massif, Germany: implications for late-stage fluid flow during the Variscan orogeny. J Geol Soc Lond 160:299–308

Walenta K, Binder W (1980) Kidwellit und Dufrenit aus den Mines de Montmins bei Echassieres (französisches Zentralmassiv). Aufschluss 31:51–54

Walter R (1992) Die Geologie von Mitteleuropa, 5th edn. Schweizerbart, Stuttgart, 561 pp

Walter F, Postl W, Taucher J (1990) Weinebeneit: Paragenese und Morphologie eines neuen Ca-Be-Phosphates von der Spodumenpegmatit-Lagerstätte Weinebene, Koralpe, Kärnten. Mitt Abt Miner Landesmuseum Joanneum 58:37–43

Walton L (2004) Exploration criteria for colored gemstone deposits in the Yukon: Yukon Geological Survey, Open File 2004-10, 184 pp

Warin R, Jacques B (2003) Le beryl-Cs d'Ambatovita, Madagascar. Morphologie et aspects macroscopiques. Le Règne Mineral 52:36–41

Wark DA, Watson EB (2006) TitaniQ: a titanium-in-quartz geothermometer. Contrib Mineral Petrol 152:743–754

Warry ND, Kramer JR (1976) Some factors affecting the synthesis of cryptocrystalline strengite from an amorphous phosphate complex. Can Mineral 14:40–46

Weber B (1977) Phosphophyllit von Hagendorf-Süd. Papier-Weber, Weiden, 8 pp

Weber B (1978a) Mineralien aus den Metapegmatiten Wilma und Gertrude bei Obersdorf und Menzlhof in der Oberpfalz. Aufschluss 29:325–329

Weber K (1978b) Das Bewegungsbild im Rhenoherzynikum – Abbild einer varistischen Subfluenz. Z Dtsch Geol Ges 129:249–281

Weber K (1995) Introduction. In: Dallmeyer RD, Franke W, Weber K (eds) Pre-permian geology of central and eastern Europe. Springer, Berlin, pp 153–154

Weber B (2013) Die Mineralien des Amphibolit-Steinbruchs Muglhof/Ödenthal bei Weiden. Aufschluss 64:268–281

Weber K, Behr JH (1983) Geodynamic interpretation of the Mid-European Variscides. In: Martin H, Eder FW (eds) Intercontinental fold belts. Springer, Berlin/Heidelberg/New York, pp 427–469

Weber K, Vollbrecht A (1989) The crustal structure at the KTB Drill Site, Oberpfalz. In: Emmermann R, Wohlenberg J (eds) The continental deep drilling program (KTB). Springer, Heidelberg, pp 5–36

Websky M (1868) Über Sarkopsid und Kochelit, zwei neue Minerale aus Schlesien. Zeitschrift der Deutschen Geologische Gesellschaft 20:245

Weger M, Loth G, Höll R, Masch L, Köhler H, Rohrmüller J (1998) The early tectonometamorphic evolution of the southern zone of Erbendorf-Vohenstrauss (ZV, NW-Bohemian Massif, Bavaria) inferred from geochemical, radiometric, petrologic, and structural characteristics. Acta Univ Carol Geol 42:357

Weiss S, Vignola P, Diella V, Meisser N, Oppizzi P, Grundmann G (2004) Die Mineralien der Pegmatite von Brissago, Tessin, Schweiz: Aussergewöhnliche Neufunde 1999-2001. Lapis 29:24–38

Wemmer K, Ahrendt H (1993) Age determination on retrograde processes in Rocks of the KTB and the surrounding area. KTB-Report 93-2, pp 129–131

Wendt I, Kreuzer H, Müller P, Schmidt H (1986) Gesteins- und Mineraldatierungen des Falkenberger Granits. Geol Jahrb E 34:5–67

Wendt I, Carl C, Kreuzer H, Müller P, Stettner G (1992) Ergänzende Messungen zum Friedenfelser Granite Steinwald und radiometrische Datierung des Ganggranite im Falkenberger Granit. Geol Jahrb A137:3–24

Wendt I, Ackermann H, Carl C, Kreuzer H, Müller P, Stettner G (1994) Rb/Sr-Gesamtgesteins- und K/Ar-Glimmerdatierungen der Granite von Flossenbürg und Bärnau. Geol Jahrb E51:3–29

Werner W, Walther HJ (1995) Metallogenesis. In: Dallmeyer RD, Franke W, Weber K (eds) Pre-permian geology of central and eastern Europe. Springer, Berlin, pp 87–95

Wignal PB (1991) Model for transgressive black shales? Geology 19:167–170

Wilk H (1959) Phosphosiderit und Strengit von Pleystein in Ostbayern. Acta Albertina Ratisbonensia 23:107–170

Wilk H (1975) Der Quarzpegmatit von Pleystein und seine Paragenese. Aufschluß, Special Volume 26, 191–206

Will TM, Schmädicke E (2001) A first report of retrogressed eclogites in the Odenwald Crystalline Complex: evidence for high-pressure metamorphism in the Mid-German Crystalline Rise, Germany. Lithos 59:109–125

Will TM, Frimmel HE, Zeh A, Le Roux P, Schmädicke E (2010) Geochemical and iso-topic constraints on the tectonic and crustal evolution of the Shackleton Range, East Antarctica, and correlation with other Gondwana crustal segments. Precambrian Res 180:85–112

Willner AP, Klemm I, Rötzler K, Maresch WV (1995) Petrologische Belege aus dem mittelkrustalen Niveau für eine variskische Krustenstapelung im Erzgebirge. Terra Nostra 95(8):138

Winchester JA, PACE TMR NETWORK TEAM (2002) Palaeozoic amalgamation of Central Europe: new results from recent geological and geophysical investigations. Tectonophysics 360:5–21

Winkler HGF (1967) Die Genese der metamorphen Gesteine. Springer, Berlin, 237 pp

Winkler HGF (1976) Petrogenesis of metamorphic rocks. Springer, New York, 348 pp

Wise MA (1999) Characterization and classification of NYF- type pegmatites. Can Mineral 37:802–803

Wooley AR, Bergman SC, Edgar AD, Le Bas MJ, Mitchell RH, Rock NS, Smith BHS (1996) Classification of lamprophyres, lamproites, kimberlites, and the kalsilitic, melilitic, and leucitic rocks. Can Mineral 34:175–186

Yakovenchuk VN, Keck E, Krivovichev SV, Pakhomovsky YA, Selivanova EA, Mikhailova JA, Chernyatieva AP, Ivanyuk GY (2012) Whiteite-(CaMnMn), $CaMnMn_2Al_2[PO_4]_4(OH)_2*8H_2O$, a new mineral from the Hagendorf-Süd granitic pegmatite, Germany. Mineral Mag 76:2761–2771

Yakubovich OV, Matviyenko YN, Simonov MA, Melínikov OK (1986) The crystalline structure of synthetic Fe^{3+}-arrojadites with the idealized formula $K_2Na_5\,Fe^{2+}{}_{14}\,Fe^{3+}\,(PO_4)_{12}(OH)_2$. Vestnik Moskovovskogo Universiteta, Seriya Geologiya 1986:36–47

York D (1969) Least squares fitting of a straight line with correlated errors. Earth Planet Sci Lett 5:320–324

Žáček V, Novák M, Raimboult L, Zachariáš J, Ackerman L (2003) Locality No. 8: Vlastějovice near Ledeč nad Sázavou. Fe-skarn, barren fluorite pegmatite. In: Novák M (ed) International symposium on light elements in rock forming minerals LERM 2003, Nové Město na Moravě, Czech Republic. Field trip guidebook, pp 61–70

Zachariáš J, Pudilová M (2002) Fluid inclusion and stable isotope study of the Kasejovice gold district, central Bohemia. Bull Czech Geol Surv 77:157–165

Zachariáš J, Stein H (2001) Re-Os ages of Variscan hydrothermal gold mineralisations, Central Bohemian metallogenetic zone, Czech Republic. In: Piestrzyński A et al (eds) Mineral deposits at the beginning of the 21st century. Swets & Zeitlinger Publishers, Lisse, pp 851–854

Zagorsky VY, Makagon VM, Shmakin BM (1999) The systematics of granitic pegmatites. Can Mineral 37:800–802

Zeh A, Will TM (2010) The mid-German crystalline rise. In: Linnemann U, Romer RL (eds) Pre-Mesozoic geology of Saxo-Thuringia – from the Cadomian active margin to the Variscan orogen. Schweizerbart, Stuttgart, pp 195–220

Zeh A, Gerdes A, Will TM, Millar IL (2005) Provenance and magmatic-metamorphic evolution of a Variscan island arc complex: constraints from U-Pb dating, petrology, and geospeedometry of the Kyffhäuser Crystalline Complex, Central Germany. J Petrol 32:1393–1420

Żelaźniewicz A (1995) Introduction (Lugicum). In: Dallmeyer RD, Franke W, Weber K (eds) Tectono-stratigraphic evolution of the Central and East European orogens. Springer, Heidelberg, pp 311–314

Zhang L (1994) Electrochemical equilibrium diagrams for sulfidization of oxide copper minerals. Miner Eng 7:927–932

Zhang RY, Liou JG, Shu JF (2002) Hydroxyl-rich topaz in high-pressure and ultrahigh-pressure kyanite quartzites, with retrograde woodhouseite, from the Sulu terrane, eastern China. Am Mineral 87:445–453

Zhang FX, Pointeau V, Shuller LC, Reaman DM, Lang M, Liu ZX, Hu J, Panero WR, Becker U, Poinssot C, Ewing RC (2009) Structural transition and electron transfer in coffinite, $USiO_4$, at high pressure. Am Mineral 94:916–920

Ziegler AM, Hansen KS, Johnson ME, Kelly MA, Scotese CR, Van der Voo R (1977) Silurian continental distributions, paleogeography, climatology, and biogeography. Tectonophysics 40:13–51

Ziehr H (1957) Uranvorkommen in Bayern. Die Atomwirtschaft 2:193–196

Zoback MD, Mastin L, Barton C (1986) In-situ stress measurements in deep boreholes using hydraulic fracturing, well breakouts, and strongly wave polarization. In: Proceedings of the international symposium on rock stress measurements, Stockholm, 1–3 September 1986, pp 289–299

Zulauf G, Schitter F, Riegler G, Finger F, Fiala J, Vejnar Z (1999) Age constraints on the Cadomian evolution of the Teplá-Barrandian unit (Bohemian Massif) through electron microprobe dating of metamorphic monazite. Z Dtsch Geol Ges 150:627–640

Index

H.G. Dill, *The Hagendorf-Pleystein Province: the Center of Pegmatites in an Ensialic Orogen*, Modern Approaches in Solid Earth Sciences 15,
DOI 10.1007/978-3-319-18806-5

C

T

MIX
Papier aus verantwortungsvollen Quellen
Paper from responsible sources
FSC® C105338

If you have any concerns about our products,
you can contact us on
ProductSafety@springernature.com

In case Publisher is established outside the EU,
the EU authorized representative is:
Springer Nature Customer Service Center GmbH
Europaplatz 3, 69115 Heidelberg, Germany

Printed by Libri Plureos GmbH
in Hamburg, Germany